Polyphenol Nutraceuticals and Healthcare

In a world increasingly focused on natural, plant-based wellness, polyphenols – potent antioxidant compounds derived from foods, fruits, herbs, and medicinal plants – are transforming the landscape of modern healthcare and nutrition. This comprehensive volume presents a rich, evidence-based exploration into the science and application of polyphenols in the nutraceutical industry.

Bridging the divide between phytochemical foundations and practical innovation, this book offers in-depth insights into:

- The chemistry, biosynthesis, and structural diversity of polyphenols
- Antioxidant mechanisms and cellular pathway modulation
- Therapeutic roles in combating chronic diseases and oxidative stress
- Pharmacokinetics, bioavailability, and regulatory standards
- Green extraction methods and formulation technologies
- Sustainable sourcing and ethical practices in product development
- The polyphenol market and its prospects

With contributions from experts in the fields of botany, chemistry, biotechnology, pharmaceutical sciences, and functional food research, this book serves as a critical resource for researchers and scholars, students and educators, healthcare professionals, entrepreneurs, and product developers.

Whether you're advancing academic research or pioneering next-generation nutraceuticals, *Polyphenol Nutraceuticals and Healthcare* provides the knowledge, tools, and vision needed to innovate with scientific integrity and ecological responsibility.

Polyphenol Nutraceuticals and Healthcare

Edited by

Richa Upadhyay and Chandrashekhar Singh

CRC Press
Taylor & Francis Group
Boca Raton London New York

CRC Press is an imprint of the
Taylor & Francis Group, an **informa** business

Designed cover image: Shutterstock

First edition published 2026
by CRC Press
2385 NW Executive Center Drive, Suite 320, Boca Raton FL 33431

and by CRC Press
4 Park Square, Milton Park, Abingdon, Oxon, OX14 4RN

CRC Press is an imprint of Taylor & Francis Group, LLC

ISBN: 978-1-032-96198-9 (hbk)
ISBN: 978-1-032-96199-6 (pbk)
ISBN: 978-1-003-58850-4 (ebk)

DOI: 10.1201/9781003588504

Typeset in Times
by Apex CoVantage, LLC

Contents

Abbreviations

AChE	Acetylcholinesterase
AMPK	AMP-activated protein kinase
ARE	Anti-oxidative response elements
ArOH	Aromatic OH Group
BChE	Butyrylcholinesterase
BDE	Bond dissociation energy
BHA	Butylated hydroxyanisole
BHT	Butylated hydroxytoluene
4CL	4-coumarate CoA-ligase
C4H	Cinnamic acid 4-hydroxylase
CAT	Catalase
CHI	Chalcone-flavanone isomerase
CHS	Chalcone synthase
COX2	Cyclooxygenase-2
CREB	cAMP response element-binding protein
DAHPS	3-deoxy-D-arabinohexulose-7-phosphate synthase
DHQD/SDH	Dehydratase/shikimate dehydrogenase
DHQS	3-dehydroquinate synthase
DMSO	Dimethylsulfoxide
DP	Degree of polymerization
DPPH	2, 2-Diphenyl-1-picrylhydrazyl
EG	Ethyl gallate
ELF-EMF	Low-frequency electromagnetic fields
EpRE	Electrophile responsive element
ERK	Extracellular signal-regulated kinase
EPSP	5-enolpyruvylshikimate 3-phosphate synthase
ETE	Electron transfer enthalpy
FOXO4	Forkhead Box O4
GA	Gallic acid
GPx	glutathione peroxidase
GR	glutathione reductase
GSSG	Glutathione disulfide
GST	Glutathione S-transferase
HAT	Hydrogen Atom Transfer
HO-1	Heme oxygenase
$\mathbf{H_2O_2}$	Hydrogen peroxide
IKK	IκB kinase
IL-1*β*	Interleukin 1 Beta
IP	Ionization potential
JNK	c-Jun N-terminal kinase
Keap1	Kelch-like ECH-associated protein 1

LDL	Low-density lipoprotein
Maf	Musculoaponeurotic fibrosarcoma
MAPK	Mitogen-activated protein kinase
MAPKP-1	Mitogen-Activated Protein Kinase Phosphatase-1
MDA	Malondialdehyde
MG	Methyl gallate
NAD+	Nicotine adenine dinucleotide
NF-κB	Nuclear factor kappa-light-chain-enhancer of activated B cells
NLRP3	NLR family pyrin domain containing 3
NLS	Nuclear localization sequence
Nrf2	Nuclear factor erythroid 2-related factor 2
NQO1	NAD(P)H quinone dehydrogenase 1
·OH	Hydroxyl radical
$\cdot O_2^{\cdot -}$	Superoxide anion radical
$^1O_2$	Singlet oxygen
OME	4, 5-dihydroxy- 3-methoxy benzoic acid
OS	Oxidative stress
PA	Proton ability
PAL	Phenylalanine ammonia-lyase
PB2	Procyanidin dimer B2
PCs	Phenolic compound
PC	Procyanidins
PGA	1, 2, 3, 4, 6-pentagalloylglucose
PGC-1α	Peroxisome proliferator-activated receptor-gamma coactivator (PGC)-1alpha
R	Anion Group
RNS	Reactive Nitrogen species
ROS	Reactive oxygen species
RSV	Resveratrol
RXRα	Retinoic X receptor a
S3P	Shikimate 3-phosphate
SDG	secoisolariciresinol diglucoside
SET	Single electron transfer
SET-PT	Single-electron transfer via proton transfer
SIRT	Sirtuin
SOD	Superoxide dismutase
STS	Stilbene synthase
TBARS	Thiobarbituric acid reactive substances
TFEB	Transcription factor EB
TNF-α	Tumor necrosis factor
Trx	Thioredoxin
TrxR	Thioredoxin reductase
γ-GCS	γ-glutamyl cysteine synthetase

Preface

In recent years, there has been a profound shift in global health and wellness paradigms, with an increasing emphasis on natural, plant-based solutions that promote holistic well-being. Among these, polyphenols – diverse and powerful phytochemicals found abundantly in medicinal plants – have emerged as a cornerstone of modern nutraceutical research and development. Recognized for their potent antioxidant properties and multifaceted therapeutic potential, polyphenols are being harnessed more than ever in the formulation of nutraceuticals that address chronic diseases, oxidative stress, and general health deterioration.

Polyphenol Nutraceuticals and Healthcare has been conceived as a comprehensive and scientifically rigorous exploration of this vital class of bioactive compounds. This contributed volume brings together a spectrum of scholarly insights and practical perspectives, aiming to bridge the gap between foundational phytochemistry and applied nutraceutical science. It provides a detailed exposition of the chemistry, biosynthesis, antioxidant mechanisms, and pharmacological benefits of polyphenols, while also addressing the practical challenges of their extraction, formulation, and integration into consumer products.

The chapters are carefully curated and contributed by experts from various disciplines – including botany, pharmaceutical sciences, biochemistry, and biotechnology – ensuring both breadth and depth. Readers will be guided through the structural intricacies of polyphenols, their role in scavenging reactive oxygen species, modulation of cellular pathways, and their influence on mitochondrial and immune functions. We also delve into the pharmacokinetics, safety, and toxicological assessments critical to the responsible use of these compounds in human health.

Equally important is our focus on sustainability and regulatory frameworks. From ethical sourcing of medicinal plants to compliance with international standards and labeling practices, this book equips researchers and professionals with a forward-looking perspective that considers ecological and commercial viability.

The motivation behind this book lies in the convergence of growing scientific evidence, technological advancements, and consumer demand for plant-derived nutraceuticals. It serves not only as a reference text for seasoned researchers and professionals but also as a guiding resource for students, educators, healthcare practitioners, and entrepreneurs seeking to innovate in the space of functional foods and dietary supplements.

We extend our deepest gratitude to the contributors for their scholarly efforts and dedication and to the institutions that support ongoing research in natural products and nutraceutical development. We also thank the reviewers and publishing team whose diligence has brought this work to fruition.

It is our hope that this volume will serve as both a foundational and inspirational resource – supporting the advancement of knowledge, guiding evidence-based practice, and fostering sustainable innovation in the field of polyphenol-based nutraceuticals.

Richa Upadhyay
Chandrashekhar Singh

About the Editors

Richa Upadhyay is an assistant professor of botany at Mihir Bhoj PG College, Dadri, Gautam Budh Nagar. She received her PhD from Banaras Hindu University, Varanasi, India. Her research interests are in natural product-based drug discovery. She has published several peer-reviewed articles, books, and patents.

Chandrashekhar Singh is a distinguished researcher and academic with expertise in plant biotechnology, clinical microbiology, and nanomedicine. He has contributed significantly to the development of bioinspired therapeutic agents targeting wound healing and antimicrobial resistance. He has held academic positions at Maharishi University of Information Technology, Lucknow, and Sri Agrasen Kanya PG College, Varanasi, and served as a postdoctoral research associate at IIT BHU. His work includes numerous peer-reviewed publications, book chapters, and three patents in nanotechnology and phytotherapy.

Contributors

Romila Rawat Bisht
Department of Botany
Hansraj College, University of Delhi
Delhi, India

Moumita Chatterjee
V Sivaram Research Foundation, Vijayanagar, Bengaluru
Karnataka, India

Srijonee Choudhury
Department of Botany
Rani Indira Debi Government Girls' College, Jhargram
West Bengal, India

Moumita Datta
Department of Botany
Rammohan College, Kolkata
West Bengal, India

Pratyusha De
Department of Botany
The University of Burdwan
Burdwan, India

Elad Eliahoo
Kimron Veterinary Institute
Ministry of Agriculture and Rural Development
Beit Dagan, Israel
The Robert H. Smith Faculty of Agriculture, Food and Environment
Hebrew University of Jerusalem
Jerusalem, Israel

Arindam Ganguly
Department of Microbiology
Bankura Sammilani College, Bankura
West Bengal, India

Rajeeva Gaur
Department of Microbiology
RML Awadh University, Ayodhya
Uttar Pradesh, India

Maneesh Kumar Gupta
Integrated Public Health Laboratory
Community Health Centre, Bhadohi
Uttar Pradesh, India

Kaushalendra Kumar
Plant Molecular Biology Lab
Department of Botany
Dayanand Angelo Vedic (PG) College, CSJM Kanpur University
Kanpur, India

Shivam Kumar
Department of Life Sciences
CSJM Kanpur University
Kanpur, India

Anna Elisabetta Maccarronello
Department of Chemical Sciences
University of Catania
Catania, Italy

Aqsa Majgaonkar
Senior Education Officer
Conservation Education Centre (CEC)
Bombay Natural History Society (BNHS)
Mumbai, India

Nidhi Minz
Department of Bioengineering and Biotechnology
BIT-Mesra, Ranchi
Jharkhand, India

Awadhesh Kumar Mishra
Department of Botany
Ramjas College, University of Delhi
Delhi, India

Vera Muccilli
Department of Chemical Sciences
University of Catania
Catania, Italy

Tripti Pal
RMIT University
Melbourne, Australia

Smaranika Pattnaik
School of Life Sciences
Sambalpur University, Sambalpur
Odisha, India

Renisha Pinto
College of Professional Studies
Northeastern University
Boston, MA

Chandra Prakash
School of Life Sciences
Jawaharlal Nehru University
New Delhi, India

Francisca Rodrigues
i4HB – Associate Laboratory Institute for Health and Bioeconomy
Faculty of Pharmacy
University of Porto
Porto, Portugal

UCIBIO – Applied Molecular Biosciences Unit
MedTech-Laboratory of Pharmaceutical Technology
Faculty of Pharmacy
University of Porto
Porto, Portugal

Rajesh Saini
Department of Botany, MMV
Banaras Hindu University, Varanasi
Uttar Pradesh, India

Deepak Sharma
School of Life Sciences
Jawaharlal Nehru University
New Delhi, India

Manoj Kumar Sharma
Department of Botany
J.V. College, Baraut, Baghpat
Uttar Pradesh, India

Vishal Sharma
Department of Pharmaceutical Engineering and Technology
IIT BHU
Varanasi, India

Madhvendra Shukla
Department of Botany, MMV
Banaras Hindu University, Varanasi
Uttar Pradesh, India

Akhil Singh
Department of Life Sciences
CSJM Kanpur University
Kanpur, India

Aparna Singh
Department of Botany, MMV
Banaras Hindu University, Varanasi
Uttar Pradesh, India

Chandrashekhar Singh
Maharishi School of Science and Humanities
Maharishi University of Information Technology, Lucknow
Uttar Pradesh, India

Ishwar Singh
Department of Botany
Hansraj College, University of Delhi
Delhi, India

Satyendra Pratap Singh
Postharvest Department, ARO
Volcani Center
Rishon LeZiyyon, Israel

Surendra Pratap Singh
Plant Molecular Biology Lab
Department of Botany
Dayanand Angelo Vedic (PG) College, CSJM Kanpur University
Kanpur, India

Dwaipayan Sinha
Department of Botany
Government General Degree College, Mohanpur, Paschim Medinipur
West Bengal, India

Filipa Teixeira
Laboratório Associado para a Química Verde, ISEP
Polytechnic of Porto
Porto, Portugal

Associate Laboratory i4HB-Institute for Health and Bioeconomy
Faculty of Pharmacy
University of Porto
Porto, Portugal

UCIBIO-Applied Molecular Biosciences Unit
MedTech-Laboratory of Pharmaceutical Technology
Faculty of Pharmacy
University of Porto
Porto, Portugal

Biomedical Sciences Institute of Abel Salazar
Porto University (ICBAS-UP)
Porto, Portugal

Kavindra Nath Tiwari
Department of Botany, MMV
Banaras Hindu University, Varanasi
Uttar Pradesh, India

Richa Upadhyay
Department of Botany
Mihir Bhoj PG College, Dadri, GB Nagar
Uttar Pradesh, India

Kajal Verma
Plant Molecular Biology Lab
Department of Botany
Dayanand Angelo Vedic (PG) College, CSJM Kanpur University
Kanpur, India

Praveen Kumar Verma
Kimron Veterinary Institute
Ministry of Agriculture and Rural Development
Beit Dagan, Israel

The Robert H. Smith Faculty of Agriculture, Food and Environment
Hebrew University of Jerusalem, Israel

Praveen Kumar Shukla
Department of Botany, MMV
Banaras Hindu University, Varanasi
Uttar Pradesh, India

Vikas
Department of Pharmaceutical Engineering and Technology
IIT BHU
Varanasi, India

Masonic Medical Research Institute
Utica, NY

Jyoti Vishwakarma
Department of Botany, MMV
Banaras Hindu University, Varanasi
Uttar Pradesh, India

Reena Yadav
Plant Science Department, ARO
Volcani Centre
Rishon LeZiyyon, Israel

1 Introduction and Overview of Polyphenols

Kajal Verma, Kaushalendra Kumar, Surendra Pratap Singh, and Reena Yadav

1.1 INTRODUCTION

Polyphenolic compounds, a vital group of secondary metabolites, are found throughout the plant kingdom (Cheynier *et al.*, 2013; Tijjani *et al.*, 2020). These bioactive molecules are synthesized in various plant organs and contribute to key physiological and ecological functions such as pigmentation, pollinator attraction, pathogen defence, UV protection, and intracellular signalling. Although found across all plant taxa, their structural diversity and abundance are especially notable in higher plants, often existing in conjugated forms like esters, amides, and glycosides. These compounds are predominantly synthesized through the shikimate and acetate–malonate pathways. Despite subclass variation, all polyphenols share a common structure of one or more hydroxylated aromatic rings. To date, over 10,000 distinct polyphenolic compounds have been identified, making them one of the most diverse groups of phytochemicals (Truzzi *et al.*, 2021). In nature, they serve as potent defensive agents, shielding plants and algae from environmental stressors such as UV exposure, microbial infections, and herbivory. Their antioxidant capacity also contributes significantly to plant resilience and human health. These compounds are abundant in plant-based foods like fruits, vegetables, nuts, seeds, flowers, and bark, as well as in beverages such as tea, coffee, wine, and beer (Abbas *et al.*, 2017; Ahmad *et al.*, 2021; Bertelli *et al.*, 2021). Their widespread inclusion in the human diet has drawn considerable scientific interest, largely due to their reported antioxidant, anti-inflammatory, antimicrobial, cardioprotective, and anticancer properties (Potì *et al.*, 2019; Tijjani *et al.*, 2020; Iqbal *et al.*, 2023). Mounting evidence suggests that they can mitigate various degenerative diseases by counteracting oxidative stress and modulating inflammatory pathways (Ahmad *et al.*, 2021). As a result, these compounds have garnered increasing attention in the fields of nutrition, medicine, and pharmacology. Quantitative studies have also emphasized their prevalence in the human diet. According to the Phenol-Explorer database, many common foods provide more than 1 mg of polyphenols per serving (Pérez–Jiménez *et al.*, 2010). A dietary study by Godos *et al.* (2017) reported that individuals in an Italian population consumed about 660 mg of polyphenols daily, mainly from citrus fruits, vegetables, tea, coffee, nuts, cherries, red wine, and chocolate. This chapter provides an in-depth overview of the chemistry, sources, extraction techniques, bioactivities, and therapeutic potential of polyphenols. It also discusses the critical challenges related to their limited

DOI: 10.1201/9781003588504-1

bioavailability and metabolic interactions, offering insights into future directions for maximizing their potential as powerful natural therapeutics.

1.2 CLASSIFICATION AND SOURCES OF POLYPHENOLS

Polyphenols are a diverse group of naturally occurring compounds, found in fruits, vegetables, teas, and herbs with more than 8,000 types identified to date. Based on their phenolic ring structure, they are classified into major groups such as flavonoids, phenolic acids, stilbenes, lignans, and others. Beyond their role in plant defence mechanisms, polyphenols are widely recognized for their significant contributions to human health, owing to their potent antioxidant and therapeutic properties. An overview of the classification of polyphenols is presented in Figure 1.1 (Cheynier *et al.* 2013; Abbas *et al.*, 2017; Rasouli *et al.*, 2017; Prabhu *et al.*, 2021).

1.2.1 Phenolic Acids

Phenolic acids represent one of the most prominent classes of plant-derived polyphenols, Chemically, they consist of an aromatic ring attached to a carboxylic acid group (-COOH). These compounds are widely utilized in the food, cosmetic, and pharmaceutical industries due to their strong antioxidant and anti-inflammatory (Teixeira *et al.*, 2013) properties. Phenolic acids are broadly divided into two major groups based on their carbon backbone: hydroxybenzoic acids (C6–C1) and hydroxycinnamic acids (C6–C3). Hydroxybenzoic acids such as gallic, protocatechuic, vanillic, syringic, and p-hydroxybenzoic acids are predominantly found in tea, grape seeds, and olive products, where they contribute to anti-inflammatory, antioxidant, and cardioprotective effects. In contrast, hydroxycinnamic acids, derived from cinnamic acid, include compounds like caffeic, ferulic, and p-coumaric acids. These are abundant in dietary sources such as coffee, apples, berries, whole grains, and kiwi fruit. Caffeic acid is among the most prevalent and is often present in its ester form, chlorogenic acid, especially in coffee. Caffeic acid itself is commonly found in fruits, such as berries and apples. Ferulic acid, another key hydroxycinnamic acid, is especially abundant in cereal grains like wheat, rice, and oats (Teixeira *et al.*, 2013; Prabhu *et al.*, 2021; Truzzi *et al.*, 2021; Iqbal *et al.*, 2023).

1.2.2 Flavonoids

Flavonoids represent a vital class of plant-derived secondary metabolites, characterized by their polyphenolic structure composed of 15 carbon atoms arranged in two aromatic rings (A and B) connected by a three-carbon bridge that forms a central heterocyclic pyran ring (C ring). This structural arrangement underpins their chemical diversity and biological activity. Based on the oxidation state and substitution pattern of the C ring, flavonoids are broadly classified into six major subclasses: flavones, flavonols, isoflavones, flavan-3-ols, flavanones, and anthocyanidins (Panche *et al.*, 2016; Abbas *et al.*, 2017). Among these, flavones such as apigenin, luteolin, baicalein, nobiletin, and chrysin are present in foods like celery, garlic, chamomile tea, parsley, and

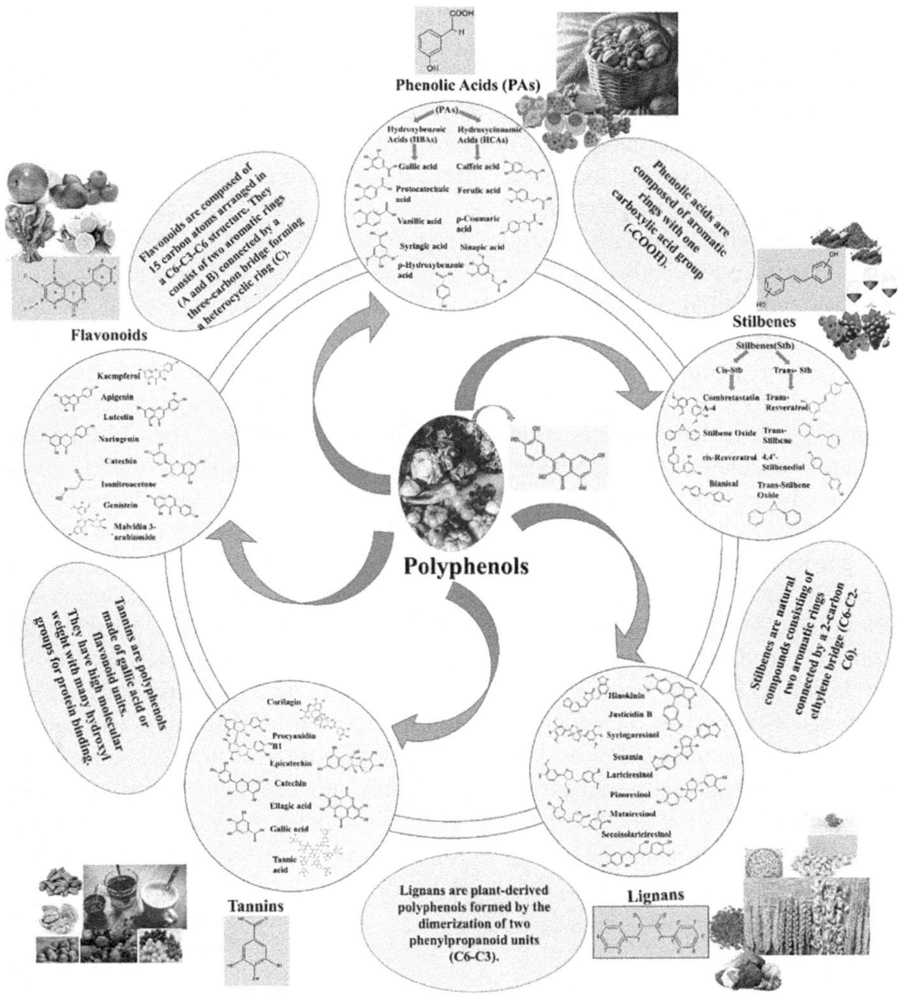

FIGURE 1.1 Classification of polyphenols based on structural types: phenolic acids, flavonoids, stilbenes, lignans, and tannins.

various traditional herbs and have shown significant anticancer potential by inducing apoptosis and inhibiting tumour proliferation (Sindhu *et al.*, 2021; Xian *et al.*, 2021). In particular, apigenin has demonstrated favourable pharmacokinetics; for instance, doses ranging from 13.82 to 100 mg/kg resulted in plasma concentrations of 0.14 to 1.33 μg/mL, indicating efficient absorption across a wide dosage range (Ding *et al.*, 2014; Dong *et al.*, 2017). Flavonols, such as quercetin and kaempferol are abundant in foods like onions, apples, green tea, broccoli, and buckwheat. Structurally distinguished by a 3-hydroxyflavone backbone, these compounds are well known for their antioxidant, anti-oestrogenic, and pro-oxidant activities. Quercetin and its glycoside form, rutin, are especially prominent in flavonoid research and have been linked to the prevention of oestrogen-dependent cancers such as breast cancer (Cui *et al.*, 2016;

Iqbal *et al.*, 2023; Sindhu *et al.*, 2021). Isoflavones, including daidzein and glycitein, are predominantly found in soybeans and soy-based products. These compounds act as phytoestrogens and have been associated with reduced risks of certain cancers, cardiovascular diseases, hypertension, and metabolic disorders. Beyond isoflavones, soy also contains other bioactive constituents such as lunasin, omega-3 fatty acids, Bowman–Birk inhibitors, and dietary fibre, all of which contribute synergistically to its health benefits (Eisen *et al.*, 2003; Sindhu *et al.*, 2021). Flavan-3-ols, like catechin, epicatechin, and gallocatechin, are present in tea, cocoa, berries, and pome fruits. They are especially noted for their vascular protective effects and ability to improve endothelial function in healthy individuals. In countries such as the UK, Australia, and Japan, tea is the primary dietary source of flavan-3-ols, whereas in other regions, pome fruits also serve as significant contributors (Ezzat *et al.*, 2018; Kuhnle, 2018). Flavanones like hesperetin, naringenin, and eriodyctiol are mainly present in citrus fruits such as oranges, lemons, and grapefruits. These compounds not only contribute to flavour but also offer health benefits and undergo modifications such as glycosylation, methylation, and hydroxylation, which enhance their biological activity. Bergamot juice (*Citrus bergamia*) is especially rich in flavanones and has garnered attention for its cardiovascular benefits (Pontifex *et al.*, 2021; Ortiz *et al.*, 2022; Skendi *et al.*, 2022). Anthocyanidins, such as cyanidin, pelargonidin, delphinidin, and peonidin, are responsible for the vibrant red, purple, and blue pigmentation seen in fruits, vegetables, and flowers, including grapes, strawberries, and purple corn. These pigments are not merely decorative; they also offer diverse pharmacological properties, including strong antioxidant, anti-inflammatory, anti-ageing, and anticancer activities. For example, anthocyanin extracts from strawberries have been shown to suppress breast cancer cell proliferation in experimental animal models (Kopustinskiene *et al.*, 2020). Collectively, flavonoids contribute significantly to human health by modulating cell signalling pathways (Cheynier *et al.*, 2013; Abbas *et al.*, 2017; Rasouli *et al.*, 2017; Prabhu *et al.*, 2021).

1.2.3 Stilbenes

Stilbenes are a small class of polyphenolic compounds structurally defined by a 1,2-diphenylethylene backbone, typically bearing hydroxyl substitutions on their aromatic rings, and can occur as monomers or oligomers. The most studied stilbene is trans-resveratrol, which features a trihydroxystilbene skeleton and has gained attention for its broad therapeutic potential. Common dietary sources of stilbenes include grapes, red wine, soy products, and peanuts, while several plants traditionally used in Asian medicine such as *Polygonum cuspidatum, Rhodomyrtus tomentosa, Rheum undulatum, Melaleuca leucadendron,* and *Euphorbia lagascae* are also known to be rich in these compounds. Pterostilbene, another notable stilbene, is mainly found in bilberries, blueberries, grapes, and their by-products, especially grape juice residues, which are increasingly used in nutraceutical formulations (Riviere *et al.*, 2012; Shaito *et al.*, 2020; Petrović *et al.*, 2024). Despite the chemical variability, stilbene production is limited to a few plant families, notably Polygonaceae, Cyperaceae, Pinaceae, and Vitaceae. Biologically, stilbenes function as phytoalexins, which are protective compounds synthesized by plants in response to environmental stresses or pathogen attacks, such as ultraviolet (UV) radiation, ozone exposure, metal toxicity, physical injury, or frost (Riviere *et al.*, 2012; Oluwole *et al.*, 2022). Among stilbenes, resveratrol has been extensively studied for its antioxidant, anti-inflammatory, anticancer, cardioprotective, neuroprotective, lipid-lowering, and immunomodulatory activities (Shaito *et al.*, 2020; Petrović *et al.*, 2024). Mechanistically, stilbenes mitigate oxidative stress by modulating cellular signalling pathways such as Nrf2 and regulating proteins like SQSTM1/p62, which influence antioxidant defences and autophagy. These properties make stilbenes valuable candidates in the development of functional foods and therapeutic agents.

1.2.4 Lignans

Lignans are a class of polyphenolic compounds formed through the dimerization of phenylpropanoid units and are abundantly present in flaxseeds, sesame seeds, whole grains, fruits, and vegetables. After ingestion, dietary lignans such as secoisolariciresinol diglucoside (SDG) are converted by gut microbiota into enterolignans like enterodiol and enterolactone, which possess weak estrogenic activity and strong antioxidant potential (Imran *et al.*, 2015; Plaha *et al.*, 2022). Flaxseeds are particularly rich in SDG, while sesame seeds contain sesamin, both contributing to various health benefits. Enterolactone is known for its anticancer effects, particularly in hormone-related cancers such as breast and prostate cancer. Other lignans, like lariciresinol and pinoresinol, help protect the cardiovascular system by neutralizing free radicals and enhancing antioxidant enzymes (Jang *et al.*, 2022). Some lignans also display anti-inflammatory effects; for instance, sesamin inhibits NF-κB and p38 MAPK signalling pathways in mast cells, reducing pro-inflammatory cytokine production (Li *et al.*, 2016), while nortrachelogenin lowers nitric oxide levels by down-regulating inducible nitric oxide synthase (Laavola *et al.*, 2017). Furthermore, lignans such as pinoresinol show antimicrobial activity by interfering with bacterial membrane receptors, metabolites, and biofilm formation (Ecevit *et al.*, 2022).

1.2.5 Other Polyphenols

1.2.5.1 Tannins

Tannins, commonly referred to as tannic acids, are high-molecular-weight, water-soluble polyphenolic compounds found in association with proteins, polysaccharides, and alkaloids in various plant-based foods. These bioactive molecules are structurally defined by two or three phenolic hydroxyl groups attached to a large phenyl ring, contributing to their strong biological functions (Okuda & Ito, 2011). Based on their chemical characteristics, tannins are mainly classified into two types: hydrolysable tannins (HTs) and condensed tannins (CTs) (García *et al.*, 2016). HTs are esters of gallic or ellagic acid, predominantly found in fruits, legumes, leafy vegetables, and several tree species. They have long been used in the leather tanning industry and are recognized for their antioxidant and antimicrobial properties (Shirmohammadli *et al.*, 2018; Mal & Pal, 2020). In contrast, CTs, also known as proanthocyanidins, consist of flavan-3-ol polymer units and are widely found in berries, cocoa, and beverages such as tea, wine, and beer (García *et al.*, 2016; Mal & Pal, 2020). Tannins exert several biological functions, including potent antioxidant and anticarcinogenic effects. They have been shown to neutralize mutagenic agents and inhibit the proliferation of fungi, bacteria, viruses, and yeasts (Shirmohammadli *et al.*, 2018). For instance, epigallocatechin-3-gallate (EGCG), a major green tea polyphenol, suppresses matrix metalloproteinases (MMP-2/-9), reducing tumour cell invasiveness (Tanimura *et al.*, 2005). In plants, tannins act as defensive compounds, particularly during fruit ripening, by hydrolyzing ester bonds and deterring microbial attack (Chung *et al.*, 1998). They also contribute to photoprotection by absorbing UV radiation and scavenging reactive oxygen species (Shirmohammadli *et al.*, 2018). Tannic acid has been shown to inhibit UV-induced enzymes like collagenase and elastase, while exhibiting broad UV absorption (Dare *et al.*, 2020). Additionally, tannin-rich extracts from *Quercus robur* have enhanced antioxidant enzyme levels such as catalase (CAT) and superoxide dismutase (SOD), highlighting their therapeutic potential in managing oxidative stress (Horvathova *et al.*, 2014).

1.2.5.2 Lignin

Lignin, a complex natural phenolic macromolecule, consists of phenylpropanoid units such as p-hydroxyphenyl (H), guaiacyl (G), and syringyl (S), which polymerize through various linkages on the aromatic rings. First described by Swiss botanist A.P. de Candolle in 1813, lignin plays a crucial structural role in plants, particularly by strengthening xylem tissues, enhancing rigidity, and providing resistance against microbial invasion and water loss (Singla *et al.*, 2019; Popoola-Akinola *et al.*, 2022; Okonkwo *et al.*, 2023). It is a heterogeneous polymer derived from a few core precursors and is found in diverse sources such as fruit stones, vegetables like garden beans, cereals, and soft- and hardwoods (Singla *et al.*, 2019; Popoola-Akinola *et al.*, 2022). Its antioxidant activity is closely linked to oxygen-containing functional groups like hydroxyl, methoxy, carboxyl, and carbonyl moieties that influence its polyphenolic reactivity (Ponomarenko *et al.*, 2015). In human health, lignin functions as a potent free radical scavenger, mitigating oxidative DNA damage induced by agents like hydrogen peroxide. It has shown promising antimutagenic and anticancer potential and, when incorporated into sunscreens, enhances photoprotection by reducing UV-induced radical formation (Li *et al.*, 2024).

1.3 EXTRACTION AND CHARACTERIZATION TECHNIQUES

The extraction and characterization of polyphenols are crucial steps in evaluating their therapeutic potential and enhancing their application in nutraceuticals. Traditionally, polyphenols have been extracted using conventional techniques such as solid–liquid and liquid–liquid extraction, with solvents like ethanol or methanol. However, these methods often have drawbacks, including low efficiency and lengthy processing times. To address these limitations, advanced techniques such as ultrasound-assisted extraction (UAE), microwave-assisted extraction (MAE), pressurized liquid extraction (PLE), and supercritical fluid extraction (SFE) have shown great promise in improving both the yield and efficiency of polyphenol extraction (Rajbhar *et al.*, 2015; Ameer *et al.*, 2017). These methods are designed to leverage the structural properties of polyphenols, facilitating their efficient separation and recovery. Alcohol-based solvents, particularly methanol and ethanol, remain widely used due to their effectiveness in extracting various polyphenol classes, including stilbenes and flavonoids (Rasouli *et al.*, 2017). For instance, Yue *et al.* (2012)

reported that UAE significantly increased the yield of (-)-epicatechin and procyanidins from unripe apples. Similarly, Dahmoune *et al.* (2015) reported that MAE of *Myrtus communis* leaves not only accelerated the process but also achieved a higher phenolic recovery compared to traditional methods. After extraction, polyphenols are characterized using analytical techniques such as UV-visible spectroscopy (e.g., Folin–Ciocalteu assay for total phenolics), high-performance liquid chromatography (HPLC) for individual compound profiling, and mass spectrometry (HPLC-MS/MS) for molecular identification. Additional methods like Fourier-transform infrared spectroscopy (FTIR) and nuclear magnetic resonance (NMR) help determine functional groups and detailed structural features (Wu *et al.*, 2024). Notably, Nkhili *et al.* (2009) demonstrated the effective quantification of catechins from green tea using HPLC after MAE, emphasizing improved yield and antioxidant potential. Together, these integrated methods enable the development of standardized, polyphenol-rich nutraceutical products (Figure 1.2).

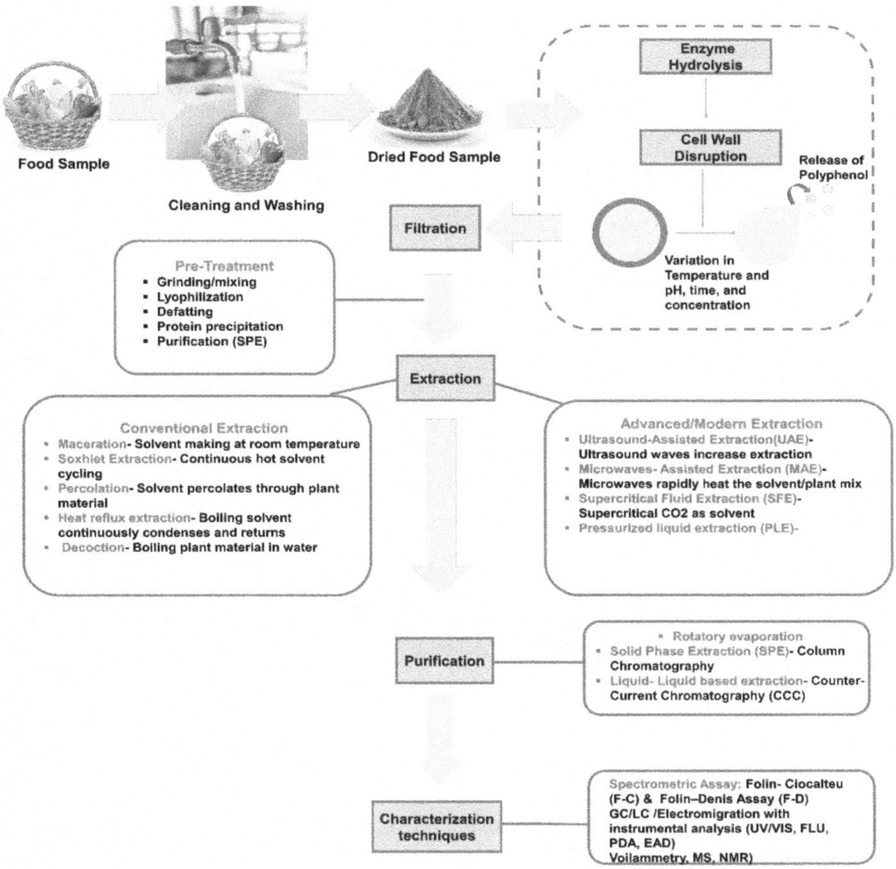

FIGURE 1.2 Overview of polyphenol extraction, purification, and characterization employing both traditional and advanced techniques.

1.4 BIOLOGICAL ACTIVITIES OF POLYPHENOLS

Polyphenols exhibit a wide range of biological functions that contribute to the prevention and management of chronic diseases. These compounds protect cellular systems by neutralizing oxidative stress, regulating inflammatory processes, and influencing molecular pathways associated with cardiovascular health, immune responses, ageing, and cancer (Figure 1.3).

1.4.1 Antioxidant Properties

Polyphenols are potent antioxidants known for their ability to donate electrons or hydrogen atoms, thereby neutralizing reactive oxygen species (ROS) and protecting

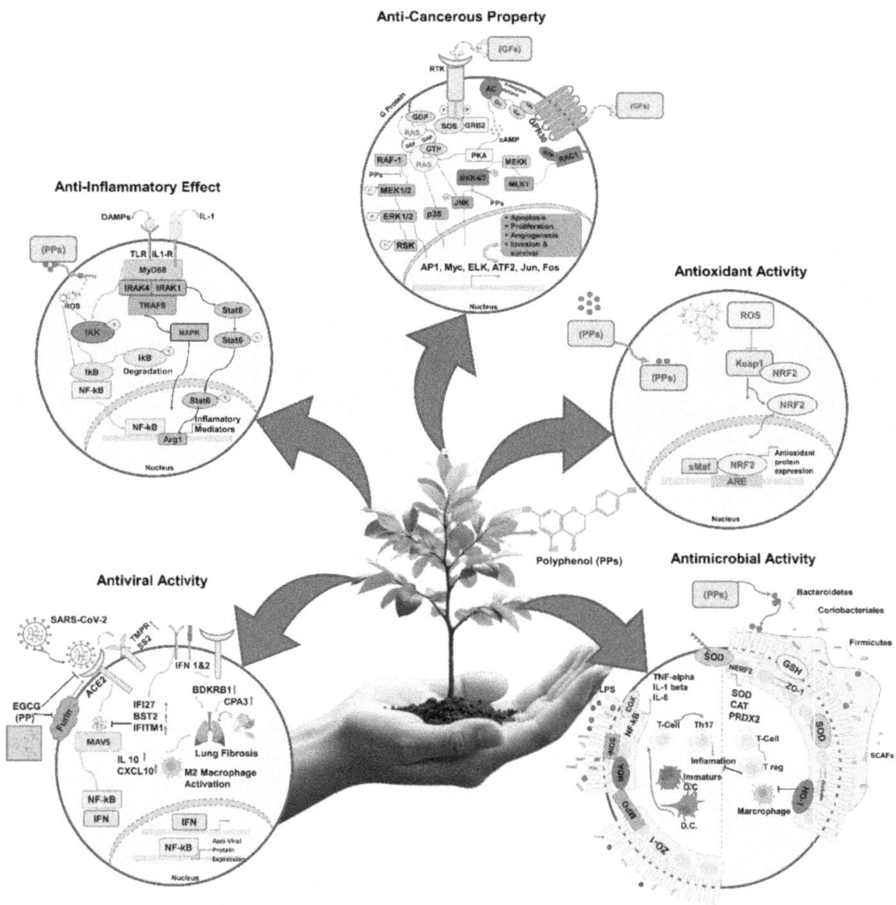

FIGURE 1.3 Illustration of polyphenol-driven modulation of biological signalling pathways. Polyphenols exert regulatory effects on oxidative stress, inflammation, immune function, and apoptosis by targeting pathways such as NF-κB, MAPK, Nrf2, and JAK/STAT, supporting their role in disease prevention and cellular protection.

cells from oxidative damage. They scavenge free radicals, chelate metal ions like Fe^{2+} and Cu^{2+}, and boost antioxidant enzymes such as SOD, catalase, and glutathione peroxidase (Heim *et al.*, 2002; Lang *et al.*, 2024). Furthermore, polyphenols reduce ROS generation by inhibiting mitochondrial adenosine triphosphate (ATP) synthesis through the blockage of the electron transport chain. They also downregulate lipopolysaccharide (LPS)-induced inducible nitric oxide synthesis (iNOS) expressions in macrophages, limiting oxidative injury. With low toxicity and broad pharmacological effects, polyphenols play a crucial role in managing oxidative stress–related diseases (Sugamura *et al.*, 2011; Raedschelders *et al.*, 2012). Beyond their antioxidant properties, polyphenols influence various intracellular signalling pathways involved in numerous diseases. For instance, curcumin has been shown to exert diverse pharmacological effects (Patel *et al.*, 2020). Gallic acid, another notable polyphenol, reduces oxidative tissue damage by neutralizing free radicals generated in Fenton reactions and decreasing lipid peroxidation markers like MDA while enhancing antioxidant enzyme activity (Saidi *et al.*, 2021). Additionally, anthocyanins were reported to activate the Nrf2 pathway and inhibit NF-κB, thereby boosting antioxidant responses in ruminants (Jeong *et al.*, 2013). Tea polyphenols have been shown to reduce the ageing of glomerular mesangial cells in diabetes by modulating the miR-126/Akt-p53-p21 signalling pathway (Cao *et al.*, 2019). Green tea, rich in polyphenols like catechin, epicatechin, and gallate, has demonstrated a dose-dependent increase in plasma antioxidant capacity. In a study with 15 healthy volunteers, consuming 500 mL of green tea with varying concentrations (1.4–2.0 g/L) revealed that higher polyphenol content led to stronger antioxidant effects lasting up to 4 hours (Pecorari *et al.*, 2010). Similarly, resveratrol, a well-known polyphenol, offers a wide range of health benefits, including antioxidant, anti-inflammatory, anticancer, anti-ageing, anti-obesity, antidiabetic, cardioprotective, and neuroprotective effects (Iqbal *et al.*, 2023).

1.4.2 Anti-Inflammatory Effects

Excessive inflammation plays a pivotal role in the pathogenesis of numerous chronic diseases such as obesity, type II diabetes, cardiovascular conditions, neurodegenerative disorders, and ageing. Polyphenols have shown strong anti-inflammatory potential in both in vivo and in vitro studies (Santangelo *et al.*, 2007). Their beneficial actions are largely attributed to their ability to modulate molecular signalling pathways that govern inflammatory responses (González *et al.* 2011). Molecules such as VCAM-1, ICAM-1, and E-selectin promote leukocyte adhesion to vascular endothelium, triggering signal cascades that recruit immune cells and release pro-inflammatory cytokines, including TNF-α, IFN-γ, and chemokines like MCP-1. These events activate NF-κB, leading to the production of interleukins such as IL-6 and IL-8, along with MMPs, all of which contribute to inflammation and atherosclerotic progression (Hansson & Libby, 2006; Libby, 2006). For instance, resveratrol inhibits prostaglandin synthesis and cyclooxygenase (COX) activity while activating endothelial nitric oxide synthase (eNOS) and suppressing peroxisome proliferator-activated receptor gamma (PPARγ), contributing to its anti-inflammatory profile (Mohar & Malik, 2012; Biasutto *et al.*, 2012). Moderate consumption of red wine, rich in resveratrol, has been shown to reduce monocyte adhesion to endothelial cells (Mellor *et al.*, 2010). Furthermore, quince polyphenols reduce inflammatory responses by increasing IL-1β levels and inhibiting LPS-induced activation of NF-κB, p38MAPK, and AKT signalling pathways (Essafi-Benkhadir *et al.*, 2012). Similarly, other polyphenols like gingerol and quercetin can elevate adiponectin, a hormone with known anti-inflammatory properties (Leiherer *et al.*, 2013). Likewise, EGCG inhibits NF-κB activation, reduces nitric oxide production via iNOS suppression, and exhibits immunomodulatory effects (Singh *et al.*, 2011; Landis-Piwowar *et al.*, 2013). Compounds like curcumin, kaempferol-3-O-sophoroside, lycopene, and oleanolic acid inhibit the pro-inflammatory high mobility group box 1 (HMGB1) protein, further supporting their role in inflammatory regulation (Bae, 2012).

1.4.3 Antimicrobial and Antiviral Activities

Polyphenols exhibit broad-spectrum antimicrobial and antiviral properties. Their mechanisms of action include disrupting microbial membranes, altering permeability, and inhibiting nucleic acid synthesis, making them effective against a range of pathogens such as bacteria, fungi, and viruses. For instance, a polyphenol-rich mix extracted from almond skins has shown significant antibacterial and antiherpetic activity against methicillin-resistant *Staphylococcus aureus* (MRSA) and herpes simplex virus type I (HSV-I) (Musarra-Pizzo *et al.*, 2019). Similarly, walnut leaves possess anti-inflammatory and antimicrobial properties and are approved for treating mild skin inflammation. Ethanolic extracts of immature *Juglans regia* fruits inhibited MRSA growth (128–512 μg/mL) and reduced biofilm formation (16 μg/mL), suggesting potential for skin infection treatment (Quave *et al.*, 2008). Among flavonoids, apigenin stands out for its antibacterial and synergistic effects against oral pathogens, while luteolin, baicalein, and tangetin also exhibit strong antimicrobial properties. Notably, luteolin disrupts nucleic acid synthesis, impairs bacterial membranes, alters cell morphology, and inhibits biofilm formation (Gutiérrez-Venegas *et al.*, 2019; Chagas *et al.*, 2022). Polyphenols also show significant antifungal activity. For example, extracts from *Acacia nilotica, Cinnamomum zeylanicum,* and *Syzygium aromaticum* were effective against *Candida albicans* (Manso et al., 2021). Compounds such as gallic acid, ellagic acid, and corilagin have demonstrated inhibitory effects against *Candida* species and *Cryptococcus neoformans*, with minimum inhibitory concentrations (MICs) of gallic acid ranging from 12 to 100 μg/mL (Li *et al.*, 2017; Harir *et al.*, 2019). Red seaweed (*Jania rubens*) and flavonoids like quercetin, rutin, morin, and myricetin have demonstrated broad antimicrobial efficacy, including antifungal and antiviral effects (Sadeghi *et al.*, 2024). Compounds such as curcumin, catechin, and resveratrol have shown the ability to interfere with viral replication by targeting ACE2 receptors and viral proteases, effectively halting the synthesis of single-stranded RNA. Polyphenols like EGCG and quercetagetin inhibit SARS-CoV-2 RNA polymerase, while others, including luteolin and naringenin, block the viral spike (S) protein, preventing entry into host cells (Montenegro-Landívar *et al.*, 2021; Mandal & Domb, 2024; Sadeghi *et al.*, 2024).

1.4.4 Anticancer Potential

Polyphenols have emerged as promising agents in cancer prevention and treatment due to their ability to target various hallmarks of cancer. They modulate critical cellular processes such as apoptosis induction, cell proliferation inhibition, angiogenesis suppression, and prevention of metastasis by influencing key signalling pathways like PI3K/Akt, MAPK, and Wnt/β-catenin. Among the most studied polyphenols, curcumins have shown strong anticancer activity in multiple cancer types, including breast, colon, and endometrial cancers. In endometrial carcinoma (Ishikawa cells), curcumin induces apoptosis and S-phase cell cycle arrest while downregulating ERK and Jun mRNA expression. It also suppresses the p-ERK-2/c-Jun pathway, thereby reducing AP-1 activity and decreasing MMP2/9 expression, which ultimately hinders cell invasion (Cháirez-Ramírez *et al.*, 2021). Similarly, resveratrol demonstrates its anticancer effects by inhibiting proliferation and promoting apoptosis in multiple myeloma cells through the downregulation of antiapoptotic proteins such as Bcl-2, survivin, and XIAP (Bhardwaj *et al.*, 2007). Moreover, it interferes with PD-L1 glycosylation, promoting its endoplasmic retention and impairing its stability, thereby enhancing cytotoxic T-cell activity and immune-mediated tumour suppression (Verdura *et al.*, 2020). Quercetin, another potent flavonoid, directly inhibits the PI3K/Akt pathway and suppresses H-RAS–mediated invasion in human epithelial cells. Intravenous administration of quercetin was also found to suppress melanoma metastasis in mice by increasing eNOS expression and nitric oxide (NO) levels, which in turn triggered ceramide generation and apoptosis (Ferrer *et al.*, 2007; Kim *et al.*, 2023). Anthocyanins like delphinidin have been reported to induce apoptosis and arrest the cell cycle by inhibiting the NF-κB signalling pathway, showing efficacy across various cancer cell types (Yun *et al.*, 2009). Additionally, polyphenol-rich extracts from hawthorn fruit, both peel and flesh, were shown to induce apoptosis in MCF-7 breast cancer cells via the mitochondrial pathway (Li *et al.*, 2013).

1.5 PHARMACOLOGICAL AND THERAPEUTIC APPLICATIONS

Polyphenols exhibit a wide range of health-promoting effects and have been explored for their roles in preventing and managing various chronic diseases. Their therapeutic potential is evident across cardiovascular, metabolic, neurological, and oncological conditions. Various polyphenols have shown targeted therapeutic effects against specific diseases, acting through antioxidant, anti-inflammatory, anticancer, antimicrobial, and antiviral mechanisms. Their biological roles, sources, delivery methods, and pathways are summarized in Table 1.1.

1.5.1 Cardiovascular Health

Polyphenols are increasingly recognized for their significant contribution to cardiovascular health, primarily through their ability to enhance endothelial function, promote vasodilation, lower arterial blood pressure, and reduce low-density lipoprotein (LDL) cholesterol levels. Among these, compounds like resveratrol and

TABLE 1.1
Overview of Polyphenols, Their Sources, Delivery Modes, Target Diseases, and Mechanisms

Polyphenol	Source	Mode of Delivery	Target Disease/ Condition	Primary Activity	Mechanism/Target Pathway	Reference
Apigenin	Chamomile	Dietary	Oral infections	Antibacterial	Synergistic with antibiotics	Cháirez-Ramírez *et al.*, 2021
Baicalein	Scutellaria	Herb extract	MRSA, dental biofilm	Antibacterial	Antibiofilm, membrane disruption	Chagas *et al.*, 2022
Tangetin	Citrus fruits	Citrus peel	Dental caries	Antibacterial	Alters bacterial shape, biofilm	Davidova *et al.*, 2024; Mandal & Domb, 2024
Luteolin	Celery, carrot	Natural foods	Oral pathogens, SARS-CoV-2	Antibacterial, antiviral	Membrane disruption, DNA inhibition	Adamczak *et al.*, 2019
Delphinidin	Berries	Berry extract	Colon, Liver, Breast cancer	Anticancer	Inhibits NF-κB, induces apoptosis	Husain *et al.*, 2022
Genistein	Soybean	Soy products	Prostate cancer	Anticancer	Inhibits tyrosine kinases	Tuli *et al.*, 2019
Curcumin	Turmeric	Oral/supplement	Endometrial and breast cancer	Anticancer, antioxidant	Downregulates ERK/c-Jun, AP-1, MMPs	Cháirez-Ramírez *et al.*, 2021
Chlorogenic Acid	Coffee, apples	Coffee/leaf extract	Type 2 diabetes	Antidiabetic	Inhibits glucose-6-phosphatase	Lu *et al.*, 2020
Corilagin	Phyllanthus	Herbal extract	Candida, Cryptococcus	Antifungal	Inhibits cell growth	Manso *et al.*, 2021
Ellagic Acid	Berries	Pomegranate/fruits	*Candida*, oxidative stress	Antifungal, Antioxidant	Cell wall disruption, ROS scavenging	Mendes *et al.*, 2024
Kaempferol	Tea, broccoli	Foods/supplement	Autoimmune conditions	Anti-inflammatory	HMGB1 inhibition, IL suppression	Cháirez-Ramírez *et al.*, 2021
Gingerol	Ginger	Raw/extract	Obesity, metabolic syndrome	Anti-inflammatory	↑Adiponectin, ↓IL-6, TNF-α	Leiherer *et al.*, 2013
Oleanolic Acid	Olive	Olive extract	Rheumatoid arthritis	Anti-inflammatory	Inhibits HMGB1, blocks cytokines	Choi *et al.*, 2016

Fisetin	Strawberries	Fruit extract	Alzheimer's disease	Anti-inflammatory	Suppresses NF-κB, COX-2	Naeimi & Alizadeh, 2017
Hesperidin	Oranges	Citrus extract	Venous insufficiency	Anti-inflammatory	Downregulates cytokines	Li & Schluesener, 2017
Resveratrol	Grapes	Red wine/oral	Multiple myeloma, obesity	Anti-inflammatory, anticancer	Blocks PD-L1 glycosylation, activates T-cells	Verdura *et al.*, 2020; Cháirez-Ramírez *et al.*, 2021
Catechin	Tea, cocoa	Tea, extract	Cardiovascular diseases	Antioxidant	ROS scavenging, mitochondrial support	Gadkari & Balaraman, 2015
Rutin	Buckwheat	Citrus, buckwheat	Vascular dysfunction	Antioxidant	Enhances plasma antioxidant potential	Nazir *et al.*, 2018
Pterostilbene	Blueberries	Oral	Neurodegeneration	Antioxidant	Modulates oxidative stress genes	Tsai *et al.*, 2017
Naringenin	Grapefruit	Citrus	Liver diseases	Antioxidant	Activates AMPK, reduces ROS	Thirumalaisamy *et al.*, 2022
Quercetin	Onion, apple	Intravenous/oral	Melanoma, atherosclerosis	Antioxidant, anticancer	Inhibits PI3K/Akt, upregulates eNOS	Ferrer *et al.*, 2007; Chagas *et al.*, 2022
Gallic Acid	Tea, gallnuts	Natural extract	Oxidative stress, candidiasis	Antioxidant, antifungal	Free radical scavenging, ↓TBARS, ↑SOD	Xu *et al.*, 2021
Anthocyanins	Blueberries, black rice	Berries/juice	Ageing, diabetes	Antioxidant, anti-inflammatory	Activates Nrf2, inhibits NF-κB	Jeong *et al.*, 2013
EGCG	Green tea	Green tea/capsule	Inflammation, SARS-CoV-2	Antioxidant, immunomodulatory	Inhibits NF-κB, ↓NO production	Singh *et al.*, 2011; Cháirez-Ramírez *et al.*, 2021
Morin	Guava	Fruit extract	SARS-CoV-2, flu	Antiviral	Broad-spectrum inhibition	Nazir *et al.*, 2018; Davidova *et al.*, 2024
Myricetin	Berries	Dietary	SARS-CoV-2	Antiviral	Inhibits viral entry (spike protein)	Davidova *et al.*, 2024
Caffeic Acid	Coffee	Coffee/supplement	Hepatitis B, HSV	Antiviral	Inhibits virus replication	Vicente-Zurdo *et al.*, 2025

quercetin have gained considerable attention for their ability to enhance vascular performance and mitigate cardiovascular risk factors (Quiñones *et al.*, 2013; Iqbal *et al.*, 2023). Although the precise mechanisms underlying these cardioprotective actions are not yet fully elucidated, a growing body of evidence suggests that the cardiovascular benefits of polyphenol-rich diets, especially those high in fruits and vegetables, are closely linked to their antioxidant and anti-inflammatory properties (Iqbal *et al.*, 2023). Among the many polyphenolic subclasses, phenolic acids stand out due to their robust antioxidant potential. These molecules can donate electrons or hydrogen atoms to neutralize free radicals, thereby limiting oxidative stress and preventing endothelial injury and the progression of atherosclerosis (Prabhu *et al.*, 2021). Indeed, oxidative stress and ROS are widely implicated in cardiovascular pathologies, including endothelial dysfunction, sustained myocardial infarction, and ischemia-reperfusion injury (Sugamura *et al.*, 2011; Raedschelders *et al.*, 2012). Polyphenols have also been shown to improve flow-mediated dilation (FMD), a key indicator of vascular health, likely through the enhancement of NO availability. This effect is attributed to increased eNOS activity and a resulting NO-mediated vasodilation, which, along with modulation of the renin–angiotensin system, contributes to blood pressure regulation (Quiñones *et al.*, 2013; Potì *et al.*, 2019). Remarkably, such vascular improvements have been observed even at typical dietary polyphenol intakes. Resveratrol, for instance, not only stimulates eNOS but also inhibits NF-κB pathway activation, thereby reducing the expression of vascular adhesion molecules like ICAM-1 and VCAM-1 (Bonnefont-Rousselot, 2016).

1.5.2 Diabetes Management

Polyphenols play a significant role in diabetes management by improving insulin sensitivity and regulating glucose metabolism. Chlorogenic acid from coffee, for instance, reduces post-meal blood sugar by slowing intestinal glucose absorption. Studies in animals and some human trials show that polyphenols help lower blood glucose and enhance insulin response through various mechanisms, such as inhibiting carbohydrate digestion, decreasing glucose absorption, stimulating insulin secretion, reducing hepatic glucose output, and promoting glucose uptake in tissues (Kobayashi *et al.*, 2000; Johnston *et al.*, 2005; Aryaeian *et al.*, 2017). Additionally, polyphenols such as catechins and epicatechins from green tea, along with chlorogenic, ferulic, caffeic, and tannic acids and flavonoids like quercetin and naringenin, can interfere with intestinal glucose transport by inhibiting sodium-dependent glucose transporters SGLT1 and SGLT2 (Kobayashi *et al.*, 2000; Johnston *et al.*, 2005). Further studies have demonstrated that these compounds modulate postprandial glucose levels by enhancing insulin responsiveness and influencing gut hormones such as glucagon-like peptide-1 (GLP-1) and glucose-dependent insulinotropic polypeptide (GIP), which play roles in glucose regulation (Dao *et al.*, 2011). Experimental evidence indicates that supplementation with citrus bioflavonoids like hesperidin and naringin in diabetic rats increases hepatic glucokinase activity and glycogen storage, while reducing gluconeogenesis via downregulation of glucose-6-phosphatase and phosphoenolpyruvate carboxykinase, thereby improving glycaemic control (Jung *et al.*, 2006). Interestingly, co-treatment with quercetin and resveratrol significantly improved serum glucose, insulin levels, lipid profiles, oxidative stress biomarkers, and electrolyte balance in streptozotocin-induced diabetic rats, suggesting their potential as a combined therapeutic strategy (Yang & Kang, 2018). Moreover, quercetin's antioxidant and anti-inflammatory properties influence several molecular pathways, including TNFα, NF-κB, AMPK, AKT, and NRF2, that are crucial in the development of insulin resistance and type 2 diabetes (Yan *et al.*, 2023).

1.5.3 Neuroprotection

Polyphenols have emerged as promising natural compounds with significant neuroprotective properties, primarily due to their potent antioxidant and anti-inflammatory effects. Neurodegenerative diseases such as Alzheimer's are marked by the gradual deterioration of neurons, often resulting in cognitive decline and dementia among the elderly (Spagnuolo *et al.*, 2016). Various naturally occurring polyphenols, including EGCG, resveratrol, curcumin, and quercetin, have demonstrated their capacity to protect neuronal cells by alleviating oxidative stress, inhibiting inflammatory processes, and preventing the formation of harmful amyloid-beta plaques (Spagnuolo *et al.*, 2016). Compounds like mangiferin and morin have been found to defend the brain against ischemic damage by restoring mitochondrial stability and modulating critical signalling pathways such as NF-κB, Bax, Akt, and Erk1/2, thereby mitigating glutamate-induced toxicity (Campos-Esparza *et al.*, 2009). Furthermore, catechin and epicatechin gallate have shown the ability to inhibit the activation of microglia and astrocytes, which are commonly involved in neuroinflammation and the apoptotic death of neurons (Li *et al.*, 2004). Similarly, catechins may help prevent brain damage linked to neurotoxins involved in Parkinson's disease (Faria *et al.*, 2011). These catechin derivatives act through multiple mechanisms, such as scavenging free radicals, chelating metal ions, and regulating genes that promote cell survival, thus potentially delaying the progression of neurodegenerative conditions. Caffeic acid also contributes to neural protection by reducing the toxicity of amyloid-beta and limiting oxidative injury (Sul *et al.*, 2009). Among these, resveratrol is particularly notable for its ability to cross the blood–brain barrier and enhance neuronal resilience by promoting the breakdown of intracellular amyloid-beta, ultimately supporting neural function in Alzheimer's disease models (Richard *et al.*, 2011).

1.5.4 Cancer Prevention and Therapy

Polyphenols have emerged as promising agents in cancer prevention and therapy due to their ability to modulate multiple cellular pathways involved in tumour development and progression. These bioactive compounds exhibit anticancer effects by inhibiting proliferation, inducing apoptosis, suppressing metastasis, and enhancing the efficacy of conventional chemotherapeutic agents (Bhardwaj *et al.*, 2007). Curcumin, one of the most extensively studied polyphenols, has shown promise in combination with conventional treatments such as paclitaxel, enhancing its therapeutic efficacy in breast cancer models (Calaf *et al.*, 2018). In a clinical study conducted by Dhillon *et al.* (2008), high-dose curcumin (8 g/day) was administered to patients with advanced pancreatic cancer, where 2 out of 21 patients showed stable disease, accompanied by marked suppression of NF-κB activity, indicating potential therapeutic effects. Recent research highlights curcumin's role in modulating sphingolipid metabolism, particularly in colorectal cancer (CRC). In combination with standard CRC therapies, curcumin has been shown to induce apoptosis in chemoresistant CRC cells and reduce their invasive capabilities. These effects are mediated through enhanced caspase-3 activation and downregulation of the TGF-β/Smad2/3 signalling pathway in both in vitro and in vivo models (Yin *et al.*, 2019; Bracci *et al.*, 2021). Catechins, known for their antimicrobial activity, have demonstrated anticancer potential by inhibiting carcinogenesis in organs such as the skin, lungs, stomach, liver, and colon (Gadkari & Balaraman, 2015). Pomegranate-derived polyphenols, including ellagic acid and flavonoid glycosides like quercetin, kaempferol, and luteolin, show antiproliferative, anti-invasive, and pro-apoptotic effects by regulating Bcl-2 proteins and disrupting the cyclin-CDK network (Turrini *et al.*, 2015). Other notable phenolic compounds such as gallic acid, chlorogenic acid, caffeic acid, carnosol, capsaicin, 6-shogaol, and 6-gingerol have also demonstrated antimetastatic and anti-invasive properties (Weng & Yen, 2012). Moreover, recent studies emphasize the advantages of nanoformulations of polyphenols, such as liposomes, gold nanoparticles, solid lipid nanoparticles, and polymeric carriers, which improve anticancer efficacy by enhancing cellular uptake, enabling sustained release, and increasing pro-apoptotic activity in tumour cells (Davatgaran-Taghipour *et al.*, 2017).

1.6 BIOAVAILABILITY AND METABOLISM OF POLYPHENOLS

Bioavailability refers to the extent and rate at which a nutrient or bioactive compound is absorbed and becomes available at the site of physiological activity. Although polyphenols are abundant in plant-based foods, their systemic availability is generally limited due to poor intestinal absorption, rapid metabolism, and efficient excretion (Arora *et al.*, 2019). Their structural complexity governs their metabolic fate, influencing both absorption and biotransformation. After ingestion, polyphenols undergo enzymatic transformations throughout the digestive tract, involving both host enzymes and gut microbiota. In the small intestine and liver, they are further modified through phase I and phase II metabolic reactions such as oxidation, methylation, sulfation, and glucuronidation, which often produce metabolites with different bioactivities than the parent compounds (Manach *et al.*, 2004; De Souza *et al.*, 2015; Di Lorenzo *et al.*, 2021) (Figure 1.4). For instance, quercetin predominantly exists in glycosylated forms, which require hydrolysis before absorption. Similarly, curcumin exhibits low bioavailability due to rapid metabolism and elimination; however, co-administration with piperine can enhance its bioavailability by up to 20-fold by inhibiting glucuronidation (Shoba *et al.*, 1998; Bertoncini-Silva *et al.*, 2024). Moreover, gut microbiota plays a critical role in polyphenol metabolism.

Unabsorbed polyphenols that reach the colon are transformed by microbial enzymes into bioactive phenolic acids. For example, ellagitannins found in pomegranate are converted by gut bacteria into urolithins, which exhibit anti-inflammatory and anticancer properties (García-Villalba *et al.*, 2013). Human studies highlight significant variability in polyphenol bioavailability, largely influenced by their chemical structure and individual physiological differences. For instance,

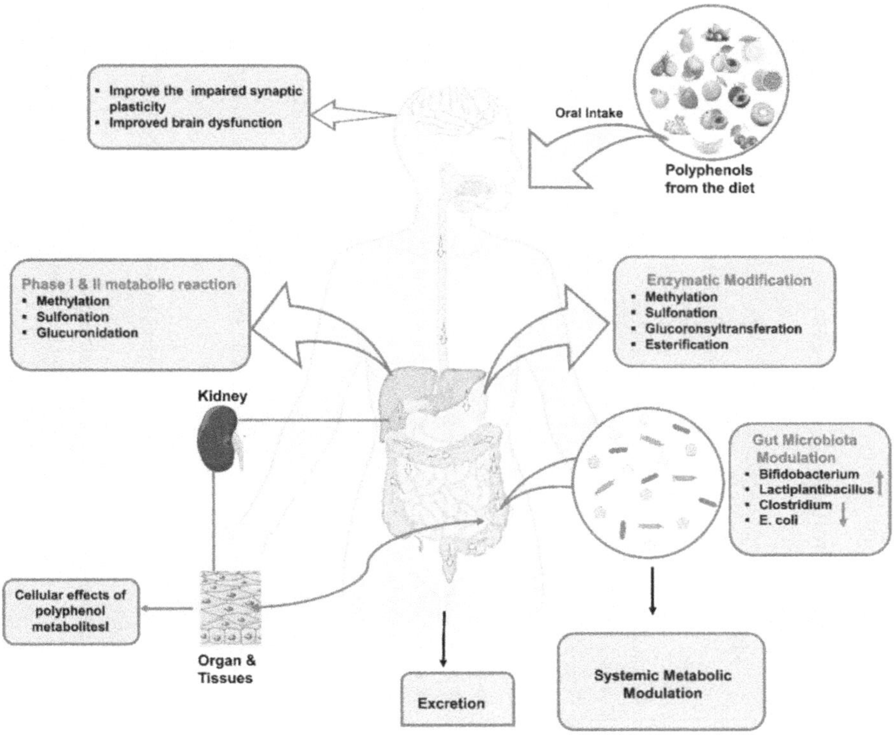

FIGURE 1.4 Systemic representation of polyphenol bioavailability and metabolism following oral intake, highlighting enzymatic modification, gut microbiota interaction, metabolic pathways, cellular effects, and excretion.

urinary excretion is low for quercetin and its glycoside rutin (0.3–1.4%), but much higher for catechins, isoflavones, flavanones, and anthocyanidins (3–26%). Naringin absorption from grapefruit juice varies significantly among individuals (5–57%) due to genetic factors and gut microbiota composition (Scalbert & Williamson, 2000). Additional factors such as the food matrix, dosage, age, gastrointestinal health, and genetic polymorphisms also contribute to interindividual differences in polyphenol bioavailability. Despite their promising in vitro effects, the limited and variable in vivo bioavailability of polyphenols poses a major hurdle in realizing their clinical potential. Consequently, improving their bioavailability through advanced delivery systems remains a key objective in nutraceutical development (Del Rio *et al.*, 2013).

1.7 RECENT ADVANCES IN POLYPHENOL-BASED NUTRACEUTICAL FORMULATIONS

In recent years, significant advancements have been made in the development of polyphenol-based nutraceuticals, with a focus on enhancing their therapeutic efficacy. Despite their potent antioxidant, anti-inflammatory, and disease-preventing

TABLE 1.2
Overview of Key Polyphenolic Compounds, Dietary Sources, Bioavailability Limitations, Enhancement Approaches, and Associated Therapeutic Benefits

Compound	Source	Bioavailability Effect	Mechanism of Low Bioavailability	Enhancement Strategies	Health Relevance	References
Apigenin	Chamomile	Low solubility	Extensive metabolism	Nanocrystal formulations	Enhances neuroprotection via improved absorption	Adel *et al.*, 2022
Catechin	Tea, cocoa	Moderate, quickly metabolized	Intestinal degradation	Cyclodextrin complexation	Improves cardiovascular bioactivity	Cai *et al.*, 2018
Chlorogenic acid	Coffee, apples	Moderate absorption	Mildly hydrolyzed	Stabilization by proteins	Maintains glucose-modulating and antioxidant roles	Lu *et al.*, 2020
Curcumin	Turmeric	Very low absorption	Glucuronidation and poor solubility	Piperine co-administration	Improves therapeutic potential by slowing metabolism	Anand *et al.*, 2007
Cyanidin-3-glucoside	Berries	Unstable under pH changes	Degradation in intestine	Encapsulation in protein/polysaccharide	Preserves antioxidant properties in food applications	Zannou *et al.*, 2023
Delphinidin	Blackcurrant, eggplant	Low intestinal stability	pH-sensitive degradation	Encapsulation in liposomes	Enhances anti-angiogenic properties	da Silva *et al.*, 2023
EGCG	Green tea	Low systemic absorption	Methylation, sulfation, glucuronidation	Nanoencapsulation, piperine	Enhances antioxidant action through better stability and absorption	Singh *et al.*, 2011
Ellagic acid	Pomegranate, berries	Poor water solubility, gut conversion to urolithins	Microbial metabolism	Probiotics, formulated with fibre	Urolithins provide superior bioactivity	Leng *et al.*, 2025
Ferulic acid	Whole grains	Good permeability	Partial degradation	Co-formulation with flavonoids	Boosts antioxidant and dermal benefits	Pyrzynska, 2024
Gallic acid	Grapes, berries	Highly water soluble	Passive absorption only	Use of stabilizing agents	Improves anti-inflammatory and liver-protective functions	Ferruzzi *et al.*, 2009

Genistein	Soybeans	Moderate, higher in aglycone form	Glycoside form less absorbed	Enzymatic treatment of soy	Improves bioavailability and anticancer potential	Tang *et al.*, 2011
Hesperidin	Oranges	Poorly absorbed glycoside	Needs intestinal conversion	Fermentation, hydrolysis techniques	Gut microbes essential for vascular benefits	Rapisarda *et al.*, 2022
Kaempferol	Kale, spinach	Limited oral absorption	P-glycoprotein efflux	Nanoemulsion, SMEDDS	Supports cardiovascular protection via nanodelivery	Kothari *et al.*, 2020
Luteolin	Parsley, celery	Lipophilic, poorly absorbed	First-pass metabolism	Nano-liposomes, solid carriers	Strengthens systemic anti-inflammatory effects	Taban *et al.*, 2023
Naringenin	Citrus fruits	Low absorption	Enzymatic hydrolysis	Inclusion complexes	Supports metabolic health and anti-obesity action	Yang *et al.*, 2022
Quercetin	Onions, apples	Poor solubility	Limited absorption in small intestine	Phytosomes, micelles	Enhanced antioxidant and anticancer efficacy	Manach *et al.*, 2004
Resveratrol	Grapes, red wine	Low bioavailability	Rapid hepatic metabolism	Liposomes, solid lipid nanoparticles	Activates SIRT1, useful in anti-ageing despite low levels	Cháirez-Ramírez *et al.*, 2021
Rutin	Buckwheat, citrus peels	Needs deglycosylation to be active	Gut enzyme dependency	Enzyme-assisted extraction	Enhances vascular benefits via increased aglycone levels	de Araújo *et al.*, 2013
Tannic acid	Tea, wine	Poor mineral bioavailability	Forms complexes with proteins	Reduce concentration in formulations	Excess may hinder mineral absorption	Jaramillo *et al.*, 2015
Theaflavin	Black tea	Poor bioavailability	pH sensitive, metabolized rapidly	Nanoformulation in beverages	Improves antioxidant and lipid-lowering effects	Pereira-Caro *et al.*, 2017

properties, the clinical application of polyphenols is often hindered by poor water solubility, low chemical stability, and limited bioavailability (Dini & Grumetto, 2022). To address these limitations, various nanotechnology-based delivery systems have been developed, including organic (protein- and polysaccharide-based) and inorganic (gold, silver, silica) nanoparticles (Salah *et al.*, 2020). Techniques such as nanoencapsulation, ionic gelation, and liposomal delivery have shown promise in protecting polyphenols from degradation and enabling sustained release (Pereira *et al.*, 2018; Ayala-Fuentes & Chavez-Santoscoy, 2021). Nanoscale platforms like nano-emulsions, nanocrystals, liposomes, phytosomes, ethosomes, and solid lipid nanoparticles have enhanced the pharmacokinetic properties of these bioactives (Pimentel-Moral *et al.*, 2016; Rudrapal *et al.*, 2022). For instance, nano-lipid carriers loaded with EGCG or curcumin have demonstrated increased blood concentrations and prolonged therapeutic action (Yan *et al.*, 2019). Additionally, co-delivery with other bioactives, the use of biopolymer-based coatings, and the 3D printing of nutraceuticals are also being explored to create tailor-made health solutions (Melocchi *et al.*, 2018; Rocha *et al.*, 2022). These advancements not only enhance the therapeutic potential of polyphenols but also support the development of next-generation nutraceuticals aligned with consumer demand for safe, effective, and natural health products.

1.8 REGULATORY FRAMEWORK, SAFETY CONSIDERATIONS, AND EXISTING CHALLENGES

Polyphenol-based nutraceuticals have attracted considerable interest owing to their potent antioxidant, anti-inflammatory, and disease-modulating properties. Despite their natural origin and growing market acceptance, critical concerns persist regarding product safety, standardization, and clinical efficacy. While often assumed to be inherently safe, these compounds exhibit considerable variability in bioavailability and may cause adverse effects if improperly processed or consumed inappropriately. Regulatory frameworks governing these products differ globally, complicating their development and approval. Agencies such as the Food and Drug Administration (FDA, USA), European Food Safety Authority (EFSA, Europe), and the Food Safety and Standards Authority of India (FSSAI, India) mandate transparency in health claims, dosage precision, and labelling to safeguard consumer welfare and ensure regulatory compliance (Girdhar *et al.*, 2017; Patanè *et al.*, 2019). However, the absence of globally harmonized definitions and regulatory standards presents a major obstacle. Additional challenges include compositional variability of raw materials, inconsistencies in extraction and formulation methods, and a paucity of large-scale clinical trials validating efficacy. To ensure the safe and effective deployment of polyphenol-based nutraceuticals, it is imperative to integrate rigorous scientific validation with globally coherent regulatory frameworks. Such alignment will foster innovation while securing consumer confidence (Dwyer *et al.*, 2018).

1.9 FUTURE PROSPECTS AND CONCLUSION

Polyphenols have emerged as powerful natural compounds known for their antioxidant, anti-inflammatory, and disease-preventive effects. Their widespread presence in plants and significant therapeutic potential have made them a key area of focus across various scientific disciplines, including nutrition, medicine, and agriculture. Despite advances in research, limitations such as low bioavailability and poor stability still hinder their full potential in clinical and nutraceutical applications. Future efforts should aim to improve extraction methods and develop advanced delivery systems, such as nanoformulations and sustainable technologies, to enhance their effectiveness. The discovery of new plant sources and further clinical validation are also essential to ensure safety, efficacy, and standardized use. As research expands, polyphenols are expected to play a vital role in next-generation functional foods and nutraceuticals, offering promising solutions for managing oxidative stress, inflammation, and chronic diseases. Overall, polyphenols represent a key bridge between plant science and human health, and with continued innovation, they are set to become integral to future healthcare and wellness strategies.

REFERENCES

Abbas, M., Saeed, F., Anjum, F. M., Afzaal, M., Tufail, T., Bashir, M. S., . . . Suleria, H. A. R. (2017). Natural polyphenols: An overview. *International Journal of Food Properties*, 20(8), 1689–1699.

Adamczak, A., Ożarowski, M., & Karpiński, T. M. (2019). Antibacterial activity of some flavonoids and organic acids widely distributed in plants. *Journal of Clinical Medicine*, 9(1), 109.

Adel, M., Zahmatkeshan, M., Akbarzadeh, A., Rabiee, N., Ahmadi, S., Keyhanvar, P., . . . Seifalian, A. M. (2022). Chemotherapeutic effects of Apigenin in breast cancer: Preclinical evidence and molecular mechanisms; enhanced bioavailability by nanoparticles. *Biotechnology Reports*, 34, e00730.

Ahmad, A., Ahmad, V., Zamzami, M. A., Chaudhary, H., Baothman, O. A., Hosawi, S., . . . Khan, M. J. (2021). Introduction and classification of natural polyphenols. *Polyphenols-Based Nanotherapeutics for Cancer Management*, 1–16.

Ameer, K., Shahbaz, H. M., & Kwon, J. H. (2017). Green extraction methods for polyphenols from plant matrices and their byproducts: A review. *Comprehensive Reviews in Food Science and Food Safety*, 16(2), 295–315.

Anand, P., Kunnumakkara, A. B., Newman, R. A., & Aggarwal, B. B. (2007). Bioavailability of curcumin: Problems and promises. *Molecular Pharmaceutics*, 4(6), 807–818.

Arora, I., Sharma, M., & Tollefsbol, T. O. (2019). Combinatorial epigenetics impact of polyphenols and phytochemicals in cancer prevention and therapy. *International Journal of Molecular Sciences*, 20(18), 4567.

Aryaeian, N., Sedehi, S. K., & Arablou, T. (2017). Polyphenols and their effects on diabetes management: A review. *Medical Journal of the Islamic Republic of Iran*, 31, 134.

Ayala-Fuentes, J. C., & Chavez-Santoscoy, R. A. (2021). Nanotechnology as a key to enhance the benefits and improve the bioavailability of flavonoids in the food industry. *Foods*, 10(11), 2701.

Bae, J. S. (2012). Role of high mobility group box 1 in inflammatory disease: Focus on sepsis. *Archives of Pharmacal Research*, 35, 1511–1523.

Bertelli, A., Biagi, M., Corsini, M., Baini, G., Cappellucci, G., & Miraldi, E. (2021). Polyphenols: From theory to practice. *Foods*, 10(11), 2595.

Bertoncini-Silva, C., Vlad, A., Ricciarelli, R., Giacomo Fassini, P., Suen, V. M. M., & Zingg, J. M. (2024). Enhancing the bioavailability and bioactivity of curcumin for disease prevention and treatment. *Antioxidants*, 13(3), 331.

Bhardwaj, A., Sethi, G., Vadhan-Raj, S., Bueso-Ramos, C., Takada, Y., Gaur, U., . . . Aggarwal, B. B. (2007). Resveratrol inhibits proliferation, induces apoptosis, and overcomes chemoresistance through down-regulation of STAT3 and nuclear factor-κB–regulated antiapoptotic and cell survival gene products in human multiple myeloma cells. *Blood*, 109(6), 2293–2302.

Biasutto, L., Mattarei, A., & Zoratti, M. (2012). Resveratrol and health: the starting point. *Chembiochem: A European Journal of Chemical Biology*, 13(9), 1256–1259.

Bonnefont-Rousselot, D. (2016). Resveratrol and cardiovascular diseases. *Nutrients*, 8(5), 250.

Bracci, L., Fabbri, A., Del Corno, M., & Conti, L. (2021). Dietary polyphenols: Promising adjuvants for colorectal cancer therapies. *Cancers*, 13(18), 4499.

Cai, Z. Y., Li, X. M., Liang, J. P., Xiang, L. P., Wang, K. R., Shi, Y. L., . . . Liang, Y. R. (2018). Bioavailability of tea catechins and its improvement. *Molecules*, 23(9), 2346.

Calaf, G. M., Ponce-Cusi, R., & Carrión, F. (2018). Curcumin and paclitaxel induce cell death in breast cancer cell lines. *Oncology Reports*, 40(4), 2381–2388.

Campos-Esparza, M. R., Sánchez-Gómez, M. V., & Matute, C. (2009). Molecular mechanisms of neuroprotection by two natural antioxidant polyphenols. *Cell Calcium*, 45(4), 358–368.

Cao, D., Zhao, M., Wan, C., Zhang, Q., Tang, T., Liu, J., . . . Jiang, C. (2019). Role of tea polyphenols in delaying hyperglycemia-induced senescence in human glomerular mesangial cells via miR-126/Akt–p53–p21 pathways. *International Urology and Nephrology*, 51, 1071–1078.

Chagas, M. D. S. S., Behrens, M. D., Moragas-Tellis, C. J., Penedo, G. X., Silva, A. R., & Gonçalves-de-Albuquerque, C. F. (2022). Flavonols and flavones as potential anti-inflammatory, antioxidant, and antibacterial compounds. *Oxidative Medicine and Cellular Longevity*, 2022(1), 9966750.

Cháirez-Ramírez, M. H., de la Cruz-López, K. G., & García-Carrancá, A. (2021). Polyphenols as antitumor agents targeting key players in cancer-driving signaling pathways. *Frontiers in Pharmacology*, 12, 710304.

Cheynier, V., Comte, G., Davies, K. M., Lattanzio, V., & Martens, S. (2013). Plant phenolics: Recent advances on their biosynthesis, genetics, and ecophysiology. *Plant Physiology And Biochemistry*, 72, 1–20.

Choi, J. K., Kim, S. W., Kim, D. S., Lee, J. Y., Lee, S., Oh, H. M., . . . Kim, S. H. (2016). Oleanolic acid acetate inhibits rheumatoid arthritis by modulating T cell immune responses and matrix-degrading enzymes. *Toxicology and Applied Pharmacology*, 290, 1–9.

Chung, K. T., Wong, T. Y., Wei, C. I., Huang, Y. W., & Lin, Y. (1998). Tannins and human health: A review. *Critical Reviews in Food Science and Nutrition*, 38(6), 421–464.

Cui, L., Liu, X., Tian, Y., Xie, C., Li, Q., Cui, H., & Sun, C. (2016). Flavonoids, flavonoid subclasses, and esophageal cancer risk: A meta-analysis of epidemiologic studies. *Nutrients*, 8(6), 350.

da Silva, A. P. G., Zia, S., John, O. D., de Souza, M. C., da Silva, L. C., & Sganzela, W. G. (2023). Delphinidin: Sources, biosynthesis, bioavailability, bioactivity, and pharmacology. In *Handbook of Dietary Flavonoids* (pp. 1–31). Cham: Springer International Publishing.

Dahmoune, F., Nayak, B., Moussi, K., Remini, H., & Madani, K. (2015). Optimization of microwave-assisted extraction of polyphenols from Myrtus communis L. leaves. *Food Chemistry*, 166, 585–595.

Dao, T. M. A., Waget, A., Klopp, P., Serino, M., Vachoux, C., Pechere, L., . . . Seree, E. (2011). Resveratrol increases glucose induced GLP-1 secretion in mice: A mechanism which contributes to the glycemic control. *PLoS One*, 6(6), e20700.

Dare, R. G., Nakamura, C. V., Ximenes, V. F., & Lautenschlager, S. O. (2020). Tannic acid, a promising anti-photoaging agent: Evidences of its antioxidant and anti-wrinkle potentials, and its ability to prevent photodamage and MMP-1 expression in L929 fibroblasts exposed to UVB. *Free Radical Biology and Medicine*, 160, 342–355.

Davatgaran-Taghipour, Y., Masoomzadeh, S., Farzaei, M. H., Bahramsoltani, R., Karimi-Soureh, Z., Rahimi, R., & Abdollahi, M. (2017). Polyphenol nanoformulations for cancer therapy: Experimental evidence and clinical perspective. *International Journal of Nanomedicine*, 2689–2702.

Davidova, S., Galabov, A. S., & Satchanska, G. (2024). Antibacterial, antifungal, antiviral activity, and mechanisms of action of plant polyphenols. *Microorganisms*, 12(12), 2502.

de Araújo, M. E. M. B., Franco, Y. E. M., Alberto, T. G., Sobreiro, M. A., Conrado, M. A., Priolli, D. G., . . . de Oliveira Carvalho, P. (2013). Enzymatic de-glycosylation of rutin improves its antioxidant and antiproliferative activities. *Food Chemistry*, 141(1), 266–273.

De Souza, J. E., Casanova, L. M., & Costa, S. S. (2015). Bioavailability of phenolic compounds: A major challenge for drug development? *Revista Fitos*, 9(1), 55–67.

Del Rio, D., Rodriguez-Mateos, A., Spencer, J. P., Tognolini, M., Borges, G., & Crozier, A. (2013). Dietary (poly) phenolics in human health: Structures, bioavailability, and evidence of protective effects against chronic diseases. *Antioxidants & Redox Signalling*, 18(14), 1818–1892.

Dhillon, N., Aggarwal, B. B., Newman, R. A., Wolff, R. A., Kunnumakkara, A. B., Abbruzzese, J. L., . . . Kurzrock, R. (2008). Phase II trial of curcumin in patients with advanced pancreatic cancer. *Clinical Cancer Research*, 14(14), 4491–4499.

Di Lorenzo, C., Colombo, F., Biella, S., Stockley, C., & Restani, P. (2021). Polyphenols and human health: The role of bioavailability. *Nutrients*, 13(1), 273.

Ding, S. M., Zhang, Z. H., Song, J., Cheng, X. D., Jiang, J., & Jia, X. B. (2014). Enhanced bioavailability of apigenin via preparation of a carbon nanopowder solid dispersion. *International Journal of Nanomedicine*, 2327–2333.

Dini, I., & Grumetto, L. (2022). Recent advances in natural polyphenol research. *Molecules*, 27(24), 8777.

Dong, X., Lan, W., Yin, X., Yang, C., Wang, W., & Ni, J. (2017). Simultaneous determination and pharmacokinetic study of quercetin, luteolin, and apigenin in rat plasma after oral administration of Matricaria chamomilla L. extract by HPLC-UV. *Evidence-Based Complementary and Alternative Medicine*, 2017(1), 8370584.

Dwyer, J. T., Coates, P. M., & Smith, M. J. (2018). Dietary supplements: regulatory challenges and research resources. *Nutrients*, 10(1), 41.

Ecevit, K., Barros, A. A., Silva, J. M., & Reis, R. L. (2022). Preventing microbial infections with natural phenolic compounds. *Future Pharmacology*, 2(4), 460–498.

Eisen, B., Ungar, Y., & Shimoni, E. (2003). Stability of isoflavones in soy milk stored at elevated and ambient temperatures. *Journal of Agricultural and Food Chemistry*, 51(8), 2212–2215.

Essafi-Benkhadir, K., Refai, A., Riahi, I., Fattouch, S., Karoui, H., & Essafi, M. (2012). Quince (Cydonia oblonga Miller) peel polyphenols modulate LPS-induced inflammation in human THP-1-derived macrophages through NF-κB, p38MAPK and Akt inhibition. *Biochemical and Biophysical Research Communications*, 418(1), 180–185.

Ezzat, S. M., Shouman, S. A., Elkhoely, A., Attia, Y. M., Elsesy, M. S., El Senousy, A. S., . . . El Tanbouly, N. (2018). Anticancer potentiality of lignan rich fraction of six Flaxseed cultivars. *Scientific Reports*, 8(1), 544.

Faria, A., Pestana, D., Teixeira, D., Couraud, P. O., Romero, I., Weksler, B., . . . Calhau, C. (2011). Insights into the putative catechin and epicatechin transport across blood-brain barrier. *Food & Function*, 2(1), 39–44.

Ferrer, P., Asensi, M., Priego, S., Benlloch, M., Mena, S., Ortega, A., . . . Estrela, J. M. (2007). Nitric oxide mediates natural polyphenol-induced Bcl-2 down-regulation and activation of cell death in metastatic B16 melanoma. *Journal of Biological Chemistry*, 282(5), 2880–2890.

Ferruzzi, M. G., Lobo, J. K., Janle, E. M., Cooper, B., Simon, J. E., Wu, Q. L., . . . Pasinetti, G. M. (2009). Bioavailability of gallic acid and catechins from grape seed polyphenol extract is improved by repeated dosing in rats: Implications for treatment in Alzheimer's disease. *Journal of Alzheimer's Disease*, 18(1), 113–124.

Gadkari, P. V., & Balaraman, M. (2015). Catechins: Sources, extraction and encapsulation: A review. *Food and Bioproducts Processing*, 93, 122–138.

García, D. E., Glasser, W. G., Pizzi, A., Paczkowski, S. P., & Laborie, M. P. (2016). Modification of condensed tannins: From polyphenol chemistry to materials engineering. *New Journal of Chemistry*, 40(1), 36–49.

García-Villalba, R., Beltrán, D., Espín, J. C., Selma, M. V., & Tomás-Barberán, F. A. (2013). Time course production of urolithins from ellagic acid by human gut microbiota. *Journal of Agricultural and Food Chemistry*, 61(37), 8797–8806.

Girdhar, S., Pandita, D., Girdhar, A., & Lather, V. (2017). Safety, quality and regulatory aspects of nutraceuticals. *Applied Clinical Research, Clinical Trials and Regulatory Affairs*, 4(1), 36–42.

Godos, J., Marventano, S., Mistretta, A., Galvano, F., & Grosso, G. (2017). Dietary sources of polyphenols in the mediterranean healthy eating, aging and lifestyle (MEAL) study cohort. *International Journal of Food Sciences and Nutrition*, 68(6), 750–756.

González, R., Ballester, I., López-Posadas, R., Suárez, M. D., Zarzuelo, A., Martínez-Augustin, O., & Medina, F. S. D. (2011). Effects of flavonoids and other polyphenols on inflammation. *Critical Reviews in Food Science and Nutrition*, 51(4), 331–362.

Gutiérrez-Venegas, G., Gómez-Mora, J. A., Meraz-Rodríguez, M. A., Flores-Sánchez, M. A., & Ortiz-Miranda, L. F. (2019). Effect of flavonoids on antimicrobial activity of microorganisms present in dental plaque. *Heliyon*, 5(12).

Hansson, G. K., & Libby, P. (2006). The immune response in atherosclerosis: A double-edged sword. *Nature Reviews Immunology*, 6(7), 508–519.

Harir, M., Bendif, H., Yahiaoui, M., Bellahcene, M., Zohra, F., & Rodríguez-Couto, S. (2019). Evaluation of antimicrobial activity of Terfezia arenaria extracts collected from Saharan desert against bacteria and filamentous fungi. *3 Biotech*, 9(7), 281.

Heim, K. E., Tagliaferro, A. R., & Bobilya, D. J. (2002). Flavonoid antioxidants: Chemistry, metabolism and structure-activity relationships. *The Journal of Nutritional Biochemistry*, 13(10), 572–584.

Horvathova, M., Orszaghova, Z., Laubertova, L., Vavakova, M., Sabaka, P., Rohdewald, P., . . . Muchova, J. (2014). Effect of the French oak wood extract Robuvit on markers of oxidative stress and activity of antioxidant enzymes in healthy volunteers: A pilot study. *Oxidative Medicine and Cellular Longevity*, 2014(1), 639868.

Husain, A., Chanana, H., Khan, S. A., Dhanalekshmi, U. M., Ali, M., Alghamdi, A. A., & Ahmad, A. (2022). Chemistry and pharmacological actions of delphinidin, a dietary purple pigment in anthocyanidin and anthocyanin forms. *Frontiers in Nutrition*, 9, 746881.

Imran, M., Ahmad, N., Anjum, F. M., Khan, M. K., Mushtaq, Z., Nadeem, M., & Hussain, S. (2015). Potential protective properties of flax lignan secoisolariciresinol diglucoside. *Nutrition Journal*, 14, 1–7.

Iqbal, I., Wilairatana, P., Saqib, F., Nasir, B., Wahid, M., Latif, M. F., . . . Mubarak, M. S. (2023). Plant polyphenols and their potential benefits on cardiovascular health: A review. *Molecules*, 28(17), 6403.

Jang, W. Y., Kim, M. Y., & Cho, J. Y. (2022). Antioxidant, anti-inflammatory, anti-menopausal, and anti-cancer effects of lignans and their metabolites. *International Journal of Molecular Sciences*, 23(24), 15482.

Jaramillo, Á., Briones, L., Andrews, M., Arredondo, M., Olivares, M., Brito, A., & Pizarro, F. (2015). Effect of phytic acid, tannic acid and pectin on fasting iron bioavailability both in the presence and absence of calcium. *Journal of Trace Elements in Medicine and Biology*, 30, 112–117.

Jeong, J. W., Lee, W. S., Shin, S. C., Kim, G. Y., Choi, B. T., & Choi, Y. H. (2013). Anthocyanins downregulate lipopolysaccharide-induced inflammatory responses in BV2 microglial cells by suppressing the NF-κB and Akt/MAPKs signaling pathways. *International Journal of Molecular Sciences*, 14(1), 1502–1515.

Johnston, K., Sharp, P., Clifford, M., & Morgan, L. (2005). Dietary polyphenols decrease glucose uptake by human intestinal Caco-2 cells. *FEBS Letters*, 579(7), 1653–1657.

Jung, U. J., Lee, M. K., Park, Y. B., Kang, M. A., & Choi, M. S. (2006). Effect of citrus flavonoids on lipid metabolism and glucose-regulating enzyme mRNA levels in type-2 diabetic mice. *The International Journal of Biochemistry & Cell Biology*, 38(7), 1134–1145.

Kim, K. H., Ki, M. R., Min, K. H., & Pack, S. P. (2023). Advanced delivery system of polyphenols for effective cancer prevention and therapy. *Antioxidants*, 12(5), 1048.

Kobayashi, Y., Suzuki, M., Satsu, H., Arai, S., Hara, Y., Suzuki, K., . . . Shimizu, M. (2000). Green tea polyphenols inhibit the sodium-dependent glucose transporter of intestinal epithelial cells by a competitive mechanism. *Journal of Agricultural and Food Chemistry*, 48(11), 5618–5623.

Kopustinskiene, D. M., Jakstas, V., Savickas, A., & Bernatoniene, J. (2020). Flavonoids as anticancer agents. *Nutrients*, 12(2), 457.

Kothari, D., Lee, W. D., & Kim, S. K. (2020). Allium flavonols: Health benefits, molecular targets, and bioavailability. *Antioxidants*, 9(9), 888.

Kuhnle, G. G. (2018). Nutrition epidemiology of flavan-3-ols: The known unknowns. *Molecular Aspects of Medicine*, 61, 2–11.

Laavola, M., Leppänen, T., Eräsalo, H., Hämäläinen, M., Nieminen, R., & Moilanen, E. (2017). Anti-inflammatory effects of nortrachelogenin in murine J774 macrophages and in carrageenan-induced paw edema model in the mouse. *Planta Medica*, 234(06), 519–526.

Landis-Piwowar, K., Chen, D., Foldes, R., Chan, T. H., & Dou, Q. P. (2013). Novel epigallocatechin gallate analogs as potential anticancer agents: A patent review (2009–present). *Expert Opinion on Therapeutic Patents*, 23(2), 189–202.

Lang, Y., Gao, N., Zang, Z., Meng, X., Lin, Y., Yang, S., . . . Li, B. (2024). Classification and antioxidant assays of polyphenols: A review. *Journal of Future Foods*, 4(3), 193–204.

Leiherer, A., Mündlein, A., & Drexel, H. (2013). Phytochemicals and their impact on adipose tissue inflammation and diabetes. *Vascular Pharmacology*, 58(1–2), 3–20.

Leng, P., Wang, Y., & Xie, M. (2025). Ellagic acid and gut microbiota: Interactions, and implications for health. *Food Science & Nutrition*, 13(4), e70133.

Li, C., & Schluesener, H. (2017). Health-promoting effects of the citrus flavanone hesperidin. *Critical Reviews in Food Science and Nutrition*, 57(3), 613–631.

Li, K., Xu, H., Liu, Y., Zhong, W., Jin, Y., & Wu, W. (2024). Exploring the relationship between lignin structure and antioxidant property using lignin model compounds. *International Journal of Biological Macromolecules*, 282, 136786.

Li, L. C., Piao, H. M., Zheng, M. Y., Lin, Z. H., Li, G., & Yan, G. H. (2016). Sesamin attenuates mast cell-mediated allergic responses by suppressing the activation of p38 and nuclear factor-κB. *Molecular Medicine Reports*, 13(1), 536–542.

Li, R., Huang, Y. G., Fang, D., & Le, W. D. (2004). (−)-Epigallocatechin gallate inhibits lipopolysaccharide-induced microglial activation and protects against inflammation-mediated dopaminergic neuronal injury. *Journal of Neuroscience Research*, 78(5), 723–731.

Li, T., Zhu, J., Guo, L., Shi, X., Liu, Y., & Yang, X. (2013). Differential effects of polyphenols-enriched extracts from hawthorn fruit peels and fleshes on cell cycle and apoptosis in human MCF-7 breast carcinoma cells. *Food Chemistry*, 141(2), 1008–1018.

Li, Z. J., Liu, M., Dawuti, G., Dou, Q., Ma, Y., Liu, H. G., & Aibai, S. (2017). Antifungal activity of gallic acid in vitro and in vivo. *Phytotherapy Research*, 31(7), 1039–1045.

Libby, P. (2006). Inflammation and cardiovascular disease mechanisms. *The American Journal of Clinical Nutrition*, 83(2), 456S–460S.

Lu, H., Tian, Z., Cui, Y., Liu, Z., & Ma, X. (2020). Chlorogenic acid: A comprehensive review of the dietary sources, processing effects, bioavailability, beneficial properties, mechanisms of action, and future directions. *Comprehensive Reviews in Food Science and Food Safety*, 19(6), 3130–3158.

Mal, S., & Pal, D. (2020). Tannins and polyphenols extracted from natural plants and their versatile application. *Bioactive Natural Products for Pharmaceutical Applications*, 715–757.

Manach, C., Scalbert, A., Morand, C., Rémésy, C., & Jiménez, L. (2004). Polyphenols: Food sources and bioavailability. *The American Journal of Clinical Nutrition*, 79(5), 727–747.

Mandal, M. K., & Domb, A. J. (2024). Antimicrobial activities of natural bioactive polyphenols. *Pharmaceutics*, 16(6), 718.

Manso, T., Lores, M., & de Miguel, T. (2021). Antimicrobial activity of polyphenols and natural polyphenolic extracts on clinical isolates. *Antibiotics*, 11(1), 46.

Mellor, D. D., Sathyapalan, T., Kilpatrick, E. S., Beckett, S., & Atkin, S. L. (2010). High-cocoa polyphenol-rich chocolate improves HDL cholesterol in Type 2 diabetes patients. *Diabetic Medicine*, 27(11), 1318–1321.

Melocchi, A., Parietti, F., Maccagnan, S., Ortenzi, M. A., Antenucci, S., Briatico-Vangosa, F., . . . Zema, L. (2018). Industrial development of a 3D-printed nutraceutical delivery platform in the form of a multicompartment HPC capsule. *AAPS PharmSciTech*, 19, 3343–3354.

Mendes, A. G. G., Campos, C. D. L., Pereira-Filho, J. L., Pereira, A. P. A., Reis, G. S. A., Araújo, Á. W. D. M. S., . . . Monteiro-Neto, V. (2024). Ellagic acid potentiates the inhibitory effects of fluconazole against candida albicans. *Antibiotics*, 13(12), 1174.

Mohar, D. S., & Malik, S. (2012). The sirtuin system: The holy grail of resveratrol? *Journal of Clinical & Experimental Cardiology*, 3(11), 216.

Montenegro-Landívar, M. F., Tapia-Quirós, P., Vecino, X., Reig, M., Valderrama, C., Granados, M., . . . Saurina, J. (2021). Polyphenols and their potential role to fight viral diseases: An overview. *Science of the Total Environment*, 801, 149719.

Musarra-Pizzo, M., Ginestra, G., Smeriglio, A., Pennisi, R., Sciortino, M. T., & Mandalari, G. (2019). The antimicrobial and antiviral activity of polyphenols from almond (Prunus dulcis L.) skin. *Nutrients*, 11(10), 2355.

Naeimi, A. F., & Alizadeh, M. (2017). Antioxidant properties of the flavonoid fisetin: An updated review of in vivo and in vitro studies. *Trends in Food Science & Technology*, 70, 34–44.

Nazir, N., Karim, N., Abdel-Halim, H., Khan, I., Wadood, S. F., & Nisar, M. (2018). Phytochemical analysis, molecular docking and antiamnesic effects of methanolic extract of Silybum marianum (L.) Gaertn seeds in scopolamine induced memory impairment in mice. *Journal of Ethnopharmacology*, 210, 198–208.

Nkhili, E., Tomao, V., El Hajji, H., El Boustani, E. S., Chemat, F., & Dangles, O. (2009). Microwave-assisted water extraction of green tea polyphenols. *Phytochemical Analysis*, 20(5), 408–415.

Okonkwo, C. E., Hussain, S. Z., Onyeaka, H., Adeyanju, A. A., Nwonuma, C. O., Bashir, A. A., . . . Shittu, T. D. (2023). Lignin polyphenol: From biomass to innovative food applications, and influence on gut microflora. *Industrial Crops and Products*, 206, 117696.

Okuda, T., & Ito, H. (2011). Tannins of constant structure in medicinal and food plants – hydrolyzable tannins and polyphenols related to tannins. *Molecules*, 16(3), 2191–2217.

Oluwole, O., Fernando, W. B., Lumanlan, J., Ademuyiwa, O., & Jayasena, V. (2022). Role of phenolic acid, tannins, stilbenes, lignans and flavonoids in human health–a review. *International Journal of Food Science and Technology*, 57(10), 6326–6335.

Ortiz, A. D. C., Fideles, S. O. M., Reis, C. H. B., Bellini, M. Z., Pereira, E. D. S. B. M., Pilon, J. P. G., . . . Buchaim, R. L. (2022). Therapeutic effects of citrus flavonoids neohesperidin, hesperidin and its aglycone, hesperetin on bone health. *Biomolecules*, 12(5), 626.

Panche, A. N., Diwan, A. D., & Chandra, S. R. (2016). Flavonoids: An overview. *Journal of Nutritional Science*, 5, e47.

Patanè, P., Laganà, P., Devi, P., Vig, A., Haddad, M. A., Natalello, S., . . . Hashim, H. A. (2019). Polyphenols and functional foods from the regulatory viewpoint. *Journal of AOAC International*, 102(5), 1373–1377.

Patel, S. S., Acharya, A., Ray, R. S., Agrawal, R., Raghuwanshi, R., & Jain, P. (2020). Cellular and molecular mechanisms of curcumin in prevention and treatment of disease. *Critical Reviews in Food Science and Nutrition*, 60(6), 887–939.

Pecorari, M., Villaño, D., Francesca Testa, M., Schmid, M., & Serafini, M. (2010). Biomarkers of antioxidant status following ingestion of green teas at different polyphenol concentrations and antioxidant capacity in human volunteers. *Molecular Nutrition & Food Research*, 54(S2), S278–S283.

Pereira, M. C., Oliveira, D. A., Hill, L. E., Zambiazi, R. C., Borges, C. D., Vizzotto, M., . . . Gomes, C. L. (2018). Effect of nanoencapsulation using PLGA on antioxidant and antimicrobial activities of guabiroba fruit phenolic extract. *Food Chemistry*, 240, 396–404.

Pereira-Caro, G., Moreno-Rojas, J. M., Brindani, N., Del Rio, D., Lean, M. E., Hara, Y., & Crozier, A. (2017). Bioavailability of black tea theaflavins: Absorption, metabolism, and colonic catabolism. *Journal of Agricultural and Food Chemistry*, 65(26), 5365–5374.

Pérez-Jiménez, J., Neveu, V., Vos, F., & Scalbert, A. (2010). Identification of the 100 richest dietary sources of polyphenols: An application of the Phenol-explorer database. *European Journal of Clinical Nutrition*, 64(3), S112–S120.

Petrović, A., Živković, N., Torović, L., Bukarica, A., Nikolić, V., Cvejić, J., & Gojković-Bukarica, L. (2024). Comparative study of the stilbenes and other phenolic compounds in cabernet sauvignon wines obtained from two different vinifications: Traditional and co-inoculation. *Processes*, 12(5), 1020.

Pimentel-Moral, S., Verardo, V., Robert, P., Segura-Carretero, A., & Martínez-Férez, A. (2016). Nanoencapsulation strategies applied to maximize target delivery of intact polyphenols. In *Encapsulations* (pp. 559–595). Academic Press.

Plaha, N. S., Awasthi, S., Sharma, A., & Kaushik, N. (2022). Distribution, biosynthesis and therapeutic potential of lignans. *3 Biotech*, 12(10), 255.

Ponomarenko, J., Lauberts, M., Dizhbite, T., Lauberte, L., Jurkjane, V., & Telysheva, G. (2015). Antioxidant activity of various lignins and lignin-related phenylpropanoid units with high and low molecular weight. *Holzforschung*, 69(6), 795–805.

Pontifex, M. G., Malik, M. M., Connell, E., Müller, M., & Vauzour, D. (2021). Citrus polyphenols in brain health and disease: current perspectives. *Frontiers in Neuroscience*, 15, 640648.

Popoola-Akinola, O. O., Raji, T. J., & Olawoye, B. (2022). Lignocellulose, dietary fibre, inulin and their potential application in food. *Heliyon*, 8(8).

Potì, F., Santi, D., Spaggiari, G., Zimetti, F., & Zanotti, I. (2019). Polyphenol health effects on cardiovascular and neurodegenerative disorders: A review and meta-analysis. *International Journal of Molecular Sciences*, 20(2), 351.

Prabhu, S., Molath, A., Choksi, H., Kumar, S., & Mehra, R. (2021). Classifications of polyphenols and their potential application in human health and diseases. *Int. J. Physiol. Nutr. Phys. Educ.*, 6(1), 293–301.

Pyrzynska, K. (2024). Ferulic acid – a brief review of its extraction, bioavailability and biological activity. *Separations*, 11(7), 204.

Quave, C. L., Plano, L. R., Pantuso, T., & Bennett, B. C. (2008). Effects of extracts from Italian medicinal plants on planktonic growth, biofilm formation and adherence of methicillin-resistant Staphylococcus aureus. *Journal of ethnopharmacology*, 118(3), 418–428.

Quiñones, M., Miguel, M., & Aleixandre, A. (2013). Beneficial effects of polyphenols on cardiovascular disease. *Pharmacological Research*, 68(1), 125–131.

Raedschelders, K., Ansley, D. M., & Chen, D. D. (2012). The cellular and molecular origin of reactive oxygen species generation during myocardial ischemia and reperfusion. *Pharmacology & Therapeutics*, 133(2), 230–255.

Rajbhar, K., Dawda, H., & Mukundan, U. (2015). Polyphenols: Methods of extraction. *Sci. Revs. Chem. Commun.*, 5(1), 1–6.

Rapisarda, P., Amenta, M., Ballistreri, G., Fabroni, S., & Timpanaro, N. (2022). Distribution, antioxidant capacity, bioavailability and biological properties of anthocyanin pigments in blood oranges and other citrus species. *Molecules*, 27(24), 8675.

Rasouli, H., Farzaei, M. H., & Khodarahmi, R. (2017). Polyphenols and their benefits: A review. *International Journal of Food Properties*, 20(sup2), 1700–1741.

Richard, T., Pawlus, A. D., Iglésias, M. L., Pedrot, E., Waffo-Teguo, P., Mérillon, J. M., & Monti, J. P. (2011). Neuroprotective properties of resveratrol and derivatives. *Annals of the New York Academy of Sciences*, 1215(1), 103–108.

Riviere, C., Pawlus, A. D., & Merillon, J. M. (2012). Natural stilbenoids: Distribution in the plant kingdom and chemotaxonomic interest in Vitaceae. *Natural Product Reports*, 29(11), 1317–1333.

Rocha, A. A., Santos, G. A. S., Ribeiro Jr, J. E., Paixão, L. C., & Santana, A. A. (2022). Recent progress in biopolymer-based delivery systems and coatings for improving stability, bioavailability and efficacy of nutraceutical products. *Biopolymers in Nutraceuticals and Functional Foods*, 36–53.

Rudrapal, M., Mishra, A. K., Rani, L., Sarwa, K. K., Zothantluanga, J. H., Khan, J., . . . Sahoo, P. K. (2022). Nanodelivery of dietary polyphenols for therapeutic applications. *Molecules*, 27(24), 8706.

Sadeghi, A., Rajabiyan, A., Nabizade, N., Nezhad, N. M., & Zarei-Ahmady, A. (2024). Seaweed-derived phenolic compounds as diverse bioactive molecules: A review on identification, application, extraction and purification strategies. *International Journal of Biological Macromolecules*, 266, 131147.

Saidi, I., Guesmi, F., Kharbech, O., Hfaiedh, N., & Djebali, W. (2021). Gallic acid improves the antioxidant ability against cadmium toxicity: Impact on leaf lipid composition of sunflower (Helianthus annuus) seedlings. *Ecotoxicology and Environmental Safety*, 210, 111906.

Salah, M., Mansour, M., Zogona, D., & Xu, X. (2020). Nanoencapsulation of anthocyanins-loaded β-lactoglobulin nanoparticles: Characterization, stability, and bioavailability in vitro. *Food Research International*, 137, 109635.

Santangelo, C., Vari, R., Scazzocchio, B., Di Benedetto, R., Filesi, C., & Masella, R. (2007). Polyphenols, intracellular signalling and inflammation. *Annali-istituto superiore di Sanita*, 43(4), 394.

Scalbert, A., & Williamson, G. (2000). Dietary intake and bioavailability of polyphenols. *The Journal of Nutrition*, 130(8), 2073S–2085S.

Shaito, A., Posadino, A. M., Younes, N., Hasan, H., Halabi, S., Alhababi, D., . . . Pintus, G. (2020). Potential adverse effects of resveratrol: A literature review. *International Journal of Molecular Sciences*, 21(6), 2084.

Shirmohammadli, Y., Efhamisisi, D., & Pizzi, A. (2018). Tannins as a sustainable raw material for green chemistry: A review. *Industrial Crops and Products*, 126, 316–332.

Shoba, G., Joy, D., Joseph, T., Majeed, M., Rajendran, R., & Srinivas, P. S. S. R. (1998). Influence of piperine on the pharmacokinetics of curcumin in animals and human volunteers. *Planta Medica*, 64(04), 353–356.

Sindhu, R. K., Verma, R., Salgotra, T., Rahman, M. H., Shah, M., Akter, R., . . . Al-Kuraishy, H. M. (2021). Impacting the remedial potential of nano delivery-based flavonoids for breast cancer treatment. *Molecules*, 26(17), 5163.

Singh, B. N., Shankar, S., & Srivastava, R. K. (2011). Green tea catechin, epigallocatechin-3-gallate (EGCG): Mechanisms, perspectives and clinical applications. *Biochemical Pharmacology*, 82(12), 1807–1821.

Singla, R. K., Dubey, A. K., Garg, A., Sharma, R. K., Fiorino, M., Ameen, S. M., . . . Al-Hiary, M. (2019). Natural polyphenols: Chemical classification, definition of classes, subcategories, and structures. *Journal of AOAC International*, 102(5), 1397–1400.
Skendi, A., Irakli, M., Chatzopoulou, P., Bouloumpasi, E., & Biliaderis, C. G. (2022). Phenolic extracts from solid wastes of the aromatic plant essential oil industry: Potential uses in food applications. *Food Chemistry Advances*, 1, 100065.
Spagnuolo, C., Napolitano, M., Tedesco, I., Moccia, S., Milito, A., & Luigi Russo, G. (2016). Neuroprotective role of natural polyphenols. *Current Topics in Medicinal Chemistry*, 16(17), 1943–1950.
Sugamura, K., & Keaney Jr, J. F. (2011). Reactive oxygen species in cardiovascular disease. *Free Radical Biology and Medicine*, 51(5), 978–992.
Sul, D., Kim, H. S., Lee, D., Joo, S. S., Hwang, K. W., & Park, S. Y. (2009). Protective effect of caffeic acid against beta-amyloid-induced neurotoxicity by the inhibition of calcium influx and tau phosphorylation. *Life Sciences*, 84(9–10), 257–262.
Taban, K., İlhan, M., & Süntar, I. (2023). Luteolin: Advances on resources, biosynthesis pathway, bioavailability, bioactivity, and pharmacology. In *Handbook of Dietary Flavonoids* (pp. 1–37). Cham: Springer International Publishing.
Tang, J., Xu, N., Ji, H., Liu, H., Wang, Z., & Wu, L. (2011). Eudragit nanoparticles containing genistein: formulation, development, and bioavailability assessment. *International Journal of Nanomedicine*, 2429–2435.
Tanimura, S., Kadomoto, R., Tanaka, T., Zhang, Y. J., Kouno, I., & Kohno, M. (2005). Suppression of tumor cell invasiveness by hydrolyzable tannins (plant polyphenols) via the inhibition of matrix metalloproteinase-2/-9 activity. *Biochemical and Biophysical Research Communications*, 330(4), 1306–1313.
Teixeira, J., Gaspar, A., Garrido, E. M., Garrido, J., & Borges, F. (2013). Hydroxycinnamic acid antioxidants: An electrochemical overview. *BioMed Research International*, 2013(1), 251754.
Thirumalaisamy, R., Bhuvaneswari, M., Haritha, S., Jeevarathna, S., Janani, K. S., & Suresh, K. (2022). Curcumin, naringenin and resveratrol from natural plant products hold promising solutions for modern world diseases–a recent review. *South African Journal of Botany*, 151, 567–580.
Tijjani, H., Zangoma, M. H., Mohammed, Z. S., Obidola, S. M., Egbuna, C., & Abdulai, S. I. (2020). Polyphenols: Classifications, biosynthesis and bioactivities. *Functional Foods and Nutraceuticals: Bioactive Components, Formulations and Innovations*, 389–414.
Truzzi, F., Tibaldi, C., Zhang, Y., Dinelli, G., & D' Amen, E. (2021). An overview on dietary polyphenols and their biopharmaceutical classification system (BCS). *International Journal of Molecular Sciences*, 22(11), 5514.
Tsai, H. Y., Ho, C. T., & Chen, Y. K. (2017). Biological actions and molecular effects of resveratrol, pterostilbene, and 3'-hydroxypterostilbene. *Journal of Food and Drug Analysis*, 25(1), 134–147.
Tuli, H. S., Tuorkey, M. J., Thakral, F., Sak, K., Kumar, M., Sharma, A. K., . . . Bishayee, A. (2019). Molecular mechanisms of action of genistein in cancer: Recent advances. *Frontiers in Pharmacology*, 10, 1336.
Turrini, E., Ferruzzi, L., & Fimognari, C. (2015). Potential effects of pomegranate polyphenols in cancer prevention and therapy. *Oxidative Medicine and Cellular Longevity*, 2015(1), 938475.
Verdura, S., Cuyàs, E., Cortada, E., Brunet, J., Lopez-Bonet, E., Martin-Castillo, B., . . . Menendez, J. A. (2020). Resveratrol targets PD-L1 glycosylation and dimerization to enhance antitumor T-cell immunity. *Aging (Albany NY)*, 12(1), 8.
Vicente-Zurdo, D., Gómez-Mejía, E., Morante-Zarcero, S., Rosales-Conrado, N., & Sierra, I. (2025). Analytical strategies for green extraction, characterization, and bioactive evaluation of polyphenols, tocopherols, carotenoids, and fatty acids in agri-food bio-residues. *Molecules*, 30(6), 1326.

Weng, C. J., & Yen, G. C. (2012). Chemopreventive effects of dietary phytochemicals against cancer invasion and metastasis: Phenolic acids, monophenol, polyphenol, and their derivatives. *Cancer Treatment Reviews*, 38(1), 76–87.

Wu, C., Ma, B., McClements, D. J., Lai, Z., Hou, J., Wang, S., . . . Wang, P. (2024). Fractionation of phenolic compounds from hickory by-products using solid phase extraction-sonication: Chemical composition, antioxidant and antimicrobial activity. *Food Chemistry*, 460, 140633.

Xian, D., Guo, M., Xu, J., Yang, Y., Zhao, Y., & Zhong, J. (2021). Current evidence to support the therapeutic potential of flavonoids in oxidative stress-related dermatoses. *Redox Report*, 26(1), 134–146.

Xu, Y., Tang, G., Zhang, C., Wang, N., & Feng, Y. (2021). Gallic acid and diabetes mellitus: Its association with oxidative stress. *Molecules*, 26(23), 7115.

Yan, L., Vaghari-Tabari, M., Malakoti, F., Moein, S., Qujeq, D., Yousefi, B., & Asemi, Z. (2023). Quercetin: An effective polyphenol in alleviating diabetes and diabetic complications. *Critical Reviews in Food Science and Nutrition*, 63(28), 9163–9186.

Yan, X., Zhang, X., McClements, D. J., Zou, L., Liu, X., & Liu, F. (2019). Co-encapsulation of epigallocatechin gallate (EGCG) and curcumin by two proteins-based nanoparticles: Role of EGCG. *Journal of Agricultural and Food Chemistry*, 67(48), 13228–13236.

Yang, D. K., & Kang, H. S. (2018). Anti-diabetic effect of cotreatment with quercetin and resveratrol in streptozotocin-induced diabetic rats. *Biomolecules & Therapeutics*, 26(2), 130.

Yang, Y., Trevethan, M., Wang, S., & Zhao, L. (2022). Beneficial effects of citrus flavanones naringin and naringenin and their food sources on lipid metabolism: An update on bioavailability, pharmacokinetics, and mechanisms. *The Journal of Nutritional Biochemistry*, 104, 108967.

Yin, J., Wang, L., Wang, Y., Shen, H., Wang, X., & Wu, L. (2019). Curcumin reverses oxaliplatin resistance in human colorectal cancer via regulation of TGF-β/Smad2/3 signaling pathway. *OncoTargets and Therapy*, 3893–3903.

Yue, T., Shao, D., Yuan, Y., Wang, Z., & Qiang, C. (2012). Ultrasound-assisted extraction, HPLC analysis, and antioxidant activity of polyphenols from unripe apple. *Journal of Separation Science*, 35(16), 2138–2145.

Yun, J. M., Afaq, F., Khan, N., & Mukhtar, H. (2009). Delphinidin, an anthocyanidin in pigmented fruits and vegetables, induces apoptosis and cell cycle arrest in human colon cancer HCT116 cells. *Molecular Carcinogenesis: Published in Cooperation with the University of Texas MD Anderson Cancer Center*, 48(3), 260–270.

Zannou, O., Oussou, K. F., Chabi, I. B., Awad, N. M., Aïssi, M. V., Goksen, G., . . . Kayodé, A. P. (2023). Nanoencapsulation of Cyanidin 3-O-glucoside: Purpose, technique, bioavailability, and stability. *Nanomaterials*, 13(3), 617.

2 Biochemistry and Antioxidant Mechanisms of Polyphenols

Praveen Kumar Verma and Elad Eliahoo

2.1 INTRODUCTION

Polyphenols contain one or more phenolic hydroxyl groups (OH-) on their aromatic rings and are exclusively synthesized by plants as secondary metabolites. Polyphenols are generally found in the entire plant kingdom in plants' fruits, leaves, roots and peels and show potent antioxidant properties (Scalbert et al., 2005). These secondary metabolites are formed from the phenylpropanoid and the shikimic acid pathways. Initially, they are present in conjugated forms with sugar residues attached to the hydroxyl group and a direct attachment with aromatic carbon. Linkages of phenol with other chemical residues, such as carboxylic acid, amines and lipids, are very common. More than 8000 structurally diverse polyphenols have been identified around the world, and these polyphenols play essential roles in plants including colour, odour, flavour, microbe infection, seed-dispersing animals, formation of nitrogen-fixing root nodules and attractants for pollinators and ultraviolet (UV) protectants (Singh et al., 2021). In addition, they can defend crops from the plague disease and can help in seed germination. Polyphenols are ingested as glycosides, which are hydrolyzed in the gut by enzymes or gut microbiota to release the active aglycone forms. Polyphenols are absorbed in the small intestine and colon. Once absorbed, they are metabolized in the liver through conjugation reactions (glucuronidation, sulfation and methylation), producing bioactive metabolites. These metabolites circulate in the bloodstream and are eventually excreted via urine or bile. In humans, these polyphenols are suggested to help treat cancer by inhibiting oxidative stress and intake of cytokines into cancer cells. Furthermore, they increase anti-inflammatory, antiviral, antitumor and antioxidant effects by different mechanisms, for example, lowering the levels of blood lipids (Kim et al., 2012; Tapiero et al., 2002).

Reactive oxygen species (ROS) such as hydroxyl radical (·OH), superoxide anion radical ($^{\cdot}O_2^{\cdot-}$), singlet oxygen ($^1O_2$) and hydrogen peroxide (H_2O_2) display high reactivity with cell organelles. On one hand, increased ROS levels inhibit cancer cell proliferation and lead to cell death. On the other hand, low concentrations of ROS levels elevate proliferation and angiogenesis (Ushio-Fukai et al., 2008). Some papers reported that the hydroxyl group (OH-) of polyphenols scavenges the ROS free radicals and that these free radicals form rough self-oxidation radicals that present toxicity and excellent stability in the cell (Alov et al., 2015). Moreover, it was found

DOI: 10.1201/9781003588504-2

that natural antioxidant polyphenols have less toxicity and carcinogenic effects on the human body than synthetic antioxidants in long-term uses (Di Meo et al., 2013).

The antioxidant mechanisms of polyphenol take place through following ways: (a) donating a hydrogen ion to free radicals and inhibition of the $\bullet O_2^-$, H_2O_2, $^{\bullet}OH$, $ROO^{\bullet}$, $^1O_2$ activities in the cell; (b) preventing generation of radicals by inhibiting the enzyme that is responsible for free radical generation, such as cyclooxygenase, lipoxygenase, NADPH oxidase, etc.; (c) chelating with metal ions to reduce the production of free radicals through resonance; (d) inhibiting oxidation reactions by inducing the antioxidant enzyme activity or antioxidant protein expression; and (e) acting as antioxidants and scavenging free radicals formed in the cells (Lv et al., 2021). The antioxidation mechanisms of various polyphenolic compounds are detailed in Table 2.1.

TABLE 2.1
Antioxidant Mechanisms of Various Polyphenolic Compounds

Species	Representative Substance	Source	Antioxidation Mechanism
Flavonoids compound	Flavonols	Citrus, apples, berries, etc.	a. Hydrogen bond formed after reaction with free radicals b. Chelation with metal ions c. Activity of antioxidant enzymes increased, such as SOD
	Procyanidins	Cocoa, wine, tea, nuts, etc.	a. Reaction of hydrogen atoms with free radicals and chelation with metal ions b. MAPK pathways down-regulated and increased ERK phosphorylation as well as Nrf2 and HO-1 expression enhanced c. PI3K/Akt pathways protect cell from H_2O_2 damage
Phenolic acid compounds	Chlorogenic acid	Ferns, dicotyledonous plants	a. Binding of hydrogen atom with free radical b. DNA fragmentation was inhibited by preventing NH_2Cl c. FOXO family gene expression was reduced and increased phosphorylation of PI2K/Akt
	Gallic acid	Tea leaves, grapes, nuts, etc.	a. Decrease in the level of MDA and NO b. Reduction in stress through activation of Nrf2/HO–1
Stilbene compounds	Resveratrol	Grape, knotweed, blueberry, peanut, blackberry	a. FOXO 4 attached with SOD and CAT promotor site and increased expression of antioxidant enzymes b. Activation of Nrf2/HO-1 pathway and increased the expression of antioxidant enzymes c. Chelate formation with metal ions
Lignan compounds	Lignans	Sesame and flaxseed	a. Chelating with metal ions (Cu^{2+}) b. Activation of Nrf2/HO-1 pathway and increased expression of antioxidant enzymes c. Reduced ROS and MDA levels, increasing antioxidant enzyme activity and inhibiting liver lipid peroxidation

Polyphenols include a high range of complex compounds that usually contain normal phenols and high-molecular-weight polymers and their derivatives. Polyphenols are classified based on their botanical source into grape polyphenols, tea polyphenols, etc. Alternatively, they can be classified based on the constituent chemical group into flavonoids, phenolic acids, stilbenes and lignans.

2.2 BIOSYNTHESIS OF POLYPHENOLS IN PLANTS

The biosynthesis of polyphenols is highly important component of plant secondary metabolites. The synthesis of different polyphenol structures occurs through two main metabolic pathways: the shikimate pathway and the phenylpropanoid pathway (Figure 2.1).

2.2.1 The Shikimate Pathways

The name shikimate originated from shikimic acid, which is the initial precursor in the biosynthesis of aromatic acids (L-phenylalanine, L-tyrosine and L-tryptophan) and polyphenols. The pathway is divided into seven steps. In its first step, phosphoenolpyruvate from glycolysis and erythrose-4-phosphate from the pentose phosphate pathway are used as substrates of primary metabolism. The condensation process of phosphoenolpyruvate and erythrose-4-phosphate leads to the formation of 3-deoxy-D-arabinohexulose-7-phosphate by DAHP synthase (DAHPS; EC 2.5.1.54) (Tzin & Galili, 2010). Next, DAHP forms 3-dehydroquinate (DHQ) by enzyme DHQ synthase (DHQS; EC 4.2.3.4). This reaction requires oxidation, phosphate elimination, reduction and ring condensation (Mittelstädt et al., 2013). In the third step, DHQ goes through two reactions that are catalysed by the bifunctional enzyme 3-dehydroquinate dehydratase/shikimate dehydrogenase (DHQD/SDH; EC 4.2.1.10 and EC 1.1.1.25, respectively) to transform into shikimate in plastids (Herrmann & Weaver, 1999). Then, it forms several hydroxybenzoic acids, such as p-hydroxybenzoic acid, protocatechuic acid and gallic acid (Ghosh et al., 2012). In its fifth step, shikimate

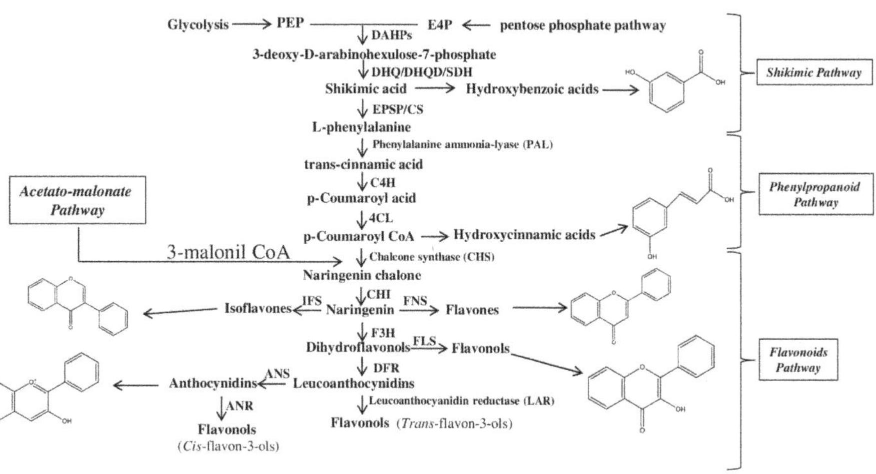

FIGURE 2.1 Biosynthesis pathway of polyphenolic compounds.

3-phosphate (S3P) is activated from shikimate through shikimate kinase (SK; EC 2.7.1.71), which requires divalent cations such as Mg^{2+} and Mn^{2+} (Maeda & Dudareva, 2012). The sixth step is condensation of S3P with PEP to produce EPSP, which is catalysed by 5-enolpyruvylshikimate 3-phosphate (EPSP) synthase (EPSPS; EC 2.5.1.19) (Schönbrunn et al., 2001). In its last step, chorismate is produced through dephosphorylated EPSP by chorismate synthase (CS; EC 4.2.3.5) (Kitzing et al., 2004). The major purpose of chorismate is to act as a precursor for the synthesis of aromatic acids such as tryptophan, tyrosine and phenylalanine.

2.2.2 The Phenylpropanoid Pathway

In this pathway, phenylalanine ammonia-lyase (PAL) acts as a key phenolic metabolic enzyme for the formation of trans-cinnamic acid through deamination of L-phenylalanine, and this step is considered the foundation of the phenylpropanoid pathway of Polyphenol componds (PCs) biosynthesis. Trans-cinnamic acid is observed as the initial product of phenolic metabolites in plant cells, which act as a bridge between aromatic amino acid and polyphenol metabolism. The transformation of L-tyrosine to trans-cinnamic acid has also been observed via tyrosine ammonia-lyase enzyme activity (Barros et al., 2016). Trans-cinnamic acid is an early phenolic metabolite generated in plant cells, which acts as bridge between aromatic and PCs metabolism. In monocot, the PAL enzyme (L-tyrosine ammonia-lyase) acts as a bifunctional for both L-phenylalanine and L-tyrosine and converts to p-coumaric acid (without the 4-hydroxylation step), but with less efficiency (Yadav et al., 2020). However, formation of different hydroxycinnamic acids takes place through a series of hydroxylations of trans-cinnamic acid. This compound cannot be stored in its free form so it undergoes another transformation through bi-oxidation, resulting in hydroxybenzoic acids that lead to cinnamic alcohols participating in lignan biosynthesis, acyl derivatives and complex esters (chlorogenic acid) and the synthesis of coumarin. This step finalizes the formation of the major constituents in the phenylpropanoid pathway, which is a central element in the biosynthesis of natural polyphenols (Sharma et al., 2019; Dong & Lin, 2021).

Another important pathway is the aceto-malonate pathway, which has a major role in the biosynthesis of phenolic compounds in higher plants. The aceto-malonate pathway couples with the phenylpropanoid pathway, which leads to the formation of flavonoids. Ring B and A in their structures are formed through the shikimate pathway and aceto-malonate pathway, respectively. p-Coumaroyl-CoA and three molecules of malonyl-CoA are used as a substrate where the condensation reaction leads to the formation of chalone through chalone synthase, which is a key enzyme in the biosynthesis of flavonoids (Nabavi et al., 2020; Wang et al., 2018). Chalcone-flavanone isomerase (CHI) actions lead to the transformation of chalcone naringenin into flavanone naringenin; then naringenin acts as a precursor for other classes of flavonoids (flavan-3-ols, flavanones, flavones, anthocyanins, etc.) except for chalcones and dihydrochalcones due to a change in their oxidation state in the central heterocyclic ring (redox reaction) (Yin et al., 2019). Also, the last products of phenolic metabolism are not in monomer forms but form more complex oligomeric and polymeric structures like proanthocyanidins (byproducts of flavan-3-ols), which are widely available in higher plant tissues. The final stage of the flavonoid pathway is the condensation of flavan-3-ols into proanthocyanidins, but this still remains unclear, and it is thought that this condensation takes place through enzymatic pathways involving peroxidase, polyphenol oxidase and laccase (Rauf et al., 2019).

Some other factors are required for controlled metabolic biosynthesis of polyphenols at gene expression levels like transcription, post-translational modifications and micro-regulators, such as non-coding RNAs. A transcription factor (TF), which is a DNA-binding protein, binds to the promoter region and starts transcription of the target genes. These TFs can influence both structural and regulatory enzyme genes in response to environmental signals (both internal and external, containing phytohormones and abiotic factors) and affect the growth of secondary metabolites (Yadav et al., 2020). Various families of TFs are well known for the metabolism of polyphenols; for example, the MYB TF family is one that is used for phenolic metabolism and helps in plant defence against several stresses. It has been shown that phenylpropanoid and flavonoid pathways are regulated by various MYB TFs, which may activate or repress the genes' expression of their enzymes (Kołton et al., 2022). For example, TF MusaMYB31 in bananas (*Musa* cultivar Rasthal) has shown reduced transcription levels within the phenylpropanoid and flavonoid pathways due to overexpression of the gene (Tak et al., 2017). Another study showed that TF PtMYB115 in poplar trees (*Populus* spp.) binds to the promoter regions of ANR1 and LAR3 genes leading to increased expression of these genes, ultimately inducing deposition of proanthocyanidins. Moreover, VvMYB5a TF in grapevines (*Vitis vinifera*) promotes flavonoid synthesis by inhibiting lignin synthesis, which helps to support the balance in the carbon flow between lignins and flavonoids in the plant (Zhang et al., 2021; Wang et al., 2017).

Polyphenols provide a wide range of biological effects, many of which contribute to their potential health benefits as follows:

- **Cardiovascular Health**: Polyphenols improve endothelial function, reduce blood pressure and prevent oxidation of low-density lipoprotein (LDL) cholesterol (Aviram et al., 2000).
- **Neuroprotection**: They protect neurons by reducing oxidative damage and modulating signalling pathways, potentially delaying the onset of neurodegenerative diseases (Letenneur et al., 2007).
- **Anti-Inflammatory Properties**: Polyphenols inhibit pro-inflammatory mediators like cytokines and enzymes such as cyclooxygenase (COX) (Yahfoufi et al., 2018).
- **Anticancer Activity**: By modulating cell proliferation, apoptosis and angiogenesis, polyphenols exhibit potential as chemopreventive agents (García-Lafuente et al., 2009).
- **Glucose Metabolism**: Certain polyphenols enhance insulin sensitivity and regulate glucose uptake, aiding in diabetes management (Aryaeian et al., 2017).
- **Antioxidant Properties:** One of the most studied properties of polyphenols is their ability to act as antioxidants. Polyphenols neutralize ROS by donating hydrogen atoms or electrons, thereby reducing oxidative stress, or through Fenton reactions. Polyphenols also regulate the activity of enzymes involved in oxidative stress, such as superoxide dismutase and catalase (Perron & Brumaghim, 2009).

2.3 ANTIOXIDANT EFFECTS AND MECHANISMS OF ACTION OF POLYPHENOLS

Scientific articles have reported the identification of various factors through which free radical accumulation occurs (including age, environment, irregular diet, stress, etc.) that leads to increased cancer development, ageing and degenerative health issues (Xin-yue et al., 2021). Hence, nature provides several plant-based beneficial components, like polyphenols, which are categorized by their low toxic effects on the body and a wide range of pharmacological activities. Therefore, researchers have studied the use of polyphenols as novel drugs against the development of several diseases. Since these polyphenols have one or more hydroxyl groups on their aromatic ring, they can neutralize the harmful effect of free radicals in the body by blocking free radical chain reactions and helping to treat many diseases (Bernatoniene & Kopustinskiene, 2018). The interactions between polyphenols and their active components have encouraged scientists to study the antioxidant activity of polyphenols (Figure 2.2).

2.3.1 Antioxidant Effects of Flavonoids

Flavonoids are polyphenolic phytochemicals found in fruits and vegetables. Their consumption as food and as nutraceuticals provides a broad range of health benefits. More than 4000 species of flavonoids have been discovered, including flavonols, procyanidins and anthocyanins (Panche et al., 2016).

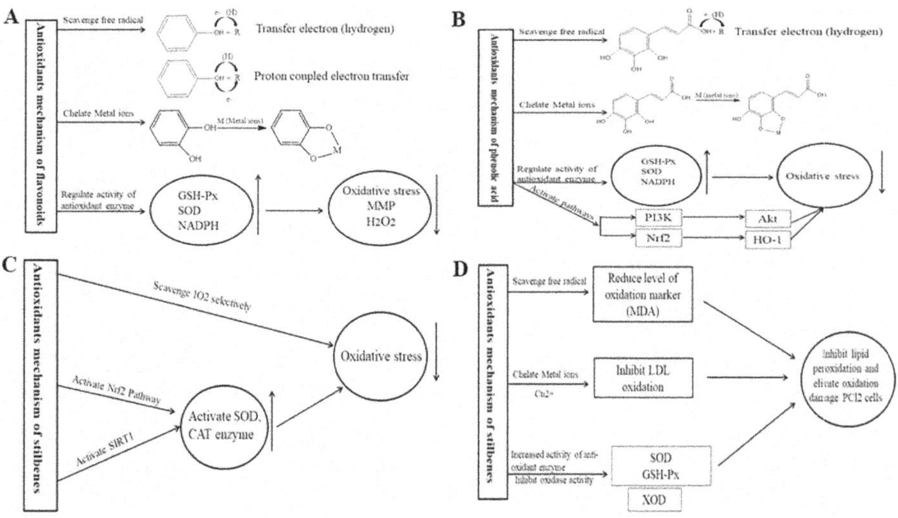

FIGURE 2.2 Schematic representation of antioxidant mechanism of polyphenols. (A) Antioxidant mechanisms of flavonoids. (B) Antioxidant mechanisms of phenolic acids. (C) Antioxidant mechanisms of stilbenes. (D) Antioxidant mechanisms of Lignans.

From a chemical approach, flavonoid has the structural backbone of the flavan nucleus, which contains two benzene rings connected through three carbon atoms to form a pyran heterocyclic ring. The whole structure contains multiple reactions and substitutions to produce various classes of flavonoids such as flavonols, flavones, flavanones, flavanols and anthocyanidins. Flavone and isoflavone are formed by a substitution reaction between phenylchromen-4-one and a phenyl ring at C_2 and C_3, whereas, flavanones, flavanols, flavonols, flavones, anthocyanidins and isoflavones are formed by a substitution reaction with some modification at the C_2-C_3 (saturated or olefinic) and C_4 (a ketone group) site (Valenzuela et al., 2016; Chagas et al., 2022). Thus, the hydroxyl groups in these molecules release hydrogen atoms during the reaction and block the oxidation of lipid molecules. In addition, it was found that the flavonoid subclass has structural features where the positions and the number of hydroxyl groups comprise large physicochemical characteristics and numerous biological functions in the body. The presence of large structural accuracies of flavonoid have shown antioxidant, anti-allergic, anti-atherogenic, blood vessel–dilating, anti-hyperglycaemic actions and so on (Vauzour, 2012; Barrajón-Catalán et al., 2014; Oteiza et al., 2018).

Among all these biological functions, all flavonoids share antioxidant activities. These activities efficiently reduce the rate of ROS generation in the body through different mechanisms in which the flavan nucleus plays a crucial role in neutralizing the free radicals by providing hydrogen atoms in the aromatic rings. However, another mechanism has been described in which flavonoids chelate themselves, and this prevents them from being available for oxidation. Various studies have revealed that humans and animals show increased markers of biological oxidation provoked by ROS, such as 8-hydroxy-20-deoxyguanosine, nitrotyrosine, carbonylated amino acids, etc., that can be inhibited by the administration of flavonoid-rich plant foods (Umeno et al., 2016).

2.3.2 ROS and Oxidative Stress

Reactive oxygen compounds are denoted by one or more unpaired electrons produced in the cell during the mitochondrial oxidative process in response to UV radiation, cytokines, environmental pollution, etc. Together, they are known as oxidizing agents. These ROS and the oxidative stress start cascade reactions that can result in oxidative damage to the cells. Likewise, normal cell metabolism produces reactive molecules essential for regular physiological functions such as cell signalling, immune response and gene expression. Some antioxidant-reactive molecules donate an electron to unpair free radicals to neutralize them (Halliwell & Gutteridge, 2015; Edeas, 2011). Other antioxidant-reactive molecules neutralize the free radicals by directly scavenging or by inhibiting lipid peroxidation. Still, the antioxidant system can take place through enzymatic or non-enzymatic pathways, where the non-enzymatic system blocks the beginning of oxidation, using compounds such as ascorbate (AsA), hydroxybenzoic acids, xanthophylls, carotene, betaxanthins, vitamin E and betanidine, and the enzymatic system takes place by several antioxidant enzymes such as superoxide dismutase (SOD), catalase (CAT), glutathione peroxidase (GPx) and glutathione reductase (GR) (Yang et al., 2020; Bhattacharya, 2015).

Furthermore, antioxidant activity is essential to the cellular systems; any imbalance in ROS production leads to a dysfunction in gene expression, protein maturation, membrane integrity, etc. This imbalance was suggested to be involved in the development of various diseases, such as Parkinson's disease, cardiovascular disease and diabetes. Excessive deposition of ROS in cells leads to the inhibition of several signalling pathways, including glutathionylation and proteasome degradation of MAP kinase phosphatase-1, which in turn inhibits insulin signalling. For example, an excess of ROS inhibits AKT signalling for glucose uptake, which contributes to type 2 diabetes. In addition, ROS in excess provokes NLRP3 and IL-1B production, leading to inflammation in atherosclerosis. Equally, down-regulation of the Nrf2 TF causes phosphorylation of tau protein and Aβ protein, which increases ROS production that was suggested to lead to Alzheimer's disease (Faria & Persaud, 2017; Zuo et al., 2022).

2.3.3 Chelation Process of Flavonoids

Flavonoids interact with metal ions, which leads to chelate formation. This is important for the inhibition of free radicals in the cells and also protects against oxidative stress. Metal-flavonoids are much stronger in terms of free radical scavenging than the free flavonoids. Flavonoids interact with a reduced form of transition metals, mainly iron and copper ions. Three potent sites for chelation have been identified (Mladěnka et al., 2010):

a) Between the 5-hydroxy and 4-carbonyl group.
b) Between the 3-hydroxy and 4-carbonyl group.
c) Between the 3',4'-hydroxy group in the B ring.

Rutin and epicatechin form complexes with iron (II), iron (III), copper (II) or zinc (II). These complexes consistently present higher potential to scavenge free radicals than free flavonoids. In vitro experiments provided supporting evidence that the complexes have increased efficiency of protection against oxidative damage to the cells. In addition, a copper complex with rutin has high potent antioxidant activity against lipid peroxidation in vivo (Kostyuk et al., 2001). Correspondingly, iron (III) complexes with flavonoids show high protection of hepatocytes against hypoxia in rats (Moridani et al., 2003). Likewise, the hydroxyl group in quercetin (a plant pigment flavonoid) is able to form complexes with several ions of metals through the 3 and 4 phenolic group to take advantage of its antioxidant activity. Nevertheless, it has been found that Cu-quercetin, Fe-quercetin and Al-quercetin complexes reduce oxidative stress much better than free quercetin (Bukhari et al., 2009).

Chelation capabilities depend on several factors such as reducing power, pH, solvent (polarity and ionic component) and stoichiometry. While rutin and negletein are potent inhibitors of the Fenton reaction when used at very low ratios, they are ineffective at the 1:1 ratio. However, in the presence of flavonoids, pure ferric salt or ferric citrate significantly increases the Fenton reaction (Barbusiński, 2009).

Several reports have observed that the number and positions of the hydroxyl groups play an essential role in increasing flavonoids, such as 6, 7-dihydoxyl, and function at neutral and acidic pH. Furthermore, the 3-hydroxyl group, 4-keto group and 2,3 double bond at the B ring display strong chelator activity in neutral and acidic pH conditions. But in neutral pH conditions, the 5-hydroxyl-4-keto chelation site in flavonoids acts as a weak chelator. Still, due to the resonance effect, Fe (II) may also bind other deprotonation sites and may exhibit a strong antioxidation effect (Barbusiński, 2009; Fernandez et al., 2002).

Another important factor is electron density, which plays a crucial role in the interaction of flavonoids with metal ions. In an aprotic solvent like dimethyl sulfoxide (DMSO), neutral and monoanionic quercetin with Fe (II) ion at a 1:1 ratio exhibits a small influence on function. However, in a protic solvent such as water, it can interact with the phenyl and the carbonyl groups of flavonoids. Thus, its influence depends more on pH and on dissociation and on its interaction with metal ions. For example, biochanin A and genistein combine at 1:2 stoichiometry in their complexes with the Fe (III) metal ion in acidic or neutral pH solvent, whereas they combine in a 1:1 ratio with Fe(III) in basic pH due to deprotonation by the phenyl group (Bodini et al., 1999; Dowling et al., 2010).

Many scientific investigations have confirmed the relationships between flavonoid structure and antioxidant activity. Flavonoids' unique chemical structure allows them to establish antioxidant activity through the hydroxyl group (ring C) and through the ortho-dihydroxy arrangement in the B ring, where the C_2-C_3 unsaturated bond combines with the C_4 carbonyl group in the C skeleton and with O-methylation (Borges Bubols et al., 2013).

In the beginning, a hydrogen atom is donated to radical molecules from the free hydroxyl group, producing a moderately stable flavonoid phenoxyl radical. Next, this molecule reacts with a radical alkoxyl (RO*) and forms a stable quinone structure. The number and positions of the hydroxyl group significantly dominate and influence its antioxidant activity. The hydroxyl structure, as well as the twisting angle of the B-ring, has the ability to scavenge both oxygen and nitrogen-centred free radicals, while hydroxyl, peroxyl and peroxynitrite radicals are stabilized by the flavan nucleus, which donates hydrogen and an electron to them (Dugas et al., 2000).

It has been reported that high antioxidant activity has been shown by the presence of the linearity and number of hydroxyl groups in the B-ring. Among quercetin and its derivatives, quercetin exhibited vigorous antioxidant activity. Since quercetin is a robust scavenger of both ROS and reactive nitrogen species (RNS), this observation supports the hypothesis that the hydroxyl group in the ring led to the ability of flavonoids to be antioxidants, therefore explaining the impact of the hydroxyl group on the ability of flavonoids to increase antioxidant activity (Moalin et al., 2011). Similarly to quercetin, genistein significantly induces antioxidant activity either through the level of hepatic mRNA of GST or NQO1 activity in rats via binding to the promoter region (Wiegand et al., 2009).

Another analysis has shown that Hepa-1c1c7 mouse hepatoma cells have increased luciferase activity in the presence of flavonoids that are comprised of 3-OH at the C-ring compared to the no-hydroxyl group and initiation of an electrophile responsive element (EpRE). EpRE is a regulatory sequence of genes encoding for phase II enzymes, which are highly linked with the antioxidant properties of flavonoids (Lee-Hilz et al., 2006). Transfer of electrons from one ring to another, due to a 2,3 double bond that is joined with a keto group together with any depletion in one or both structures, significantly reduces cellular antioxidant activity. This analysis established that the binding of rings A and B induces a resonance effect of the aromatic nucleus that produces flavonoid radicals (Wolfe & Liu, 2008). Moreover, the hydrophobicity and molecular planarity of structures are the main causes of differences in antioxidant activity of flavonoids. Wen et al. disclosed the differences in methylated and unmethylated flavones. They showed that metabolically steady and high intestinal absorption through human colon adenocarcinoma (Caco-2) cells were found in methylated flavones and preserved the compound from hepatic conversion (Wen & Walle, 2006). In addition, in vitro studies have found that both 5,7-dimethoxyflavone and 5,7,4-trimethoxyflavone compounds have high inhibitory activity, greater hepatic metabolic stability and enhanced intestinal absorption in comparison to the unmethylated compounds. These differences were due to the hindering free hydroxyl group by expulsion of metabolizing enzymes that later enhance their antioxidant activity (Walle et al., 2007).

2.3.4 Antioxidant Effects of Flavonols and Flavones

Flavonols contain 2-phenyl chromone molecules in the parent nucleus (flavonoids) and form derivatives such as quercetin, kaempferol and myricetin. These derivatives are present in fruits, vegetables and flowers and play important roles in the human body. In normal conditions, the cell maintains a balance between ROS production and its antioxidant activities. Thus, the antioxidant enzymes SOD and CAT directly neutralize the generated ROS or use glutathione (GSH), which is ubiquitous in nature, and cofactors for glutathione peroxidase in the detoxification of hydrogen peroxide and of free radicals. Yet CAT is the enzyme that is accountable for the breakdown of peroxide in the cell (Rice-Evans, 2001). Next, these intracellular superoxides are converted into water and oxygen by SOD1 and SOD2 enzymes, whereas extracellular superoxides are converted by SOD3. However, the extracellular H_2O_2 is carried to the cells by the aquaporin (AQP) channel and converted to H_2O through the enzymes CAT, peroxidase (PRX) and GPX (Wu C. et al., 2022).

For example, quercetin shows antioxidant activity and prevents ROS generation by directly acting on GSH. It affects signal transduction and increases GSH's enzymatic activity, which retains the oxidative balance. Thus, GSH acts as a reducing agent regulated by quercetin where conversion of H_2O_2 to H_2O takes place through the transfer of a hydrogen donor. In addition, quercetin inhibits the enzyme activities of acetylcholinesterase (AChE) and butyrylcholinesterase (BChE) that are related to the oxidative mechanisms through signal transduction (Gao et al., 2018). Moreover, it increases the levels of antioxidant enzymes such as GSH, CAT and SOD. In addition, ROS can be removed by quercetin through various mechanisms including maintaining the balance of ROS production and adjusting antioxidant-related gene expression, showing antitumour activity and preventing metastasis (Hur et al., 2021). Recent studies have reported that quercetin obtained from *Flaveria bidentis* (L.) can prevent ROS production in mononuclear and polymorphic cells better than vitamin C and can reverse the $FeSO_4$ ion oxidative effect in bovine sperm cells. In conclusion, they have suggested that flavonols increase the expression of HO-1 through activation of the Keapl-Nrf2 signalling pathway, and they can inhibit NADPH oxidase activity by induction of HO-1, which leads to suppression of the oxidative stress (Luo et al., 2019).

2.3.5 Structure and Regulation of Nrf2 for Antioxidant Activity

The transcription factor Nrf2 is composed of 605 amino acids and contains seven highly conserved domains known as Nrf2-ECH homology (Neh) 1 to 7. Neh1 protein has an essential leucine zipper (bZIP) motif and can bind with small muscle Apo neurosis fibromatous (sMaf) protein. These heterodimer protein complexes control the binding of Nrf2-ARE in the nucleus and help to activate transcription of antioxidant enzymes (Saha et al., 2020). Neh2 protein contains DLG and ETGE motifs, which bind with Nrf2 and help Keapl1 act as a negative regulator of Nrf2. The Neh protein domains can transcriptionally act as co-activators, which modulate ARE-dependent genes. For instance, Neh3 binds with CHD6, while Neh4 and Neh5 bind with cyclic adenosine monophosphate (cAMP) response element-binding protein (CREB)–binding protein (CBP) and Neh6 and Neh7 are involved in Nrf2-ARE degradation and suppression through binding with ß-transducin repeat-containing protein (β-TrCP) and via binding to the retinoic X receptor a (RXRα), respectively (Wang et al., 2013).

The activation and regulation of Nrf2 occur through two different pathways, i.e., canonical (Keapl-dependent) and non-canonical (Keapl-independent). Under normal conditions, the Keapl1-Nrf2 complex forms when Keapl protein binds with ETGE and DLG motifs of Nrf2. Then, this complex binds with cullin-based (Cul3) E_3 ligase to form a Keapl-Cul3-RBX1 complex with the help of the Neh2 domain of Nrf2. Lastly, the Keapl-Cul3-RBX1 complex stimulates proteasomal degradation of Nrf2 by the 26S proteasome (Sivandzade et al., 2019). However, in stress conditions, overproduction of ROS and electrophiles leads to modification in the cysteine residue of Keapl1, causing Nrf2 to separate from the Keapl-Cul3-RBX1 complex. Consequently, Nrf2 goes into the nucleus, where it binds with target genes and enhances the expression of several antioxidant genes (Bellezza et al., 2018). Nonetheless, several other factors regulate Nrf2 activity. For example, some kinases regulate Nrf2 through a Keapl1-independent manner, while others, such as glycogen synthase kinase-3 (GSK-3), phosphorylate Nrf2 through the Neh6 domain. It was also found that GSK-3 negatively regulates Nrf2 through phosphorylation at Ser9 by PI3K/AKT and thus promotes Nrf2 activity (Wu S. et al., 2022). Other factors like c-Jun N-terminal kinase (JNK), a member of mitogen-activated protein kinase (MAPK) and ERK, as well as p38 MAPK, can modulate Nrf2 activity through phosphorylation at several sites. However, ERK and JNK regulate Nrf2 positively, while p38 MAPK was found to regulate Nrf2 both negatively and positively (Matzinger et al., 2018). AMP-activated kinase (AMPK) is another trimeric serine/threonine kinase that increases Nrf2 responses by phosphorylating the Neh1 domain, which inhibits Nrf2 transfer from the nucleus (Liu et al., 2021). Likewise, protein kinase C (PKC) can phosphorylate Nrf2 at Ser-40 of the Neh2 domain, separating the Nrf2 domain from Keapl1 and stimulating Nrf2 translocation into the nucleus. Conversely, protein sequestosome 1 (p62) prevents binding of Nrf2 to Keapl1 by binding directly to Keapl1, leading to autophagic degradation of Keapl1 but not Nrf2 (Shaw & Chattopadhyay, 2020). Nonetheless, there are other ways for Nrf2 activation and regulation to occur through transcriptionally mechanisms. For example, miRNAs such as mi-34a, -144 and -28 regulate Nrf2 post-transcriptionally. Still, these are not the primary mechanisms for flavones to regulate Nrf2 protein. DNA methylation and histone modification are other ways to regulate Nrf2 expression. For instance, the sirtuin (SIRT) protein family is composed of conserved proteins of class III histone deacetylases that share a nicotine adenine dinucleotide (NAD) + binding site and a catalytic domain, which deacetylases the Nrf2 gene in early transcription, leading to an increment in the activity of the Nrf2/ARE pathways. Hence, SIRT1 plays a vital role in histone deacetylation that decreases the ubiquitination of Nrf2 and increases the ability of Nrf2 to bind with the ARE site (Liu et al., 2022; Wang et al., 2019).

2.3.6 Mechanism of Nrf2 and NF-κB for Antioxidant Activity

Nrf2 and NF-kB are the main factors that control ROS and RNS. The NF-kB family contains five members: RelA (p65), RelB, c-Rel, p50 (NF-κB1) and p52 (NF-κB2). Of these members, the P65/50 dimer reflects the most active form that regulates transcription. In the latent phase, NF-kB dimers are kept in their inactive form in the cytoplasm by the IκB protein inhibitor. The IκB protein family contains three members: IκBα, IκBβ and IκBε, and all three inhibit the nuclear translocation of NF-kB by wrapping around its nuclear localization sequence (NLS) (Thiruvengadam et al., 2021).

When oxidative stress (OS) stimulates IκB kinase (IKK), this kinase gets activated and phosphorylates IκB proteins. IκB phosphorylation leads to ubiquitination and proteasomal degradation of the IκB protein. Subsequently, NF-kB is released and translocated into the nucleus, where it binds to the DNA to raise the expression of several interleukins such as COX-2 and inducible nitric oxide synthase (iNOS). Some research reports have shown that Nrf2 negatively controls the NF-kB pathways by various mechanisms. Firstly, activation of Nrf2 leads to up-regulating of the antioxidant gene expression to increase ROS or free radical scavenging and inhibit OS-induced NF-kB activation (Thiruvengadam et al., 2021). Secondly, NF-κB nuclear translocation is inhibited through blocking its release by the Keapl1/Nrf2 pathways while proteasomal degradation and ubiquitination of IKKβ and IκB-α take place. Furthermore, up-regulation of Nrf2 increases the expression of HO-1 and catalyses haem to Co^+ and Fe^+, which leads to converting biliverdin to bilirubin, resulting in NF-κB inhibition. However, it was found that NF-κB activation stimulates the production of inflammation factors, whereas Nrf2/Keapl1 inactivates NF-κB through IKKβ and IκB-degradation. Thirdly, Nrf2 and NF-κB inhibit each other by striving for the transcriptional co-activator CBP-p300 complex. Taken together, Nrf2 and

NF-κB are linked under OS conditions in a cell type–dependent manner (Zinatizadeh et al., 2021; Socha et al., 2021).

2.3.7 Procyanidins

Proanthocyanidins copiously appear as polyphenols in plants and are considered secondary metabolites that are found in fruits, cereals, beans, nuts, apples, grapes, kiwis and soybeans. Proanthocyanidins are categorized into propelargonidins, prodelphinidins and procyanidins based on their substitution pattern and their oxidation reaction. Proanthocyanidins are very common in plants and show anticancer, antioxidant and hypoglycaemic properties (Gu et al., 2003). Procyanidins, also called condensed tannins, have a structural unit of flavone-3-ol, and their basic structure consists of a C6-C3-C6 phenolic flavonoid ring (Kelm et al., 2004). Furthermore, the degree of polymerization (DP) and interflavonoid binding site determine the function of different procyanidins. The molecular weight and the total number of hydroxyl groups of procyanidins increase exponentially as the DP increases (Abe et al., 2008). In addition, it was found that the identification of procyanidins is difficult due to the exponential increase in different potential isomers in correlation with DP. Hence, procyanidins are divided into oligomers (low DP) and polymers (high DP), whereas dimeric procyanidins are divided into types A and B depending on their stereo configuration and monomer linkage (Bittner et al., 2013), where one intraflavan bond is present between procyanidins monomers (C4-C8) is considered to be type B. In contrast, two intraflavan and ether bonds present between procyanidins monomers (C4-C8) are considered type A procyanidins (Rue et al., 2018).

Procyanidins endogenously and exogenously exhibit potent antioxidant and free radical scavenging activity due to their special molecular stereo-structure and many phenyl hydroxyl groups (Kehrer & Klotz, 2015). For example, expression of SOD, catalase and other antioxidant enzymes was significantly increased in mice after treatment with lotus-seeded procyanidins and prevented OS after exposure to low-frequency electromagnetic fields (ELF-EMF) (Luo et al., 2016). In addition, it was found that procyanidins control antioxidant enzymes and alcohol metabolism in rats by MAPK. Thus, MAPK reduces the expression of aspartate transaminase and alanine transaminase enzymes in the cell, presenting antioxidant and hepatoprotective effects (Bak et al., 2016). For instance, procyanidin dimer B2 (PB2) exhibits several crucial functions such as scavenging superoxide anion radicals, inhibiting NADPH oxidase and protecting neurons through the removal of O_2 and N_2 molecules. PB2 reduces free fatty acids and induces lipid degradation through regulation of the transcription factor EB (TFEB)–mediated lysosomal pathway. Alternatively, procyanidin B1 (PB1) elevates the antioxidant protein expression by activating the Nrf2 antioxidant response elements (AREs) (Steffen et al., 2008; Li et al., 2019). Furthermore, Li et al. has shown that A-type oligomeric procyanidins have higher scavenging abilities than the (–)-epicatechin, (+)-catechin in the pericarp of *Litchi chinensis*. Thus, antioxidant activity directly depends on the mDP of procyanidins, whereas, a high number of hydroxyl groups in the structure was found to be a more potent antioxidant than a monomeric phenolic (Liu et al., 2017; Hagerman et al.,

1998). In conclusion, these studies revealed that antioxidant activity and free radical scavenging capacity closely depend on the structure of procyanidins and the DP.

2.3.7.1 MAPK Signalling

Some kinase families, such as MAPK family kinases p38 MAPKs, c-Jun amino-terminal kinases (JNKs) and extracellular signal-regulated kinases, are responsible for arbitrating stress in cells (Chen et al., 2001). OS activates various signalling pathways that are induced by ROS through phosphorylation, resulting in nuclear translocation and expression of stress factors, apoptosis and tissue alteration (Bachelor & Bowden, 2004). Hence, procyanidins show the ability to downregulate MAPK, ERK, p38 and JNK signalling pathways, inhibiting ROS production and OS damage. In addition, it increases Nrf2 expression and AREs via the PI3K/Akt pathways in HepG2 cells and induces phase II/antioxidant enzyme expression (Bak et al., 2016; Bak et al., 2012).

2.3.7.2 NF-κB Signalling

Nuclear factor-κB (NF-κB) is triggered by multiple cellular oxidants and many antigens, which leads to the expression of many genes in response to OS, apoptosis and inflammation (Siomek, 2012). Its target genes include cytokines, chemokines, antioxidant enzymes and anti-apoptotic proteins. NF-κB remains in the cytoplasm with its inhibitory proteins (IκBs). However, OS phosphorylates IκBs that dissociate from NF-κB, leading to its translocation into the nucleus where the MAPK signal regulates its response pathways. Additionally, procyanidins (PC) inhibit OS damage by restraining NF-κB, ERK, p38 and JNK phosphorylation. In conclusion, PC reduce the mRNA expression of TNF-α, IL-1β (proinflammatory cytokines) and COX2 through inhibition of both the NF-κB and the MAPK pathways (Schoonbroodt & Piette, 2000; Lee et al., 2017).

2.3.7.3 Nrf2 Signalling

Nrf2 is a nuclear transcriptional factor that promotes the expression of protective genes in response to OS. It remains inactive in the cytoplasm through binding with Kelch-like ECH associating protein 1 (Keapl), which induces degradation of Nrf2 (Giudice et al., 2010). Under OS, Nrf2 dissociates from Keapl1 and translocates into the nucleus, where it binds with ARE. Subsequently, it induces or promotes activation of antioxidant enzymes such as phase II peroxiredoxin 1, glutamate-cysteine ligase catalytic subunit and NAD(P)H quinone dehydrogenase 1 (NQO1) (MacLeod et al., 2009). It was found that in order to reduce OS, PC induce the elevation of Nrf2 expression and ARE-mediated transcription. In addition, oligomeric PC enhance Nrf2 translocation and elevate the expression of HO-1 and thioredoxin reductase 1 by suppressing H_2O_2-induced OS damage in A549 cells (Sun et al., 2017).

2.3.7.4 JAK/STAT Signalling

The Janus kinase-signal transducer and activator of transcription (JAK/STAT) pathway is associated with OS through inflammatory factors and cytokines. This pathway is activated by ROS and by osmotic stress when specific cytokines bind to their receptors (Dudley et al., 2004). The PI3K/AKT pathway is essential for cell proliferation and regulation of ROS; however, irregularities in this signalling pathway can lead to cancer induction. Procyanidins inhibit apoptosis and reduce neurological deformities in type 2 diabetes by lowering STAT1 expression and preventing the JAK/STAT signalling (Song et al., 2014). This also prevents high glucose vascular smooth muscle cell growth by hindering the PI3k/Akt-dependent signalling pathway and ROS overproduction. In addition, PI3K/MAPK signalling pathways regulate Nrf2 transcription and increase phase II antioxidant enzyme expression (Wang et al., 2010).

2.4 ANTIOXIDANT EFFECTS OF PHENOLIC ACIDS

Phenolic acid compounds are natural antioxidants found in plant foods such as vegetables, fruits, cereals, seeds, berries, tea, onion bulbs, wine and vegetable oils. These compounds are synthesized from phenylalanine or tyrosine by the shikimic acid pathway, ranging from simple to complex compounds (Dimitrios, 2006). The hydroxyl group on the benzene ring is the main factor for antioxidant activity in phenolic compounds. Polyhydroxy phenolic compounds contain more than two hydroxyl groups, whereas polyphenolic compounds contain more than one phenol group (García-Lafuente et al., 2009).

Phenolic acids are classified into hydroxybenzoic acid and hydroxycinnamic acid derivatives. Furthermore, hydroxycinnamic acids are categorized into chlorogenic acid, caffeic acid, coumaric acid and ferulic acid, whereas hydroxybenzoic acids are divided into gallic acid and protocatechuic acid. It has been found that phenolic acids can activate antioxidant enzymes in the cells to eliminate free radicals (Clifford, 1999). The antioxidant mechanisms of phenolic compounds take place through the HAT mechanism, single-electron transfer via proton transfer (SET-PT) or transition metal mechanism.

2.4.1 HAT Mechanism

AH represents the antioxidant H-atom, which reacts with free radicals and stabilizes to form neutral species and an antioxidant free radical (A*). Subsequently, non-radical substrates like RH, ROH and ROOH are formed. The antioxidant properties of phenolic compounds originate from their structures that contain a benzene ring and the number of the OH group and their positions. While reacting with free radicals, the benzene ring stabilizes the antioxidant molecules, and free radicals are stabilized due to the resonance effect of the ring (Rajan & Muraleedharan, 2017).

In contrast to flavonol (i.e., quercetin), which has high antioxidant activity due to the presence of the hydroxyl group in the structure, phenolic acids can increase their activity if the COOH group has more space in the benzene ring. This increase in their activity is because the larger space influences the electron-withdrawing property of the carboxylic group (COOH), as was demonstrated in the case of cinnamic acid and its derivatives that have higher antioxidant activity than those of benzoic acid derivatives (Tsao, 2010). The hydroxyl group position on the benzene ring is vital to the antioxidant effect. However, the ortho and para positions show higher antioxidant activity than meta and unsubstituted phenols. Substitution of ethyl or n-butyl at the para position increases the activity of antioxidants compared to a methyl group. Also, the alkyl, bulky and branched groups decrease antioxidant activity at the para position and reduce the circulation of free radicals in the ring (Göçer, 2011).

The bond dissociation energy (BDE) of the hydrogen atom is also crucial in the antioxidant mechanism. A low BDE value suggests quick OH bond dissociation and reaction with free radicals. Therefore, the presence of the hydroxyl group on the ortho position of the benzene ring leads to a lower BDE value than that of the phenol group because an intramolecular hydrogen bond is formed during the reaction between a free radical and a hydrogen atom (Leopoldini et al., 2011; Wright et al., 2001).

2.4.2 Single Electron Transfer

In SET, the R group (anion) has an even number of electrons and it is a very stable species, whereas the Ar-OH (the cation radical) is a less stable and less reactive species. The reaction between the R group and the cation radical creates an odd number of electrons distributed over the aromatic ring. In the SET mechanism, the various atoms' ionization potential (IP) is essential for estimating scavenging activity. If the IP value is low, there is a quick electron removal after reacting with radicals, thus increasing antioxidant activity (Platzer et al., 2021; Rhile et al., 2006). In sequential proton loss electron transfer (SPLET), the phenolic group provides a proton to the free radical, forms an anion and transfers electrons again to create a stable product. The antioxidant activity is determined by the proton ability (PA) and the electron transfer enthalpy. PA values represent the intensity of phenolic hydroxyl dephosphorylation, whereas the electron transfer enthalpy (ETE) values denote the ability to donate electrons to phenolic ions. The electron-donating group at the ortho position has a higher PA value than the NO_2 group (electron-releasing group). At the same time, the ETE value is increased by the NO_2 group and is decreased by the methoxy group at the ortho position (Apak et al., 2016; Lee et al., 2020).

2.4.3 Transition Metal Chelation

The phenolic group can chelate with transition metal ions for its antioxidant activity. Transition metals, such as copper, cobalt, etc., catalyse reactions and chelate with polyphenols to form stable products. Chelation inhibits Fe^{3+} reduction, thus decreasing the reactive OH free radical formation by the Fenton reaction (Vo et al., 2020). Metal chelation mainly depends on the reduction potential of the phenolic acids. The reduction potential of PCs is as follows: propyl gallate > gallic acid > caffeic acid > pyrogallol > protocatechuic acid > sinapic acid > BHA > vanillyl alcohol > ferulic acid > guaiacol > vanillic acid > coumaric acid (Perron & Brumaghim, 2009). Phenolic acids and flavonoids show vigorous antioxidant activity and free radical scavenging potential. However, the chelating power and reducing properties may differ based on their structural characteristics. Flavonoid radical scavengers show higher chelating activity due to the catechol moiety (ring B), conjugated 4-4-carboxylic group in ring C and a hydroxyl group at positions 3 and 5. It was found that polyphenol's ligand effectively stabilizes Fe^{3+} because the Fe^{2+} complex is rapidly oxidized to form Fe^{3+} polyphenol complexes, known as auto-oxidation (Mathew et al., 2015). It was seen that Fe^{2+} oxidation in the presence of O_2 is very slow; however, when polyphenol ligands are attached to Fe^{2+}, there is a sudden drop in its reduction potential, which increases the rate of Fe oxidation. The oxidation rate of Fe for polyphenols is different, i.e., gallate shows a higher oxidation rate than catecholate complexes. However, electron-giving solid properties of oxygen ligands provide higher stability to the Fe^{3+} ion (Chvátalová et al., 2008).

2.4.4 Chlorogenic Acid

During aerobic respiration, the shikimic acid pathway produces chlorogenic acid in coffee, sunflower and dicot plants, known as phenylpropanoid or plant gold. Condensation of quinic acid and trans-cinnamic acid produces chlorogenic acid in plants. Chlorogenic acid has shown numerous biological functions such as antihypertensive, antitumor, stimulating the central nervous system (CNS), antioxidant, anti-inflammatory and decreased lipid peroxidation (Maalik et al., 2016). For its antioxidant activity, chlorogenic acid can scavenge free radicals by providing hydrogen atoms to DPPH, O_2-, and ONOO-ions, which assist their oxidation in order to reach their resonance stage. Furthermore, chlorogenic acid can chelate with metal ions. Hence, it helps to decrease the oxidation damage that is produced by free radicals through blocking the OH production in the cells (Cha et al., 2014). For example, Yang, R et al. found that chlorogenic acid chelates with Fe^{3+} ions to reduce oxidative damage (Yang et al., 2021), whereas Shibata et al. showed that chlorogenic acid causes plasmid DNA fragmentation in white blood cells, such as neutrophils, by inhibiting NH_2Cl function through a redox reaction (Shibata et al., 1999). Moreover, chlorogenic acid decreases the expression of the FOXO4 family of genes mesenchymal stem cells by PI3K/Akt phosphorylation, which leads to the prevention of OS. Finally, chlorogenic acid activates Nrf2 pathways, improves the expression of cellular antioxidant enzymes like NADPH and helps reduce OS-mediated injury (Li et al., 2012; Wang et al., 2021).

2.4.5 Gallic Acid

The chemical name of gallic acid (GA) is 3,4,5-trihydroxybenzoic acid, and it is abundantly present in the plant kingdom. It is present in a free form or as a derivative in various food sources like nuts, grapes and vegetables (Samanidou et al., 2012). The deposition of ROS in tissue is reduced and removed by GA through the Fenton reaction (Lima et al., 2023). The BDE and IP values of GA derivatives are high among four phenolic acids: GA, methyl gallate (MG), ethyl gallate (EG), and 4,5-dihydroxy-3-methoxy benzoic acid (OME) (Kalita et al., 2012). Compared with ester derivatives, the BDE value of GA was lower, but GA antioxidant activity was high due to the donating carboxylate group. In addition, unpaired electron spin density helps GA delocalize the electron over the entire structure of the radical, with a large area centred around the three carbon atoms of the aromatic ring, which is bound to the hydroxyl group (Šolc et al., 2014). For that reason, it is presumed that the electron donating group (OH) and the withdrawing group (-COOH) present at the ortho and para positions stabilize free radicals through resonance (Leopoldini et al., 2004).

2.4.6 Free Radical Scavenging Activity

The antioxidant potential is mainly measured by the 2,2-diphenyl-1-picrylhydrazyl (DPPH) assay, which contains stable free radicals that vanish upon reduction by an antiradical agent. GA has shown high antiradical potential from all phenolic compounds, where it can reduce six DPPHs per molecule (Yen et al., 2002). Research indicates that GA and 1,2,3,4,6-pentagalloylglucose (PGA), which can scavenge DPPH radicals by donating hydrogen, are more efficient than vitamin E (Abdelwahed et al., 2007). The phenolic acid compound has different capacities to quench and react with OH·, O^{2+} and DPPH. Moreover, it was determined that the number and positions of the hydroxyl groups in the phenolic rings affect free radical scavenging potential, i.e., monophenol shows less reducing power than polyphenols. The CHO and OH benzene derivatives present less antioxidant activity in comparison to the COOH derivatives because localizing electrons in the aromatic ring is difficult. However, the bulky group and double bond play an important role in the ring to stabilize the free radical by resonance (Ji et al., 2006).

The mode of action of GA can restore the low levels of endogenous antioxidant defence systems. Antioxidant enzymes like CAT, SOD and GST are the first line of defence against free radicals (Priscilla & Prince, 2009). But free radical accumulation can hinder endogenous antioxidant activity and increase OS. Superoxide radicals may reduce the SOD and CAT activities, resulting in low antioxidant response. In contrast, the inactivation of GRX leads to the production of excess GSSG, which ultimately inhibits protein synthesis (Hundal et al., 1997). GA and propyl gallate refurbished the activities of GPx and CAT. They help decrease the harmful effects of hydrogen peroxide and lipid peroxide and reduce malondialdehyde (MDA) concentration in some organs. Furthermore, GA can also reduce OS by decreasing the deposition of lipid peroxidation products like MDA and TBARS and enhancing antioxidant enzyme activity such as SOD, CAT and GSH-Px (Nabavi et al., 2013). Lastly, GA reduces the expression of the proinflammatory cytokine caused by chromium and reduces lipid peroxidation and NO_2 levels through the induction of SOD and GST antioxidant enzymes, as well as OS, by triggering the Nrf2/HO-1 pathways (Mohamed et al., 2016).

2.5 ANTIOXIDANT EFFECTS OF STILBENES

Stilbenes are phenolic compounds that can be found in several plant families, and their different structures in different plant families provide nutraceutical value. Stilbenes contain an essential class of non-flavonoid polyphenols, a standard structure distinguished by 14 carbons composed of two benzene rings linked by an ethylene bridge. Furthermore, stilbenes have two stereoisomers, namely cis and trans, due to the occurrence of a central ethylene moiety between the aromatic rings. Usually, trans-form stilbenes are present in nature and form polyketides when combined with polyphenol-like flavonoids and isoflavonoids (Roupe et al., 2006). Around 400 stilbene compounds are presently known and derived from resveratrol (3,5,4'-trihydroxy-trans-stilbene). Stilbenes are synthesized through the shikimate pathway or phenylpropanoid pathway, in which L-phenylalanine is the bio-precursor for the phenylpropanoid pathway (El Khawand et al., 2018). Moreover, stilbenes have a dual role in plants such as constitutive protection against both biotic (phytopathogen, herbivorous animal) and abiotic (UV radiation) stress. Hence, they protect cells from oxidative damage and act as antioxidants (Jeandet et al., 2010). Furthermore, they act as antibacterial, antifungal and insecticidal agents and can prevent vertebrate herbivory. The basic function of stilbenes is the same as other polyphenols, mainly counteracting OS by activating pathways for defence (Liu et al., 2013).

Resveratrol is a well-known stilbenoid polyphenol having two phenol rings linked to each other through an ethylene bond. As mentioned, two isomeric forms of resveratrol have been discovered: cis- and trans-resveratrol. The trans-form of resveratrol (E-5–(4-hydroxyethyl) benzene-1, 3-diol) is essential in terms of occurrence and its various biological functions such as differentiation, antiproliferative activity, etc. (Akinwumi et al., 2018; Orallo, 2006). Additionally, the trans-form is very stable under 70–75% humidity and 35–40°C in the presence of air. The geometric isomeric structure of the trans-resveratrol form can change into the cis-form under UV radiation exposure. Due to the low bioavailability of resveratrol in nature, researchers give special attention to synthesized modified resveratrol structures for therapeutic purposes, such as halogenated derivatives, which have shown promising therapeutic potential (Cichewicz & Kouzi, 2002). Glycosylated forms of resveratrol, like piceid, are present in our dietary foods, though humans, plants and some pathogens contain enzymes that oxidize these polyphenols. Nevertheless, glycosylation inhibits resveratrol's enzymatic oxidation, consequently preserving and improving its biological effect, stability and bioavailability. Although three different glycosylated resveratrol analogues that are present in nature, i.e., piceid, piceatannol glucoside and resveratroloside, show antibacterial activity, strong inflammatory, immune-modulatory and antiproliferative effects, only piceatannol glucoside demonstrated an anti-leishmanial effect (Fan et al., 2009).

Resveratrol has antioxidant properties and the ability to scavenge free radicals like nitric oxide (NO), superoxide radical ($O_2^{\cdot-}$) and hydrogen peroxide (H_2O_2) because of the presence of a hydroxyl group and conjugate double bond system (Stivala et al., 2001). It was found that antioxidant functions are dependent on the substitution of hydrogen in the hydroxyl group with CH_3 or elimination of the hydroxyl group. In addition, the 4′ hydroxyl activity has a crucial role in the hydrogen atom transfer (HAT), leading to SPLET in the resveratrol scavenging function (Benayahoum et al., 2014). HAT is a process in which three hydrogen atoms are formed and transferred to a nearby phenolic group for delocalization of unpaired electrons on the ring through a dynamic flip-flop motion, tautomerism rearrangement and intracellular nucleophilic attack (Caruso et al., 2004).

Leonard et al. has shown that resveratrol has the aptitude to scavenge free radicals using the ESR spin trap technique. They found that resveratrol (RSV) neutralizes the hydroxyl group produced during the Fenton reaction and the superoxide radical reaction (Leonard et al., 2003). Moreover, under in vitro conditions, RSV showed scavenging activity lower than butylated hydroxytoluene (BHT), butylated hydroxyanisole (BHA), tocopherol and Trolox compounds. However, the antioxidant efficiency of RSV was increased after chelating with metal ions, which inhibits hydroxyl radical production and their oxidation. It was also seen that RSV uses antioxidant enzymes to scavenge free radicals, which were produced by xanthine oxidase, mitochondrial respiratory chain enzymes, eNOS, etc. (Gülçin, 2010). Thus, in normal conditions, SOD, catalase and antioxidant enzymes can remove or destroy ROS generation during mitochondrial oxidative respiration. However, in abnormal conditions, chemical reagents and UV rays interfere with deposition of H_2O_2 and $O_2^{\cdot-}$ free radicals were reduced by RSV as observed after successful administration in diabetic rats (Palsamy & Subramanian, 2011).

2.5.1 Antioxidant Stress Mechanisms of RSV

All organisms possess complex antioxidant functions, which makes it very tough to recognize the exact molecular mechanism of RSV. Several research papers have shown that RSV employs an antioxidant effect through various signalling pathways and antioxidant enzymes (Zhang & Tsao, 2016). Still, RSV's antioxidant effect has three main pathways: Nrf2, NF-κB and SIRT1 signalling.

2.5.1.1 Nrf2 Signalling

Nrf2 is an important transcription factor for the regulation and expression of antioxidant genes and protection against oxidative damage. The antioxidant gene contains AREs, which help activate these pathways. Still, the activity of Nrf2 is under the control of a regulatory protein called Keapl (Teixeira et al., 2017). Nrf2 is present in the cytoplasm with inactivated Keapl in the absence of stimulus. But in the presence of ROS, conformational changes take place in Keapl leading to its disassociation from Nrf2 that translocates into the nucleus. Subsequently, Nrf2 forms a heterodimer with musculoaponeurotic fibrosarcoma (Maf) protein and then binds with AREs to increase the expression of antioxidant genes generating the antioxidant enzymes (Menshchikova et al., 2013). In addition, the activation of Nrf2/ARE leads to protein production, which helps remove ROS and many other harmful substances from the cells (He et al., 2020). Several studies demonstrate that RSV activates Nrf2 downstream of pathways such as the PI3K/AKT pathway. RSV reduces T-cell apoptosis by induction of an Nrf2/ARE-dependent oxidation gene like haeme oxygenase-1 (HO-1) through phosphorylation of Ser9 glycogen synthase kinase 3β (GSK3β) in AMPK pathways (Iwasaki et al., 2013). RSV also reduces cell death of PC12 cells induced by H_2O_2 through ERK and Akt pathway activation, where Nrf2 translocates into the nucleus and induces the expression of HO-1 (Chen et al., 2005). Finally, it was also reported that RSV activates Nrf2, causing up-regulation of glutamate-cysteine ligase (GCL) expression and induction of GSH expression to protect cells from OS in alveolar epithelial cells. Yet further work is required to investigate the mechanism of Nrf2 activation by RSV that induces downstream antioxidant enzyme genes.

2.5.1.2 NF-κB Signalling

NF-κB is a nuclear protein that exists in inactive form and normally associates with IκBs, including IκBα and IκBβ. This protein plays a vital role in the regulation of the expression of genes during inflammation such as IL-1, IL-6, IL-10 and tumour necrosis factor (TNF) in lipopolysaccharide (LPS)-stimulated cells (Tak & Firestein, 2001). However, the capability of resveratrol is much higher than naringenin and naringin to inhibit NO synthase (NOS) and NF-κB activation in macrophages. Likewise, resveratrol inhibits PMA-induced, LPS H_2O_2, okadaic acid, TNF-α-, IL-1β and UVB. Resveratrol reduces TNF-induced NF-κB activation in a dose- and time-dependent fashion (Adhami et al., 2003). Zhang et al. revealed that resveratrol inhibits the ERK1/2/NF-κB pathway to reduce OS and inflammation (Zhang J. et al., 2013). Expression of proinflammatory cytokines COX-2 and TNF-α was enhanced by NF-κB activation, while RSV inhibits those cytokines in both a dose-dependent and time-dependent manner in different cell lines (Manna et al., 2000). Furthermore, it was reported that administration of RSV (5 mg/kg) can reduce the GSSH/GSH ratio and NF-κB expression by inducing SOD activity in diabetic male Wistar rats (Soufi et al., 2012).

2.5.1.3 SIRT1 Signalling

Sirtuins play an essential role in various cellular functions, such as metabolic processes and OS. These functions mainly depend on NAD+ molecules that cause the deacetylation of several proteins such as FOXO, p53, etc. (Chen et al., 2015). To date, seven sirtuins have been found in mammals that are involved in DNA damage repair and cellular stress defence, mainly SIRT1, SIRT2 and SIRT6. However, RSV and SIRT1 do not interact directly. Instead, RSV initiates a cascade in which it activates AMPK (phosphodiesterase, non-degradable) through inactivation or inhibition of cAMP, which leads to induction of NAD+ that increases SIRT1 activation (Park et al., 2012). Moreover, RSV's effect on hyperglycaemia-induced mitochondrial OS takes place through increased SIRT1 activation, GSH content and MnSOD expression levels, while reduction in mtROS generation as well as H_2O_2 levels in human coronary artery endothelial cells (CAECs) takes place (Ungvari et al., 2009). In addition, it was reported that RSV shows a protective effect in low temperatures in tilapia where RNA expression levels of sirtuin homologs (sirtl, sirt2, sirt3, sirt5a and sirt6), SOD levels and catalase levels were found to increase (Wang et al., 2020). Lastly, SIRT1 deacetylates the transcription factor FOXO that would increase the ability of FOXO to ultimately scavenge ROS by increasing the expression of several antioxidant enzymes. Hence, SIRT1 signalling regulates the expression of the FOXO gene for its antioxidant activity. Still, the interaction between SIRT1, NF-κB and Nrf2 signalling pathways has not been sufficiently clarified, and there might be some redundancy between them. Therefore, additional works needs to be done to use them properly as a solution for disease treatments (Olmos et al., 2013).

2.6 ANTIOXIDANT EFFECTS OF LIGNANS

Lignans are dimers of phenol having 2,3-dibenzylbutane structures that are primarily collected from the bran layer of cereal grain to form the building block of lignin in the plant cell wall. Lignans are synthesized through the shikimic acid pathway in the plants and act as secondary metabolites (Rodríguez-García et al., 2019). In addition, lignans are found in all parts of the plants and classified according to the carbon chain skeleton and the integration of oxygen molecules into the carbon chain. Their structure is formed by polymerization of two phenyl-propyl groups at the β-positions of the side chains and mainly present in glycosidic form. Lignans are categorized into seven types: secoisolariciresionol (Seco), pinoresinol (Pino), matairesinol (Mat), medioresinol (Med), sesamin (Ses), syringaresinol (Syr) and lariciresinol (Lari). The two main mammalian lignans are enterodiol and enterolactone, which are products of secoisolariciresinol diglucoside (SDG) (Adlercreutz, 2007). Flax and sesame seeds contain the SDG conjugate where the gut microbiota converts SDG to Seco through deglycosylation. Two faecal bacterial strains in the colon, *Peptostreptococcus* and *Eggerthella lenta*, do this conversion that leads to the absorption of lignans in the gastrointestinal tract (Clavel et al., 2006). However, few studies have revealed that these metabolites perform as antioxidants and free radical scavengers to reduce the risk of various human diseases (Teodor et al., 2020). These studies have found that ROS containing free radicals like hydrogen peroxide (H_2O_2) and superoxide (O_2) are the main factors for death of normal cells and may cause several severe inflammatory diseases, such as cancer and cardiovascular disease in humans. For example, inflammatory bowel disease is caused by the excessive release of ROS in enteric cells. This ROS release induces the expression of pro-inflammatory molecules and some antioxidant reagents, such as riboflavin, that are used as therapeutic agents against this inflammatory disease. On the other hand, enterolactone presents antioxidant activity against human LDL oxidation under in vitro conditions. Hence, it inhibits the growth of colon cancer cells by inducing NADPH and quinone reductase enzymes (Parikh et al., 2019).

SDG was found to be the most crucial flaxseed phenolic compound for antioxidant activity through direct radical scavenging or through inhibition of lipid peroxidation (You & Cho, 2021). The glycosylated forms of SDG induce the expression of SOD and CAT enzymes, which ultimately scavenge free radicals that have been produced in the cells (Ge et al., 2022; Villavicencio Tejo & Quintanilla, 2021). In addition, it was found that glutathione concentration increased after SDG treatment. In contrast, MDA concentration was reduced due to Nrf2 activation, where an ARE binds with the antioxidant genes and induces the expression of HO-1 (Zhang M. et al., 2013). Lignans have the ability to chelate with Cu^{2+} metal ions and scavenge free radicals and to induce lipid peroxidation metabolism. For instance, sesamin with a Cu^{2+} metal ion presents potent antioxidative activity as shown by DPPH and TBARS assays on LDL (Saleem et al., 2005).

Nrf2 plays a crucial role in defence against OS and is known as the redox-sensitive transcription factor. Under normal conditions, Nrf2 binds to Keapl and remains inactive in the cytosol via ubiquitination and the 26S proteasome-mediated degradation process. Nrf2 activation is regulated by several factors, including ROS, which dissociate Nrf2 from Keapl through phosphorylation and leads to Nrf2 activation; other kinases that might help in its activation are protein kinase B (Akt), extracellular signal-regulated kinase (ERK), p38 mitogen-activated protein kinase (p38 MAPK) and protein kinase C (PKC) (Checa & Species, 2020).

After activation, Nrf2 enters the nucleus and forms a heterodimer with musculoaponeurotic fibrosarcoma (Maf) protein. Next, the complex binds to specific DNA sequences, i.e., ARE consensus, and starts the downstream transcription of cytoprotective genes of ARE-element genes such as redox-balancing proteins and phase II enzymes. These redox-balancing proteins include HO-1, NAD(P)H:quinone oxidoreductase 1 (NQO1), γ-glutamyl cysteine synthetase (γ-GCS), thioredoxin (Trx) and thioredoxin reductase (TrxR). The phase II enzymes such as GSH-utilizing enzymes include SOD, GPx, CAT and glutathione S-transferase (GST). These AREs are involved in ROS elimination and maintain cellular redox activity as well as remove toxicants (Vinogradov & Grivennikova, 2016).

Mitochondrial degradation leads to increased levels of ROS and of oxidized lipids through Bax/Bak pathways, which initiate NF-κB–mediated inflammatory and oxidative responses. Subsequently, induction of AMP/ATP and NAD+/NADH activates PGC-1α through different pathways, i.e., SIRT1 deacetylation and AMPK-mediated phosphorylation. Next, elevation of the PGC-1α level leads to increased Nrf2 activation that was initially inhibited by GSK3ß (blocks the translocation of Nrf2 into the nucleus through phosphorylation). Moreover, under OS, p38 AMPK, which is a metabolic regulator protein, keeps GSK3ß inactivated and phosphorylates Nrf2 at Ser50 to promote its translocation. There, PGC-1α assists in Nrf2 binding to the GCLC gene ARE, increasing the content of SOD and GCL proteins and decreasing the effect of ROS/RNS through phosphorylation (Rabinovitch et al., 2017; Wu et al., 2017).

It was found that Nrf2 and PGC-1α are concurrently activated by Erk1/2 signalling pathways through phosphorylation of LKB1, which in succession phosphorylates AMPK (Hota et al., 2012). Some medicines, such as erythropoietin (EPO), help in mitochondrial biogenesis through the Akt/eNOS/PGC-1α and AMPK/PGC-1α pathways. EPO can activate both Nrf2 and PGC-1α pathways through a common point, i.e., Akt phosphorylation can activate Nrf2 via GSK3b phosphorylation and activate antioxidant defence (Piantadosi et al., 2008).

The high amount of free radicals leads to ageing through the buildup of OS. Tissue degeneration by ROS productions takes place due to mtDNA mutations and ETC dysfunction. It was suggested that the imbalance between mitochondrial biogenesis and mitophagy is the reason for ageing due to the changes in ROS metabolism (Harman, 2009). In addition, the mtDNA mutations are caused by defilement in the mtDNA repair and mitophagic dysfunction. These Nrf2/ARE/ PGC-1α pathways play an essential role in the reduction and balance of ROS and may help in the decrease of age-related damages. Although some regulators, such as AMPK and SIRT1, are necessary for the liver and skeletal muscles, normal functions are found to decrease in old age (Cho et al., 2015).

In age-related neurodegenerative disease, i.e., Parkinson's disease, in which mitochondrial biogenesis is inhibited, activation of Nrf2/ARE/PGC-1α pathways may help in targeted therapy to slow them down. Since PARKIN and PINK1 (PTEN-induced kinase 1) proteins play an important role in Parkinson's disease, mutations in the gene that translates these proteins mediate mitophagy. However, Nrf2 and PINK1 have a direct relation; Nrf2 regulates PINK1 expression, although four ARE regions were deleted in the promoter of this gene. Nrf2 also regulates the expression of p62/SQSTM1 (adapter molecule), which helps interface ubiquitinated molecules straight to the autophagosome and can help in Parkinson's disease therapy (Murata et al., 2015; Jain et al., 2010; Thomas et al., 2012).

2.7 CONCLUSION

It is not difficult to conclude that polyphenols provide many health benefits and protection against the growth and development of many chronic pathological conditions such as diabetes, cardiovascular disease, Parkinson's disease and cancer. The mechanism of action of the antioxidant properties of polyphenols is now much better understood. Many types of research have shown that polyphenols have a reducing agent property, where they can scavenge free radicals in many mechanisms. It can reduce the $\bullet O^{2-}$, H_2O_2, $\bullet OH$, $ROO\bullet$, $^1O_2$ active radical directly by providing a hydrogen atom. In addition, it inhibits enzyme activity, reduces the production of free radicals and induces antioxidant enzyme activity such as SOD, CAT and GST. Furthermore, polyphenolic compounds chelate metal ions and reduce the production of free radicals. They also produce synergistic antioxidant activity with other molecules including Nrf2, SIRT1 and MAPK (summarized in Figure 2.2). Finally, it was found that the antioxidant activity of polyphenols can be used in many fields, such as medicine, agriculture, food products, chemicals, etc. Taken together, polyphenols have health benefits and antioxidant properties, which are the reason they are considered the seventh major nutrition component.

REFERENCES

Abdelwahed A, Bouhlel I, Skandrani I, Valenti K, Kadri M, Guiraud P, Steiman R, Mariotte AM, Ghedira K, Laporte F, Dijoux-Franca MG. Study of antimutagenic and antioxidant activities of Gallic acid and 1, 2, 3, 4, 6-pentagalloylglucose from Pistacia lentiscus: Confirmation by microarray expression profiling. *Chemico-Biological Interactions*. 2007 Jan 5;165(1):1–3.

Abe Y, Shoji T, Kawahara N, Kamakura H, Kanda T, Goda Y, Ozeki Y. Structural characterization of a procyanidin tetramer and pentamer from the apple by low-temperature NMR analysis. *Tetrahedron Letters*. 2008 Nov 3;49(45):6413–8.

Adhami VM, Afaq F, Ahmad N. Suppression of ultraviolet B exposure-mediated activation of NF-κB in normal human keratinocytes by resveratrol. *Neoplasia*. 2003 Jan 1;5(1):74–82.

Adlercreutz H. Lignans and human health. *Critical Reviews in Clinical Laboratory Sciences*. 2007 Jan 1;44(5–6):483–525.

Akinwumi BC, Bordun KA, Anderson HD. Biological activities of stilbenoids. *International Journal of Molecular Sciences*. 2018 Mar 9;19(3):792.

Alov P, Tsakovska I, Pajeva I. Computational studies of free radical-scavenging properties of phenolic compounds. *Current Topics in Medicinal Chemistry*. 2015 Jan 1;15(2):85–104.

Apak R, Özyürek M, Güçlü K, Çapanoğlu E. Antioxidant activity/capacity measurement. 1. Classification, physicochemical principles, mechanisms, and electron transfer (ET)-based assays. *Journal of Agricultural and Food Chemistry*. 2016 Feb 10;64(5):997–1027.

Aryaeian N, Sedehi SK, Arablou T. Polyphenols and their effects on diabetes management: A review. *Medical Journal of the Islamic Republic of Iran*. 2017 Dec 26;31:134.

Aviram M, Dornfeld L, Rosenblat M, Volkova N, Kaplan M, Coleman R, Hayek T, Presser D, Fuhrman B. Pomegranate juice consumption reduces oxidative stress, atherogenic modifications to LDL, and platelet aggregation: Studies in humans and in atherosclerotic apolipoprotein E–deficient mice. *The American Journal of Clinical Nutrition*. 2000 May 1;71(5):1062–76.

Bachelor MA, Bowden GT. UVA-mediated activation of signaling pathways involved in skin tumor promotion and progression. In *Seminars in Cancer Biology* (Vol. 14, No. 2, pp. 131–138). Academic Press; 2004 Apr 1.

Bak MJ, Jun M, Jeong WS. Procyanidins from wild grape (Vitis amurensis) seeds regulate ARE-mediated enzyme expression via Nrf2 coupled with p38 and PI3K/Akt pathway in HepG2 cells. *International Journal of Molecular Sciences*. 2012 Jan 13;13(1):801–18.

Bak MJ, Truong VL, Ko SY, Nguyen XN, Ingkasupart P, Jun M, Shin JY, Jeong WS. Antioxidant and hepatoprotective effects of procyanidins from wild grape (Vitis amurensis) seeds in ethanol-induced cells and rats. *International Journal of Molecular Sciences.* 2016 May 18;17(5):758.

Barbusiński K. Henry John Horstman Fenton-short biography and brief history of Fenton reagent discovery. *Chemistry-Didactics-Ecology-Metrology.* 2009;14(1–2):101–5.

Barrajón-Catalán E, Herranz-López M, Joven J, Segura-Carretero A, Alonso-Villaverde C, Menéndez JA, Micol V. Molecular promiscuity of plant polyphenols in the management of age-related diseases: Far beyond their antioxidant properties. *Oxidative Stress and Inflammation in Non-Communicable Diseases-Molecular Mechanisms and Perspectives in Therapeutics.* 2014:141–59.

Barros J, Serrani-Yarce JC, Chen F, Baxter D, Venables BJ, Dixon RA. Role of bifunctional ammonia-lyase in grass cell wall biosynthesis. *Nature Plants.* 2016 May 9;2(6):1–9.

Bellezza I, Giambanco I, Minelli A, Donato R. Nrf2-Keap1 signaling in oxidative and reductive stress. *Biochimica et Biophysica Acta (BBA)-Molecular Cell Research.* 2018 May 1;1865(5):721–33.

Benayahoum A, Amira-Guebailia H, Houache O. Homolytic and heterolytic O–H bond cleavage in trans-resveratrol and some phenantrene analogs: A theoretical study. *Computational and Theoretical Chemistry.* 2014 Jun 1;1037:1–9.

Bernatoniene J, Kopustinskiene DM. The role of catechins in cellular responses to oxidative stress. *Molecules.* 2018 Apr 20;23(4):965.

Bhattacharya S. Reactive oxygen species and cellular defense system. *Free Radicals in Human Health and Disease.* 2015:17–29.

Bittner K, Rzeppa S, Humpf HU. Distribution and quantification of flavan-3-ols and procyanidins with low degree of polymerization in nuts, cereals, and legumes. *Journal of Agricultural and Food Chemistry.* 2013 Sep 25;61(38):9148–54.

Bodini ME, Copia G, Tapia R, Leighton F, Herrera L. Iron complexes of quercetin in aprotic medium. Redox chemistry and interaction with superoxide anion radical. *Polyhedron.* 1999 Jul 16;18(17):2233–9.

Borges Bubols G, da Rocha Vianna D, Medina-Remon A, von Poser G, Maria Lamuela-Raventos R, Lucia Eifler-Lima V, Cristina Garcia S. The antioxidant activity of coumarins and flavonoids. *Mini Reviews in Medicinal Chemistry.* 2013 Mar 1;13(3):318–34.

Bukhari SB, Memon S, Mahroof-Tahir M, Bhanger MI. Synthesis, characterization and antioxidant activity copper–quercetin complex. *Spectrochimica Acta Part A: Molecular and Biomolecular Spectroscopy.* 2009 Jan 1;71(5):1901–6.

Caruso F, Tanski J, Villegas-Estrada A, Rossi M. Structural basis for antioxidant activity of trans-resveratrol: ab initio calculations and crystal and molecular structure. *Journal of Agricultural and Food Chemistry.* 2004 Dec 1;52(24):7279–85.

Cha JW, Piao MJ, Kim KC, Yao CW, Zheng J, Kim SM, Hyun CL, Ahn YS, Hyun JW. The polyphenol chlorogenic acid attenuates UVB-mediated oxidative stress in human HaCaT keratinocytes. *Biomolecules & Therapeutics.* 2014 Mar;22(2):136.

Chagas MD, Behrens MD, Moragas-Tellis CJ, Penedo GX, Silva AR, Gonçalves-de-Albuquerque CF. Flavonols and flavones as potential anti-inflammatory, antioxidant, and antibacterial compounds. *Oxidative Medicine and Cellular Longevity.* 2022;2022(1):9966750.

Checa J, Species JA. *Drivers of Physiological and Pathological Processes,* 2020, 13. https://doi.org/10.2147/JIR.S.;275595:1057-73.

Chen B, Zang W, Wang J, Huang Y, He Y, Yan L, Liu J, Zheng W. The chemical biology of sirtuins. *Chemical Society Reviews.* 2015;44(15):5246–64.

Chen CY, Jang JH, Li MH, Surh YJ. Resveratrol upregulates heme oxygenase-1 expression via activation of NF-E2-related factor 2 in PC12 cells. *Biochemical and Biophysical Research Communications.* 2005 Jun 17;331(4):993–1000.

Chen Z, Gibson TB, Robinson F, Silvestro L, Pearson G, Xu BE, Wright A, Vanderbilt C, Cobb MH. MAP kinases. *Chemical Reviews*. 2001 Aug 8;101(8):2449–76.

Cho SH, Chen JA, Sayed F, Ward ME, Gao F, Nguyen TA, Krabbe G, Sohn PD, Lo I, Minami S, Devidze N. SIRT1 deficiency in microglia contributes to cognitive decline in aging and neurodegeneration via epigenetic regulation of IL-1β. *Journal of Neuroscience*. 2015 Jan 14;35(2):807–18.

Chvátalová K, Slaninová I, Březinová L, Slanina J. Influence of dietary phenolic acids on redox status of iron: Ferrous iron autoxidation and ferric iron reduction. *Food Chemistry*. 2008 Jan 15;106(2):650–60.

Cichewicz RH, Kouzi SA. Resveratrol oligomers: Structure, chemistry, and biological activity. *Studies in Natural Products Chemistry*. 2002 Jan 1;26:507–79.

Clavel T, Henderson G, Engst W, Doré J, Blaut M. Phylogeny of human intestinal bacteria that activate the dietary lignan secoisolariciresinol diglucoside. *FEMS Microbiology Ecology*. 2006 Mar 1;55(3):471–8.

Clifford MN. Chlorogenic acids and other cinnamates–nature, occurrence and dietary burden. *Journal of the Science of Food and Agriculture*. 1999 Mar 1;79(3):362–72.

Di Meo F, Lemaur V, Cornil J, Lazzaroni R, Duroux JL, Olivier Y, Trouillas P. Free radical scavenging by natural polyphenols: Atom versus electron transfer. *The Journal of Physical Chemistry A*. 2013 Mar 14;117(10):2082–92.

Dimitrios B. Sources of natural phenolic antioxidants. *Trends in Food Science & Technology*. 2006 Sep 1;17(9):505–12.

Dong NQ, Lin HX. Contribution of phenylpropanoid metabolism to plant development and plant–environment interactions. *Journal of Integrative Plant Biology*. 2021 Jan;63(1):180–209.

Dowling S, Regan F, Hughes H. The characterisation of structural and antioxidant properties of isoflavone metal chelates. *Journal of Inorganic Biochemistry*. 2010 Oct 1;104(10):1091–8.

Dudley AC, Thomas D, Best J, Jenkins A. The STATs in cell stress-type responses. *Cell Communication and Signaling*. 2004 Dec;2:1–5.

Dugas Jr AJ, Castaneda-Acosta J, Bonin GC, Price KL, Fischer NH, Winston GW. Evaluation of the total peroxyl radical-scavenging capacity of flavonoids: Structure– activity relationships. *Journal of Natural Products*. 2000 Feb 26;63(3):327–31.

Edeas M. Strategies to target mitochondria and oxidative stress by antioxidants: Key points and perspectives. *Pharmaceutical Research*. 2011 Nov;28(11):2771–9.

El Khawand T, Courtois A, Valls J, Richard T, Krisa S. A review of dietary stilbenes: Sources and bioavailability. *Phytochemistry Reviews*. 2018 Oct;17(5):1007–29.

Fan P, Marston A, Hay AE, Hostettmann K. Rapid separation of three glucosylated resveratrol analogues from the invasive plant Polygonum cuspidatum by high-speed countercurrent chromatography. *Journal of Separation Science*. 2009 Sep;32(17):2979–84.

Faria A, Persaud SJ. Cardiac oxidative stress in diabetes: Mechanisms and therapeutic potential. *Pharmacology & Therapeutics*. 2017 Apr 1;172:50–62.

Fernandez MT, Mira ML, Florêncio MH, Jennings KR. Iron and copper chelation by flavonoids: An electrospray mass spectrometry study. *Journal of Inorganic Biochemistry*. 2002 Nov 11;92(2):105–11.

Gao W, Pu L, Chen M, Wei J, Xin Z, Wang Y, Yao Z, Shi T, Guo C. Glutathione homeostasis is significantly altered by quercetin via the Keap1/Nrf2 and MAPK signaling pathways in rats. *Journal of Clinical Biochemistry and Nutrition*. 2018;62(1):56–62.

García-Lafuente A, Guillamón E, Villares A, Rostagno MA, Martínez JA. Flavonoids as anti-inflammatory agents: Implications in cancer and cardiovascular disease. *Inflammation Research*. 2009 Sep;58(9):537–52.

Ge J, Hao R, Rong X, Dou QP, Tan X, Li G, Li F, Li D. Secoisolariciresinol diglucoside mitigates benzo [a] pyrene-induced liver and kidney toxicity in mice via miR-101a/

MKP-1-mediated p38 and ERK pathway. *Food and Chemical Toxicology*. 2022 Jan 1;159:112733.

Ghosh S, Chisti Y, Banerjee UC. Production of shikimic acid. *Biotechnology Advances*. 2012 Nov 1;30(6):1425–31.

Giudice A, Arra C, Turco MC. Review of molecular mechanisms involved in the activation of the Nrf2-ARE signaling pathway by chemopreventive agents. *Transcription Factors: Methods and Protocols*. 2010 May 10:37–74.

Göçer H, Gülçin İ. Caffeic acid phenethyl ester (CAPE): Correlation of structure and antioxidant properties. *International Journal of Food Sciences and Nutrition*. 2011 Dec 1;62(8):821–5.

Gu L, Kelm MA, Hammerstone JF, Beecher G, Holden J, Haytowitz D, Prior RL. Screening of foods containing proanthocyanidins and their structural characterization using LC-MS/MS and thiolytic degradation. *Journal of Agricultural and Food Chemistry*. 2003 Dec 3;51(25):7513–21.

Gülçin İ. Antioxidant properties of resveratrol: A structure–activity insight. *Innovative Food Science & Emerging Technologies*. 2010 Jan 1;11(1):210–8.

Hagerman AE, Riedl KM, Jones GA, Sovik KN, Ritchard NT, Hartzfeld PW, Riechel TL. High molecular weight plant polyphenolics (tannins) as biological antioxidants. *Journal of Agricultural and Food Chemistry*. 1998 May 18;46(5):1887–92.

Halliwell B, Gutteridge JM. *Free Radicals in Biology and Medicine*. Oxford University Press; 2015.

Harman D. Origin and evolution of the free radical theory of aging: A brief personal history, 1954–2009. *Biogerontology*. 2009 Dec;10(6):773–81.

He F, Ru X, Wen T. NRF2, a transcription factor for stress response and beyond. *International Journal of Molecular Sciences*. 2020 Jul 6;21(13):4777.

Herrmann KM, Weaver LM. The shikimate pathway. *Annual Review of Plant Biology*. 1999 Jun;50(1):473–503.

Hota KB, Hota SK, Chaurasia OP, Singh SB. Acetyl-L-carnitine-mediated neuroprotection during hypoxia is attributed to ERK1/2-Nrf2-regulated mitochondrial biosynthesis. *Hippocampus*. 2012 Apr;22(4):723–36.

Hundal MK, Jnagal S, Khanduja KL. Influence of gallic acid and propylgallate on antioxidant defense systems and lipid peroxidation in mice. *Journal of Clinical Biochemistry and Nutrition*. 1997;22(3):155–61.

Hur S, Jang E, Lee JH. Beneficial actions of Orostachys japonica and its compounds against tumors via MAPK signaling pathways. *Nutrients*. 2021 Feb 8;13(2):555.

Iwasaki K, Ray PD, Huang BW, Sakamoto K, Kobayashi T, Tsuji Y. Role of AMP-activated protein kinase in ferritin H gene expression by resveratrol in human T cells. *Biochemistry*. 2013 Jul 30;52(30):5075–83.

Jain A, Lamark T, Sjøttem E, Larsen KB, Awuh JA, Øvervatn A, McMahon M, Hayes JD, Johansen T. p62/SQSTM1 is a target gene for transcription factor NRF2 and creates a positive feedback loop by inducing antioxidant response element-driven gene transcription. *Journal of Biological Chemistry*. 2010 Jul 16;285(29):22576–91.

Jeandet P, Delaunois B, Conreux A, Donnez D, Nuzzo V, Cordelier S, Clément C, Courot E. Biosynthesis, metabolism, molecular engineering, and biological functions of stilbene phytoalexins in plants. *Biofactors*. 2010 Sep;36(5):331–41.

Ji HF, Zhang HY, Shen L. Proton dissociation is important to understanding structure–activity relationships of gallic acid antioxidants. *Bioorganic & Medicinal Chemistry Letters*. 2006 Aug 1;16(15):4095–8.

Kalita D, Kar R, Handique JG. A theoretical study on the antioxidant property of gallic acid and its derivatives. *Journal of Theoretical and Computational Chemistry*. 2012 Apr;11(02):391–402.

Kehrer JP, Klotz LO. Free radicals and related reactive species as mediators of tissue injury and disease: Implications for health. *Critical Reviews in Toxicology*. 2015 Oct 21;45(9):765–98.

Kelm MA, Hammerstone JF, Beecher G, Holden J, Haytowitz D, Gebhardt S, Gu L, Prior RL. Concentrations of proanthocyanidins in common foods and estimations of normal consumption. *The Journal of Nutrition*. 2004 Mar 1;134(3):613–7.

Kim DK, Jeong SC, Gorinstein S, Chon SU. Total polyphenols, antioxidant and antiproliferative activities of different extracts in mungbean seeds and sprouts. *Plant Foods for Human Nutrition*. 2012 Mar;67:71–5.

Kitzing K, Auweter S, Amrhein N, Macheroux P. Mechanism of chorismate synthase: Role of the two invariant histidine residues in the active site. *Journal of Biological Chemistry*. 2004 Mar 2;279(10):9451–61.

Kołton A, Długosz-Grochowska O, Wojciechowska R, Czaja M. Biosynthesis regulation of folates and phenols in plants. *Scientia Horticulturae*. 2022 Jan 3;291:110561.

Kostyuk VA, Potapovich AI, Vladykovskaya EN, Korkina LG, Afanas' ev IB. Influence of metal ions on flavonoid protection against asbestos-induced cell injury. *Archives of Biochemistry and Biophysics*. 2001 Jan 1;385(1):129–37.

Lee CY, Sharma A, Semenya J, Anamoah C, Chapman KN, Barone V. Computational study of ortho-substituent effects on antioxidant activities of phenolic dendritic antioxidants. *Antioxidants*. 2020 Feb 25;9(3):189.

Lee JW, Kim YI, Im CN, Kim SW, Kim SJ, Min S, Joo YH, Yim SV, Chung N. Grape seed proanthocyanidin inhibits mucin synthesis and viral replication by suppression of AP-1 and NF-κB via p38 MAPKs/JNK signaling pathways in respiratory syncytial virus-infected A549 cells. *Journal of Agricultural and Food Chemistry*. 2017 Jun 7;65(22):4472–83.

Lee-Hilz YY, Boerboom AM, Westphal AH, van Berkel WJ, Aarts JM, Rietjens IM. Pro-oxidant activity of flavonoids induces EpRE-mediated gene expression. *Chemical Research in Toxicology*. 2006 Nov 20;19(11):1499–505.

Leonard SS, Xia C, Jiang BH, Stinefelt B, Klandorf H, Harris GK, Shi X. Resveratrol scavenges reactive oxygen species and effects radical-induced cellular responses. *Biochemical and Biophysical Research Communications*. 2003 Oct 3;309(4):1017–26.

Leopoldini M, Marino T, Russo N, Toscano M. Antioxidant properties of phenolic compounds: H-atom versus electron transfer mechanism. *The Journal of Physical Chemistry A*. 2004 Jun 3;108(22):4916–22.

Leopoldini M, Russo N, Toscano M. The molecular basis of working mechanism of natural polyphenolic antioxidants. *Food Chemistry*. 2011 Mar 15;125(2):288–306.

Letenneur L, Proust-Lima C, Le Gouge A, Dartigues JF, Barberger-Gateau P. Flavonoid intake and cognitive decline over a 10-year period. *American Journal of Epidemiology*. 2007 Jun 15;165(12):1364–71.

Li S, Bian H, Liu Z, Wang Y, Dai J, He W, Liao X, Liu R, Luo J. Chlorogenic acid protects MSCs against oxidative stress by altering FOXO family genes and activating intrinsic pathway. *European Journal of Pharmacology*. 2012 Jan 15;674(2–3):65–72.

Li T, Li Q, Wu W, Li Y, Hou DX, Xu H, Zheng B, Zeng S, Shan Y, Lu X, Deng F. Lotus seed skin proanthocyanidin extract exhibits potent antioxidant property via activation of the Nrf2–ARE pathway. *Acta Biochimica et Biophysica Sinica*. 2019 Jan;51(1):31–40.

Lima JP, Tabelini CH, Aguiar A. A review of gallic acid-mediated Fenton processes for degrading emerging pollutants and dyes. *Molecules*. 2023 Jan 24;28(3):1166.

Liu HJ, Pan XX, Liu BQ, Gui X, Hu L, Jiang CY, Han Y, Fan YX, Tang YL, Liu WT. Grape seed-derived procyanidins alleviate gout pain via NLRP3 inflammasome suppression. *Journal of Neuroinflammation*. 2017 Dec;14:1–10.

Liu S, Pi J, Zhang Q. Signal amplification in the KEAP1-NRF2-ARE antioxidant response pathway. *Redox Biology*. 2022 Aug 1;54:102389.

Liu T, Lv YF, Zhao JL, You QD, Jiang ZY. Regulation of Nrf2 by phosphorylation: Consequences for biological function and therapeutic implications. *Free Radical Biology and Medicine*. 2021 May 20;168:129–41.

Liu YQ, Li XJ, Zhao CY, Lu Y, Li WQ, Liu ZL, Feng G, Yang L. Synthesis and insect antifeedant activity of stilbene derivatives against Brontispa longissima larvae. *Medicinal Chemistry Research*. 2013 May;22:2196–206.

Luo M, Tian R, Yang Z, Peng YY, Lu N. Quercetin suppressed NADPH oxidase-derived oxidative stress via heme oxygenase-1 induction in macrophages. *Archives of Biochemistry and Biophysics*. 2019 Aug 15;671:69–76.

Luo X, Chen M, Duan Y, Duan W, Zhang H, He Y, Yin C, Sun G, Sun X. Chemoprotective action of lotus seedpod procyanidins on oxidative stress in mice induced by extremely low-frequency electromagnetic field exposure. *Biomedicine & Pharmacotherapy*. 2016 Aug 1;82:640–8.

Lv QZ, Long JT, Gong ZF, Nong KY, Liang XM, Qin T, Huang W, Yang L. Current state of knowledge on the antioxidant effects and mechanisms of action of polyphenolic compounds. *Natural Product Communications*. 2021 Aug;16(7). https://doi.org/10.1177/1934578X211027745.

Maalik A, Bukhari SM, Zaidi AS, Shah KH, Khan FA. Chlorogenic acid: A pharmacologically potent molecule. *Acta Pol Pharm*. 2016 Jul 1;73(4):851–4.

MacLeod AK, McMahon M, Plummer SM, Higgins LG, Penning TM, Igarashi K, Hayes JD. Characterization of the cancer chemopreventive NRF2-dependent gene battery in human keratinocytes: Demonstration that the KEAP1–NRF2 pathway, and not the BACH1–NRF2 pathway, controls cytoprotection against electrophiles as well as redox-cycling compounds. *Carcinogenesis*. 2009;30(9):1571–80.

Maeda H, Dudareva N. The shikimate pathway and aromatic amino acid biosynthesis in plants. *Annual Review of Plant Biology*. 2012 Jun 2;63(1):73–105.

Manna SK, Mukhopadhyay A, Aggarwal BB. Resveratrol suppresses TNF-induced activation of nuclear transcription factors NF-κB, activator protein-1, and apoptosis: potential role of reactive oxygen intermediates and lipid peroxidation. *The Journal of Immunology*. 2000 Jun 15;164(12):6509–19.

Mathew S, Abraham TE, Zakaria ZA. Reactivity of phenolic compounds towards free radicals under in vitro conditions. *Journal of Food Science and Technology*. 2015 Sep;52:5790–8.

Matzinger M, Fischhuber K, Heiss EH. Activation of Nrf2 signaling by natural products-can it alleviate diabetes?. *Biotechnology Advances*. 2018 Nov 1;36(6):1738–67.

Menshchikova EB, Zenkov NK, Tkachev VO, Lemza AE, Kandalintseva NV. Protective effect of ARE-inducing phenol antioxidant TS-13 in chronic inflammation. *Bulletin of Experimental Biology and Medicine*. 2013 Jul;155:330–4.

Mittelstädt G, Negron L, Schofield LR, Marsh K, Parker EJ. Biochemical and structural characterisation of dehydroquinate synthase from the New Zealand kiwifruit Actinidia chinensis. *Archives of Biochemistry and Biophysics*. 2013 Sep 15;537(2):185–91.

Mladěnka P, Zatloukalová L, Filipský T, Hrdina R. Cardiovascular effects of flavonoids are not caused only by direct antioxidant activity. *Free Radical Biology and Medicine*. 2010 Sep 15;49(6):963–75.

Moalin M, van Strijdonck GP, Beckers M, Hagemen GJ, Borm PJ, Bast A, Haenen GR. A planar conformation and the hydroxyl groups in the B and C rings play a pivotal role in the antioxidant capacity of quercetin and quercetin derivatives. *Molecules*. 2011 Nov 21;16(11):9636–50.

Mohamed HM, Abd El-Twab SM. Gallic acid attenuates chromium-induced thyroid dysfunction by modulating antioxidant status and inflammatory cytokines. *Environmental Toxicology and Pharmacology*. 2016 Dec 1;48:225–36.

Moridani MY, Pourahmad J, Bui H, Siraki A, O'Brien PJ. Dietary flavonoid iron complexes as cytoprotective superoxide radical scavengers. *Free Radical Biology and Medicine*. 2003 Jan 15;34(2):243–53.

Murata H, Takamatsu H, Liu S, Kataoka K, Huh NH, Sakaguchi M. NRF2 regulates PINK1 expression under oxidative stress conditions. *PLoS One*. 2015 Nov 10;10(11):e0142438.

Nabavi SM, Habtemariam S, Nabavi SF, Sureda A, Daglia M, Moghaddam AH, Amani MA. Protective effect of gallic acid isolated from Peltiphyllum peltatum against sodium fluoride-induced oxidative stress in rat's kidney. *Molecular and Cellular Biochemistry*. 2013 Jan;372:233–9.

Nabavi SM, Šamec D, Tomczyk M, Milella L, Russo D, Habtemariam S, Suntar I, Rastrelli L, Daglia M, Xiao J, Giampieri F. Flavonoid biosynthetic pathways in plants: Versatile targets for metabolic engineering. *Biotechnology Advances*. 2020 Jan 1;38:107316.

Olmos Y, Sánchez-Gómez FJ, Wild B, García-Quintans N, Cabezudo S, Lamas S, Monsalve M. SirT1 regulation of antioxidant genes is dependent on the formation of a FoxO3a/PGC-1α complex. *Antioxidants & Redox Signaling*. 2013 Nov 1;19(13):1507–21.

Orallo F. Comparative studies of the antioxidant effects of cis-and trans-resveratrol. *Current Medicinal Chemistry*. 2006 Jan 1;13(1):87–98.

Oteiza PI, Fraga CG, Mills DA, Taft DH. Flavonoids and the gastrointestinal tract: Local and systemic effects. *Molecular Aspects of Medicine*. 2018 Jun 1;61:41–9.

Palsamy P, Subramanian S. Resveratrol protects diabetic kidney by attenuating hyperglycemia-mediated oxidative stress and renal inflammatory cytokines via Nrf2–Keap1 signaling. *Biochimica et Biophysica Acta (BBA)-Molecular Basis of Disease*. 2011 Jul 1;1812(7):719–31.

Panche AN, Diwan AD, Chandra SR. Flavonoids: An overview. *Journal of Nutritional Science*. 2016 Jan;5:e47.

Parikh M, Maddaford TG, Austria JA, Aliani M, Netticadan T, Pierce GN. Dietary flaxseed as a strategy for improving human health. *Nutrients*. 2019 May 25;11(5):1171.

Park SJ, Ahmad F, Philp A, Baar K, Williams T, Luo H, Ke H, Rehmann H, Taussig R, Brown AL, Kim MK. Resveratrol ameliorates aging-related metabolic phenotypes by inhibiting cAMP phosphodiesterases. *Cell*. 2012 Feb 3;148(3):421–33.

Perron NR, Brumaghim JL. A review of the antioxidant mechanisms of polyphenol compounds related to iron binding. *Cell Biochemistry and Biophysics*. 2009 Mar;53:75–100.

Piantadosi CA, Carraway MS, Babiker A, Suliman HB. Heme oxygenase-1 regulates cardiac mitochondrial biogenesis via Nrf2-mediated transcriptional control of nuclear respiratory factor–1. *Circulation Research*. 2008 Nov 21;103(11):1232–40.

Platzer M, Kiese S, Herfellner T, Schweiggert-Weisz U, Miesbauer O, Eisner P. Common trends and differences in antioxidant activity analysis of phenolic substances using single electron transfer based assays. *Molecules*. 2021 Feb 25;26(5):1244.

Priscilla DH, Prince PS. Cardioprotective effect of gallic acid on cardiac troponin-T, cardiac marker enzymes, lipid peroxidation products and antioxidants in experimentally induced myocardial infarction in Wistar rats. *Chemico-Biological Interactions*. 2009 May 15;179(2–3):118–24.

Rabinovitch RC, Samborska B, Faubert B, Ma EH, Gravel SP, Andrzejewski S, Raissi TC, Pause A, Pierre JS, Jones RG. AMPK maintains cellular metabolic homeostasis through regulation of mitochondrial reactive oxygen species. *Cell Reports*. 2017 Oct 3;21(1):1–9.

Rajan VK, Muraleedharan K. A computational investigation on the structure, global parameters and antioxidant capacity of a polyphenol, *Gallic Acid Food Chemistry*. 2017 Apr 1;220:93–9.

Rauf A, Imran M, Abu-Izneid T, Patel S, Pan X, Naz S, Silva AS, Saeed F, Suleria HA. Proanthocyanidins: A comprehensive review. *Biomedicine & Pharmacotherapy*. 2019 Aug 1;116:108999.

Rhile IJ, Markle TF, Nagao H, DiPasquale AG, Lam OP, Lockwood MA, Rotter K, Mayer JM. Concerted proton– electron transfer in the oxidation of hydrogen-bonded phenols. *Journal of the American Chemical Society*. 2006 May 10;128(18):6075–88.

Rice-Evans C. Flavonoid antioxidants. *Current Medicinal Chemistry*. 2001 Jun 1;8(7):797–807.

Rodríguez-García C, Sánchez-Quesada C, Toledo E, Delgado-Rodríguez M, Gaforio JJ. Naturally lignan-rich foods: A dietary tool for health promotion?. *Molecules*. 2019 Mar 6;24(5):917.
Roupe KA, Remsberg CM, Yáñez JA, Davies NM. Pharmacometrics of stilbenes: Seguing towards the clinic. *Current Clinical Pharmacology*. 2006 Jan 1;1(1):81–101.
Rue EA, Rush MD, van Breemen RB. Procyanidins: A comprehensive review encompassing structure elucidation via mass spectrometry. *Phytochemistry Reviews*. 2018 Feb;17:1–6.
Saha S, Buttari B, Panieri E, Profumo E, Saso L. An overview of Nrf2 signaling pathway and its role in inflammation. *Molecules*. 2020 Nov 23;25(22):5474.
Saleem M, Kim HJ, Ali MS, Lee YS. An update on bioactive plant lignans. *Natural Product Reports*. 2005;22(6):696–716.
Samanidou V, Tsagiannidis A, Sarakatsianos I. Simultaneous determination of polyphenols and major purine alkaloids in Greek Sideritis species, herbal extracts, green tea, black tea, and coffee by high-performance liquid chromatography-diode array detection. *Journal of Separation Science*. 2012 Feb;35(4):608–15.
Scalbert A, Manach C, Morand C, Rémésy C, Jiménez L. Dietary polyphenols and the prevention of diseases. *Critical Reviews in Food Science and Nutrition*. 2005 Jun 1;45(4):287–306.
Schönbrunn E, Eschenburg S, Shuttleworth WA, Schloss JV, Amrhein N, Evans JN, Kabsch W. Interaction of the herbicide glyphosate with its target enzyme 5-enolpyruvylshikimate 3-phosphate synthase in atomic detail. *Proceedings of the National Academy of Sciences*. 2001 Feb 13;98(4):1376–80.
Schoonbroodt S, Piette J. Oxidative stress interference with the nuclear factor-κB activation pathways. *Biochemical Pharmacology*. 2000 Oct 15;60(8):1075–83.
Sharma A, Shahzad B, Rehman A, Bhardwaj R, Landi M, Zheng B. Response of phenylpropanoid pathway and the role of polyphenols in plants under abiotic stress. *Molecules*. 2019 Jul 4;24(13):2452.
Shaw P, Chattopadhyay A. Nrf2–ARE signaling in cellular protection: Mechanism of action and the regulatory mechanisms. *Journal of Cellular Physiology*. 2020 Apr;235(4):3119–30.
Shibata H, Sakamoto Y, Oka M, Kono Y. Natural antioxidant, chlorogenic acid, protects against DNA breakage caused by monochloramine. *Bioscience, Biotechnology, and Biochemistry*. 1999 Jan 1;63(7):1295–7.
Singh S, Kaur I, Kariyat R. The multifunctional roles of polyphenols in plant-herbivore interactions. *International Journal of Molecular Sciences*. 2021 Feb 1;22(3):1442.
Siomek A. NF-κB signaling pathway and free radical impact. *Acta Biochimica Polonica*. 2012 Aug 1;59(3):323–31.
Sivandzade F, Prasad S, Bhalerao A, Cucullo L. NRF2 and NF-κB interplay in cerebrovascular and neurodegenerative disorders: molecular mechanisms and possible therapeutic approaches. *Redox Biology*. 2019 Feb 1;21:101059.
Socha MW, Malinowski B, Puk O, Wartęga M, Stankiewicz M, Kazdepka-Ziemińska A, Wiciński M. The role of NF-κB in uterine spiral arteries remodeling, insight into the cornerstone of preeclampsia. *International Journal of Molecular Sciences*. 2021 Jan 12;22(2):704.
Šolc R, Gerzabek MH, Lischka H, Tunega D. Radical sites in humic acids: A theoretical study on protocatechuic and gallic acids. *Computational and Theoretical Chemistry*. 2014 Mar 15;1032:42–9.
Song CG, Yang X, Min LQ, Liu CX, Zhao CS. The effect of procyanidin on expression of STAT1 in type 2 diabetes mellitus SD rats with focal cerebral ischemia. *Neuroendocrinology Letters*. 2014 Jan 1;35(1):68–72.
Soufi FG, Mohammad-Nejad D, Ahmadieh H. Resveratrol improves diabetic retinopathy possibly through oxidative stress – nuclear factor κB – apoptosis pathway. *Pharmacological Reports*. 2012 Nov;64(6):1505–14.

Steffen Y, Gruber C, Schewe T, Sies H. Mono-O-methylated flavanols and other flavonoids as inhibitors of endothelial NADPH oxidase. *Archives of Biochemistry and Biophysics*. 2008 Jan 15;469(2):209–19.

Stivala LA, Savio M, Carafoli F, Perucca P, Bianchi L, Maga G, Forti L, Pagnoni UM, Albini A, Prosperi E, Vannini V. Specific structural determinants are responsible for the antioxidant activity and the cell cycle effects of resveratrol. *Journal of Biological Chemistry*. 2001 Jun 22;276(25):22586–94.

Sun C, Jin W, Shi H. Oligomeric proanthocyanidins protects A549 cells against H2O2-induced oxidative stress via the Nrf2-ARE pathway. *International Journal of Molecular Medicine*. 2017 Jun 1;39(6):1548–54.

Tak H, Negi S, Ganapathi TR. Overexpression of MusaMYB31, a R2R3 type MYB transcription factor gene indicate its role as a negative regulator of lignin biosynthesis in banana. *PLoS One*. 2017 Feb 24;12(2):e0172695.

Tak PP, Firestein GS. NF-κB: A key role in inflammatory diseases. *The Journal of Clinical Investigation*. 2001 Jan 1;107(1):7–11.

Tapiero H, Tew KD, Ba GN, Mathe G. Polyphenols: Do they play a role in the prevention of human pathologies? *Biomedicine & Pharmacotherapy*. 2002 Jun 1;56(4):200–7.

Teixeira TM, da Costa DC, Resende AC, Soulage CO, Bezerra FF, Daleprane JB. Activation of Nrf2-antioxidant signaling by 1, 25-dihydroxycholecalciferol prevents leptin-induced oxidative stress and inflammation in human endothelial cells. *The Journal of Nutrition*. 2017 Apr 1;147(4):506–13.

Teodor ED, Moroeanu V, Radu GL. Lignans from medicinal plants and their anticancer effect. *Mini Reviews in Medicinal Chemistry*. 2020 Jul 1;20(12):1083–90.

Thiruvengadam M, Venkidasamy B, Subramanian U, Samynathan R, Ali Shariati M, Rebezov M, Girish S, Thangavel S, Dhanapal AR, Fedoseeva N, Lee J. Bioactive compounds in oxidative stress-mediated diseases: Targeting the NRF2/ARE signaling pathway and epigenetic regulation. *Antioxidants*. 2021 Nov 23;10(12):1859.

Thomas RR, Keeney PM, Bennett JP. Impaired complex-I mitochondrial biogenesis in Parkinson disease frontal cortex. *Journal of Parkinson's Disease*. 2012 Jan 1;2(1):67–76.

Tsao R. Chemistry and biochemistry of dietary polyphenols. *Nutrients*. 2010 Dec 10;2(12):1231–46.

Tzin V, Galili G. New insights into the shikimate and aromatic amino acids biosynthesis pathways in plants. *Molecular Plant*. 2010 Nov 1;3(6):956–72.

Umeno A, Horie M, Murotomi K, Nakajima Y, Yoshida Y. Antioxidative and antidiabetic effects of natural polyphenols and isoflavones. *Molecules*. 2016 May 30;21(6):708.

Ungvari Z, Labinskyy N, Mukhopadhyay P, Pinto JT, Bagi Z, Ballabh P, Zhang C, Pacher P, Csiszar A. Resveratrol attenuates mitochondrial oxidative stress in coronary arterial endothelial cells. *American Journal of Physiology-Heart and Circulatory Physiology*. 2009 Nov;297(5):H1876–81.

Ushio-Fukai M, Nakamura Y. Reactive oxygen species and angiogenesis: NADPH oxidase as target for cancer therapy. *Cancer Letters*. 2008 Jul 18;266(1):37–52.

Valenzuela JC, Vergara-Salinas JR, Perez-Correa JR. *Advances in Technologies for Producing Food-Relevant Polyphenols*. CRC Press; 2016 Sep 19.

Vauzour D. Dietary polyphenols as modulators of brain functions: Biological actions and molecular mechanisms underpinning their beneficial effects. *Oxidative Medicine and Cellular Longevity*. 2012;2012(1):914273.

Villavicencio Tejo F, Quintanilla RA. Contribution of the Nrf2 pathway on oxidative damage and mitochondrial failure in Parkinson and Alzheimer's disease. *Antioxidants*. 2021 Jul 2;10(7):1069.

Vinogradov AD, Grivennikova VG. Oxidation of NADH and ROS production by respiratory complex I. *Biochimica et Biophysica Acta (BBA)-Bioenergetics*. 2016 Jul 1;1857(7):863–71.

Vo QV, Tam NM, Van Bay M, Thong NM, Le Huyen T, Hoa NT, Mechler A. The antioxidant activity of natural diterpenes: theoretical insights. *RSC Advances*. 2020;10(25): 14937–43.

Walle T, Ta N, Kawamori T, Wen X, Tsuji PA, Walle UK. Cancer chemopreventive properties of orally bioavailable flavonoids – methylated versus unmethylated flavones. *Biochemical Pharmacology*. 2007 May 1;73(9):1288–96.

Wang D, Hou J, Wan J, Yang Y, Liu S, Li X, Li W, Dai X, Zhou P, Liu W, Wang P. Dietary chlorogenic acid ameliorates oxidative stress and improves endothelial function in diabetic mice via Nrf2 activation. *Journal of International Medical Research*. 2021 Jan;49(1):0300060520985363.

Wang H, Liu K, Geng M, Gao P, Wu X, Hai Y, Li Y, Li Y, Luo L, Hayes JD, Wang XJ. RXRα inhibits the NRF2-ARE signaling pathway through a direct interaction with the Neh7 domain of NRF2. *Cancer Research*. 2013 May 15;73(10):3097–108.

Wang L, Ran L, Hou Y, Tian Q, Li C, Liu R, Fan D, Luo K. The transcription factor MYB115 contributes to the regulation of proanthocyanidin biosynthesis and enhances fungal resistance in poplar. *New Phytologist*. 2017 Jul;215(1):351–67.

Wang L, Zhu LH, Jiang H, Tang QZ, Yan L, Wang D, Liu C, Bian ZY, Li H. Grape seed proanthocyanidins attenuate vascular smooth muscle cell proliferation via blocking phosphatidylinositol 3-kinase-dependent signaling pathways. *Journal of Cellular Physiology*. 2010 Jun;223(3):713–26.

Wang MC, Wang YC, Peng HW, Hseu JR, Wu GC, Chang CF, Tseng YC. Resveratrol induces expression of metabolic and antioxidant machinery and protects tilapia under cold stress. *International Journal of Molecular Sciences*. 2020 May 8;21(9):3338.

Wang N, Qu C, Jiang S, Chen Z, Xu H, Fang H, Su M, Zhang J, Wang Y, Liu W, Zhang Z. The proanthocyanidin-specific transcription factor Md MYBPA 1 initiates anthocyanin synthesis under low-temperature conditions in red-fleshed apples. *The Plant Journal*. 2018 Oct;96(1):39–55.

Wang W, Sun W, Cheng Y, Xu Z, Cai L. Role of sirtuin-1 in diabetic nephropathy. *Journal of Molecular Medicine*. 2019 Mar 7;97:291–309.

Wen X, Walle T. Methylated flavonoids have greatly improved intestinal absorption and metabolic stability. *Drug Metabolism and Disposition*. 2006 Oct 1;34(10):1786–92.

Wiegand H, Wagner AE, Boesch-Saadatmandi C, Kruse HP, Kulling S, Rimbach G. Effect of dietary genistein on Phase II and antioxidant enzymes in rat liver. *Cancer Genomics & Proteomics*. 2009 Mar 1;6(2):85–92.

Wolfe KL, Liu RH. Structure– activity relationships of flavonoids in the cellular antioxidant activity assay. *Journal of Agricultural and Food Chemistry*. 2008 Sep 24;56(18):8404–11.

Wright JS, Johnson ER, DiLabio GA. Predicting the activity of phenolic antioxidants: Theoretical method, analysis of substituent effects, and application to major families of antioxidants. *Journal of the American Chemical Society*. 2001 Feb 14;123(6):1173–83.

Wu C, Xu D, Ge M, Luo J, Chen L, Chen Z, You Y, Zhu YX, Lin H, Shi J. Blocking glutathione regeneration: Inorganic NADPH oxidase nanozyme catalyst potentiates tumoral ferroptosis. *Nano Today*. 2022 Oct 1;46:101574.

Wu H, Zhao J, Chen M, Wang H, Yao Q, Fan J, Zhang M. The anti-aging effect of erythropoietin via the ERK/Nrf2-ARE pathway in aging rats. *Journal of Molecular Neuroscience*. 2017 Mar;61:449–58.

Wu S, Liao X, Zhu Z, Huang R, Chen M, Huang A, Zhang J, Wu Q, Wang J, Ding Y. Antioxidant and anti-inflammation effects of dietary phytochemicals: The Nrf2/NF-κB signalling pathway and upstream factors of Nrf2. *Phytochemistry*. 2022 Dec 1;204:113429.

Xin-Yue HO, Jian-Mei WU, Xiao-Qiao LU, Kai LI, Wei-Jie ZH, Guo-Wen ZH. Research progress of inhibitory effects of polyphenolic compounds on xanthine oxidase. *Food and Machinery*. 2021;37(2):1–8.

Yadav V, Wang Z, Wei C, Amo A, Ahmed B, Yang X, Zhang X. Phenylpropanoid pathway engineering: An emerging approach towards plant defense. *Pathogens*. 2020 Apr 23;9(4):312.

Yahfoufi N, Alsadi N, Jambi M, Matar C. The immunomodulatory and anti-inflammatory role of polyphenols. *Nutrients*. 2018 Nov 2;10(11):1618.

Yang D, Wang T, Long M, Li P. Quercetin: Its main pharmacological activity and potential application in clinical medicine. *Oxidative Medicine and Cellular Longevity*. 2020;2020(1):8825387.

Yang R, Tian J, Liu Y, Zhu L, Sun J, Meng D, Wang Z, Wang C, Zhou Z, Chen L. Interaction mechanism of ferritin protein with chlorogenic acid and iron ion: The structure, iron redox, and polymerization evaluation. *Food Chemistry*. 2021 Jul 1;349:129144.

Yen GC, Duh PD, Tsai HL. Antioxidant and pro-oxidant properties of ascorbic acid and gallic acid. *Food Chemistry*. 2002 Nov 1;79(3):307–13.

Yin YC, Zhang XD, Gao ZQ, Hu T, Liu Y. The research progress of chalcone isomerase (CHI) in plants. *Molecular Biotechnology*. 2019 Jan 15;61:32–52.

You L, Cho JY. The regulatory role of Korean ginseng in skin cells. *Journal of Ginseng Research*. 2021 May 1;45(3):363–70.

Zhang H, Tsao R. Dietary polyphenols, oxidative stress and antioxidant and anti-inflammatory effects. *Current Opinion in Food Science*. 2016 Apr 1;8:33–42.

Zhang J, Chen J, Yang J, Xu CW, Pu P, Ding JW, Jiang H. Resveratrol attenuates oxidative stress induced by balloon injury in the rat carotid artery through actions on the ERK1/2 and NF-kappa B pathway. *Cellular Physiology and Biochemistry*. 2013 Mar 1;31(2–3):230–41.

Zhang M, An C, Gao Y, Leak RK, Chen J, Zhang F. Emerging roles of Nrf2 and phase II antioxidant enzymes in neuroprotection. *Progress in Neurobiology*. 2013 Jan 1;100:30–47.

Zhang S, Yang J, Li H, Chiang VL, Fu Y. Cooperative regulation of flavonoid and lignin biosynthesis in plants. *Critical Reviews in Plant Sciences*. 2021 Mar 4;40(2):109–26.

Zinatizadeh MR, Schock B, Chalbatani GM, Zarandi PK, Jalali SA, Miri SR. The nuclear factor Kappa B (NF-kB) signaling in cancer development and immune diseases. *Genes & Diseases*. 2021 May 1;8(3):287–97.

Zuo J, Zhang Z, Luo M, Zhou L, Nice EC, Zhang W, Wang C, Huang C. Redox signaling at the crossroads of human health and disease. *MedComm*. 2022 Jun;3(2):e127.

3 Pharmaceutical, Biological, and Medical Benefits of Polyphenol-Based Nutraceuticals Derived from Foods, Fruits, Herbs, and Medicinal Plants

Vikas and Vishal Sharma

3.1 INTRODUCTION

Medicinal plants have been used as traditional remedies globally and possess a long history of efficacy in addressing various ailments (Petrovska, 2012). Medicinal plants possess a diverse array of bioactive compounds that offer nutraceutical potential and health-related benefits beyond an ordinary diet (Sorrenti et al., 2023). Nutraceuticals derived from plants can alleviate ailments and enhance overall well-being. Natural sources of medicine are appealing substitutes for pharmaceuticals because of their availability, cost-effectiveness, safety, and potential for synergistic impacts (Nasri et al., 2014). Nutraceuticals serve dual purposes as dietary supplements and medicinal agents that protect against chronic disease and provide physiological benefits. Nutraceuticals can improve body structure and function, slow ageing, prevent chronic diseases, and increase life expectancy (Villeponteau et al., 2000). Various chemical compounds, such as essential oils, terpenoids, alkaloids, flavonoids, and polyphenols, are distributed in plants and have different therapeutic activities such as anticancer, antibacterial, anti-inflammatory and antioxidant effects (Karasawa & Mohan, 2018). Medicinal plants, natural food sources, and fruits are blessed with a wide variety of phenolic compounds. Polyphenols such as phenylpropanoids, flavonoids, and phenolic chemicals are secondary metabolites typically biosynthesised by plant sources using the acetate-malonate and shikimate pathways (Zagoskina et al., 2023). Polyphenols are recognised as basic dietary components of the human diet since they are found in a wide variety of dietary products and drinks, including fruits, vegetables, nuts, seeds, and flowers (Bertelli et al., 2021). The amount of polyphenolic chemicals

DOI: 10.1201/9781003588504-3

in plant foods affects their nutritional value and sensory qualities like colour, smell, and astringency (Osakabe et al., 2024). The dietary intake of polyphenols is recommended to be 1 gram per day, higher than other nutritional antioxidants (Scalbert & Williamson, 2000). The potential involvement of dietary polyphenols as functional food elements in the production of nutraceuticals has recently garnered considerable attention. Polyphenols are categorised according to the number of phenolic rings they contain, along with the structural elements (Pandey & Rizvi, 2009). Most of the polyphenols are recognised to have therapeutic potential, including antioxidant, anticancer, and antibacterial agents. As a result, they are becoming a good alternative for management of chronic disorders (Tomar et al., 2024). These substances are broadly categorised into four major areas: phenolic acids, stilbenes, flavonoids, and lignans (Table 3.1). Polyphenols, particularly flavonoids, improve learning and memory processes and are being examined for their effect to avoid age-related cognitive decline in animals and humans (de Vries et al., 2021). Fruit peel is a highly abundant source of flavonoids, with the quantity varying depending on the species and duration of light exposure (Panche et al., 2016). More than 8,000 distinct polyphenol types have been discovered so far (Cory et al., 2018). Fruits like berries, grapes, pears, apples, and cherries typically contain about 200–300 mg of polyphenols per 100 g of fresh weight (Rosales & Fabi, 2023; Kim et al., 2016). Typically, each serving of red wine, tea, or coffee has approximately 100 mg of polyphenols (Fukushima et al., 2014). Phenolic acids are abundant in fruits and vegetables, including broccoli, kale, and onions. Consuming juice and extracts from grapes, tea, and pomegranate can assist in the reduction of atherosclerosis plaque (Habauzit & Morand, 2012). The intake of a polyphenol-based diet has been found to lower the risk of age-related disorders and chronic degenerative diseases (Grabska-Kobyłecka et al., 2023). This chapter presents a comprehensive overview of polyphenol-based nutraceuticals, including their biological sources, geographical origins, chemical constituents, and therapeutic and nutraceutical applications.

3.2 POLYPHENOL-RICH PLANTS AND THEIR NUTRACEUTICAL APPLICATIONS

The utilisation of edible and medicinal plants, including fruits and vegetables, to provide nourishment and natural remedies has been linked to improved health, decreased reliance on synthetic antibiotics, and prolonged lifespan (Rana et al., 2022). Plant antioxidants such as polyphenols are frequently employed in traditional medicine as an efficient way of providing defence against disease caused by free radicals. Phenolics in edible and medicinal plants boost antioxidant activity, making them more effective in fighting degenerative illnesses, diabetes, cancer, ageing, and cardiovascular disorders (Rudrapal et al., 2022). A vast array of medicinal plants with nutraceutical applications have been utilised in food and cooking. In this chapter, we explore the role of polyphenol-rich medicinal plants in our regular dietary intake. Table 3.2 provides a comprehensive list of frequently used medicinal plants along with their pharmacognostic, therapeutic, and nutraceutical uses.

TABLE 3.1
Polyphenol-Rich Herbs With Their Chemical Structures

Phenolic Acids	Stilbenes	Lignans	Flavonoids and Polyphenols
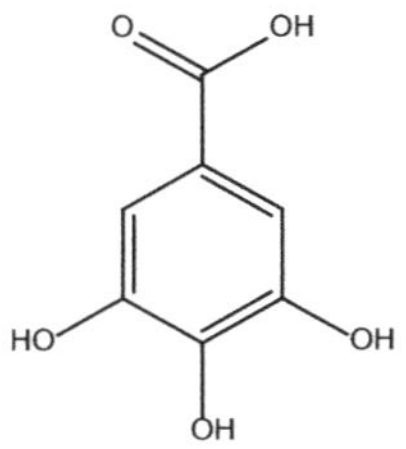 Gallic acid (tea, grapes and lemons)	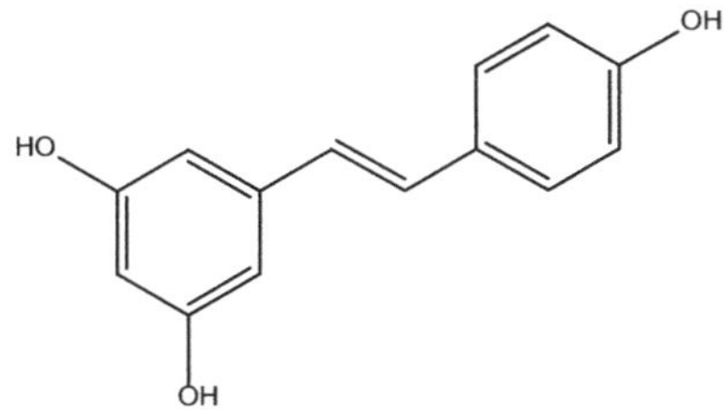 Resveratrol (blueberries and peanuts)	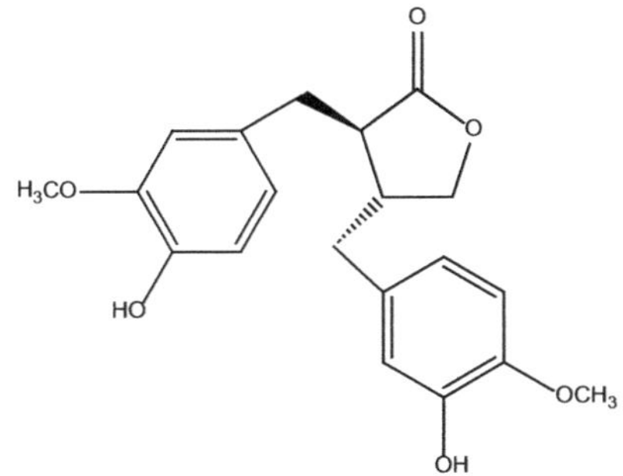 Matairesinol (linseed)	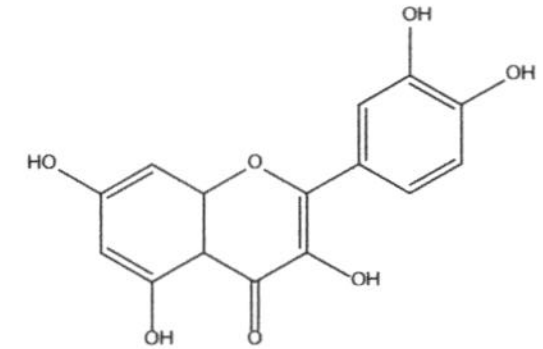 I. Flavonols Quercetin (tea, grapes, and apples)
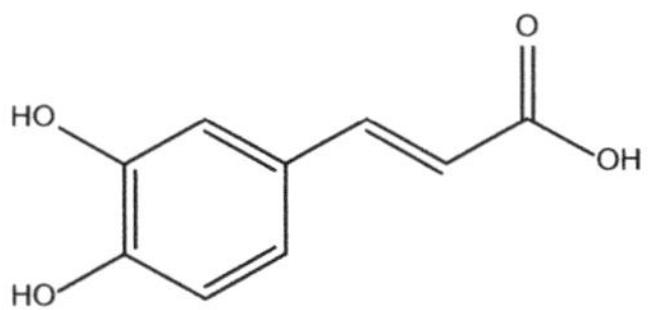 Caffeic acid (coffee and tea)	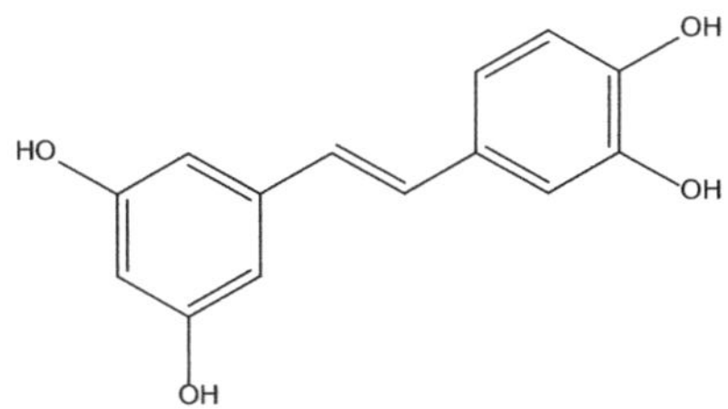 Piceatannol (Japanese knotweed, grapes, and white tea)	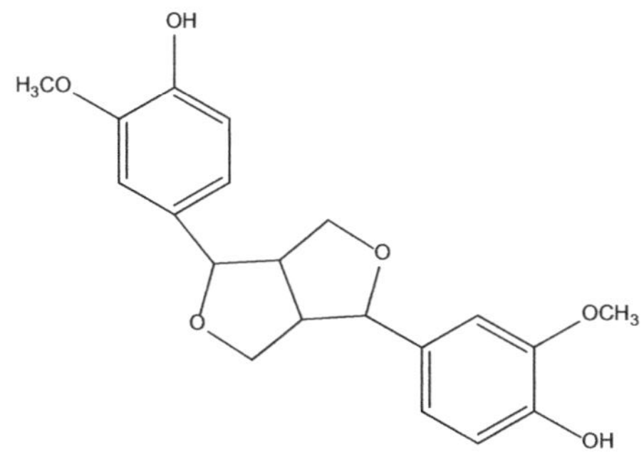 Pinoresinol (Olive oil)	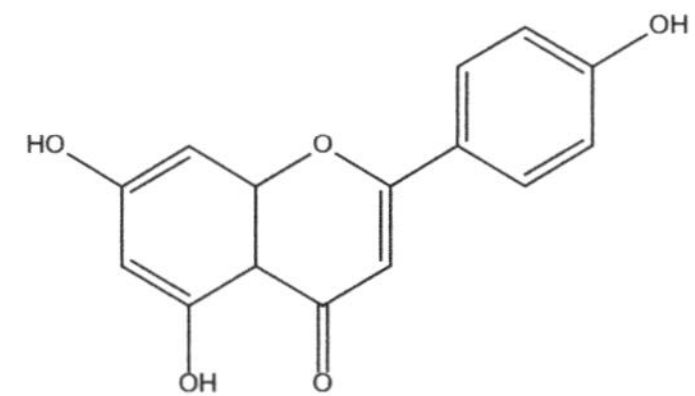 II. Flavones Apigenin (celery, cilantro, and basil)

(Continued)

TABLE 3.1 (*Continued*)
Polyphenol-Rich Herbs With Their Chemical Structures

Phenolic Acids	Stilbenes	Lignans	Flavonoids and Polyphenols
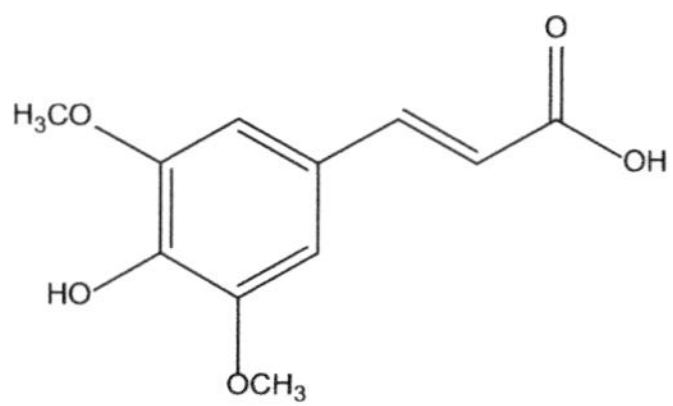 Sinapic or sinapinic acid (grapes, oranges, and cereals)	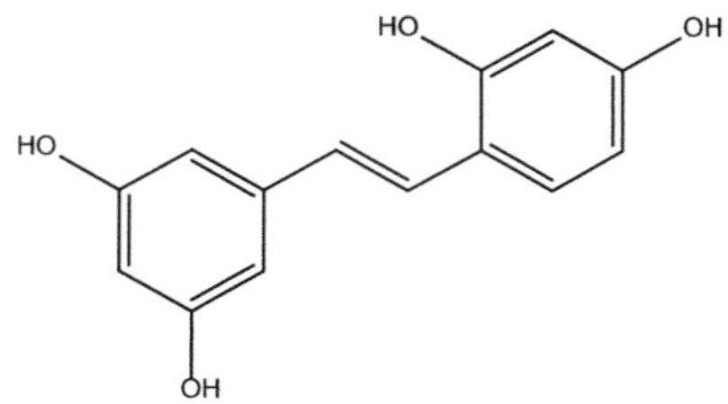 Oxyresveratrol (peanuts, grapes, and mulberries)	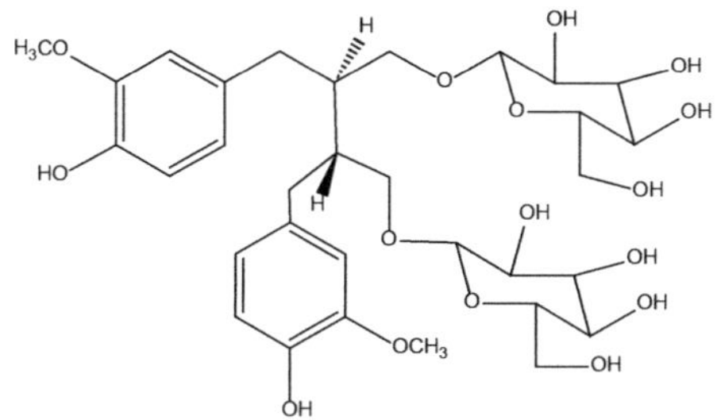 Secoisolariciresinol diglucoside (fennel and flaxseed seeds)	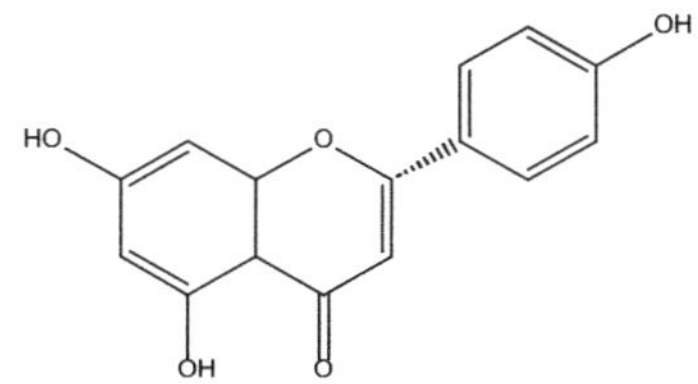 III. Flavanone Naringenin (grapefruit)
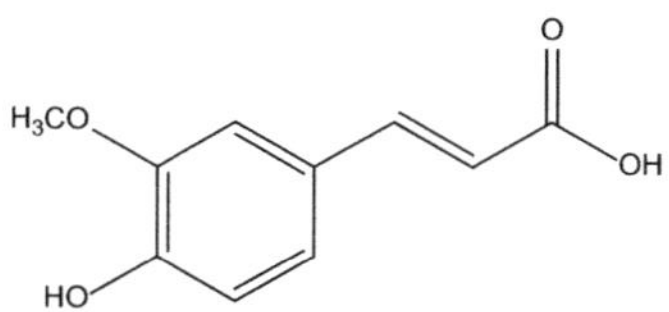 Ferulic acid (pineapple, rice, wheat, and oats)	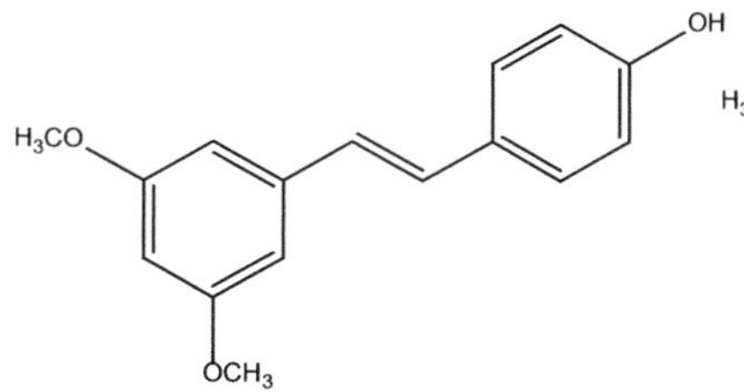 Pterostilbene (blueberries, peanuts, and grapes)	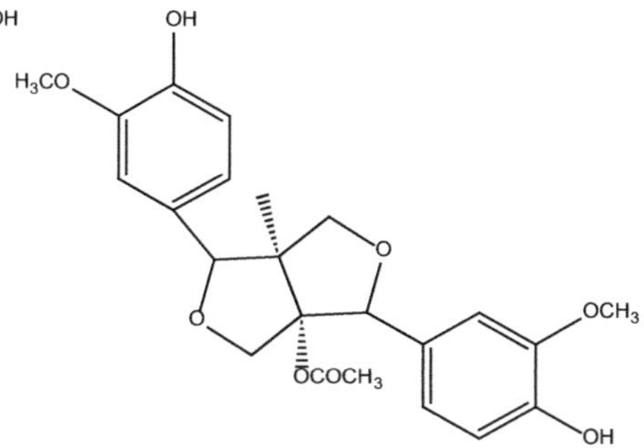 1-acetoxypinoresinol (olive oil)	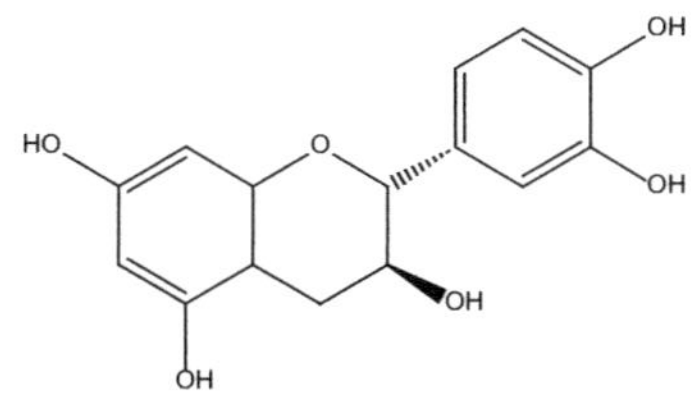 IV. Flavanols Catechin (cocoa-based products, green tea, and wine)

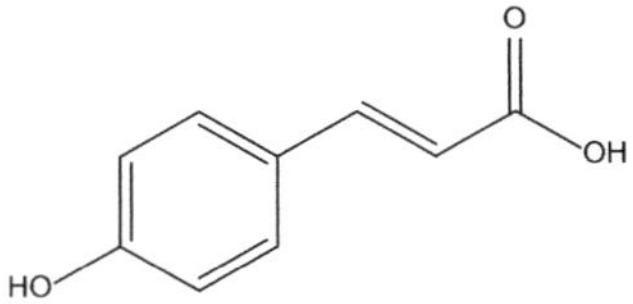

p-Coumaric acid (carrots, garlic, basil, and tomatoes)

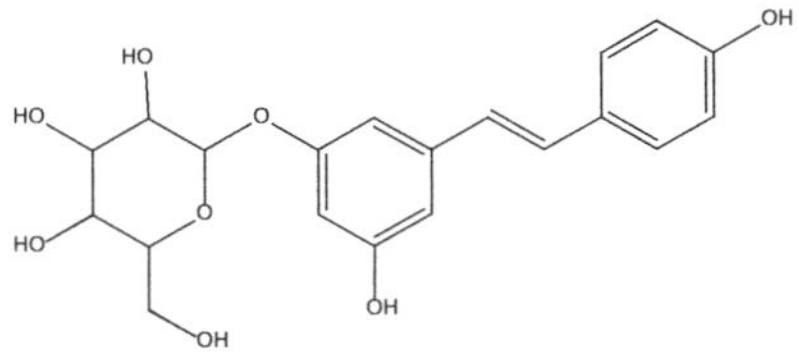

Polydatin (peanuts and grapes)

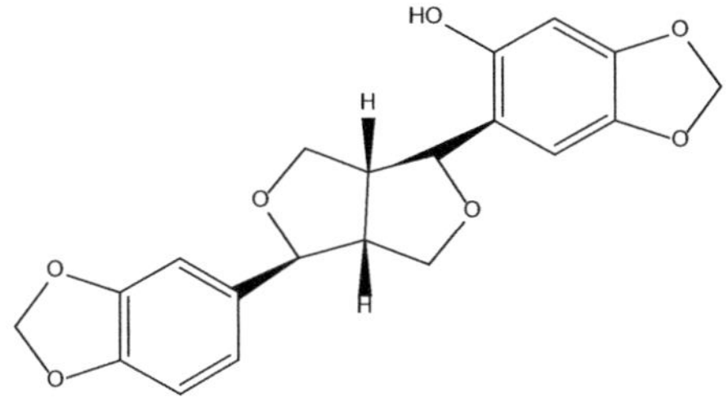

Sesaminol (sesame oil)

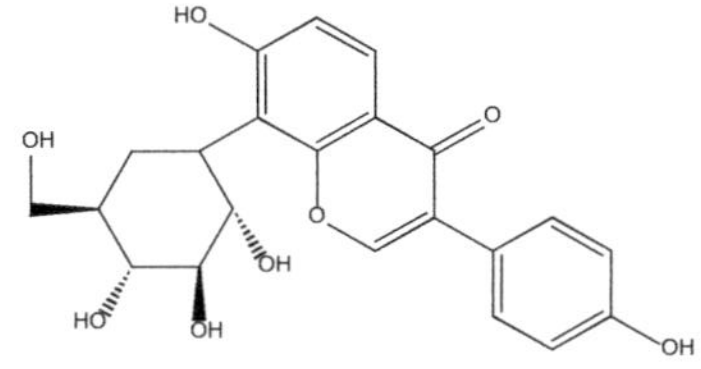

V. Isoflavones

Puerarin (*Pueraria*)

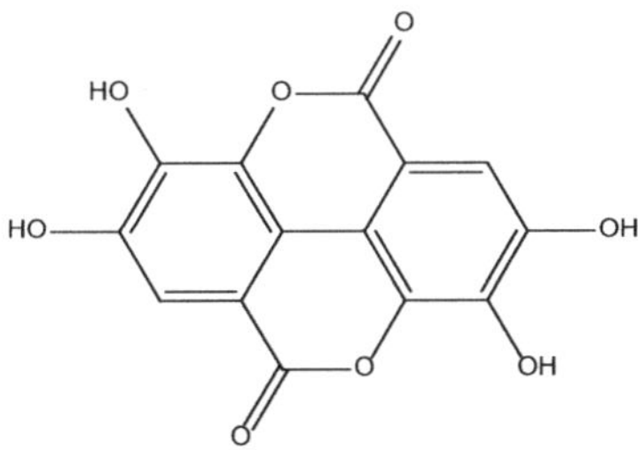

Ellagic acid (amla)

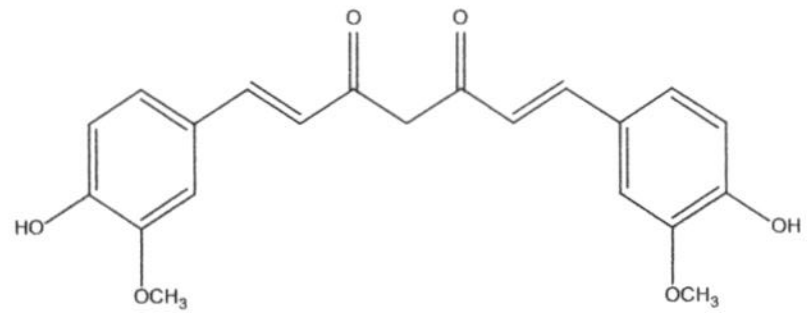

Curcumin (turmeric)

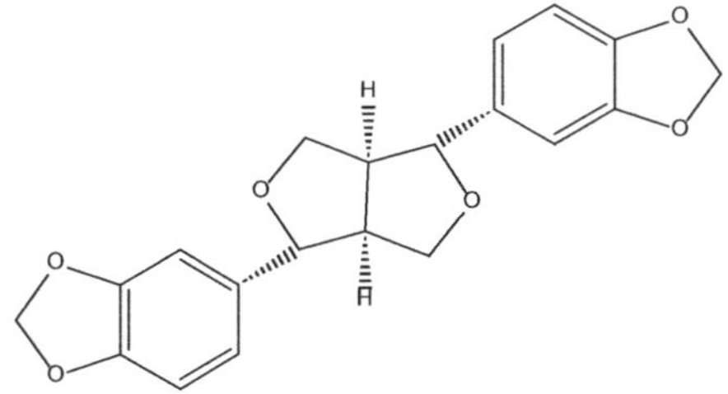

Sesamin (sesame oil)

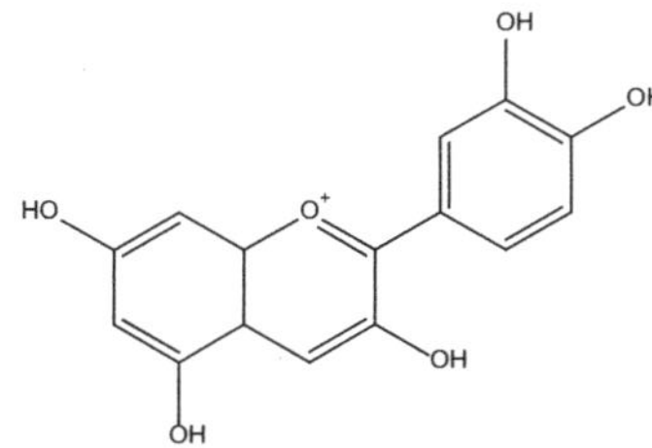

VI. Anthocyanin

Cyanidin (grapes and blackberries)

TABLE 3.2
Polyphenol-Rich Plants With Their Pharmacognostical and Nutraceutical Characteristics

Medicinal Plant	Biological Source	Geographical Source	Chemical Constituents	Nutraceutical Applications	Therapeutic Applications	References
Black chokeberry (Aronia)	Fruits obtained from the shrubs *Aronia melanocarpa*, belongs to family Rosaceae	It originated in eastern North America and has been widely planted in Europe and Asia	Chlorogenic acid, neochlorogenic acid, cyanidin 3-arabinoside, and chrysanthemin	It is used for the large-scale manufacturing of jams, juices, and wines, together with an abundant resource of natural food colours	It is used to treat inflammation, obesity, hypertension, cardiovascular, cerebrovascular, and metabolic disorders, such as hyperlipidaemia and hyperglycaemia	(Sidor et al., 2019; Sidor & Gramza-Michałowska, 2019)
Grapes: Colours vary with variety	Fruit derived from the plant *Vitis vinifera,* belongs to the family Vitaceae	Grapes are primarily cultivated in Asia, the USA, Africa, and the Middle East	Stilbenes, proanthocyanadines, anthocyanins, flavonoids, and phenolic acids	It is extensively consumed fresh as fruits, dehydrated as raisins, and utilised in the commercial production of vines, jams, and beverages	It is used to prevent and manage diabetes, cancer, and heart disease and possesses anti-ageing characteristics	(Georgiev et al., 2014; Leal et al., 2020)
Aloe vera (Ghritkumari)	Leaves derived from a perennial succulent plant *Aloe barbadensis* belonging to the Liliaceae family	It is primarily found in desert areas across Africa, Asia, Europe, and America	Tannins, phlorotannins, saponin, flavonoids, steroids, terpenoids, and cardiac glycosides	It is utilised in the production of numerous food items like milk, ice cream, and confectionaries	Its anti-inflammatory, antibacterial, wound healing, and antioxidant characteristics make it a versatile medicine	(Gao et al., 2019; Maan et al., 2018)
Coffee (Arabica)	The dried ripe seed derived from *Coffea arabica* belongs to the Rubiaceae family	Indigenous to tropical Asia and southern and tropical Africa	Caffeine, chlorogenic acid, trigonelline, theophylline, and theobromine	Coffee intake is utilised for socialising, improving work performance, and overall well-being	It provides respiratory stimulation, antimicrobial action, gastrointestinal effects, antioxidant potential, and anticancer effects	(Chen, 2019; Bondam et al., 2022)

Tea (chai)	Fresh leaves of an evergreen shrub, *Camellia sinensis*, belongs to the Theaceae family	Primarily grown in Asia and Africa but also extensively grown in the southern American and Caspian Sea regions	Catechins that include gallocatechin gallate, epicatechin, epicatechin-3-gallate, and epicatechin	Tea is an extremely adaptable beverage that works accompanying a wide range of foods	It provides various beneficial effects, including antioxidant, anti-ageing, antidiabetic, and cardioprotective properties	(Saeed et al., 2017; Zaveri, 2006)
Ginger (adrak)	It consists of the rhizome of the herbaceous *Zingiber officinale*, which belongs to the Zingiberaceae family	Cultivated in Africa, Asia, Central America, and South America	Farnesene, gingerol, gingerol, phellandrene, camphene, gingerol, gingerol, and trans-6-shogaol	It adds a spicy flavour to a range of meals, including fruit salads, beverages, curries, preserves, and baked products	It is utilised to address a diverse array of disorders, including hypertension, colds, vertigo, arthritis, and migraines	(Beristain-Bauza et al., 2019; Mahomoodally et al., 2021)
Turmeric (Indian saffron, Haldi)	It is the rhizome of a perennial plant *Curcuma longa,* belonging the family Zingiberaceae	It is native to the tropical regions of South Asia	Curcumin, curcuminoid bisdemethoxycurcumin, zingiberene, piperine, and curcumol are its primary constituents	Turmeric is a popular natural colourant and preservative used in various cuisines and commercial foods	It has been found to have positive effects on liver diseases, chronic anterior uveitis, conjunctivitis, smallpox, varicella, wound healing, urinary tract infections, and rheumatoid arthritis	(Meng et al., 2018; Ahmad et al., 2020)
Fennel (Saunf)	Fennel is dried ripe fruits of *Foeniculum vulgare*, belongs to the family Umbelliferae	It is primarily grown in Japan, India, Germany, France, Russia, and Romania	It includes α-pinene, fenchol, and anethole, while chlorogenic acid, ferulic acid, cinnamic acid, rosmarinic acid, caffeine and gallic acid are polyphenolic components	Fennel is utilised as a condiment in foods that have been stewed, cooked, grilled, or baked; in herbal teas and beverages; and is consumed fresh in salads and snacks	It exhibits carminative, antibacterial, anti-inflammatory, analgesic, diuretic, and anticancer properties	(Jadid et al., 2023; Noreen et al., 2023)

(Continued)

TABLE 3.2 (*Continued*)
Polyphenol-Rich Plants With Their Pharmacognostical and Nutraceutical Characteristics

Medicinal Plant	Biological Source	Geographical Source	Chemical Constituents	Nutraceutical Applications	Therapeutic Applications	References
Cumin (Zeera)	Cumin is derived from the dried seeds of *Cuminum cyminum*, which is an annual dicotyledonous plant belonging to the Apiaceae family	It is native to south-west Asia and the eastern Mediterranean region, with India accounting for most of the global production	Phenolic acids, polyphenols, and flavonoids are all present in cumin seeds; p-coumaric acid, caffeine, and ferulic acid are its polyphenolic components	It is commonly utilised as a food additive, widely used spice, and flavouring agent in various food preparations	It has been beneficial in treating numerous diseases, such as cancer, metabolic disorders, cardiovascular disease, digestive problems, and neurological disorders	(Mnif & Aifa, 2015; Hassanien et al., 2015)
Liquorice (Mulethi)	It is comprised of dried roots and rhizomes from *Glycyrrhiza glabra*, belongs to the Leguminosae family	It is cultivated primarily in the littoral regions of the Mediterranean and Central Asia, but also includes the USA, China, Europe, India, Iraq, Japan, Kurdistan, and Spain	Glycyrrhizin, isoliquiritin, liquiritigenin, and liquiritin, together with the phenolic compounds chalcone isoliquiri-tigenin and isoflavonoid glabridin	It is commonly found in soft drinks, food products, snacks, and herbal medicines as an additive ingredient	It exhibits beneficial therapeutic properties, including antibacterial, antioxidant, anti-athero-sclerotic, and estrogenic actions	(Karkanis, Martins, Petropoulos, & Ferreira, 2018; Ding et al., 2022)

Amla (Indian gooseberry)	It consists of the fresh and dried fruits of *Emblica officinalis*, belonging to the Euphorbiaceae family	It is indigenous to India, China, Pakistan, Iran, and Southeast Asia	It contains ascorbic acid (vitamin C), ellagitannins, emblicanin A and B, and pedunculagin; furthermore, it contains phenolic acids such as vanillic acid, 4-hydroxybenzoic acid, gallic acid, protocatechuic acid, syringic acid, and coumaric acid	It has a wide range of applications, including being incorporated into sauces, candy, and dried chips; in South Asia, the fruit is often pickled using a blend of salt, oil, and spices, which enhance its taste	It is extensively used because of its immuno-modulatory, adaptogenic, antidiabetic, anticancer, nootropic, and antibacterial properties	(Variya et al., 2016; Baliga & Dsouza, 2011)
Cloves (Laung)	It is composed of the floral buds of the evergreen plant *Eugenia caryophyllus*, which is a member of the Myrtaceae family	It is indigenous to India and the North Moluccas Islands in Indonesia	Flavonoids, hidroxibenzoic acids, hidroxicinamic eugenyl acetate, β-caryophyllene, and α-humulene acids contribute to its chemical composition; eugenol is the principal compound	It is utilised to impart flavour to rice dishes, stews, and sauces, particularly in a variety of traditional recipes; infuses foods with a peppery and warming essence	It is utilised for its therapeutic properties, which include antibacterial capabilities, cardiovascular protection, anti-inflammatory effects, and antiviral activity	(Idowu et al., 2021; Cortés-Rojas et al., 2014)

3.2.1 Black Chokeberry

The black chokeberry, also known as *Aronia melanocarpa*, is a shrub from the Rosaceae family indigenous to North America. It naturally occurs in the Appalachian Mountains and extends to the Great Lakes in North America. It contains a range of flavonoids, including anthocyanins, flavanols, and flavanol and phenolic acid compounds, including caffeic acid, chlorogenic acid, and neochlorogenic acid. In ancient North American medicine, it is widely used to treat colds (Gao et al., 2024). Recent studies have found that the functional components are associated with various therapeutic outcomes, including cardiovascular and cerebrovascular diseases, obesity, hypertension, inflammation, and metabolic disorders such as hyperlipidaemia and hyperglycaemia (Kasprzak-Drozd et al., 2021; Kim et al., 2018).

3.2.2 Grapes

Grapes are an excellent source of bioactive compounds that have an advantageous effect. Given their indispensable function in the production of wine, the ancient Romans and Greeks highly esteemed grapes (Sabra et al., 2021). A wide array of grape varieties exist, encompassing those that are employed for culinary purposes and wine production. For instance, people worldwide consume grapes in the form of raisins, grape juice, and wine. Grapes are a great source of numerous nutrients, such as minerals, vitamins, phytochemicals (including polyphenols), carbohydrates, and edible fibres. Polyphenols, the primary phytochemicals in grapes, have numerous advantageous biological effects and health benefits. Stilbenes, flavanols, flavonols, phenolic acids, and anthocyanins are among the polyphenols found in grapes. Most anthocyanins found in grape peels are (+)-catechins, (-)-epicatechins, and procyanidins (Buljeta et al., 2023). Red grapes contain a high concentration of anthocyanins, whereas white grapes contain a significant amount of flavan-3-ols (Zhou et al., 2022).

3.2.3 Aloe Vera

The aloe vera leaf contains a range of beneficial components, including tannin, phlorotannins, saponin, flavonoids, steroids, terpenoids, and cardiac glycosides (Sánchez et al., 2020). The primary active compound found in aloe vera juice is aloin, which is a glycoside of anthraquinone. The leaves and flowers of aloe vera contain a variety of polyphenols, such as gallic acid, protocatechuic acid, catechin, vanillic acid, epicatechin, syringic acid, chlorogenic acid, gentisic acid, caffeic acid, coumaric acid, ferulic acid, rutin, miricetin, quercetin, kaempferol, apigenin, kaempferol, and sinapic acid (López et al., 2013). The U.S. food and drug administration (FDA) has officially approved the use of aloe juice as a "dietary supplement." Aloe vera has become increasingly popular as a nutraceutical for cosmetics and other health-related products (Guo & Mei, 2016).

3.2.4 Coffee and Tea

Coffee and tea are the second most popular beverages in the world after water and are a rich source of polyphenols. Their appropriate and controlled consumption is potentially helpful in the treatment of diseases associated with an unhealthy lifestyle due to their antioxidant properties (Liczbiński & Bukowska, 2022). Tea is derived from the fresh leaves of an evergreen shrub identified as *Camellia sinensis*, which belongs to the Theaceae family. Tea contains bioactive compounds like polyphenols, polysaccharides, vitamins, and amino acids. The primary polyphenolic compounds found in green tea are catechins, which include gallocatechin gallate, epicatechin, epicatechin-3-gallate, epicatechin, and epigallocatechin (Aboulwafa et al., 2019; Musial et al., 2020). Such compounds have numerous therapeutic effects and can be utilized as food additives as nutraceuticals. Like tea, coffee is also a commonly used beverage in the world and is linked to a multitude of health benefits when consumed in an appropriate and controlled manner. Coffee is derived from the dried ripe seed of *Coffea arabica*, which belongs to the family Rubiaceae and comprises a total of 103 species that are distributed throughout Asia, Africa, and Latin America (Davis et al., 2006). Coffee contains various active constituents such as caffeine, chlorogenic acid, diterpenes, and trigonelline. Several studies have demonstrated that coffee and green tea augment human health. These include the ability to treat cancer, obesity, diabetes, high blood pressure, and liver damage (Bułdak et al., 2018). Therapeutic effects linked with the consumption of coffee and tea have been attributed to the chlorogenic acid in coffee and epigallocatechin gallate in green tea. It is postulated that these compounds function via a multitude of mechanisms, wherein the involvement of reactive oxygen species holds particular importance. The majority of individuals can benefit from consuming up to three to four cups of coffee or tea per day (Khan & Mukhtar, 2013; Poole et al., 2017).

3.2.5 Ginger

Ginger is the dried rhizome of the plant *Zingiber officinale*, which belongs to the Zingiberaceae family. Ginger is commonly utilised as a condiment in cuisines around the world. The chemical composition of ginger consists of raw fibres, organic acids, phenolic compounds, terpenes, polysaccharides, and lipids (Mao et al., 2019). Ginger contains phenolic chemicals like gingerols and shogaols, vital in promoting health benefits. Moreover, it also contains several important polyphenols, specifically 6-gingerol, 8-gingerol, and 10-gingerol (Yücel et al., 2022). Upon exposure to heat or prolonged duration, gingerols undergo a conversion process, forming shogaols. Ginger is a vital component of the diet that provides numerous health benefits. It functions as a carminative, thereby providing relief from flatulence and bloating (Nikkhah Bodagh et al., 2019). Additionally, ginger alleviates intestinal distress, prevents dyspepsia, and reduces pressure on the lower oesophageal sphincter. Ginger tea, a beverage prepared through the extraction of ginger, is an exceptionally effective remedy for cough and aching throats. In addition, ginger has been shown to be advantageous for sexual issues. According to some studies, ginger may improve sexual function by enhancing sexual excitement (Wen et al., 2024). Also, using ginger powder in the diet is also helpful in relieving menstruation cramps (Negi et al., 2021). Moreover, ginger also been associated with other therapeutic effects, such as antibacterial, anti-inflammatory, neuroprotective, and cardioprotective. Additionally, it aids in managing weight, regulating blood sugar, relieving nausea, and preventing vomiting (Ballester et al., 2022).

3.2.6 Turmeric

Turmeric is derived from the rhizomes of a perennial plant called *Curcuma longa*, which belongs to the Zingiberaceae family (Qadir et al., 2016). It is a frequently utilised spice in curry dishes and has its origins in India. It gained considerable attention in the last decade due to its bioactive curcuminoid components, namely curcumin, bisdemethoxycurcumin, and demethoxycurcumin (Jayaprakasha et al., 2006). Turmeric is recognised for its therapeutic attributes owing to the presence of curcumin that demonstrated great potential for bacterial infections, inflammation, cancer, and ageing (Kunnumakkara et al., 2023).

In India, the consumption of turmeric powder with milk has been found to have favourable effects on the healing process and could potentially accelerate recovery from injuries (Fuloria et al., 2022). Applying turmeric powder paste to wounds was believed to aid in healing by reducing inflammation and fighting bacteria (Gupta & Balasubrahmanyam, 1998). The therapeutic effectiveness of curcumin as an adjunct medication in minimising brain injury and improving patient outcomes has also been demonstrated (Gupta et al., 2013). Studies have indicated that the use of turmeric extract may have the potential to regulate blood sugar levels and enhance the management of diabetes. Furthermore, various studies demonstrate that curcumin can improve both cellular immunity and humoral immunity. Traditionally, turmeric was employed as an herbal remedy to address a range of health issues, encompassing skin discomfort, respiratory and liver conditions, and gastrointestinal disorders such as abdominal pain, diarrhoea, constipation, coughing, and dyspepsia (Nikkhah Bodagh et al., 2019). Besides the medicinal properties of turmeric, it is commonly utilised as a spice for both food preservation and colouring agents (Sharifi-Rad et al., 2020).

3.2.7 Fennel

Fennel is a Mediterranean herb known as *Foeniculum vulgare,* which belongs to the Apiaceae family. It is found in temperate and tropical climates. The seeds are used in the culinary and medicinal industries for their curative and flavourful effects due to the presence of phytochemicals (Badgujar et al., 2014). Fennel seeds contain a high concentration of phenolic compounds and medicinal components, including vitamins E and C. Furthermore, its seeds also contain several chemicals, including chlorogenic acid, ferulic acid, cinnamic acid, rosmarinic acid, caffeine, gallic acid, p-coumaric acid, apigenin, hesperidin, and quercimeritrin (Barakat et al., 2022). Fennel has several therapeutic characteristics that can help with digestive difficulties, including carminative, antibacterial, anti-inflammatory, analgesic, diuretic, and anticancer actions (Noreen et al., 2023).

It is believed that consuming a preparation made from dried fennel seeds can aid in the digestion of fatty foods. Fennel seeds aid in reducing weight by raising metabolism, relieving irritated throats, and promoting the removal of phlegm from the respiratory system. Fennel seeds are often regarded as effective home treatments for a variety of common diseases, notably digestive issues. To relieve discomfort and eye strain, a decoction of finely ground seed powder is used to cleanse the eyes (DeFilipps & Krupnick, 2018). The diuretic properties of this plant render it beneficial for addressing kidney and bladder ailments. It can also be used to prevent vomiting and treat nausea. In addition, it has been demonstrated to be advantageous in the management of chronic fever and the removal of blockages in the liver, digestive, respiratory, and urinary systems (Noreen et al., 2023). Additionally, it is employed as a galactagogue for the purpose of treating breastfeeding mothers (Bazzano et al., 2016).

3.2.8 Cumin

Cumin is a spice frequently used in a variety of culinary applications, and it is employed in ayurvedic medication due to its stimulating, carminative, and astringent properties. The bitter cumin (*Cuminum nigrum*) has the highest amount of antioxidant property in comparison to other types of cumin, such as cumin and black cumin (Johri, 2011). Seeds are commonly used for a numerous therapeutic applications, including healing diarrhoea, lowering acidity, energising the body, and easing bloating and pain. In addition, it is used as a medical dressing to relieve the pain associated with helminthic infection. Several studies have demonstrated its health advantages, including its capacity to reduce inflammation, fight bacteria, and even prevent cancer (Mnif & Aifa, 2015). Cumin seeds contain variety of polyphenolic compounds, including tannins, phenolic acids, and flavonoids. The phenolic acids in cummin seed are caffeic acid, ferulic acid, and p-coumaric acid (El Tannir et al., 2024). Flavonoid components of cumin such as apigenin, luteolin, and quercetin are other chemical compounds that have anti-inflammatory and antioxidant characteristics (Al-Khayri et al., 2022).

3.2.9 Liquorice

The application of liquorice in ayurvedic and Chinese medication is widespread and well-established. It is known botanically as *Glycyrrhiza glabra* and is composed of desiccated roots and rhizomes belonging to the Leguminosae family. Because of its wide variety of secondary metabolites, liquorice is highly regarded in healthcare. In past decades, it has been used to improve wound healing. Liquorice root extracts have been shown to successfully reduce skin damage caused by free radicals. The primary bioactive components of liquorice are triterpenes, flavonoids, and polysaccharides (Wahab et al., 2021). Glycyrrhizin, the saccharine element of liquorice root, is a triterpene saponin. It has anti-inflammatory, antiviral, anticancer, and antibacterial effects. The biological action of liquorice root is primarily attributed to glycyrrhizin, as well as the phenolic components and chalcone derivatives such as isoliquiritigenin and isoflavonoid derivatives including glabridin (Pastorino et al., 2018). In addition to glycyrrhizin, it also comprises salicylic acid, trans-ferulic acid, syringic acid, 2,5-dihydroxybenzoic acid, and abscisic acid (Leite et al., 2022). The FDA has approved foods containing liquorice extract or glycyrrhizin. The U.S. Flavour and Extract Manufacturers Association has recognised the general safety of liquorice (Pastorino et al., 2018).

3.2.10 Amla

Amla (*Emblica officinalis*) is a medicinal plant that belongs to the Euphorbiaceae family, commonly called Indian gooseberry. Amla fruits are extensively consumed and primarily cultivated in India, Southeast Asia, China, Iran, and Pakistan. Amla is widely employed in traditional Indian medicine due to its vast health benefits (Gul et al., 2022). It is used to treat anaemia, improve male reproductive health and fertility, aid digestion, promote liver health, and benefit the cardiovascular system (Gul et al., 2022). It contains phytochemicals in several parts, such as fruits, leaves, and roots. The amla fruit contains a substantial quantity of phenolic compounds, which makes it a valuable source of natural antioxidants and medicinal components derived from plants. The phenolic acids in fresh and processed fruit are 4-hydroxybenzoic acid, coumaric acid, gallic acid, protocatechuic acid, syringic acid, and vanillic acid (Alam et al., 2022). Amla has demonstrated diverse therapeutic qualities in animal and human studies, including antihyperglycaemic, hypoglycaemic, anti-inflammatory, antihyperlipidaemic, and antioxidant effects (Kapoor et al., 2020).

3.2.11 Cloves

Cloves are the fragrant flower buds of the plant known as *Eugenia caryophyllus*, belonging to the Myrtaceae family. It has substantial medicinal and culinary characteristics, making it a popular spice and taste enhancer. Cloves contain bioactive phenolic chemicals that have been demonstrated to possess potent antioxidant, antiviral, antibacterial, and anticancer activities. Clove has grown in favour as a natural antioxidant in culinary products in recent decades due to its high bioactivity (Han & Parker, 2017). Clove buds, oil, and oleoresins are considered generally identified as safe food additives by the FDA. The essential oil predominantly contributes to pleasant flavours, but the non-volatile polyphenols are accountable for the bitter and astringent properties (Kumar Pandey et al., 2022). Cloves have high levels of phenolic chemicals, such as hydroxybenzoic acids, flavonoids, hydroxyphenyl propene, and hydroxycinnamic acids. Eugenol is an essential bioactive chemical found in cloves, whereas gallic acid is the most abundant phenolic acid component (Gengatharan & Rahim, 2023). Cloves also contain caffeine, ferulic acid, ellagic acid, and salicylic acid. Relatively small levels of flavonoids such as kaempferol, quercetin, and glycosylated derivatives are present. Clove oil is often used as a home cure for toothaches because it contains eugenol, which has potent analgesic properties (Sharma et al., 2022).

3.3 APPLICATION OF POLYPHENOL-RICH NUTRACEUTICALS

3.3.1 Gastrointestinal

Inflammatory bowel diseases are chronic health conditions, and their exact etiology remains unknown. However, the disorders are marked by an inflammatory response, gastrointestinal dysbiosis, and oxidative stress. Additionally, disrupting the bowel epithelium is a significant aspect of these disorders. Dietary polyphenols could have beneficial effects in the prevention and cure of inflammatory bowel disease (Zhong et al., 2023). Polyphenols are believed to play a role in regulating cellular

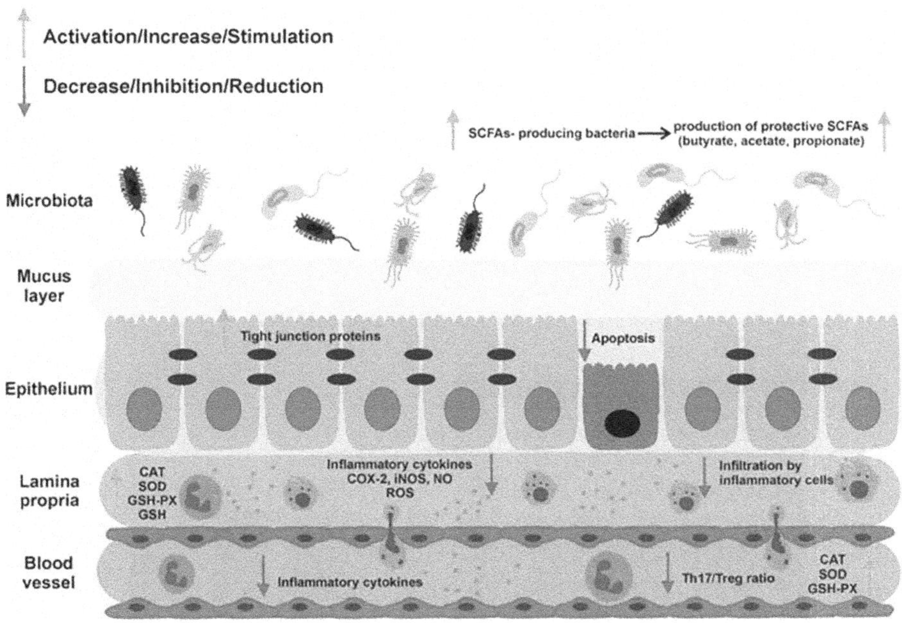

FIGURE 3.1 The fundamental mechanisms through which polyphenols contribute to alleviating symptoms associated with inflammatory bowel disease. (Reproduced from Caban & Lewandowska, 2022; under CC BY 4.0 [https://creativecommons.org/licenses/by/4.0/]; Figure 2; *Journal of Functional Foods*])

signalling pathways, enhancing the integrity of the epithelial barrier, reducing reactive oxygen species (ROS), and decreasing the activity of inflammatory enzymes such as inducible nitric oxide synthase (iNOS) and cyclooxygenase 2 (COX-2). Additionally, they may help lower elevated levels of inflammatory cells, including T helper 17 cells (Th17) and regulatory T cells (Treg), along with their associated cytokines. Additionally, they can enhance antioxidant defence mechanisms by elevating glutathione (GSH) levels and promoting antioxidant enzymes such as glutathione peroxidase (GSH-PX), catalase (CAT), and superoxide dismutase (SOD), while also maintaining a healthy microbiome, particularly for bacteria that produce short chain fatty acids (SCFAs) (Figure 3.1) (Caban & Lewandowska, 2022).

3.3.2 Anticancer

Chemotherapy and radiation, which are well-known cancer treatments, often result in significant side effects. Therefore, it is crucial to come up with additional or different approaches. Polyphenol compounds are abundant in plants, vegetables, and food products and have a broad spectrum of medicinal properties, including anticarcinogenic effects (Niedzwiecki et al., 2016). Research has demonstrated that polyphenols exert a significant influence on inflammation, apoptosis, cancer cell proliferation, tumour growth, angiogenesis, and metastasis. Furthermore, polyphenol can modulate

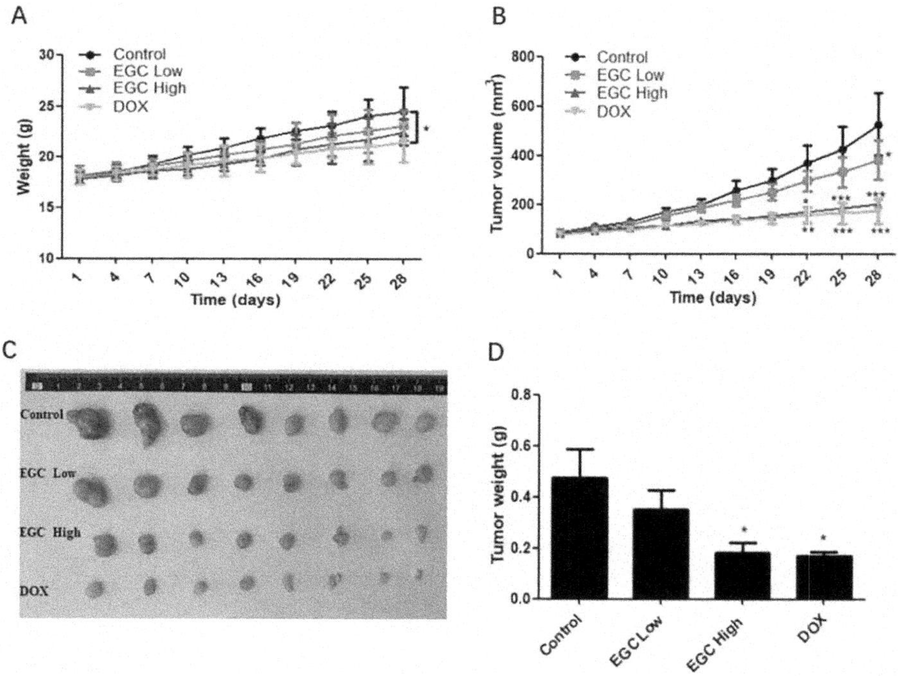

FIGURE 3.2 Antitumour effect of EGC in SW620 nude mice xenograft tumour model. (A) Body weight, (B) tumour volume, (C) representative tumour images, and (D) tumour weight across different treatment groups. (*Reprinted from* **Journal of Functional Foods**, Vol. 112, Ke-Wang Luo, Wei Ye, Ning Li, Bao-Hui Cheng, "*Tea polyphenol EGC suppresses colorectal cancer cell proliferation both in vitro and in vivo via downregulation of STAT3,*" Article 105977, January 2024, ©2024, with permission from Elsevier.)

immune system responses and protect healthy cells from free radical damage. The anticancer mechanisms of polyphenols have been extensively studied, focusing on specific compounds (Sufianova et al., 2022). Luo et al. 2024 demonstrated the anticancer properties of epigallocatechin (EGC), a primary bioactive polyphenol found in green tea, specifically for colorectal cancer (Figure 3.2). According to the findings, epigallocatechin substantially inhibited apoptosis and cell growth in three colorectal cancer cells. Western blot analysis verified the down-regulation of cell proliferation–associated signalling molecules, such as STAT3 phosphorylation and Bcl-2 expression. The study also identified the activation of apoptosis-related mediators, including caspase-3 and poly(ADP-ribose)polymerase 1. In mice with SW620 tumours, epigallocatechin (100 mg/kg) substantially reduced tumour weight and volume (63.7%). Furthermore, the examination of tumour organoids demonstrated that epigallocatechin inhibited the proliferation of colorectal organoids efficiently (Luo et al., 2024).

3.3.3 Cardioprotective

The most prevalent cause of mortality and disability worldwide is cardiovascular disorders (CVDs) (Roth et al., 2020). Polyphenols have been shown to improve endothelial function, prevent abnormal platelet aggregation, reduce inflammation, and optimize the plasma lipid profile, all of which serve to promote cardiovascular health (Iqbal et al., 2023). Food rich in polyphenols plays a vital role in preventing cardiovascular disease and promoting overall health. The main causes of heart failure are increased levels of nicotinamide adenine dinucleotide phosphate oxidases and heightened production of ROS in the mitochondria, leading to excessive oxidative stress in the heart (Moris et al., 2017). Research has shown that polyphenols can reduce levels of malondialdehyde and ROS in the heart. Furthermore, studies have shown that they increase the synthesis of GSH-PX, CAT, and SOD, among other detoxifying enzymes (Babich et al., 2011). Evidence also suggests that polyphenols may minimize the harmful effects of oxidised low-density lipoproteins on endothelial cells by inhibiting their oxidation (Rudrapal et al., 2022). Wan et al. 2023 demonstrated that epigallocatechin-3-gallate (EGCG), a major polyphenolic component in green tea, inhibited hyperglycaemia (HG)–induced cardiomyocyte hypertrophy, inflammation, oxidative stress response, and cellular senescence. To evaluate the impact of HG-induced damage on H9C2 cells in relation to green tea–derived EGCG, the study analysed the mRNA expression levels of cardiac hypertrophy–related genes, including *β-MHC*, *ANF*, and *BNP*. The analysis revealed an increase in the expression of these genes following HG treatment. However, their activation was inhibited by EGCG. As evidenced in the alterations in hypertrophic genes, EGCG significantly inhibited the increase in cardiomyocyte surface area caused by HG. The researchers also investigated changes in inflammatory markers, including IL-6, TNF-α, and IL-1β. Their findings revealed that EGCG administration reduced HG-induced inflammatory reactions in H9C2 cardiomyocytes. Moreover, they also analysed the oxidative stress–related indicators, such as malondialdehyde (MDA) and SOD levels in the culture media, and found that EGCG treatment dramatically reduced the oxidative stress response caused by HG. Dichloro-dihydro fluorescein diacetate (DCF-DA) staining experiments demonstrated that EGCG decreased the HG-induced ROS level in cardiomyocytes. All these results demonstrate that EGCG treatment greatly reduced inflammation, HG-induced cardiomyocyte hypertrophy, and oxidative stress (Wan et al., 2023).

3.3.4 Antioxidant

The progression of numerous chronic diseases, including cardiovascular disease, cancer, obesity, hypertension, and type 2 diabetes, are significantly influenced by oxidative stress (Masenga et al., 2023). An overabundance of ROS can potentially compromise the functionality of the antioxidant defence mechanism. Free radical production can cause cellular damage and cell death (Afzal et al., 2023). Fruits and vegetables are rich in antioxidants such as vitamin C, polyphenols, and β-carotene. Polyphenols represent a prevalent and extensively distributed category of secondary plant metabolites, with a multitude of compounds having been identified (Jideani et al., 2021). Incorporating foods that are high in antioxidants into the diet is the simplest and most effective way to prevent diseases that are linked to oxidative stress. Polyphenols serve as potent antioxidants, effectively neutralising free radicals by donating an electron or hydrogen atom. The primary function is to hinder the series of events that lead to lipid peroxidation by directly eliminating free radicals. The existence of phenolic hydroxyl groups in polyphenol molecules facilitates the direct scavenging of ROS. The amount and position of hydroxyl groups, as well as the structure of substituents, determine polyphenols' ability to neutralize ROS (Rudrapal et al., 2022). Furthermore, polyphenols have antioxidant properties by controlling the formation and activity of endogenous oxidase and antioxidant enzymes. Sodium oxide dismutase 1 (SOD1), located in the cytosol, and SOD2, located in the matrix of mitochondria, are two intracellular enzymes that efficiently neutralize oxidants by rapidly converting superoxide to hydrogen peroxide. Hydrogen peroxide is further broken down by GSH and CAT, which turn it into water and oxygen. Furthermore, polyphenols may increase cellular antioxidant activity by regulating the nuclear factor erythroid 2-related factor 2 (Nrf2)–mediated pathway. The transcription factor Nrf2 controls the synthesis of a variety of detoxification enzymes, including SOD, GSH, and NADP quinone oxidoreductase 1 (Tsao, 2010).

3.3.5 Anti-inflammatory

Inflammation is a typical response of the body to damage or infection. It occurs when the body produces chemicals that stimulate the immune system to combat infection or repair damaged tissue (Trowbridge, 1990). Inflammation is often required to preserve body homeostasis in response to pathogenic stimuli and situations such as tissue injury and infection. Acute and chronic inflammation are distinct; acute inflammation can last several days, whereas chronic or long-term inflammation can endure for days or even years. Acute inflammation, albeit transient, is usually helpful to the body. Diabetes, arthritis, pancreatitis, cardiovascular disease, neurological disorders, metabolic disorders, and various types of cancer have all been linked to chronic inflammation (Cain et al., 2009). Curcumin, a polyphenolic compound found in turmeric, imparts some of its numerous medicinal benefits, including anti-inflammatory effects, which extend beyond its dietary applications (Sharifi-Rad et al., 2020). In addition, a polyphenol-rich diet demonstrated potential to aid in the management of inflammation and muscle discomfort induced by physical activity, consequently enhancing recovery and performance in affected individuals. Polyphenols exert their mechanistic effects through various mechanisms, including radical scavenging, moderation of cellular activities in inflammatory cells, modulation of proinflammatory molecule production, and regulation of enzyme activities associated with arachidonic acid metabolism (phospholipase A2, COX) and arginine metabolism (nitric oxide synthases). Furthermore, they exhibit the ability to inhibit nuclear factor κB (NF-κB) and activator protein 1 (AP-1), induce phase II antioxidant detoxifying enzymes, and stimulate Nrf2-related factor, protein kinase-C, and mitogen activated protein kinase (Yahfoufi et al., 2018).

3.3.6 Antimicrobial

Polyphenols are frequently utilised as natural preservatives in the food industry because of their potent antioxidant and antibacterial characteristics (Manso et al., 2021). Polyphenols have been found to exhibit antibacterial properties against various microorganisms, including gram-positive and gram-negative bacteria and fungi (Manso et al., 2021). The effectiveness of polyphenols as an antimicrobial agent is influenced by various factors, including differences in bacterial structure, the concentration and type of polyphenol, the method of extraction, and the duration of bacterial exposure to the polyphenol. Flavone polyphenols are characterised by the presence of a solitary ketone substitution on an aromatic ring. Plants generate these chemicals as a defence mechanism against microbes. Therefore, it is anticipated that they exhibit antibacterial properties. The antibacterial activities of these compounds are most likely derived from their interactions with soluble and extracellular proteins, as well as bacterial cell walls (Kumar & Pandey, 2013).

3.4 CONCLUSION

Medicinal plants comprise an extensive variety of bioactive compounds that exhibit considerable potential as nutraceuticals, imparting health benefits that exceed those provided by a conventional diet. Polyphenols are indispensable components of the human dietary regimen due to their abundance in fruits, vegetables, nuts, seeds, flowers, and tree trunks, among other foods and beverages. A diet abundant in polyphenols may be advantageous in the prevention, management, and treatment of various disorders. These substances demonstrate advantageous therapeutic attributes, including anticancer, anti-inflammatory, antioxidant, and antibacterial properties. Moreover, a diet comprising medicinal plants and vegetables is considered safe. Thus, a diverse array of medicinal plants and food sources rich in polyphenols can be incorporated into a well-balanced diet to provide medicinal advantages. As a result, including a diet rich in polyphenols in adequate amounts is recommended for maintaining excellent health in most people and is useful for patients suffering from various ailments. Individuals with specific medical ailments are encouraged to take polyphenol-rich dietary components in appropriate levels, which have demonstrated benefits for managing those disorders, following the suggestions of dieticians and health specialists.

REFERENCES

Aboulwafa, M. M., Youssef, F. S., Gad, H. A., Altyar, A. E., Al-Azizi, M. M., & Ashour, M. L. (2019). A Comprehensive Insight on the Health Benefits and Phytoconstituents of Camellia Sinensis and Recent Approaches for Its Quality Control. *Antioxidants (Basel)*, *8*(10). https://doi.org/10.3390/antiox8100455

Afzal, S., Abdul Manap, A. S., Attiq, A., Albokhadaim, I., Kandeel, M., & Alhojaily, S. M. (2023). From Imbalance to Impairment: The Central Role of Reactive Oxygen Species in Oxidative Stress-Induced Disorders and Therapeutic Exploration. *Front Pharmacol*, *14*, 1269581. https://doi.org/10.3389/fphar.2023.1269581

Ahmad, R. S., Hussain, M. B., Sultan, M. T., Arshad, M. S., Waheed, M., Shariati, M. A., . . . Hashempur, M. H. (2020). Biochemistry, Safety, Pharmacological Activities, and Clinical Applications of Turmeric: A Mechanistic Review. *Evidence-Based Complementary and Alternative Medicine, 2020*, 7656919. https://doi.org/10.1155/2020/7656919

Alam, S., Dhar, A., Hasan, M., Richi, F. T., Emon, N. U., Aziz, M. A., . . . Rashid, M. A. (2022). Antidiabetic Potential of Commonly Available Fruit Plants in Bangladesh: Updates on Prospective Phytochemicals and Their Reported MoAs. *Molecules*, *27*(24). https://doi.org/10.3390/molecules27248709

Al-Khayri, J. M., Sahana, G. R., Nagella, P., Joseph, B. V., Alessa, F. M., & Al-Mssallem, M. Q. (2022). Flavonoids as Potential Anti-Inflammatory Molecules: A Review. *Molecules*, *27*(9). https://doi.org/10.3390/molecules27092901

Babich, H., Schuck, A. G., Weisburg, J. H., & Zuckerbraun, H. L. (2011). Research Strategies in the Study of the Pro-Oxidant Nature of Polyphenol Nutraceuticals. *J Toxicol, 2011*, 467305. https://doi.org/10.1155/2011/467305

Badgujar, S. B., Patel, V. V., & Bandivdekar, A. H. (2014). Foeniculum Vulgare Mill: A Review of Its Botany, Phytochemistry, Pharmacology, Contemporary Application, and Toxicology. *Biomed Res Int, 2014*, 842674. https://doi.org/10.1155/2014/842674

Baliga, M. S., & Dsouza, J. J. (2011). Amla (Emblica Officinalis Gaertn), a Wonder Berry in the Treatment and Prevention of Cancer. *European Journal of Cancer Prevention*, *20*(3). https://journals.lww.com/eurjcancerprev/fulltext/2011/05000/amla__emblica_officinalis_gaertn_,_a_wonder_berry.11.aspx

Ballester, P., Cerdá, B., Arcusa, R., Marhuenda, J., Yamedjeu, K., & Zafrilla, P. (2022). Effect of Ginger on Inflammatory Diseases. *Molecules*, *27*(21). https://doi.org/10.3390/molecules27217223

Barakat, H., Alkabeer, I. A., Aljutaily, T., Almujaydil, M. S., Algheshairy, R. M., Alhomaid, R. M., . . . Mohamed, A. (2022). Phenolics and Volatile Compounds of Fennel (Foeniculum vulgare) Seeds and Their Sprouts Prevent Oxidative DNA Damage and Ameliorates CCl(4)-Induced Hepatotoxicity and Oxidative Stress in Rats. *Antioxidants (Basel)*, *11*(12). https://doi.org/10.3390/antiox11122318

Bazzano, A. N., Hofer, R., Thibeau, S., Gillispie, V., Jacobs, M., & Theall, K. P. (2016). A Review of Herbal and Pharmaceutical Galactagogues for Breast-Feeding. *Ochsner J*, *16*(4), 511–524.

Beristain-Bauza, S. D. C., Hernández-Carranza, P., Cid-Pérez, T. S., Ávila-Sosa, R., Ruiz-López, I. I., & Ochoa-Velasco, C. E. (2019). Antimicrobial Activity of Ginger (Zingiber Officinale) and Its Application in Food Products. *Food Reviews International*, *35*(5), 407–426. https://doi.org/10.1080/87559129.2019.1573829

Bertelli, A., Biagi, M., Corsini, M., Baini, G., Cappellucci, G., & Miraldi, E. (2021). Polyphenols: From Theory to Practice. *Foods*, *10*(11). https://doi.org/10.3390/foods10112595

Bondam, A. F., Diolinda da Silveira, D., Pozzada dos Santos, J., & Hoffmann, J. F. (2022). Phenolic Compounds from Coffee By-Products: Extraction and Application in the Food and Pharmaceutical Industries. *Trends in Food Science & Technology*, *123*, 172–186. https://doi.org/10.1016/j.tifs.2022.03.013

Bułdak, R. J., Hejmo, T., Osowski, M., Bułdak, Ł., Kukla, M., Polaniak, R., & Birkner, E. (2018). The Impact of Coffee and Its Selected Bioactive Compounds on the Development and Progression of Colorectal Cancer In Vivo and In Vitro. *Molecules*, *23*(12). https://doi.org/10.3390/molecules23123309

Buljeta, I., Pichler, A., Šimunović, J., & Kopjar, M. (2023). Beneficial Effects of Red Wine Polyphenols on Human Health: Comprehensive Review. *Curr Issues Mol Biol*, *45*(2), 782–798. https://doi.org/10.3390/cimb45020052

Caban, M., & Lewandowska, U. (2022). Polyphenols and the Potential Mechanisms of their Therapeutic Benefits against Inflammatory Bowel Diseases. *Journal of Functional Foods*, *95*, 105181. https://doi.org/10.1016/j.jff.2022.105181

Cain, D., Kondo, M., Chen, H., & Kelsoe, G. (2009). Effects of Acute and Chronic Inflammation on B-Cell Development and Differentiation. *Journal of Investigative Dermatology*, *129*(2), 266–277. https://doi.org/10.1038/jid.2008.286

Chen, X. (2019). A Review on Coffee Leaves: Phytochemicals, Bioactivities and Applications. *Critical Reviews in Food Science and Nutrition*, *59*(6), 1008–1025. https://doi.org/10.1080/10408398.2018.1546667

Cortés-Rojas, D. F., de Souza, C. R. F., & Oliveira, W. P. (2014). Clove (Syzygium Aromaticum): A Precious Spice. *Asian Pacific Journal of Tropical Biomedicine*, *4*(2), 90–96. https://doi.org/10.1016/S2221-1691(14)60215-X

Cory, H., Passarelli, S., Szeto, J., Tamez, M., & Mattei, J. (2018). The Role of Polyphenols in Human Health and Food Systems: A Mini-Review. *Front Nutr*, *5*, 87. https://doi.org/10.3389/fnut.2018.00087

Davis, A. P., Govaerts, R., Bridson, D. M., & Stoffelen, P. (2006). An Annotated Taxonomic Conspectus of the Genus Coffea (Rubiaceae). *Botanical Journal of the Linnean Society*, *152*(4), 465–512. https://doi.org/10.1111/j.1095-8339.2006.00584.x

de Vries, K., Medawar, E., Korosi, A., & Witte, A. V. (2021). The Effect of Polyphenols on Working and Episodic Memory in Non-Pathological and Pathological Aging: A Systematic Review and Meta-Analysis. *Front Nutr*, *8*, 720756. https://doi.org/10.3389/fnut.2021.720756

DeFilipps, R. A., & Krupnick, G. A. (2018). The Medicinal Plants of Myanmar. *PhytoKeys*, *102*, 1–341. https://doi.org/10.3897/phytokeys.102.24380

Ding, Y., Brand, E., Wang, W., & Zhao, Z. (2022). Licorice: Resources, Applications in Ancient and Modern Times. *Journal of Ethnopharmacology*, *298*, 115594. https://doi.org/10.1016/j.jep.2022.115594

El Tannir, H., Houhou, D., Debs, E., Koubaa, M., Jammoul, A., Azakir, B., . . . Louka, N. (2024). Optimization of Aqueous Extraction of Polyphenols from Cuminum cyminum Seeds Using Response Surface Methodology and Assessment of Biological Activity. *BioTech (Basel)*, *13*(1). https://doi.org/10.3390/biotech13010007

Fukushima, Y., Tashiro, T., Kumagai, A., Ohyanagi, H., Horiuchi, T., Takizawa, K., . . . Kondo, K. (2014). Coffee and Beverages are the Major Contributors to Polyphenol Consumption from Food and Beverages in Japanese Middle-Aged Women. *J Nutr Sci*, *3*, e48. https://doi.org/10.1017/jns.2014.19

Fuloria, S., Mehta, J., Chandel, A., Sekar, M., Rani, N. N. I. M., Begum, M. Y., . . . Fuloria, N. K. (2022). A Comprehensive Review on the Therapeutic Potential of Curcuma longa Linn. in Relation to its Major Active Constituent Curcumin. *Frontiers in Pharmacology*, *13*. https://doi.org/10.3389/fphar.2022.820806

Gao, N., Shu, C., Wang, Y., Tian, J., Lang, Y., Jin, C., . . . Li, B. (2024). Polyphenol Components in Black Chokeberry (Aronia Melanocarpa) as Clinically Proven Diseases Control Factors – an Overview. *Food Science and Human Wellness*, *13*(3), 1152–1167. https://doi.org/10.26599/FSHW.2022.9250096

Gao, Y., Kuok, K. I., Jin, Y., & Wang, R. (2019). Biomedical Applications of Aloe Vera. *Critical Reviews in Food Science and Nutrition*, *59*(sup1), S244–S256. https://doi.org/10.1080/10408398.2018.1496320

Gengatharan, A., & Rahim, M. H. A. (2023). The Application of Clove Extracts as a Potential Functional Component in Active Food Packaging Materials and Model Food Systems: A Mini-Review. *Applied Food Research*, *3*(1), 100283. https://doi.org/10.1016/j.afres.2023.100283

Georgiev, V., Ananga, A., & Tsolova, V. (2014). Recent Advances and Uses of Grape Flavonoids as Nutraceuticals. *Nutrients*, *6*(1), 391–415. https://doi.org/10.3390/nu6010391

Grabska-Kobyłecka, I., Szpakowski, P., Król, A., Książek-Winiarek, D., Kobyłecki, A., Głąbiński, A., & Nowak, D. (2023). Polyphenols and Their Impact on the Prevention of Neurodegenerative Diseases and Development. *Nutrients*, *15*(15). https://doi.org/10.3390/nu15153454

Gul, M., Liu, Z. W., Iahtisham Ul, H., Rabail, R., Faheem, F., Walayat, N., . . . Aadil, R. M. (2022). Functional and Nutraceutical Significance of Amla (Phyllanthus emblica L.): A Review. *Antioxidants (Basel)*, *11*(5). https://doi.org/10.3390/antiox11050816

Guo, X., & Mei, N. (2016). Aloe Vera: A Review of Toxicity and Adverse Clinical Effects. *J Environ Sci Health C Environ Carcinog Ecotoxicol Rev*, *34*(2), 77–96. https://doi.org/10.1080/10590501.2016.1166826

Gupta, R. K., & Balasubrahmanyam, L. (1998). The Turmeric Effect. *World Patent Information*, *20*(3), 185–191. https://doi.org/10.1016/S0172-2190(98)00045-3

Gupta, S. C., Patchva, S., & Aggarwal, B. B. (2013). Therapeutic Roles of Curcumin: Lessons Learned from Clinical Trials. *AAPS J*, *15*(1), 195-218. https://doi.org/10.1208/s12248-012-9432-8

Habauzit, V., & Morand, C. (2012). Evidence for a Protective Effect of Polyphenols-Containing Foods on Cardiovascular Health: An Update for Clinicians. *Ther Adv Chronic Dis*, *3*(2), 87–106. https://doi.org/10.1177/2040622311430006

Han, X., & Parker, T. L. (2017). Anti-Inflammatory Activity of Clove (Eugenia Caryophyllata) Essential Oil in Human Dermal Fibroblasts. *Pharm Biol*, *55*(1), 1619–1622. https://doi.org/10.1080/13880209.2017.1314513

Hassanien, M. F. R., Assiri, A. M. A., Alzohairy, A. M., & Oraby, H. F. (2015). Health-Promoting Value and Food Applications of Black Cumin Essential Oil: An Overview. *Journal of Food Science and Technology*, *52*(10), 6136–6142. https://doi.org/10.1007/s13197-015-1785-4

Idowu, S., Adekoya, A. E., Igiehon, O. O., & Idowu, A. T. (2021). Clove (Syzygium Aromaticum) Spices: A Review on Their Bioactivities, Current Use, and Potential Application in Dairy Products. *Journal of Food Measurement and Characterization*, *15*(4), 3419–3435. https://doi.org/10.1007/s11694-021-00915-9

Iqbal, I., Wilairatana, P., Saqib, F., Nasir, B., Wahid, M., Latif, M. F., . . . Mubarak, M. S. (2023). Plant Polyphenols and Their Potential Benefits on Cardiovascular Health: A Review. *Molecules*, *28*(17). https://doi.org/10.3390/molecules28176403

Jadid, N., Widodo, A. F., Ermavitalini, D., Sa'adah, N. N., Gunawan, S., & Nisa, C. (2023). The Medicinal Umbelliferae Plant Fennel (Foeniculum vulgare Mill.): Cultivation, Traditional Uses, Phytopharmacological Properties, and Application in Animal Husbandry. *Arabian Journal of Chemistry*, *16*(3), 104541. https://doi.org/10.1016/j.arabjc.2023.104541

Jayaprakasha, G. K., Jaganmohan Rao, L., & Sakariah, K. K. (2006). Antioxidant Activities of Curcumin, Demethoxycurcumin and Bisdemethoxycurcumin. *Food Chemistry*, *98*(4), 720–724. https://doi.org/10.1016/j.foodchem.2005.06.037

Jideani, A. I. O., Silungwe, H., Takalani, T., Omolola, A. O., Udeh, H. O., & Anyasi, T. A. (2021). Antioxidant-Rich Natural Fruit and Vegetable Products and Human Health. *International Journal of Food Properties*, *24*(1), 41–67. https://doi.org/10.1080/10942912.2020.1866597

Johri, R. K. (2011). Cuminum Cyminum and Carum Carvi: An Update. *Pharmacogn Rev*, *5*(9), 63–72. https://doi.org/10.4103/0973-7847.79101

Kapoor, M. P., Suzuki, K., Derek, T., Ozeki, M., & Okubo, T. (2020). Clinical Evaluation of Emblica Officinalis Gatertn (Amla) in Healthy Human Subjects: Health Benefits and Safety Results from a Randomized, Double-Blind, Crossover Placebo-Controlled Study. *Contemp Clin Trials Commun*, *17*, 100499. https://doi.org/10.1016/j.conctc.2019.100499

Karasawa, M. M. G., & Mohan, C. (2018). Fruits as Prospective Reserves of Bioactive Compounds: A Review. *Natural Products and Bioprospecting*, *8*(5), 335–346. https://doi.org/10.1007/s13659-018-0186-6

Karkanis, A., Martins, N., Petropoulos, S. A., & Ferreira, I. C. F. R. (2018). Phytochemical Composition, Health Effects, and Crop Management of Liquorice (Glycyrrhiza glabra L.): A Medicinal Plant. *Food Reviews International*, *34*(2), 182–203. https://doi.org/10.1080/87559129.2016.1261300

Kasprzak-Drozd, K., Oniszczuk, T., Soja, J., Gancarz, M., Wojtunik-Kulesza, K., Markut-Miotła, E., & Oniszczuk, A. (2021). The Efficacy of Black Chokeberry Fruits against Cardiovascular Diseases. *Int J Mol Sci*, *22*(12). https://doi.org/10.3390/ijms22126541

Khan, N., & Mukhtar, H. (2013). Tea and Health: Studies in Humans. *Curr Pharm Des*, *19*(34), 6141–6147. https://doi.org/10.2174/1381612811319340008

Kim, N. H., Jegal, J., Kim, Y. N., Heo, J. D., Rho, J. R., Yang, M. H., & Jeong, E. J. (2018). Chokeberry Extract and Its Active Polyphenols Suppress Adipogenesis in 3T3-L1 Adipocytes and Modulates Fat Accumulation and Insulin Resistance in Diet-Induced Obese Mice. *Nutrients*, *10*(11). https://doi.org/10.3390/nu10111734

Kim, Y., Keogh, J. B., & Clifton, P. M. (2016). Polyphenols and Glycemic Control. *Nutrients*, *8*(1). https://doi.org/10.3390/nu8010017

Kumar, S., & Pandey, A. K. (2013). Chemistry and Biological Activities of Flavonoids: An Overview. *Scientific World Journal, 2013*, 162750. https://doi.org/10.1155/2013/162750

Kumar Pandey, V., Shams, R., Singh, R., Dar, A. H., Pandiselvam, R., Rusu, A. V., & Trif, M. (2022). A Comprehensive Review on Clove (Caryophyllus aromaticus L.) Essential Oil and Its Significance in the Formulation of Edible Coatings for Potential Food Applications. *Frontiers in Nutrition*, *9*. https://doi.org/10.3389/fnut.2022.987674

Kunnumakkara, A. B., Hegde, M., Parama, D., Girisa, S., Kumar, A., Daimary, U. D., . . . Aggarwal, B. B. (2023). Role of Turmeric and Curcumin in Prevention and Treatment of Chronic Diseases: Lessons Learned from Clinical Trials. *ACS Pharmacology & Translational Science*, *6*(4), 447–518. https://doi.org/10.1021/acsptsci.2c00012

Leal, C., Gouvinhas, I., Santos, R. A., Rosa, E., Silva, A. M., Saavedra, M. J., & Barros, A. I. R. N. A. (2020). Potential Application of Grape (Vitis vinifera L.) Stem Extracts in the Cosmetic and Pharmaceutical Industries: Valorization of a By-Product. *Industrial Crops and Products*, *154*, 112675. https://doi.org/10.1016/j.indcrop.2020.112675

Leite, C. D. S., Bonafé, G. A., Carvalho Santos, J., Martinez, C. A. R., Ortega, M. M., & Ribeiro, M. L. (2022). The Anti-Inflammatory Properties of Licorice (Glycyrrhiza glabra)-Derived Compounds in Intestinal Disorders. *Int J Mol Sci*, *23*(8). https://doi.org/10.3390/ijms23084121

Liczbiński, P., & Bukowska, B. (2022). Tea and Coffee Polyphenols and Their Biological Properties Based on the Latest In Vitro Investigations. *Ind Crops Prod*, *175*, 114265. https://doi.org/10.1016/j.indcrop.2021.114265

López, A., de Tangil, M. S., Vega-Orellana, O., Ramírez, A. S., & Rico, M. (2013). Phenolic Constituents, Antioxidant and Preliminary Antimycoplasmic Activities of Leaf Skin and Flowers of Aloe Vera (L.) Burm. f. (syn. A. barbadensis Mill.) from the Canary Islands (Spain). *Molecules*, *18*(5), 4942–4954. https://doi.org/10.3390/molecules18054942

Luo, K.-W., Ye, W., Li, N., & Cheng, B.-H. (2024). Tea Polyphenol EGC Suppresses Colorectal Cancer Cell Proliferation Both In Vitro and In Vivo via Downregulation of STAT3. *Journal of Functional Foods*, *112*, 105977. https://doi.org/10.1016/j.jff.2023.105977

Maan, A. A., Nazir, A., Khan, M. K. I., Ahmad, T., Zia, R., Murid, M., & Abrar, M. (2018). The Therapeutic Properties and Applications of Aloe Vera: A Review. *Journal of Herbal Medicine*, *12*, 1–10. https://doi.org/10.1016/j.hermed.2018.01.002

Mahomoodally, M. F., Aumeeruddy, M. Z., Rengasamy, K. R. R., Roshan, S., Hammad, S., Pandohee, J., . . . Zengin, G. (2021). Ginger and Its Active Compounds in Cancer Therapy: From Folk Uses to Nano-Therapeutic Applications. *Seminars in Cancer Biology*, *69*, 140–149. https://doi.org/10.1016/j.semcancer.2019.08.009

Manso, T., Lores, M., & de Miguel, T. (2021). Antimicrobial Activity of Polyphenols and Natural Polyphenolic Extracts on Clinical Isolates. *Antibiotics (Basel)*, *11*(1). https://doi.org/10.3390/antibiotics11010046

Mao, Q. Q., Xu, X. Y., Cao, S. Y., Gan, R. Y., Corke, H., Beta, T., & Li, H. B. (2019). Bioactive Compounds and Bioactivities of Ginger (Zingiber officinale Roscoe). *Foods*, *8*(6). https://doi.org/10.3390/foods8060185

Masenga, S. K., Kabwe, L. S., Chakulya, M., & Kirabo, A. (2023). Mechanisms of Oxidative Stress in Metabolic Syndrome. *Int J Mol Sci*, *24*(9). https://doi.org/10.3390/ijms24097898

Meng, F.-C., Zhou, Y.-Q., Ren, D., Wang, R., Wang, C., Lin, L.-G., . . . Zhang, Q.-W. (2018). Chapter 10 – Turmeric: A Review of Its Chemical Composition, Quality Control,

Bioactivity, and Pharmaceutical Application. In A. M. Grumezescu & A. M. Holban (Eds.), *Natural and Artificial Flavoring Agents and Food Dyes* (pp. 299–350): Academic Press.

Mnif, S., & Aifa, S. (2015). Cumin (Cuminum cyminum L.) from Traditional Uses to Potential Biomedical Applications. *Chemistry & Biodiversity*, *12*(5), 733–742. https://doi.org/10.1002/cbdv.201400305

Moris, D., Spartalis, M., Spartalis, E., Karachaliou, G. S., Karaolanis, G. I., Tsourouflis, G., . . . Theocharis, S. (2017). The Role of Reactive Oxygen Species in the Pathophysiology of Cardiovascular Diseases and the Clinical Significance of Myocardial Redox. *Ann Transl Med*, *5*(16), 326. https://doi.org/10.21037/atm.2017.06.27

Musial, C., Kuban-Jankowska, A., & Gorska-Ponikowska, M. (2020). Beneficial Properties of Green Tea Catechins. *Int J Mol Sci*, *21*(5). https://doi.org/10.3390/ijms21051744

Nasri, H., Baradaran, A., Shirzad, H., & Rafieian-Kopaei, M. (2014). New Concepts in Nutraceuticals as Alternative for Pharmaceuticals. *Int J Prev Med*, *5*(12), 1487–1499.

Negi, R., Sharma, S. K., Gaur, R., Bahadur, A., & Jelly, P. (2021). Efficacy of Ginger in the Treatment of Primary Dysmenorrhea: A Systematic Review and Meta-Analysis. *Cureus*, *13*(3), e13743. https://doi.org/10.7759/cureus.13743

Niedzwiecki, A., Roomi, M. W., Kalinovsky, T., & Rath, M. (2016). Anticancer Efficacy of Polyphenols and Their Combinations. *Nutrients*, *8*(9). https://doi.org/10.3390/nu8090552

Nikkhah Bodagh, M., Maleki, I., & Hekmatdoost, A. (2019). Ginger in Gastrointestinal Disorders: A Systematic Review of Clinical Trials. *Food Sci Nutr*, *7*(1), 96–108. https://doi.org/10.1002/fsn3.807

Noreen, S., Tufail, T., Badar Ul Ain, H., & Awuchi, C. G. (2023). Pharmacological, Nutraceutical, Functional and Therapeutic Properties of Fennel (Foeniculum vulgare). *International Journal of Food Properties*, *26*(1), 915–927. https://doi.org/10.1080/10942912.2023.2192436

Osakabe, N., Shimizu, T., Fujii, Y., Fushimi, T., & Calabrese, V. (2024). Sensory Nutrition and Bitterness and Astringency of Polyphenols. *Biomolecules*, *14*(2). https://doi.org/10.3390/biom14020234

Panche, A. N., Diwan, A. D., & Chandra, S. R. (2016). Flavonoids: An Overview. *J Nutr Sci*, *5*, e47. https://doi.org/10.1017/jns.2016.41

Pandey, K. B., & Rizvi, S. I. (2009). Plant Polyphenols as Dietary Antioxidants in Human Health and Disease. *Oxid Med Cell Longev*, *2*(5), 270–278. https://doi.org/10.4161/oxim.2.5.9498

Pastorino, G., Cornara, L., Soares, S., Rodrigues, F., & Oliveira, M. (2018). Liquorice (Glycyrrhiza glabra): A Phytochemical and Pharmacological Review. *Phytother Res*, *32*(12), 2323–2339. https://doi.org/10.1002/ptr.6178

Petrovska, B. B. (2012). Historical Review of Medicinal Plants' Usage. *Pharmacogn Rev*, *6*(11), 1–5. https://doi.org/10.4103/0973-7847.95849

Poole, R., Kennedy, O. J., Roderick, P., Fallowfield, J. A., Hayes, P. C., & Parkes, J. (2017). Coffee Consumption and Health: Umbrella Review of Meta-Analyses of Multiple Health Outcomes. *BMJ*, *359*, j5024. https://doi.org/10.1136/bmj.j5024

Qadir, M. I., Naqvi, S. T., & Muhammad, S. A. (2016). Curcumin: A Polyphenol with Molecular Targets for Cancer Control. *Asian Pac J Cancer Prev*, *17*(6), 2735–2739.

Rana, A., Samtiya, M., Dhewa, T., Mishra, V., & Aluko, R. E. (2022). Health Benefits of Polyphenols: A Concise Review. *J Food Biochem*, *46*(10), e14264. https://doi.org/10.1111/jfbc.14264

Rosales, T. K. O., & Fabi, J. P. (2023). Valorization of Polyphenolic Compounds from Food Industry By-Products for Application in Polysaccharide-Based Nanoparticles. *Front Nutr*, *10*, 1144677. https://doi.org/10.3389/fnut.2023.1144677

Roth, G. A., Mensah, G. A., Johnson, C. O., Addolorato, G., Ammirati, E., Baddour, L. M., . . . Fuster, V. (2020). Global Burden of Cardiovascular Diseases and Risk Factors,

1990–2019: Update From the GBD 2019 Study. *J Am Coll Cardiol*, *76*(25), 2982–3021. https://doi.org/10.1016/j.jacc.2020.11.010
Rudrapal, M., Khairnar, S. J., Khan, J., Dukhyil, A. B., Ansari, M. A., Alomary, M. N., . . . Devi, R. (2022). Dietary Polyphenols and Their Role in Oxidative Stress-Induced Human Diseases: Insights Into Protective Effects, Antioxidant Potentials and Mechanism(s) of Action. *Front Pharmacol*, *13*, 806470. https://doi.org/10.3389/fphar.2022.806470
Sabra, A., Netticadan, T., & Wijekoon, C. (2021). Grape Bioactive Molecules, and the Potential Health Benefits in Reducing the Risk of Heart Diseases. *Food Chem X*, *12*, 100149. https://doi.org/10.1016/j.fochx.2021.100149
Saeed, M., Naveed, M., Arif, M., Kakar, M. U., Manzoor, R., Abd El-Hack, M. E., . . . Sun, C. (2017). Green Tea (Camellia sinensis) and l-Theanine: Medicinal Values and Beneficial Applications in Humans – a Comprehensive Review. *Biomedicine & Pharmacotherapy*, *95*, 1260–1275. https://doi.org/10.1016/j.biopha.2017.09.024
Sánchez, M., González-Burgos, E., Iglesias, I., & Gómez-Serranillos, M. P. (2020). Pharmacological Update Properties of Aloe Vera and its Major Active Constituents. *Molecules*, *25*(6). https://doi.org/10.3390/molecules25061324
Scalbert, A., & Williamson, G. (2000). Dietary Intake and Bioavailability of Polyphenols. *The Journal of Nutrition*, *130*(8), 2073S–2085S. https://doi.org/10.1093/jn/130.8.2073S
Sharifi-Rad, J., Rayess, Y. E., Rizk, A. A., Sadaka, C., Zgheib, R., Zam, W., . . . Martins, N. (2020). Turmeric and Its Major Compound Curcumin on Health: Bioactive Effects and Safety Profiles for Food, Pharmaceutical, Biotechnological and Medicinal Applications. *Front Pharmacol*, *11*, 01021. https://doi.org/10.3389/fphar.2020.01021
Sharma, A., Bhardwaj, G., Sohal, H. S., & Gohain, A. (2022). Chapter 9 – Eugenol. In J. Kour & G. A. Nayik (Eds.), *Nutraceuticals and Health Care* (pp. 177–198): Academic Press.
Sidor, A., Drożdżyńska, A., & Gramza-Michałowska, A. (2019). Black Chokeberry (Aronia melanocarpa) and Its Products as Potential Health-Promoting Factors – An Overview. *Trends in Food Science & Technology*, *89*, 45–60. https://doi.org/10.1016/j.tifs.2019.05.006
Sidor, A., & Gramza-Michałowska, A. (2019). Black Chokeberry Aronia Melanocarpa L. – a Qualitative Composition, Phenolic Profile and Antioxidant Potential. *Molecules*, *24*(20). https://doi.org/10.3390/molecules24203710
Sorrenti, V., Burò, I., Consoli, V., & Vanella, L. (2023). Recent Advances in Health Benefits of Bioactive Compounds from Food Wastes and By-Products: Biochemical Aspects. *Int J Mol Sci*, *24*(3). https://doi.org/10.3390/ijms24032019
Sufianova, G., Gareev, I., Beylerli, O., Wu, J., Shumadalova, A., Sufianov, A., . . . Zhao, S. (2022). Modern Aspects of the Use of Natural Polyphenols in Tumor Prevention and Therapy. *Front Cell Dev Biol*, *10*, 1011435. https://doi.org/10.3389/fcell.2022.1011435
Tomar, R., Das, S. S., Balaga, V. K. R., Tambe, S., Sahoo, J., Rath, S. K., . . . Kesari, K. K. (2024). Therapeutic Implications of Dietary Polyphenols-Loaded Nanoemulsions in Cancer Therapy. *ACS Applied Bio Materials*, *7*(4), 2036–2053. https://doi.org/10.1021/acsabm.3c01205
Trowbridge, H. O. (1990). Immunological Aspects of Chronic Inflammation and Repair. *Journal of Endodontics*, *16*(2), 54–61. https://doi.org/10.1016/S0099-2399(06)81564-5
Tsao, R. (2010). Chemistry and Biochemistry of Dietary Polyphenols. *Nutrients*, *2*(12), 1231–1246. https://doi.org/10.3390/nu2121231
Variya, B. C., Bakrania, A. K., & Patel, S. S. (2016). Emblica Officinalis (Amla): A Review for Its Phytochemistry, Ethnomedicinal Uses and Medicinal Potentials with Respect to Molecular Mechanisms. *Pharmacological Research*, *111*, 180–200. https://doi.org/10.1016/j.phrs.2016.06.013
Villeponteau, B., Cockrell, R., & Feng, J. (2000). Nutraceutical Interventions May Delay Aging and the Age-Related Diseases. *Experimental Gerontology*, *35*(9), 1405–1417. https://doi.org/10.1016/S0531-5565(00)00182-0

Wahab, S., Annadurai, S., Abullais, S. S., Das, G., Ahmad, W., Ahmad, M. F., . . . Amir, M. (2021). Glycyrrhiza glabra (Licorice): A Comprehensive Review on Its Phytochemistry, Biological Activities, Clinical Evidence and Toxicology. *Plants (Basel)*, *10*(12). https://doi.org/10.3390/plants10122751

Wan, C., Hu, X., Li, M., Rengasamy, K. R. R., Cai, Y., & Liu, Z. (2023). Potential Protective Function of Green Tea Polyphenol EGCG against High Glucose-Induced Cardiac Injury and Aging. *Journal of Functional Foods*, *104*, 105506. https://doi.org/10.1016/j.jff.2023.105506

Wen, G., Zhang, Y., Nyman, T. J., Jern, P., & Santtila, P. (2024). Effects of Ginger on Disgust, Sexual Arousal, and Sexual Engagement: A Placebo-Controlled Experiment. *J Sex Res*, *61*(4), 658–670. https://doi.org/10.1080/00224499.2023.2175191

Yahfoufi, N., Alsadi, N., Jambi, M., & Matar, C. (2018). The Immunomodulatory and Anti-Inflammatory Role of Polyphenols. *Nutrients*, *10*(11). https://doi.org/10.3390/nu10111618

Yücel, Ç., Karatoprak, G., Açıkara Ö, B., Akkol, E. K., Barak, T. H., Sobarzo-Sánchez, E., . . . Shirooie, S. (2022). Immunomodulatory and Anti-Inflammatory Therapeutic Potential of Gingerols and Their Nanoformulations. *Front Pharmacol*, *13*, 902551. https://doi.org/10.3389/fphar.2022.902551

Zagoskina, N. V., Zubova, M. Y., Nechaeva, T. L., Kazantseva, V. V., Goncharuk, E. A., Katanskaya, V. M., . . . Aksenova, M. A. (2023). Polyphenols in Plants: Structure, Biosynthesis, Abiotic Stress Regulation, and Practical Applications (Review). *Int J Mol Sci*, *24*(18). https://doi.org/10.3390/ijms241813874

Zaveri, N. T. (2006). Green Tea and Its Polyphenolic Catechins: Medicinal Uses in Cancer and Noncancer Applications. *Life Sciences*, *78*(18), 2073–2080. https://doi.org/10.1016/j.lfs.2005.12.006

Zhong, W., Gong, J., Su, Q., Farag, M. A., Simal-Gandara, J., Wang, H., & Cao, H. (2023). Dietary Polyphenols Ameliorate Inflammatory Bowel Diseases: Advances and Future Perspectives to Maximize Their Nutraceutical Applications. *Phytochemistry Reviews*. https://doi.org/10.1007/s11101-023-09866-z

Zhou, D. D., Li, J., Xiong, R. G., Saimaiti, A., Huang, S. Y., Wu, S. X., . . . Li, H. B. (2022). Bioactive Compounds, Health Benefits and Food Applications of Grape. *Foods*, *11*(18). https://doi.org/10.3390/foods11182755

4 Extraction Techniques for Polyphenol-Rich Nutraceuticals

Anna Elisabetta Maccarronello, Filipa Teixeira, Vera Muccilli, and Francisca Rodrigues

4.1 INTRODUCTION

The demand for high-quality natural extracts to be included in nutraceutical formulations has sharply increased due to a growing consumer preference for more holistic therapeutic approaches, perceived as safer and equally effective (Pinto et al. 2020; Carpentieri et al. 2021). Natural extracts are products containing high concentrations of phytoconstituents, such as vitamins, minerals, essential oils, and active ingredients, primarily sourced from plants (Wang and Weller 2006), algae (Kadam et al. 2013), fruits (Beconcini et al. 2020), vegetables (Fernandes et al. 2023), seeds (Boussetta et al. 2012), nuts (Mandalari et al. 2022), and, more recently, even from agricultural and food waste (Maccarronello et al. 2024a; Kumar et al. 2017). They are obtained through extraction, a chemical process facilitating the transfer of biomass constituents from the cellular matrix to an appropriate solvent system (Naviglio et al. 2019). Natural extracts, valued for their functional properties and health benefits (Pinto et al. 2021a; Silva et al. 2022a; Ferreira et al. 2022), can be incorporated into cosmetics, pharmaceuticals, and nutraceuticals directly in liquid form or, more frequently, after being dried to a solid residue that, in turn, can be easily included in tablets or capsules. Polyphenol-rich nutraceuticals, in particular, are highly regarded for their safety, high biocompatibility, and effectiveness in maintaining an overall state of well-being or treating various oxidative stress–related disorders with negligible undesirable effects when compared to traditional synthetic drugs (Pandey and Rizvi 2009). Polyphenols constitute the largest group of naturally occurring bioactive phytochemicals (Pandey and Rizvi 2009; Muccilli et al. 2023), being characterized by one or more hydroxyl groups bonded to multiple aromatic rings arranged in more or less complex structures, generally of high molecular weight (Hogervorst Cvejić et al. 2017). Depending on their chemical structure, polyphenols are classified as phenolic acids, flavonoids, tannins (hydrolysable or condensed), stilbenes, or lignans (Singla et al. 2019). Their extraction from natural biomasses consists of various steps: dried sample preparation, selection and optimization of an efficient extraction method, solvent concentration or total evaporation to achieve the powdered extract,

DOI: 10.1201/9781003588504-4

and, eventually, isolation and purification of specific target molecules by chromatography (Pham 2017). Before extraction, source material must be promptly processed, since factors such as high temperature, ultraviolet (UV) radiation, or enzymatic activity can cause oxidation and subsequent degradation of bioactive constituents (Singla et al. 2019). Specifically, lowering moisture content by drying is fundamental to reduce microbial growth and prevent enzymatic degradation. Lastly, grinding and sieving the dried material to an appropriate particle size facilitate extraction efficiency by increasing the surface area (Singla et al. 2019). Numerous extraction techniques are available to obtain polyphenol-rich nutraceuticals, each playing a pivotal role in determining the yield and quality of the final product (Wang and Weller 2006).

This chapter provides a comprehensive overview of the fundamental principles on which conventional solid-liquid and cutting-edge extraction technologies are based, examining their respective advantages, drawbacks, and critical parameters necessitating optimization for maximal efficiency. Special emphasis is placed on technologies strongly aligned with the principles of green analytical chemistry, namely supercritical fluid extraction (SFE), subcritical water extraction (SWE), microwave-assisted (MAE) and ultrasound-assisted extraction (UAE), pressurized liquid extraction (PLE), enzyme-assisted extraction (EAE), and pulsed electric field (PEF). Furthermore, the chapter explores recent advancements in combined technologies strategically designed to overcome limitations of individual methods and the use of novel bio-solvent systems providing enhanced sustainability over traditional organic solvents. Finally, the potential applications and industrial extraction scaling-up perspectives for polyphenol-rich nutraceuticals are discussed.

4.2 TRADITIONAL EXTRACTION METHODS

Solvent-liquid extractions have long been used to obtain nutraceuticals from a variety of medicinal plants. These traditional methods include maceration with organic solvents, such as hexane and ethanol, Soxhlet extraction, and steam distillation (Garcia-Vaquero et al. 2020).

In maceration and Soxhlet extraction, the biomass is immersed in a solvent, which is slowly enriched with extractable solutes (Naviglio et al. 2019). These techniques are based on the combinative effect of simple diffusion and osmosis, phenomena that promote the compound transition from the biological matrix surface to the appropriate solvent system, consisting of a single organic solvent or a mixture of solvents (Naviglio et al. 2019).

Maceration represents the simplest conventional solvent-based method, allowing the extraction of soluble constituents without the necessity of dedicated equipment. Nevertheless, parameters such as solvent polarity, temperature, and extraction time strongly affect the mass transfer kinetics and, in turn, the phytochemical composition and the final quality of the extracts (Garcia-Vaquero et al. 2020). Maceration protocols can be fine-tuned by adjusting such variables, targeting the extraction of certain classes of active principles rather than others, hence achieving a certain degree of selectivity. For instance, by varying maceration conditions, Costa *et al.* obtained different extracts from coffee silverskin, a main by-product of the coffee torrefaction industry (Costa et al. 2014). The authors observed that a 50:50 (*v/v*) aqueous ethanol mixture at 50 °C for 90 min allowed reaching the highest polyphenolic content. Alternatively, a significantly higher recovery of flavonoids could be promoted by employing 75% (*v/v*) ethanol in water concentration at 60 °C for 90 min. Conversely, the recovery of tannins could be improved by using 50% (*v/v*) ethanol at 40 °C for 60 min (Costa et al. 2014).

In the so-called "Soxhlet extraction" method, multiple automatic extraction cycles are performed with a single sample loading. As shown in Figure 4.1, the biomass is put in a cellulose thimble inserted into the extraction chamber filled with condensed fresh solvent derived from a distillation flask. When the chamber is filled, a siphon aspirates the solution from the thimble, unloading it back into the distillation flask. In this way, extracted solutes are collected into the bulk liquid, and fresh solvent is returned by distillation, reaching the biological material again.

By performing multiple extraction cycles, high-quality polyphenol-rich extracts are obtained (Naviglio et al. 2019). Although the lack of a filtering stage constitutes a

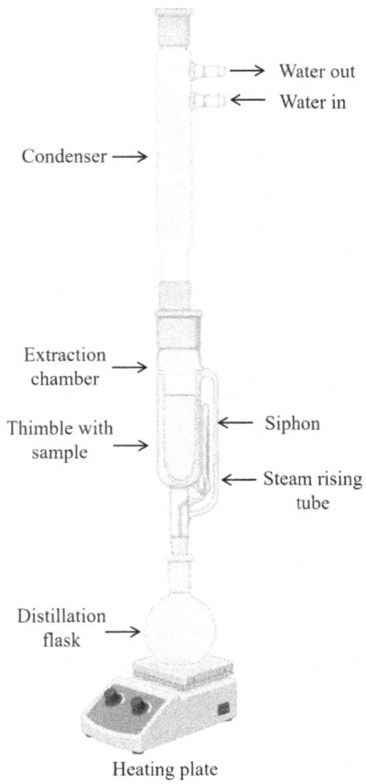

FIGURE 4.1 Schematic representation of a traditional Soxhlet extractor.

significant advantage of this method, the solvent evaporation/concentration remains necessary at the end of the process (Garcia-Vaquero et al. 2020). Consequently, as for maceration, depending on the biological matrix, the total extraction time required is notable (Garcia-Vaquero et al. 2020). To date, the European Pharmacopoeia still recommends Soxhlet extraction for the recovery of silymarin, a well-known flavolignan mixture obtained from the fruits of milk thistle (*Silybum marianum* L. Gaertner) and employed for the symptomatic relief of digestive disorders, as well as to support liver function (Wianowska and Wiśniewski 2015). The procedure consists of two steps: a preliminary defatting of the fruits for 6 hours using *n*-hexane, followed by the silymarin extraction with methanol for more 5 hours (Wianowska and Wiśniewski 2015). Considering the inclusion of *n*-hexane among hazardous air pollutants by the U.S. Environmental Protection Agency (Cravotto et al. 2022), as well as the remarkable time and energy consumption, many safer and more eco-friendly extraction alternatives started being proposed to overcome the drawbacks of this recovery process. Duan *et al.* replaced organic extraction solvents with hot water at 100–140 °C in a dynamic batch, whereas Wianowska *et al.* suggested a one-step PLE method to achieve higher silymarin yields when compared to the Soxhlet protocol (Wianowska and Wiśniewski 2015; Duan et al. 2004). The silymarin case represents just one example.

Many efforts are ongoing to combine traditional methods with the new technological trends, trying to overcome environmental concerns and satisfy the current industrial needs, which require larger volumes of extracts in a shorter time. High-pressure Soxhlet extraction, supercritical fluid-Soxhlet extractors, and automated Soxhlet extraction systems have been recently developed for these purposes (de Castro and Priego-Capote 2010).

Over the last decades, widespread criticism has been triggered regarding the significant drawbacks of traditional extraction methods, including the huge quantities of often hazardous organic solvents and the excessive time and energy required (Joshi and Adhikari 2019). Green solvents and modern technologies are catching on as valid alternatives to overcome such limitations. Nowadays, traditional methods are mostly considered reference methods to compare the performance of innovative extraction procedures to, even if, in some cases, they are still being used as the techniques of choice for recovering active ingredients (Garcia-Vaquero et al. 2020). This is the case of steam distillation.

According to the current International Organization for Standardization (ISO) standards, steam distillation remains the preferred way of producing essential oils from citrus peel (Chemat and Boutekedjiret 2015; Garcia-Vaquero et al. 2020). For ages, these mixtures of volatile molecules have been used for their healthy properties in treating irritable bowel syndrome, metabolic disorders, inflammation, immunodeficiency, neurodegenerative, and mood disorders (Matera et al. 2023). In steam distillation, aromatic plant material is heated by hot steam generated in a boiler outside the still. Hot steam vaporizes the plant material's volatile compounds that, in turn, rise and enter a condenser. After being condensed, the essential oils and the water-soluble parts of the plant – known as hydrosols – are collected into a receiver and easily separated through an upper and a lower outflow depending on their density differences (Matera et al. 2023).

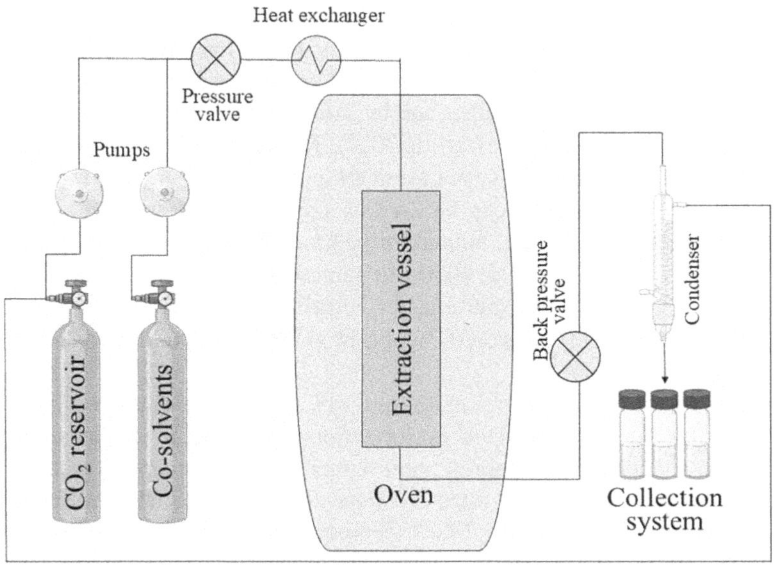

FIGURE 4.2 Schematic representation of the SFE system.

4.3 EMERGING TECHNIQUES FOR PHENOLIC COMPOUND EXTRACTION

4.3.1 Supercritical Fluid Extraction (SFE)

SFE uses a supercritical fluid as a solvent: a substance that when at a temperature and pressure above its critical point exhibits no liquid-gas phase boundary nor surface tension (Rodríguez García and Raghavan 2022; Panja 2018), therefore possessing properties of both a liquid (low surface tension, high solvent power) and a gas (decreased viscosity, high diffusivity) (Perez-Vazquez et al. 2023; Pagano et al. 2021). The critical point of a substance is a thermodynamic property mainly defined by a temperature and a pressure where the densities of liquid and vapor are the same (Rodríguez García and Raghavan 2022) – in other words, the hottest temperature allowing gases to transition into liquids under high pressures and the greatest pressure under which liquids transition to gases at high temperatures (Pagano et al. 2021).

In this extraction technique, the supercritical fluid passes through the sample in the extraction cell (Figure 4.2), where it interacts with the target compounds, diffusing into the solid matrix like a gas and dissolving active materials such as a liquid (Panja 2018). The supercritical fluid is then depressurized or cooled, causing the dissolved compounds to precipitate out, leaving behind the extracted material (Wang and Weller 2006).

To ensure safety during high-temperature operations, flammable solvents should be avoided (Panja 2018). Carbon dioxide (CO_2) is the most commonly used supercritical fluid in SFE due to its relatively low critical temperature (31 °C) and pressure (73 atm) (Panja 2018), as well as its non-toxic, chemically inert, and non-flammable nature (Uwineza and Waśkiewicz 2020). Additionally, CO_2 is readily accessible in ample quality and quantity, rendering it a cost-effective selection for extraction procedures (Uwineza and Waśkiewicz 2020), allowing minimal alteration of bioactive compounds, thereby safeguarding their functional and biological attributes (López-Hortas et al. 2022). Moreover, CO_2 can effortlessly be disposed of from samples as a gas at room temperature (Pagano et al. 2021). However, many other compounds can be used as solvents or supercritical fluids, such as propane, methanol, ethanol, butane, or pentane (Panja 2018; Rodríguez García and Raghavan 2022; Pinto et al. 2020). In some cases, the use of co-solvents may be necessary to improve the solubility of certain compounds in the supercritical fluid, depending on their polarity (Pagano et al. 2021; López-Hortas et al. 2022; Picot-Allain et al. 2021).

In order to maximize the extraction process, several variables in SFE must be optimized. For instance, pressure optimization is crucial as its elevation, while maintaining a constant temperature, increases solvent density, typically resulting in improved mass transfer rates and overall extraction yields, efficiency, and selectivity (López-Hortas et al. 2022). Temperature optimization is also crucial, as it greatly influences the solubility of the target compounds in the supercritical fluid, impacting extraction kinetics and yield (Wang and Weller 2006), along with the flow rate of the supercritical fluid through the extraction vessel, affecting the residence time of the solvent with the sample and influencing extraction efficiency (López-Hortas et al. 2022). The particle size of the sample, whether solid or semi-solid, should also be optimized, since it affects the surface area available for extraction and can influence mass transfer kinetics (López-Hortas et al. 2022; Wang and Weller 2006).

SFE offers several advantages over traditional solvent extraction methods, including higher selectivity, faster extraction rates, improved mass transfer, promoted extraction efficiency, and ability to operate under relatively mild conditions, which can preserve the integrity of heat-sensitive compounds (Rodríguez García and Raghavan 2022). Moreover, this technique requires small amounts of sample and organic solvents, allowing the supercritical fluid to be re-used (López-Hortas et al. 2022). The drawbacks of SFE include expensive equipment and maintenance, as well as high energy consumption, not being suitable for the extraction of polar compounds, and requiring the use of organic solvents, even if in low amounts (Rodríguez García and Raghavan 2022; Picot-Allain et al. 2021). Moreover, while SFE is suitable for laboratory-scale and small-scale production, scaling up to industrial levels can be challenging and costly (López-Hortas et al. 2022).

Several polyphenols have been extracted by SFE, namely anthocyanins, flavonoids, essential oils, carotenoids, and fatty acids (Uwineza and Waśkiewicz 2020; Pagano et al. 2021; Wang and Weller 2006), from multiple matrices, such as *Castanea sativa* shells and *Citrus aurantium* L. (Pinto et al. 2020; Jerković et al. 2015). The extracted compounds can then be used in various industries, including pharmaceuticals, food, beverage, cosmetic, and environmental analysis due to the absence of solvent residues in extracts and better retention of natural aromas and bioactivity (López-Hortas et al. 2022; Wang and Weller 2006).

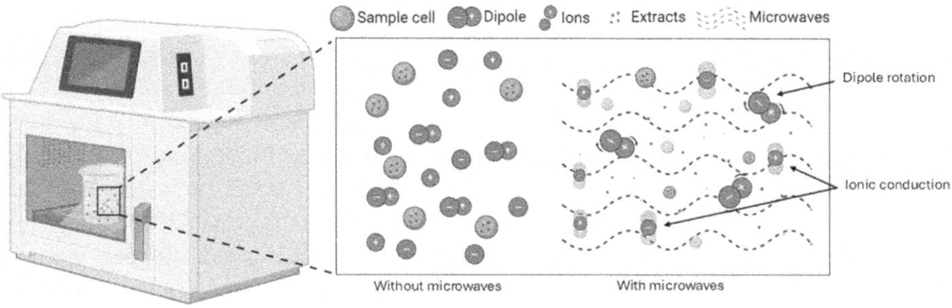

FIGURE 4.3 Schematic representation of an MAE system and mechanism.

4.3.2 Microwave-Assisted Extraction (MAE)

MAE is a modern extraction technique that utilizes microwave energy to facilitate the extraction of compounds from solid, semi-solid, or liquid samples (Cannavacciuolo et al. 2024). MAE requires oscillation of microwave wavelengths between 1 mm and 1 m and frequencies ranging from 0.3 GHz to 300 GHz (Pinto, Silva et al. 2021). Two types of systems can be used to perform MAE: (i) closed vessels under controlled pressure and temperature (susceptible to volatile compound losses) and (ii) open vessels under controlled temperature and normal atmospheric pressure (Rodríguez García and Raghavan 2022; Wang and Weller 2006).

In either system, the sample is placed in a suitable solvent within a microwave vessel, and microwave energy is applied to the mixture (Figure 4.3), being absorbed by the polar molecules within the solvent (Perez-Vazquez et al. 2023; Pinto, Silva et al. 2021). Then, rapid heating takes place through ionic conduction and dipole rotation, which involve collisions between ions and dipoles, resulting in high extraction efficiency within brief periods (Pinto, de la Luz Cádiz-Gurrea, et al. 2021). This pressure promotes solvent diffusion into the solid matrix, breaking the phenolics and cell wall linkages, and releasing the target compounds from the sample into the solvent, thereby enhancing the extraction efficiency (Macedo et al. 2021).

MAE can be tailored by adjusting multiple parameters. For instance, solvent composition needs to be carefully chosen since, in this extraction process, it needs to not only have affinity to the targeted compounds but also to be capable of absorbing microwave radiation (Wang and Weller 2006). Temperature also must be optimized because it can increase the mobility of molecules and enhance higher mass transfer rates, along with the time of extraction (ensures maximum yield while minimizing energy consumption) and the solvent-to-sample ratio (improves extraction efficiency by ensuring sufficient contact between the solvent and the sample matrix) (Putra et al. 2023; Wang and Weller 2006). Microwave power is another critical parameter in MAE: if high, it can lead to faster heating and higher extraction efficiency; if too high, it can lead to the degradation of heat-sensitive compounds (Bagade and Patil 2021; Putra et al. 2023; Wang and Weller 2006). By optimizing these parameters, this technique offers several advantages over conventional extraction techniques. For instance, MAE enables a fast and efficient extraction, reducing extraction times and providing higher yields (Silva et al. 2021; Pinto, de la Luz Cádiz-Gurrea, et al. 2021). Also, it requires smaller volumes of solvent when compared to traditional methods, contributing to lower solvent consumption and waste generation (Bagade and Patil 2021; Wang and Weller 2006). Moreover, MAE equipment is relatively low-cost, simple, and clean to operate, therefore being user-friendly (Picot-Allain et al. 2021). In addition, MAE is a straightforward extraction technique since the system is simple to set up and operate (Rodríguez García and Raghavan 2022; Panja 2018).

However, for smaller operations or research labs with limited budgets, the cost of acquiring and maintaining MAE equipment can pose a barrier to adoption (Putra et al. 2023). Another drawback of this technique is the non-selectivity of the extraction process (Picot-Allain et al. 2021). Also, MAE protocols often utilize organic solvents to enhance extraction efficiency, raising environmental and health concerns (Cannavacciuolo et al. 2024; Rodríguez García and Raghavan 2022). Additionally, MAE is more effective for polar or semi-polar compounds, while non-polar, heat-sensitive, or volatile ingredients may not be efficiently extracted or may be compromised by the intense electromagnetic radiation (Bagade and Patil 2021).

This green extraction technique has been successfully employed in the extraction of several organic compounds, namely phenols, carbohydrates, and lipids, from multiple natural matrices such as *Actinidia arguta* leaves, *C. sativa* shells, and *Beta vulgaris* L. (Silva et al. 2021; Pinto, Silva, et al. 2021; Macedo et al. 2021; Maravić et al. 2022). Many industries have benefited from the compounds recovered through this technique, including pharmaceutical, natural product extraction facilities, food and beverage analysis laboratories, environmental monitoring agencies, and forensic science organizations (Putra et al. 2023). Its ability to provide rapid and efficient extraction of target compounds makes it a valuable tool for researchers and analysts seeking to streamline their extraction processes.

4.3.3 Ultrasound-Assisted Extraction (UAE)

UAE is a green extraction method that uses ultrasound energy to facilitate better penetration of solvent and improve mass transfer (González et al. 2022; Perez-Vazquez et al. 2023; Das et al. 2021). Two different systems are available for UAE: ultrasonic probe (direct sonication; Figure 4.3) and ultrasound bath (indirect sonication) (Das et al. 2021; Rodríguez García and Raghavan 2022), both equipped with a transducer responsible for converting mechanical or electrical energy into sound energy (ultrasonic waves) (Lameirão et al. 2020). Even though an ultrasonic bath is a cheap and easy-to-operate device, it has a lower cavitation efficiency because the ultrasonic waves are not delivered to the sample uniformly, as they must pass through the medium and the container before reaching it (Kumar, et al. 2021). In contrast, the ultrasonic probe, due to direct immersion in the solution, has an increased contact area with the material, which reduces the resistance in mass transfer (Lameirão et al. 2020). Nevertheless, it should be mentioned that direct contact between the probe and the reaction mix can lead to undesired effects, such as possible contamination (Dzah et al. 2020).

The ultrasound waves, generated at a low frequency of 16–100 kHz and a power of 80–200 W, are composed of a series of compression and rarefaction cycles,

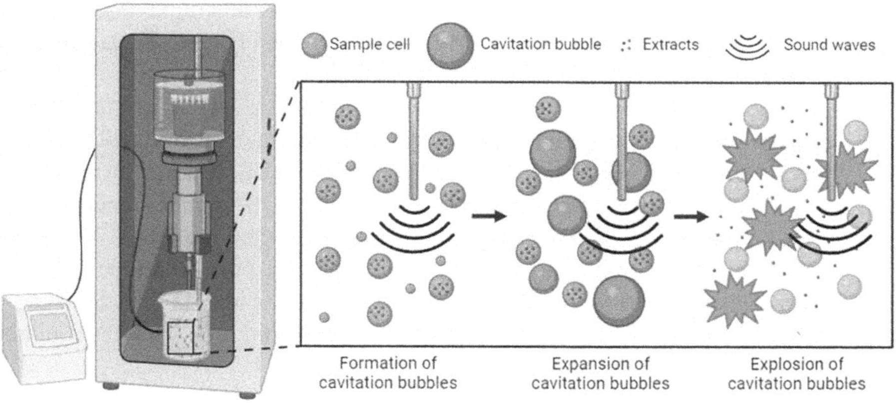

FIGURE 4.4 Schematic representation of UAE equipment and acoustic cavitation effect.

which leads the solvent to undergo acoustic cavitation (Figure 4.4): a phenomenon where small bubbles or voids (cavitation bubbles) form and collapse rapidly in the solvent (Das et al. 2021; Perez-Vazquez et al. 2023). These collapses generate intense local heating and pressure (approximately 5000 °C and 100 MPa, respectively) (Panja 2018; Lameirão et al. 2020), leading to the disruption of the sample matrix and facilitating the release of target compounds into the solvent (Pagano et al. 2021; Uwineza and Waśkiewicz 2020). This phenomenon is called erosion, and it can occur alone or in combination with other factors, such as fragmentation, sonoporation, shear force, or the increased absorption and swelling index of the cell membrane (Das et al. 2021).

Several variables can be optimized to maximize the efficiency of this extraction technique, including frequency, power/intensity, solvent type and constitution, solid-to-liquid ratio, sample particle size, and extraction time (Wang and Weller 2006; Dzah et al. 2020).

Some of the advantages offered by UAE include shorter extraction times compared to conventional methods (González et al. 2022; Silva, Almeida, et al. 2022); improved extraction efficiency with cavitation-induced disruption of the sample matrix enhancing the solubility and release of target compounds, resulting in higher extraction yields (Silva, Pinto, et al. 2022; Wang and Weller 2006); and less energy and power usage (Macedo et al. 2023; Rodríguez García and Raghavan 2022). Moreover, UAE operates at room temperature, minimizing the degradation of heat-sensitive compounds and preserving their integrity during extraction (González et al. 2022; Rodríguez García and Raghavan 2022), and it requires smaller amounts of solvent and sample, contributing to lower solvent consumption and waste generation (Panja 2018; Uwineza and Waśkiewicz 2020). In addition, UAE is an easy method to implement since it does not require any complex maintenance or expertise (González et al. 2022; Lameirão et al. 2020). However, there is an increased risk of potential contamination when the probe is in direct contact with the extraction mixture (Dzah et al. 2020), and the likelihood of free radical generation at high sonication powers is high. Moreover, although this is a non-thermal technique, heating of the system is unavoidable due to the high energy levels in use, which can lead to the degradation of some more sensitive compounds (Kumar et al. 2021).

UAE is widely used in various fields, including pharmaceutical, cosmetic, food and beverage industries, environmental monitoring, and biofuel production (Wang and Weller 2006; Silva, Garcia, et al. 2022; Pinto, de la Luz Cádiz-Gurrea, et al. 2021). Its ability to provide rapid and efficient extraction of target compounds has been proven in countless studies using *Actinidia arguta* fruit and leaves, *C. sativa* shells, waste coffee grounds, olive pomace, and different potato varieties (*Bintje*, *Challenger*, *Daisy*, *Innovator*, and *Fontane*) (Silva, Pinto, et al. 2022; Macedo et al. 2023; Lameirão et al. 2020; Al-Dhabi, et al. 2017; Goldsmith et al. 2018; Riciputi et al. 2018) as sample matrices.

4.3.4 Pressurized Liquid Extraction (PLE)

PLE, also referred to as accelerated solvent extraction, operates on the principle that the boiling point temperature increases with pressure (Panja 2018; Uwineza and Waśkiewicz 2020). Thus, the extraction system's pressure is initially raised (10 and 15 MPa) to maintain the solvent near supercritical regions – and therefore in a liquid state – before gradually increasing the temperature (50–200 °C) (Picot-Allain et al. 2021; Wang and Weller 2006). In such circumstances, the solvents' surface tension and viscosity are diminished, which promotes penetration into the solid matrix (Picot-Allain et al. 2021; Barp et al. 2023).

The sample is placed in the extraction cell and then sealed and pressurized with an inert gas, such as nitrogen, to increase the pressure and temperature within the system (Figure 4.5) (Perez-Vazquez et al. 2023). As the temperature and pressure

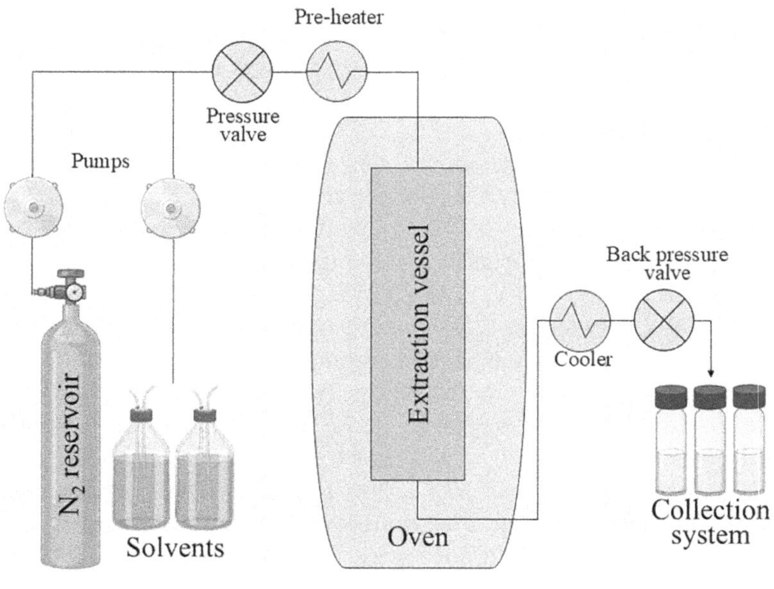

FIGURE 4.5 Schematic representation of PLE system.

rise, the solvent pumped through the sample becomes superheated, facilitating the penetration of the solvent into the sample matrix and achieving a higher solubility for the target compounds present in the sample (Pagano et al. 2021). Solvents more commonly used consist of water and aqueous alcohols, such as aqueous methanol, ethanol, 1-propanol, and acetone (Panja 2018). When the aim is to extract compounds with different polarities, a mixture of solvents with high and low polarity can be used (Perez-Vazquez et al. 2023).

When water is used as a solvent, this extraction technique may be called pressurized hot water extraction (PHWE) or subcritical water extraction (SWE). With the increased pressure and temperature, water's polarity becomes comparable to organic solvents: the dielectric constant decreases, weakening the hydrogen bonds (Barp et al. 2023), therefore enabling the extraction of less polar compounds like carotenoids, curcuminoids, lipids, and essential oils (Pagano et al. 2021).

The choice of extraction conditions increases the selectivity of PLE, which ideally allows the desired compounds to dissolve in the chosen solvent and the rest to remain insoluble in the solid matrix. For this effect to take place, some parameters must be optimized, including temperature and pressure, which interfere with the diffusivity, viscosity, and density of the solvent, directly affecting the extraction kinetics and efficiency (Pagano et al. 2021; Barp et al. 2023). Furthermore, the solvent type and composition can impact the selectivity and efficiency of extraction by optimizing interactions with target compounds (Barp et al. 2023), as well as controlling the sample particles' size since the smaller the particles, the greater the surface area available for penetration of the solvent into the sample matrix (Perez-Vazquez et al. 2023; Barp et al. 2023). The extraction time, flow rate, and number of cycles must also be optimized in order to improve the extraction efficiency since they can influence the mass transfer kinetics and maximize contact between the solvent and the sample (Perez-Vazquez et al. 2023).

PLE offers several advantages over conventional extraction techniques, including shorter extraction times, increased extraction efficiencies (mostly due to higher permeability induction in cell membranes), and reduced solvent consumption and energy input (Panja 2018). Solvents used in PLE are mainly water and aqueous alcohols; therefore, they are environmentally friendly, non-toxic, and low-cost (Panja 2018). Nevertheless, considering the mild condition in which PLE operates, heat-sensitive chemicals retain their integrity throughout extraction, and the final products' structure and quality are preserved (Picot-Allain et al. 2021). Furthermore, PLE systems offer automation compatibility, facilitating increased reproducibility, efficiency, and ease of application in high-throughput processes, while requiring minimal manual intervention. The main limitations of PLE include the high cost of the equipment, as well as rigorous optimization of extraction parameters, namely, temperature and pressure (Picot-Allain et al. 2021). For instance, high temperatures may degrade active components and dissolve undesired compounds (Wang and Weller 2006), making this technique unfitting to extract heat-sensitive compounds (Uwineza and Waśkiewicz 2020; Putra et al. 2023). Also, the choice of solvents is relatively limited, since they should have high auto-ignition temperatures and reduced metal corrosiveness at high temperatures and pressures (Panja 2018).

Many studies have proven the usefulness of this technique in recovering compounds from natural matrixes such as *Passiflora ligularis* and *B. vulgaris* L.

(Vardanega et al. 2023; Maravić et al. 2022). Moreover, PLE is widely used in various fields such as environmental analysis, food and beverage testing, pharmaceutical analysis, and natural product extraction (Barp et al. 2023; Wang and Weller 2006).

4.3.5 Subcritical Water Extraction (SWE)

As mentioned, SWE is also known as PHWE. This extraction technique uses water as a solvent below its critical point (Tc = 374.15 °C, Pc = 22.1 MPa) but above its boiling point (100 °C) and under sufficient pressure to maintain the water in liquid state (around 1–10 MPa) (Correia et al. 2022; Pinto, de la Luz Cádiz-Gurrea, et al. 2021). Therefore, water is heated under moderate pressure to temperatures typically ranging from 50 °C to 200 °C, depending on the desired application. While at ambient conditions water is considered an extremely polar solvent (dielectric constant of around 80), in subcritical conditions, water's dielectric constant decreases to approximately 30, weakening the hydrogen bonds and resembling less polar organic solvents such as methanol and ethanol (Barp et al. 2023; Cvjetko Bubalo et al. 2018; Tomšik et al. 2017). In addition, in a subcritical state, surface tension, viscosity, and density of water are also significantly reduced, which facilitates faster water wetting and dissolution of targeted compounds and enhances penetration into the material and improves diffusion rates (Cvjetko Bubalo et al. 2018; Pinto, de la Luz Cádiz-Gurrea, et al. 2021).

The extraction process begins by loading the sample into an extraction vessel, which is then pressurized with water and heated to the desired temperature (Figure 4.6). The elevated temperature and pressure help to disrupt the sample matrix, allowing water to penetrate and solubilize the target compounds (Pinto, de

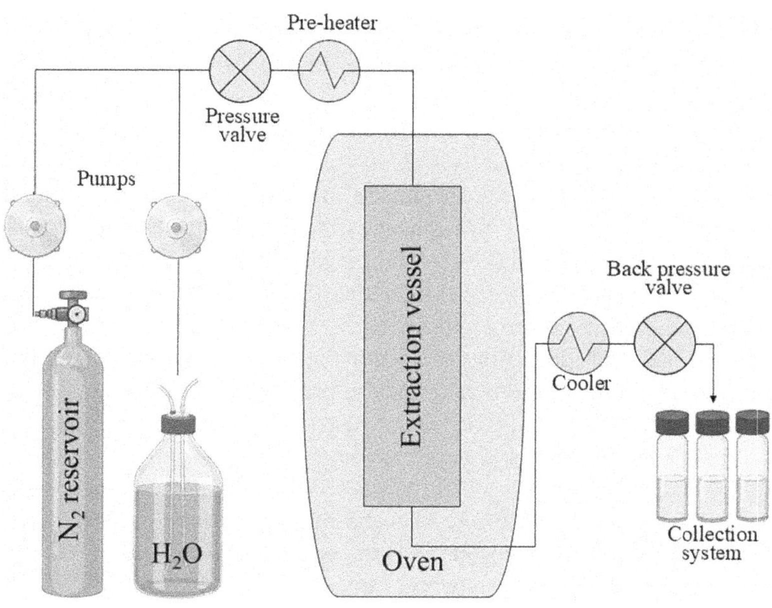

FIGURE 4.6 Schematic representation of SWE system.

la Luz Cádiz-Gurrea, et al. 2021). The extraction duration can vary depending on the desired extraction efficiency and the nature of the sample. After extraction, the mixture of water and dissolved compounds is typically collected and separated from the solid residue using filtration or centrifugation.

Several SWE variables can be studied to optimize the extraction process. For example, by carefully adjusting temperature and pressure, the solvent can attain the optimal density, viscosity, and solubility required to efficiently extract target compounds, leading to enhanced extraction efficiency and yield (Cvjetko Bubalo et al. 2018; Silva, Luís, et al. 2022). Optimizing the duration of the extraction process is also important as it ensures sufficient contact between the sample and the water solvent to extract target compounds effectively, as well as controlling the sample size and solvent flow rate (Cvjetko Bubalo et al. 2018).

The main advantage of SWE is that by using a non-toxic, non-flammable, highly available, environmentally friendly, and safe solvent (water) (Cvjetko Bubalo et al. 2018), it is still possible to extract both polar and non-polar compounds (Razak et al. 2023; Pinto, de la Luz Cádiz-Gurrea, et al. 2021). Moreover, this technique is highly efficient and selective, as the high diffusivity and solvent power of subcritical water enable rapid extraction and increase mass-transfer properties (Cvjetko Bubalo et al. 2018; Correia et al. 2022). It is also a low-cost process, with reduced solvent consumption and extraction times and low energy requirements (Correia et al. 2022; Silva, Luís, et al. 2022). In addition, SWE allows the preservation of the integrity of heat-sensitive compounds during extraction given the use of mild extraction conditions (Silva, Luís, et al. 2022).

However, SWE also has some drawbacks, such as the possibility of compound degradation at elevated temperatures, undesirable reactions (caramelization, for instance), and the loss of selectivity due to an increase of solubility of other matrix compounds (Cvjetko Bubalo et al. 2018; Pinto, Vieira, et al. 2021). Furthermore, this technique involves a high investment in equipment acquisition and maintenance.

Nevertheless, a wide range of compounds, such as phenolic compounds (phenols and flavonoids), essential oils, polysaccharides, fatty acids, sugars, and protein (Correia et al. 2022; Cvjetko Bubalo et al. 2018), have been efficiently and sustainably extracted from complex matrices, namely *Salicornia ramosissima*, *Allium ursinum* L., *Zingiber officinale* Roscoe, and *Actinidia arguta* leaves (Correia et al. 2022; Tomšik et al. 2017; Razak et al. 2023; Silva, Luís, et al. 2022), by SWE. The application of this technique has also been explored in various fields, including environmental analysis, food and beverage testing, pharmaceuticals, cosmetics, and natural product extraction industry (Correia et al. 2022; Pinto, de la Luz Cádiz-Gurrea, et al. 2021).

4.3.6 Enzyme-Assisted Extraction (EAE)

EAE is a method used to extract compounds from solid, semi-solid, or liquid samples with the assistance of enzymes, i.e., biological catalysts that facilitate chemical reactions without being consumed in the process (Das et al. 2021). Specific enzymes are added to the sample to catalyse the breakdown of cell walls, membranes, or other structural components (Figure 4.7), thereby facilitating the release of target

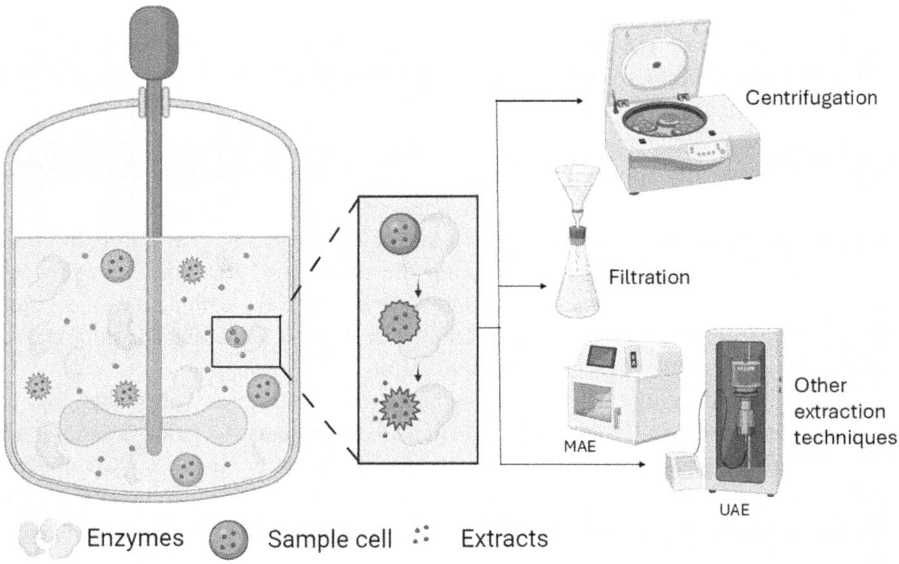

FIGURE 4.7 Schematic representation of EAE system and mechanism.

compounds into the extraction solvent (Rodríguez García and Raghavan 2022; Łubek-Nguyen et al. 2022). After the enzymatic reaction is complete, the solvent and enzyme separation from the extracted compounds may be performed following different procedures, namely centrifugation, filtration, or even another extraction technique such as UAE, MAE, or SFE (Das et al. 2021).

To achieve outstanding performance with this extraction technique, it is essential to optimize several extraction parameters. The selection of the appropriate enzymes and their optimal concentration, along with the appropriate solvent, is crucial for maximizing the breakdown of cellular structures and releasing target compounds from the sample matrix (Rodríguez García and Raghavan 2022; Panja 2018). The choice of enzymes mostly depends on the nature of the sample and the compounds of interest (Habeebullah et al. 2020). For example, cellulase and pectinase enzymes are used to extract compounds from plant matrices due to their capacity to break down plant cell walls (González et al. 2022; Macedo et al. 2021). As for the solvent, it mostly depends on the type of enzyme used and the polarity of the targeted metabolites (Cannavacciuolo et al. 2024). Enzyme activity is highly dependent on pH and temperature (Picot-Allain et al. 2021), underscoring the importance of optimizing these two parameters for maximum efficiency. Additionally, the duration of enzymatic treatment, which impacts substrate degradation and the release of target compounds, is important, as well as the particle size of the sample, such that when smaller it provides a larger surface area for enzyme-substrate interactions, potentially enhancing extraction efficiency, yield, and quality of extracted compounds (Cannavacciuolo et al. 2024; Łubek-Nguyen et al. 2022); however, it must be carefully optimized.

EAE offers several advantages over conventional extraction methods, namely enhanced extraction efficiency, since enzymes facilitate the breakdown of structural barriers within the sample, allowing for more efficient release of target compounds into the solvent (Das et al. 2021; Rodríguez García and Raghavan 2022). The possibility of performing the extraction process under lower temperatures and pH levels and using non-organic solvents such as water is also an advantage, since it helps to preserve the integrity of more sensitive compounds (Łubek-Nguyen et al. 2022; Macedo et al. 2021; Cannavacciuolo et al. 2024). Furthermore, specific enzymes can be chosen to target compounds or classes of compounds, providing greater selectivity in extraction (Łubek-Nguyen et al. 2022), and since enzymes are biodegradable and the extraction solvent to be used does not necessarily have to be an organic solvent, the extraction technique is environmentally friendly (Panja 2018; González et al. 2022; Habeebullah et al. 2020). In addition, the use of enzymes reduces the need for mechanical agitation or high temperatures, which results in lower energy requirements for the extraction process (Łubek-Nguyen et al. 2022; González et al. 2022).

Nevertheless, this is a relatively highly time-consuming (1–8 hours) (Panja 2018) and delicate technique as enzymes are susceptible to changes in temperature, pH, and other environmental factors (Picot-Allain et al. 2021; González et al. 2022). Enzymes may deactivate or denature over time, leading to inconsistent or ineffective extraction processes (Das et al. 2021). Also, EAE equipment and infrastructure are costly and technically demanding to implement, mostly because enzymes can be relatively expensive, especially when large volumes of samples are involved (Picot-Allain et al. 2021; Das et al. 2021), along with the fact that maintaining consistent enzyme activity, controlling process parameters, and ensuring uniform extraction across large volumes of samples can be daunting. In addition, enzymes are not always able to completely break through the cell matrix, thus requiring additional processing to separate and purify the products (Das et al. 2021). Due to all these drawbacks, EAE is often applied in combination with other green extraction techniques.

Even so, EAE finds application in various industries, including food processing, pharmaceuticals, biofuels, cosmetics, and natural product extraction. Its ability to improve extraction efficiency while operating under mild conditions makes it a valuable tool for researchers and manufacturers seeking to optimize their extraction processes.

4.3.7 Electrical Technologies: Pulsed Electric Field (PEF)

PEF has recently emerged as an advanced electrically assisted technology to enhance the recovery of phytochemicals from various medicinal plants, algae, and agri-food by-products (Morales de la Peña et al. 2023; Salgado-Ramos et al. 2023; Kashyap et al. 2024). This non-thermal technique offers numerous benefits over other extraction methods, including brief processing times (microseconds to milliseconds), possible use of green solvents, low energy input, and high extraction efficiencies with a minimal thermal impact on bioactive components, satisfying the requirements for a future-oriented bioeconomy and final products of superior quality (Poojary et al. 2020).

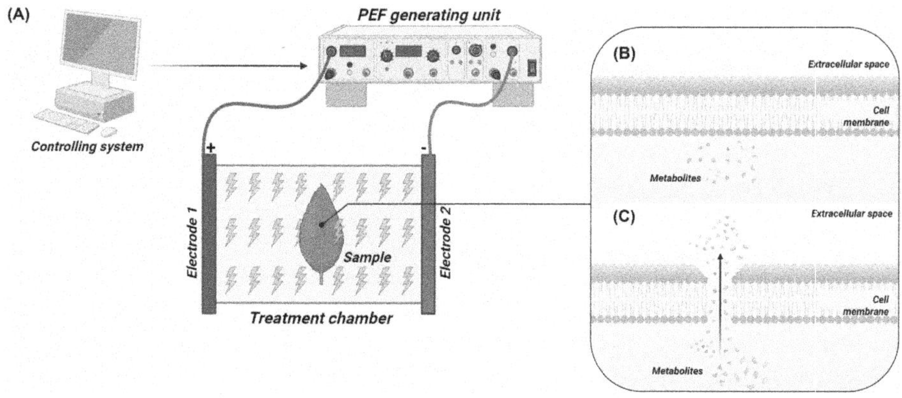

FIGURE 4.8 (A) Schematic representation of PEF setup: the sample in an idoneous conductive medium is put inside the treatment chamber connected to a moderate-to-high electric pulse generator. (B) Cell membrane before PEF-induced electroporation; (C) electroporation in cell membranes, resulting in loss of cell integrity and diffusion of intracellular metabolites. PEF, pulsed electric field.

In a typical PEF treatment, the membrane electroporation principle promotes the mass transfer of intracellular secondary metabolites (Naliyadhara et al. 2022). Medium- to high-voltage electric pulses are applied to the biomass of interest, which is placed in a suitable conductive solvent medium (e.g., distilled water or water/ethanol binary mixtures). According to the transmembrane potential breakdown theory proposed by Sale and Hamilton (Sale and Hamilton 1968), the external electric field generated increases the cell transmembrane potential, prompting pore formation in the lipid bilayer (Poojary et al. 2020). Particularly, when the transmembrane potential reaches its critical value (around 1 V), the membrane starts losing its structural integrity by forming micropores. A further voltage increase leads to larger pores and irreversible electroporation. Generally, field strength varies between 0.1 and 20 kV/cm without significantly affecting the extraction medium temperature and, consequently, the stability of target compounds (Poojary et al. 2020).

As shown in Figure 4.8, PEF equipment, ranging from laboratory-scale to pilot-scale batches, consists of three basic components: a high-energy pulse generator; a square, easily cleanable treatment chamber equipped with an insulator placed between two electrodes; and a controlling system to monitor various process parameters, including electric field intensity, pulse number, pulse width, and exposure time (Arshad et al. 2020).

Recent studies demonstrated that PEF treatment allows for the achievement of higher polyphenol recovery compared to other methods (Boussetta et al. 2012; Salgado-Ramos et al. 2023). Furthermore, Morales de la Peña et al. proposed PEF technology as a viable approach to obtain healthy polyphenol-enriched beverages from red prickly pears and soymilk (Morales de la Peña et al. 2023). However, it must be underlined that scaled-up industrial applications are still in their early stage due to the lack of technology readiness. Continuous reactors and appropriate, more cost-effective protocols need to be developed to affirm PEF as an efficient technology for the high-throughput production of food additives and nutraceuticals.

4.3.8 Combined Extraction Technologies

Coupled synergic technologies represent a relatively recent field of study. As mentioned in the previous sections, all extraction techniques inevitably have distinct strengths and weaknesses (Table 4.1), as well as specific practical issues that must be considered to preserve bioactive compounds from alterations and ensure the final quality of the nutraceutical extract.

However, the shortcomings of a single method can be overcome by coupling two or more extraction techniques based on different principles and mechanisms, generating a synergistic effect and boosting the overall extraction efficiency.

Appropriate combinations have been created between two modern advanced processes, such as SFE and PLE, or a conventional solvent-based method with

TABLE 4.1
Advantages and Disadvantages of Different Innovative Extraction Techniques

	Advantages	Disadvantages	References
SFE	• Time saving • Use of little amount of sample and organic solvents • Enhanced mass transfer, selectivity, and extraction yield • Recyclability of supercritical fluid • Use of mild conditions	• Energy consuming • Use of organic solvents • Expensive equipment and maintenance • Not suitable for polar molecules	(Rodríguez García and Raghavan 2022; López-Hortas et al. 2022; Picot-Allain et al. 2021)
MAE	• Energy and time saving • Easy and clean to operate • Use of little amount of sample and solvent • High extraction yields • Better extract quality	• Uneven heating and/or overheating of sample • Use of organic solvents • Difficult to scale up due to expensive equipment and maintenance • Not suitable for non-polar or sensitive compounds	(Maccarronello et al. 2024b; Rodríguez García and Raghavan 2022; Panja 2018; Picot-Allain et al. 2021; Putra et al. 2023; Bagade and Patil 2021)
UAE	• Use of eco-friendly solvents • Non-thermal technique • Energy and time saving • High extraction yields • Use of little amount of sample and solvent • Appropriate for sensitive ingredients	• Expensive equipment and maintenance • Possible free radical generation at high sonication powers • Risk of potential contamination when using ultrasonic probe	(Wang and Weller 2006; Rodríguez García and Raghavan 2022; Picot-Allain, et al. 2021)

TABLE 4.1 (*Continued*)
Advantages and Disadvantages of Different Innovative Extraction Techniques

	Advantages	Disadvantages	References
PLE	• Energy and time saving • Use of little amount of sample and solvent • High extraction yields	• Expensive equipment and maintenance • Not suitable for sensitive compounds	(Wang and Weller 2006; Panja 2018; Uwineza and Waśkiewicz 2020; Picot-Allain, et al. 2021; Putra et al. 2023; Vardanega et al. 2023)
SWE	• Energy and time saving • High selectivity and extraction yields • Use of eco-friendly solvents • Possible to extract both polar and non-polar compounds • Use of little amount of sample and solvent	• Uneven heating and/or overheating of sample • Inefficient in large-scale production • Expensive equipment and maintenance • Not suitable for sensitive compounds	(Razak et al. 2023; Pinto, de la Luz Cádiz-Gurrea et al. 2021)
EAE	• Energy saving • Use of eco-friendly solvents • Use of mild conditions • High selectivity and extraction yields • Suitable for separating bound compounds • Appropriated for sensitive ingredients	• Enzyme sensitivity • Time consuming • Difficult to separate the extracted compounds • Inefficient in large-scale production • Expensive price of enzyme	(Rodríguez García and Raghavan 2022; Picot-Allain et al. 2021; Cannavacci-uolo et al. 2024; Macedo et al. 2021; González et al. 2022; Łubek-Nguyen et al. 2022)
PEF	• Non-thermal technique • Energy and time saving • Use of eco-friendly solvents • High extraction yields	• Expensive equipment and maintenance • Difficult to scale up due to lack of technology readiness	(Morales de la Peña et al. 2023; Salgado-Ramos et al. 2023; Arshad et al. 2020)

an innovative one, such as hydro-distillation coupled with microwaves (Picot-Allain et al. 2021). Moreover, hybrid extraction methods are referred to as either "sequential" when the two processes are performed in series or "simultaneous" when more treatments occur in a single operating device (Capaldi et al. 2024). An example of hybrid sequential technology is the combination of PEF and PLE (Capaldi et al. 2024). Zhou et al. have studied the synergistic effect of a PEF pre-treatment followed by PLE, obtaining a higher quantity and variety of bioactive molecules extracted from *Spirulina* biomass compared to the not-combined methods, considerably shortening the extraction times (Zhou et al. 2022). Despite the promising results, sequential processes have superior operation costs compared to the simultaneous ones, which generally require less time and energy. Simultaneous microwave-ultrasound-assisted extraction (USMAE) has been applied to various biological sources of therapeutic ingredients (Ordoñez-Torres et al. 2021; Quan et al. 2023; Luo et al. 2017). For instance, Sillero et al. employed USMAE to obtain a polyphenol-rich *Larix decidua* bark extract with improved antioxidant capacity, doubling the extraction yield and reducing the time 47-fold compared to conventional extraction (Sillero et al. 2020). Apart from the number of literature studies, ultrasonic microwave combined systems are commercially available in the form of a single extractor in which the capacity of ultrasound technology to disrupt cellular matrix by cavitation, in conjunction with the dielectric heating generated by microwave irradiation, synergistically acts to improve both the extraction rate and polyphenolic yield (Ojha et al. 2020).

Combined synergic technologies also include enzyme-assisted extraction coupled with microwave or ultrasound technologies. Treating the biomass with suitable enzymes able to cause the depolymerization of cell wall constituents such as pectin, cellulose, and hemicellulose is possible to improve the cell matrix permeation to the extraction solvent. As a result, solvent consumption, extraction times, and overall processing costs are notably reduced. Although coupled synergic technologies are a relatively recent field of study, their effective employment in the nutraceutical and cosmetic fields may lead to more promising results than those obtained by applying a single method.

4.4 BIOSOLVENTS AND NATURAL DEEP EUTECTIC SOLVENTS (NaDES)

Nowadays, increasingly larger volumes of solvents are consumed by the nutraceutical industry, leading to the necessity of finding greener and more sustainable solutions to limit the enormous environmental impact as well as the notable disposal costs connected with the employment of conventional organic solvents. Ideally, solvents should be inexpensive, stable, safe, non-flammable, biodegradable, and easily separable from extraction products. Deep eutectic solvents (DES) and natural DES (NaDES) present many of these characteristics, and, for this reason, widespread research has been focused on them over the last few years (Abbott et al. 2004; Hammond et al. 2017).

DES and NaDES are considered the fourth generation of non-toxic ionic liquids with exceptional physicochemical characteristics in the face of a poor environmental footprint (Abbott et al. 2004). Unlike traditional organic solvents, these eutectic mixtures have a very low vapour tension, making them barely flammable and non-volatile (Usmani et al. 2023). Moreover, their chemical diversity, biodegradability, and safety profile to humans make them suitable for biomedical and nutraceutical applications (Ruesgas-Ramón et al. 2017; Usmani et al. 2023).

4.4.1 DES/NaDES Composition

Eutectic biosolvents consist of special mixtures that are formed when immiscible solid compounds, mixed at specific molar ratios, are weakly heated until they become

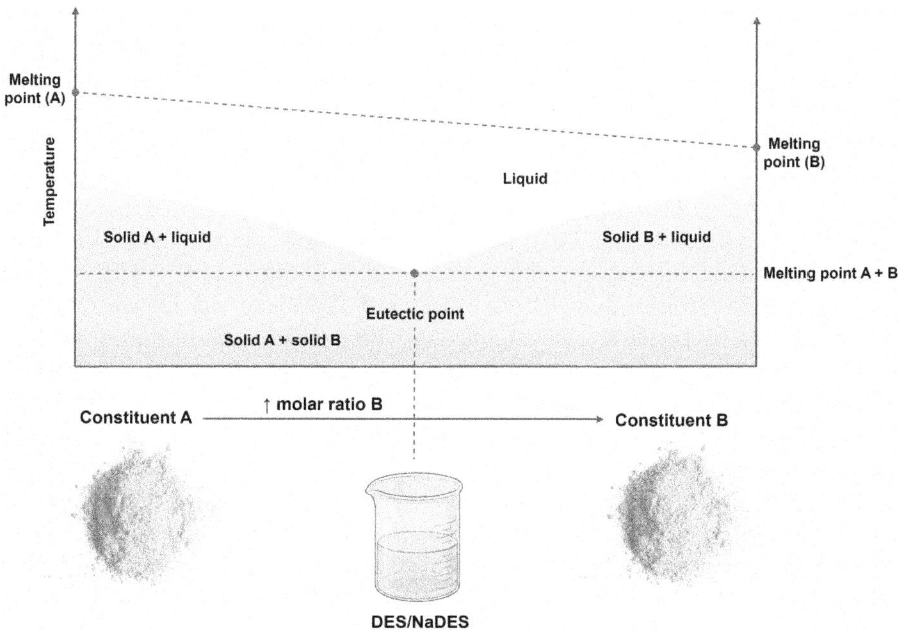

FIGURE 4.9 Schematic representation of the phase diagram of a eutectic system composed of two components.

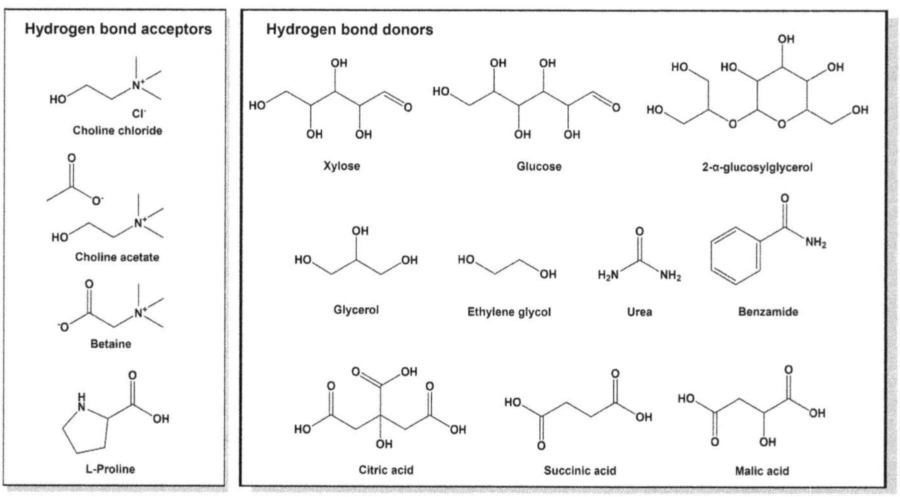

FIGURE 4.10 Chemical structure of the most common DES and NaDES constituents.

liquid (Pena-Pereira and Namiesnik 2014). The transition to the liquid phase occurs as the mixture has a lower melting point than each single component (Figure 4.9). DES/NaDES represent simple physical systems of this type (Usmani et al. 2023; Craveiro et al. 2016).

They are formed by combining neutral hydrogen bond donors (HBDs) species with ionic salts acting as hydrogen bond acceptors (HBAs) (Usmani et al. 2023). The extensive hydrogen bond network, van der Waals interactions, and charge delocalization are responsible for the melting point depression and, consequently, for the eutectic solvent formation (Hammond et al. 2017; Craveiro et al. 2016). Figure 4.10 illustrates some of the most common molecules employed as DES/NaDES constituents.

Generally, quaternary ammonium salts, such as choline chloride (ChCl), are combined with sugars, polyalcohols, amines, or carboxylic acids (Usmani et al. 2023). Born from the idea that similar mixtures may already exist in nature to transport poorly water-soluble compounds inside all living cells, NaDES represent a further evolution of DES (Choi et al. 2011). NaDES are basically composed of amino acids and sugars, and, for this reason, they are even more biocompatible with nearly all living organisms (Cvjetko Bubalo et al. 2018).

4.4.2 Synthesis and Physicochemical Properties of Eutectic Solvents

Despite the differences in labelling, these sustainable media can be easily synthesized by heating, evaporation, or freeze-drying methods (Ruesgas-Ramón et al. 2017). Independently from the technique employed, no waste product formation is generally observed, further highlighting the completely green nature of these solvent systems (Usmani et al. 2023).

In the heating method, starting constituents are magnetically stirred at temperatures ranging from 60 to 80 °C until a clear liquid appears (Ruesgas-Ramón et al. 2017). Alternatively, in the evaporating method, constituents are first dissolved in an appropriate water percentage that then is removed in a rotavapor at 50 °C. The remaining DES/NaDES is placed in a dryer until a constant weight is achieved to ensure complete dehydration (Ruesgas-Ramón et al. 2017). Lastly, freeze-drying the aqueous solutions of the individual components has been recognized as another possible method for the achievement of clear viscous eutectic mixtures (Gutiérrez et al. 2009).

If, on the one hand, the supramolecular structure endows eutectic solvents with unique physicochemical characteristics, such as a high ionic charge, high conductivity, a polarity higher than that of water, and, consequently, a high solvation capacity towards a large variety of molecules (Craveiro et al. 2016), on the other hand, it can negatively impact mass transfer phenomena occurring during the extraction of natural compounds. The dense hydrogen bond network results in poor mobility of solvent constituents, leading to high viscosity, high surface tension, and low diffusivity coefficient (Usmani et al. 2023). Noteworthy, a high solvent viscosity and a low fluidity constitute an obstacle during the extraction process (Morales de la Peña, et al. 2023). On a practical level, some strategies can be adopted to bypass such limitations. The first is through the proper addition of water (or other co-solvents) to partially dissolve DES/NaDES components, reducing their chemical interactions and decreasing the impact of the previously mentioned physicochemical characteristics on the extraction process (Ruesgas-Ramón et al. 2017; Usmani et al. 2023). However, in this case, it must be taken into account that excessive additions can completely dissolve the intramolecular forces, nullifying all the liquid properties (Ruesgas-Ramón et al. 2017). The second strategy that can be used to improve the solvent fluidity consists of heating, since, as for all liquids, an increase in temperature corresponds to a decrease in viscosity. Among the extraction methods, MAE fits particularly well when coupled with the use of these polar solvents, given their high dielectric constant (Lévêque and Cravotto 2006). Lastly, mass transfer and diffusion phenomena have been improved even at room temperature by conjugating eutectic solvents with UAE technology (Lévêque and Cravotto 2006). Nevertheless, extracting bioactive compounds using DES/NaDES as alternative biosolvents requires a careful optimization process, not only of the parameters related to the selected extraction method (e.g., temperature, pressure, time) but also of the factors intrinsically related to NaDES, namely the constituents' affinity towards target phenolic compounds, the molar ratio between mixture components, the percentage of water added to reduce the viscosity of the solvent system, and the biomass/solvent ratio.

4.4.3 DES/NaDES for Sustainable Extraction of Polyphenol-Rich Nutraceuticals

Thanks to their eco-friendly characteristics, including low-cost, easy and completely green synthesis, biodegradability, safety profile to humans, and reusability after being recovered by a chromatographic procedure, eutectic solvents have great application potential in the extraction of polyphenol-rich nutraceuticals.

They have been successfully employed as bioextractants to recover different classes of phenolic compounds from various natural sources (Table 4.2). The possibility of modifying their physicochemical properties by varying the HBA/HBD composition or their respective molar ratio makes them highly customizable and versatile, allowing for the extraction of a broad range of bioactive molecules (Hammond et al. 2017; Ruesgas-Ramón et al. 2017).

Apart from their use as extraction biosolvents, NaDES themselves can act as ingredients in functional food and nutraceuticals (Secretan et al. 2023; Nystedt et al. 2023). DES/NaDES constituted by poorly water-soluble active ingredients and idoneous HBA such as ChCl, sometimes referred to as vitamin B_4, have been investigated as promising formulations in which a compound's stability, dissolution, and bioavailability can be significantly improved (Abbott et al. 2017; Secretan et al. 2023). Morrison et al. reported that benzoic acid, danazol, griseofulvin, and itraconazole solubility in urea-ChCl and malonic acid-ChCl was 5- to 22.000-fold higher than in water (Morrison et al. 2009). Similarly, Secretan et al. formulated liquid-filled capsules containing a deep eutectic mixture composed of epigallocatechin gallate and ChCl to treat transthyretin cardiac amyloidosis. The authors reported that the active substance's stability was still maintained after 18 months in storage (Secretan et al. 2023).

Although the physicochemical properties of the individual constituents of DES/NaDES are well known in the literature, many studies are still necessary to provide a complete vision of their eco-toxicological properties, which can substantially differ from those of the individual components (Usmani et al. 2023).

TABLE 4.2
Examples of DES and NaDES Used as Bioextractants for the Recovery of Active Ingredients by Different Extraction Methods

Target Bioactive Compound	Natural Source	DES/NaDES Combination	Molar Ratio	Extraction Method	Reference
Anthocyanins	*Euterpe edulis* Mart. fruits	ChCl:xylitol	5:2	UAE	(Vannuchi et al. 2022)
Triterpenes	*Fomes officinalis* mushrooms	Betaine: 2-α-glucosylglycerol	1:2	UAE	(Zhang et al. 2024)
Pectins	*Citrus sinensis* peel	ChCl:formic acid	1:2	MAE	(Turan et al. 2024)
Stilbenes	*Vitis* sp. shoots	ChCl:levulinic acid	1:2	UAE	(Duarte et al. 2024)
Cinnamic alcohols	*Rhodiola rosea* L.	ChCl:malonic acid	1:1	UAE	(Tsvetov et al. 2023)
Flavonoids	Propolis	Betaine:malic acid:proline	1:1:1	UAE	(Alpat et al. 2023)
Rutin	*Sophora japonica* L.	ChCl:ethylene glycol	1:2	Maceration	(Le et al. 2023)
Methylxanthines	*Theobroma cacao* L. shells	ChCl:lactic acid	1:2	Agitation	(Benítez-Correa et al. 2023)

When DES/NaDES cannot be directly incorporated into the final formulation, it necessitates separation from the extract. Due to their non-volatility, simple evaporation cannot remove eutectic solvents. The recovery of the extracted components is generally carried out by means of a chromatographic column using resins or molecular sieves or through precipitation by adding anti-solvents such as water (Ruesgas-Ramón et al. 2017). The recovery of bioactive metabolites from *Carthamus tinctorius* L. after extraction with NaDES was carried out by Dai et al. through a weak polar resin HP-20 consisting of styrene-divinylbenzene (Dai et al. 2013). The phenolic compounds were adsorbed onto the resin, while the NaDES was removed by water elution. Finally, the polyphenols were collected in ethanol (Dai et al. 2013). Although this method involves an overall increase in extraction times, the NaDES can be renewed by evaporating the water and therefore reused, limiting the total costs of the extraction process (Usmani et al. 2023).

Overall, DES/NaDES have been identified as bioextractants representing an eco-sustainable alternative to organic solvents and toxic ionic liquids. The development of bioactive eutectic systems containing a bioactive molecule as one of the DES/NaDES components further increases the potential of these systems and opens a broad spectrum for future developments in the pharmaceutical and nutraceutical fields.

4.5 SCALING UP OF EXTRACTION PROTOCOLS

Scaling up a method constitutes the basis for translating academic knowledge development into business development (Belwal et al. 2020). While lab-scale extractions are carried out using a maximum of a few grams of the starting biomass and relatively limited solvent quantities, industrial extraction processes require thousands of solvent litres per hour to ensure productivity in line with the demand of the ever-increasing nutraceutical market (Gutiérrez et al. 2009). Generally, scaling-up studies are performed to ensure that operating conditions efficiently adapt to industrial needs and to ascertain comparable results to those achieved at a lab scale. The poor availability of appropriate pilot equipment, the costly initial investment, and the necessity of careful process optimization that guarantees a balance between energy expenditure and economic profit are some of the challenges that must be faced with special attention during the scale-up of an extraction protocol (Belwal et al. 2020). UAE and MAE have proven to be the most mature extraction technologies to be applied at an industrial level, offering superior yields in much less time than traditional methods (Belwal et al. 2020; Tapia-Quirós et al. 2024). Paniwnyk et al. evaluated the effect of scaling-up the UAE of rosmarinic and carnosic acids from *Rosmarinus officinalis* leaves (Paniwnyk et al. 2009). The authors conducted initial experiments using 1 kg of dried rosemary leaves in 20 L of ethanol at 38 kHz. Afterward, 6.25 kg of biomass was extracted with 125 L of solvent, employing an ultrasonic bath (25 kHz) equipped with six vertical transducers. Rosmarinic acid content was 66% superior to the one achieved by a thermal process at 35 °C for 60 min (Paniwnyk et al. 2009). The promising results presented in numerous papers strongly encourage the development of industrial reactors specifically designed to allow extraction process intensification with large solvent volumes (Belwal et al. 2020). GMC (http://www.gmariani.it/),

GIOTTI (http://www.giotti.it/), and PELLAS (http://pellasnature.com/) are some of the companies specializing in the extraction of food and pharmaceutical ingredients that use MAE and UAE industrial equipment for the production of polyphenol-rich nutraceuticals (Belwal et al. 2020).

4.6 CONCLUSIONS

This chapter presented the basic principles of obtaining bioactive compounds, such as polyphenols, for nutraceutical applications. It discussed traditional (maceration, Soxhlet, and distillation); emerging (SFE, SWE, MAE, UAE, PLE, EAE, PEF); and combined (e.g. PEF + PLE, USMAE) extraction techniques, highlighting their strengths and weaknesses, along with strategies that can be adopted to maximize the desired product yield. Every natural source has its own physical characteristics and peculiar chemical composition. Hence, the selection of the most appropriate extraction technique must be based on the type of starting biomass and the target bioactive molecules of interest. To reduce the environmental impact and simultaneously enhance the yield and quality of the final extracts, new green technologies and biosolvents have emerged. These technologies have been demonstrated to be a sustainable alternative to conventional extraction, even if further efforts are required to turn the acquired scientific knowledge into a business-level process. Continuous process monitoring and total adherence to good manufacturing practices (GMP) are fundamental for obtaining standardized nutraceutical products that always have the same desired characteristics, enabling consumers to trust these treatments with confidence.

4.7 DECLARATION OF COMPETING INTEREST

The authors declare that there is no conflict of interests.

4.8 FUNDING

This work was supported by MUR ITALY PRIN 2022 PNRR (Project No. P2022MWY3P). The authors also acknowledge the support from the Aliança para a Investigação e Ensino Superior, and the funding received by FCT, in the scope of the project UIDP/04378/2020 (DOI:10.54499/UIDP/04378/2020) and UIDB/04378/2020 DOI:10.54499/UIDB/04378/2020) of the Research Unit on Applied Molecular Biosciences – UCIBIO and the project LA/P/0140/2020 (DOI:10.54499/LA/P/0140/2020) of the Associate Laboratory Institute for Health and Bioeconomy – i4HB.

ACKNOWLEDGEMENTS

Anna Elisabetta Maccarronello is thankful for the PhD grant (D.M. no. 1061, PON R&I 2014/2020, 2020/2021 – Cycle 37) financed by the Italian Ministry for Education, University and Research MUR. Filipa Teixeira (2024.01202.BD) is thankful for her PhD grant financed by POPH-QREN and subsidized by the European Science Foundation and Ministério da Ciência, Tecnologia e Ensino Superior. Francisca

Rodrigues was supported by FCT – Fundação para a Ciência e a Tecnologia, I.P under the TENURE program [2023.14718.TENURE.016].

REFERENCES

Abbott, A.P., et al. 2004. Deep eutectic solvents formed between choline chloride and carboxylic acids: versatile alternatives to ionic liquids. *Journal of the American Chemical Society*. 126 (29): 9142–9147. https://doi.org/10.1021/ja048266j

Abbott, A.P., et al. 2017. Liquid pharmaceuticals formulation by eutectic formation. *Fluid Phase Equilibria*. 448: 2–8. https://doi.org/10.1016/j.fluid.2017.05.009

Al Dhabi, N.A., et al. 2017. Development and validation of ultrasound assisted solid-liquid extraction of phenolic compounds from waste spent coffee grounds. *Ultrasonics Sonochemistry*. 34: 206–213. https://doi.org/10.1016/j.ultsonch.2016.05.005

Alpat, U., et al. 2023. Optimization of propolis extraction with natural deep eutectic solvents using central composite design. *Phytochemistry Letters*. 58: 49–59. https://doi.org/10.1016/j.phytol.2023.09.013

Arshad, R.N., et al. 2020. Electrical systems for pulsed electric field applications in the food industry: An engineering perspective. *Trends in Food Science & Technology*. 104: 1–13. https://doi.org/10.1016/j.tifs.2020.07.008

Bagade, S.B., Patil, M. 2021. Recent advances in microwave assisted extraction of bioactive compounds from complex herbal samples: a review. *Critical Reviews in Analytical Chemistry*. 51(2): 138–149. https://doi.org/10.1080/10408347.2019.1686966

Barp, L., et al. 2023. Pressurized liquid extraction: a powerful tool to implement extraction and purification of food contaminants. *Foods*. 12 (10): 2017. https://doi.org/10.3390/foods12102017

Beconcini, D., et al. 2020. Antioxidant and anti-inflammatory properties of cherry extract: nanosystems based strategies to improve endothelial function and intestinal absorption. *Foods*. 9(2): 207. https://doi.org/10.3390/foods9020207

Belwal, T., et al. 2020. Recent advances in scaling-up of non-conventional extraction techniques: learning from successes and failures. *TrAC Trends in Analytical Chemistry*. 127: 115895. https://doi.org/10.1016/j.trac.2020.115895

Benítez-Correa, E., et al. 2023. Effect of choline chloride based deep eutectic solvents on polyphenols extraction from cocoa (*Theobroma cacao* L.) bean shells and antioxidant activity of extracts. *Current Research in Food Science*. 7: 100179. https://doi.org/10.1016/j.crfs.2023.100614

Boussetta, N., et al. 2012. Application of electrical treatments in alcoholic solvent for polyphenols extraction from grape seeds. *LWT Food Science and Technology*. 46 (1): 127–134. https://doi.org/10.1016/j.lwt.2011.10.016

Cannavacciuolo, C., et al. 2024. Critical analysis of green extraction techniques used for botanicals: Trends, priorities, and optimization strategies – A review. *Trends in Analytical Chemistry*. 173: 117627. https://doi.org/10.1016/j.trac.2024.117627

Capaldi, G., et al. 2024. New trends in extraction process intensification: Hybrid and sequential green technologies. *Industrial Crops and Products*. 209: 117906. https://doi.org/10.1016/j.indcrop.2023.117906

Carpentieri, S., et al. 2021. Emerging green techniques for the extraction of antioxidants from agri food by products as promising ingredients for the food industry. *Antioxidants*. 10 (9): 1417. https://doi.org/10.3390/antiox10091417

Chemat, F. and C. Boutekedjiret. 2015. Extraction // steam distillation. *Reference Module in Chemistry, Molecular Sciences and Chemical Engineering*. https://doi.org/10.1016/B978-0-12-409547-2.11557-4

Choi, Y.H., et al. 2011. Are natural deep eutectic solvents the missing link in understanding cellular metabolism and physiology? *Plant Physiology*. 156 (4): 1701–1705. https://doi.org/10.1104/pp.111.178426

Correia, A., et al. 2022. *Salicornia ramosissima*: A new green cosmetic ingredient with promising skin effects. *Antioxidants*. 11 (12): 2449. https://doi.org/10.3390/antiox11122449

Costa, A.S.G., et al. 2014. Optimization of antioxidants extraction from coffee silverskin, a roasting by product, having in view a sustainable process. *Industrial Crops and Products*. 53: 350–357. https://doi.org/10.1016/j.indcrop.2014.01.006

Craveiro, R., et al. 2016. Properties and thermal behavior of natural deep eutectic solvents. *Journal of Molecular Liquids*. 215: 534–540. https://doi.org/10.1016/j.molliq.2016.01.038

Cravotto, C., et al. 2022. Towards substitution of hexane as extraction solvent of food products and ingredients with no regrets. *Foods*. 11 (21): 3412. https://doi.org/10.3390/foods11213412

Cvjetko Bubalo, M., et al. 2018. New perspective in extraction of plant biologically active compounds by green solvents. *Food and Bioproducts Processing*. 109: 52–73. https://doi.org/10.1016/j.fbp.2018.03.001

Dai, Y.T., et al. 2013. Natural deep eutectic solvents as a new extraction media for phenolic metabolites in *Carthamus tinctorius* L. *Analytical Chemistry*. 85 (13): 6272–6278. https://doi.org/10.1021/ac400432p

Das, S., et al. 2021. Integrated strategies for enzyme assisted extraction of bioactive molecules: a review. *International Journal of Biological Macromolecules*. 191: 899–917. https://doi.org/10.1016/j.ijbiomac.2021.09.060

de Castro, M.D.L. and F. Priego-Capote. 2010. Soxhlet extraction: past and present panacea. *Journal of Chromatography A*. 1217 (16): 2383–2389. https://doi.org/10.1016/j.chroma.2009.11.027

Duan, L.J., D.J. Carrier and E.C. Clausen. 2004. Silymarin extraction from milk thistle using hot water. In *Proceedings of the Twenty-Fifth Symposium on Biotechnology for Fuels and Chemicals. Biotechnology for Fuels and Chemicals,* ed. Finkelstein, M., McMillan, J.D., Davison, B.H., Evans, B. Humana Press, Totowa, NJ. https://doi.org/10.1007/978-1-59259-837-3_46

Duarte, H., et al. 2024. Sustainable extraction of polyphenols from vine shoots using deep eutectic solvents: influence of the solvent, *Vitis* sp., and extraction technique. *Talanta*. 267: 125135. https://doi.org/10.1016/j.talanta.2023.125135

Dzah, C.S., et al. 2020. The effects of ultrasound assisted extraction on yield, antioxidant, anticancer and antimicrobial activity of polyphenol extracts: a review. *Food Bioscience*. 35: 100547. https://doi.org/10.1016/j.fbio.2020.100547

Fernandes, A., et al. 2023. Polyphenol dietary fiber conjugates from fruits and vegetables: nature and biological fate in a food and nutrition perspective. *Foods*. 12 (5): 1052. https://doi.org/10.3390/foods12051052

Ferreira, A.S., et al. 2022. New perspectives on the sustainable employment of chestnut shells as active ingredient against oral mucositis: a first screening. *International Journal of Molecular Sciences*. 23 (23): 14956. https://doi.org/10.3390/ijms232314956

Garcia Vaquero, M., et al. 2020. Chapter 7 – Conventional extraction techniques: solvent extraction. In *Sustainable Seaweed Technologies. Advances in Green and Sustainable Chemistry*, ed. Torres, M.D., Kraan, S., Dominguez, H., 171–189. https://doi.org/10.1016/B978-0-12-817943-7.00006-8

Goldsmith, C.D., et al. 2018. Ultrasound increases the aqueous extraction of phenolic compounds with high antioxidant activity from olive pomace. *LWT*. 89: 284–290. https://doi.org/10.1016/j.lwt.2017.10.065

González, J.A., et al. 2022. A comparison study between ultrasound-assisted and enzyme-assisted extraction of anthocyanins from blackcurrant (*Ribes nigrum* L.). *Food Chemistry: X*. 13: 100192. https://doi.org/10.1016/j.fochx.2021.100192

Gutiérrez, M.C., et al. 2009. Freeze drying of aqueous solutions of deep eutectic solvents: a suitable approach to deep eutectic suspensions of self assembled structures. *Langmuir*. 25 (10): 5509–5515. https://doi.org/10.1021/la900552b

Habeebullah, S.F.K., et al. 2020. Enzyme-assisted extraction of bioactive compounds from brown seaweeds and characterization. *Journal of Applied Phycology*. 32 (1): 615–629. https://doi.org/10.1007/s10811-019-01906-6

Hammond, O.S., et al. 2017. Resilience of malic acid natural deep eutectic solvent nanostructure to solidification and hydration. *Journal of Physical Chemistry B*. 121 (31): 7473–7483. https://doi.org/10.1021/acs.jpcb.7b05454

Hogervorst Cvejić, J., et al. 2017. Chapter 7 – Polyphenols. In *Nutraceutical and Functional Food Components*, ed. Galanakis, C.M., 203–258. Academic Press. https://doi.org/10.1016/B978-0-12-805257-0.00007-7

Jerković, I., et al. 2015. GC-FID/MS profiling of supercritical CO2 extracts of peels from *Citrus aurantium, C. sinensis* cv. *Washington navel, C. sinensis* cv. *Tarocco* and *C. sinensis cv.* Doppio Sanguigno from Dubrovnik area (Croatia). *Natural Product Communications*. 10(7): 1315–1318.

Joshi, D.R. and N. Adhikari. 2019. An overview on common organic solvents and their toxicity. *Journal of Pharmaceutical Research International*. 28 (3): 1–18. https://doi.org/10.9734/jpri/2019/v28i330203

Kadam, S.U., et al. 2013. Application of novel extraction technologies for bioactives from marine algae. *Journal of Agricultural and Food Chemistry*. 61 (20): 4667–4675. https://doi.org/10.1021/jf400819p

Kashyap, M., et al. 2024. High voltage pulsed electric field and electroporation technologies for algal biomass processing. *Journal of Applied Phycology*. 36 (1): 29–39. https://doi.org/10.1007/s10811-023-03145-2

Kumar, K., et al. 2017. Food waste: a potential bioresource for extraction of nutraceuticals and bioactive compounds. *Bioresources and Bioprocessing*. 4: 18. https://doi.org/10.1186/s40643-017-0148-6

Kumar, K., et al. 2021. Ultrasound assisted extraction (UAE) of bioactive compounds from fruit and vegetable processing by-products: a review. *Ultrasonics Sonochemistry*. 70: 105325. https://doi.org/10.1016/j.ultsonch.2020.105325

Lameirão, F., et al. 2020. Green sustainable recovery of phenolic and antioxidant compounds from industrial chestnut shells using ultrasound assisted extraction: optimization and evaluation of biological activities *in vitro*. *Antioxidants*. 9(3): 267. https://doi.org/10.3390/antiox9030267

Le, N.T., et al. 2023. Green solvents based rutin extraction from *Sophora japonica* L. *Journal of Applied Research on Medicinal and Aromatic Plants*. 36: 100508. https://doi.org/10.1016/j.jarmap.2023.100508

Lévêque, J.M. and G. Cravotto. 2006. Microwaves, power ultrasound, and ionic liquids: a new synergy in green organic synthesis. *Chimia*. 60(6): 313–320. https://doi.org/10.2533/000942906777836255

López-Hortas, L., et al. 2022. Supercritical fluid extraction as a suitable technology to recover bioactive compounds from flowers. *The Journal of Supercritical Fluids*. 188: 105652. https://doi.org/10.1016/j.supflu.2022.105652

Łubek-Nguyen, A., et al. 2022. Application of enzyme-assisted extraction for the recovery of natural bioactive compounds for nutraceutical and pharmaceutical applications. *Applied Sciences*. 12(7): 3232. https://doi.org/10.3390/app12073232

Luo, Y., et al. 2017. Optimization of simultaneous microwave/ultrasonic assisted extraction of phenolic compounds from walnut flour using response surface methodology. *Pharmaceutical Biology*. 55(1): 1999–2004. https://doi.org/10.1080/13880209.2017.1347189

Maccarronello, A.E., et al. 2024a. From waste to bioactive compounds: a response surface methodology approach to extract antioxidants from *Pistacia vera* shells for postprandial hyperglycaemia management. *Food Chemistry*. 443: 138504. https://doi.org/10.1016/j.foodchem.2024.138504

Maccarronello, A.E., et al. 2024b. Unlocking the nutraceutical potential of *Corylus avellana* L. shells: microwave-assisted extraction of phytochemicals with antiradical and anti-diabetic properties. *Journal of Agricultural and Food Chemistry*. 104: 9472–9485. https://doi.org/10.1002/jsfa.13770

Macedo, C., et al. 2023. Insights into the polyphenols extraction from *Actinidia arguta* fruit (kiwiberry): a source of pro-healthy compounds. *Scientia Horticulturae*. 313: 111910. https://doi.org/10.1016/j.scienta.2023.111910

Macedo, G.A., et al. 2021. Integrated microwave and enzyme assisted extraction of phenolic compounds from olive pomace. *LWT*. 138: 110621. https://doi.org/10.1016/j.lwt.2020.110621

Mandalari, G., et al. 2022. Pistachio nuts (*Pistacia vera* L.): production, nutrients, bioactives and novel health effects. *Plants-Basel*. 11(1). https://doi.org/10.3390/plants11010018

Maravić, N., et al. 2022. From agricultural waste to antioxidant rich extracts: green techniques in extraction of polyphenols from sugar beet leaves. *Sustainable Chemistry and Pharmacy* 28: 100728. https://doi.org/10.1016/j.scp.2022.100728

Matera, R., et al. 2023. Plant essential oils as healthy functional ingredients of nutraceuticals and diet supplements: a review. *Molecules*. 28(2): 901. https://doi.org/10.3390/molecules28020901

Morales de la Peña, et al. 2023. Application of moderate intensity pulsed electric fields in red prickly pears and soymilk to develop a plant-based beverage with potential health related benefits. *Innovative Food Science & Emerging Technologies*. 88: 103421. https://doi.org/10.1016/j.ifset.2023.103421

Morrison, H.G., et al. 2009. Characterization of thermal behavior of deep eutectic solvents and their potential as drug solubilization vehicles. *International Journal of Pharmaceutics*. 378(1–2): 136–139. https://doi.org/10.1016/j.ijpharm.2009.05.039

Muccilli, V., et al. 2023. Green^3: a green extraction of green additives for green plastics. *Heliyon*. 10(2): e24469. https://doi.org/10.1016/j.heliyon.2024.e24469

Naliyadhara, N., et al. 2022. Pulsed electric field (PEF): avant garde extraction escalation technology in food industry. *Trends in Food Science & Technology*. 122: 238–255. https://doi.org/10.1016/j.tifs.2022.02.019

Naviglio, D., et al. 2019. Rapid solid liquid dynamic extraction (RSLDE): a powerful and greener alternative to the latest solid liquid extraction techniques. *Foods*. 8(7): 245. https://doi.org/10.3390/foods8070245

Nystedt, H.L., et al. 2023. Neutral natural deep eutectic solvents as anti-biofilm agents. *Biofilm*. 5: 100114. https://doi.org/10.1016/j.bioflm.2023.100114

Ojha, K.S., et al. 2020. Ultrasound technology for the extraction of biologically active molecules from plant, animal and marine sources. *TRAC Trends in Analytical Chemistry*. 122: 115663. https://doi.org/10.1016/j.trac.2019.115663

Ordoñez-Torres, A., et al. 2021. Ultrasound microwave assisted extraction of polyphenolic compounds from Mexican "Ataulfo" mango peels: antioxidant potential and identification by HPLC/ESI/MS. *Phytochemical Analysis*. 32 (4): 495–502. https://doi.org/10.1002/pca.2997

Pagano, I., et al. 2021. Green non-conventional techniques for the extraction of polyphenols from agricultural food by products: a review. *Journal of Chromatography A*. 1651: 462295. https://doi.org/10.1016/j.chroma.2021.462295

Pandey, K.B. and S.I. Rizvi. 2009. Plant polyphenols as dietary antioxidants in human health and disease. *Oxidative Medicine and Cellular Longevity* 2 (5): 270–278. https://doi.org/10.4161/oxim.2.5.9498

Paniwnyk, L., et al. 2009. The enhancement and scale up of the extraction of antioxidants from *Rosmarinus officinalis* using ultrasound. *Ultrasonics Sonochemistry* 16 (2): 287–292. https://doi.org/10.1016/j.ultsonch.2008.06.007

Panja, P. 2018. Green extraction methods of food polyphenols from vegetable materials. *Current Opinion in Food Science* 23: 173–182. https://doi.org/10.1016/j.cofs.2017.11.012

Pena-Pereira, F. and J. Namiesnik. 2014. Ionic liquids and deep eutectic mixtures: sustainable solvents for extraction processes. *Chemsuschem.* 7 (7): 1784–1800. https://doi.org/10.1002/cssc.201301192

Perez-Vazquez, A., et al. 2023. Pressurized liquid extraction for the recovery of bioactive compounds from seaweeds for food industry application: A review. *Antioxidants* 12 (3): 612. https://doi.org/10.3390/antiox12030612

Pham, H.N.T. 2017. Chapter 4 – Recovering bioactive compounds from fruit and vegetable wastes. In *Recovering Bioactive Compounds from Agricultural Wastes. Agricultural Wastes*, ed. Nguyen, V.T., 81–99. https://doi.org/10.1002/9781119168850.ch4

Picot-Allain, C., et al. 2021. Conventional *versus* green extraction techniques – a comparative perspective. *Current Opinion in Food Science.* 40: 144–156. https://doi.org/10.1016/j.cofs.2021.02.009

Pinto, D., et al. 2020. Valorisation of underexploited *Castanea sativa* shells bioactive compounds recovered by supercritical fluid extraction with CO2: a response surface methodology approach. *Journal of CO_2 Utilization.* 40: 101194. https://doi.org/10.1016/j.jcou.2020.101194

Pinto, D., de la Luz Cádiz-Gurrea, M., et al. 2021a. *Castanea sativa* shells: a review on phytochemical composition, bioactivity and waste management approaches for industrial valorization. *Food Research International.* 144: 110364. https://doi.org/10.1016/j.foodres.2021.110364

Pinto, D., Silva, A.M., et al. 2021. Microwave assisted extraction as a green technology approach to recover polyphenols from *Castanea sativa* shells. *ACS Food Science & Technology.* 1(2): 229–241. https://doi.org/10.1021/acsfoodscitech.0c00055

Pinto, D., Vieira, E.F., et al. 2021. Optimizing the extraction of phenolic antioxidants from chestnut shells by subcritical water extraction using response surface methodology. *Food Chemistry* 334: 127521. https://doi.org/10.1016/j.foodchem.2020.127521

Poojary, M.M., et al. 2020. Chapter 4 – Pulsed electric field (PEF) as an efficient technology for food additives and nutraceuticals development. In *Pulsed Electric Fields to Obtain Healthier and Sustainable Food for Tomorrow*, ed. Barba, F.J., Parniakov, O., Wiktor, A., 65–99. Academic Press. https://doi.org/10.1016/B978-0-12-816402-0.00004-5

Putra, N.R., et al. 2023. Advancements and challenges in green extraction techniques for Indonesian natural products: A review. *South African Journal of Chemical Engineering.* 46: 88–98. https://doi.org/10.1016/j.sajce.2023.08.002

Quan, N., et al. 2023. Ultrasound microwave combined extraction of novel polysaccharide fractions from *Lycium barbarum* leaves and their *in vitro* hypoglycemic and antioxidant activities. *Molecules.* 28 (9): 3880. https://doi.org/10.3390/molecules28093880.

Razak, A.M., et al. 2023. A subcritical water extract of soil grown *Zingiber officinale* Roscoe: comparative analysis of antioxidant and anti-inflammatory effects and evaluation of bioactive metabolites. *Frontiers in Pharmacology.* 14: 2023. https://doi.org/10.3389/fphar.2023.1006265

Riciputi, Y., et al. 2018. Establishment of ultrasound-assisted extraction of phenolic compounds from industrial potato by-products using response surface methodology. *Food Chemistry.* 269: 258–263. https://doi.org/10.1016/j.foodchem.2018.06.154

Rodríguez García, S.L. and V. Raghavan. 2022. Green extraction techniques from fruit and vegetable waste to obtain bioactive compounds – a review. *Critical Reviews in Food Science and Nutrition.* 62(23): 6446–6466. https://doi.org/10.1080/10408398.2021.1901651

Ruesgas Ramón, M., et al. 2017. Application of deep eutectic solvents (DES) for phenolic compounds extraction: Overview, challenges, and opportunities. *Journal of Agricultural and Food Chemistry.* 65 (18): 3591–3601. https://doi.org/10.1021/acs.jafc.7b01054

Sale, A.J.H. and W.A. Hamilton. 1968. Effects of high electric fields on micro-organisms: III. Lysis of erythrocytes and protoplasts. *Biochimica et Biophysica Acta (BBA) – Biomembranes.* 163 (1): 37–43. https://doi.org/10.1016/0005-2736(68)90030-8

Salgado-Ramos, M., et al. 2023. Winemaking derived by-products: in-depth characterization and sustainable, advanced pulsed electric field (PEF) processing to a zero-waste-based approach. *Journal of Environmental Chemical Engineering.* 11 (5): 110535. https://doi.org/10.1016/j.jece.2023.110535

Secretan, P.H., et al. 2023. 3D printed, liquid-filled capsules of concentrated and stabilized polyphenol epigallocatechin gallate, developed in a clinical trial. *Antioxidants.* 12 (2): 424. https://doi.org/10.3390/antiox12020424

Sillero, L., et al. 2020. Simultaneous microwave–ultrasound-assisted extraction of bioactive compounds from bark. *Chemical Engineering and Processing – Process Intensification.* 156: 108100. https://doi.org/10.1016/j.cep.2020.108100

Silva, A.M., et al. 2021. An insight into kiwiberry leaf valorization: phenolic composition, bioactivity and health benefits. *Molecules.* 26 (8): 2314. https://doi.org/10.3390/molecules26082314

Silva, A.M., Almeida, A., *et al.* 2022a. Insights into the 3D *in vitro* permeability and *in vivo* antioxidant protective effects of kiwiberry leaf extract: a step forward to human nutraceutical use. *International Journal of Molecular Sciences.* 23 (22). https://doi.org/10.3390/ijms232213509

Silva, A.M., Garcia, J., et al. 2022. Eco-friendly insights on kiwiberry leaves valorization through *in vitro* and *in vivo* studies. *Industrial Crops and Products.* 184: 115090. https://doi.org/10.1016/j.indcrop.2022.115090

Silva, A.M, Luís, A.S., et al. 2022. Influence of temperature on the subcritical water extraction of *Actinidia arguta* leaves: a screening of pro-healthy compounds. *Sustainable Chemistry and Pharmacy.* 25: 100593. https://doi.org/10.1016/j.scp.2021.100593

Silva, A.M., Pinto, D., et al. 2022. Valorization of kiwiberry leaves recovered by ultrasound-assisted extraction for skin application: a response surface methodology approach. *Antioxidants.* 11 (4): 763. https://doi.org/10.3390/antiox11040763

Singla, R.K., et al. 2019. Natural polyphenols: chemical classification, definition of classes, subcategories, and structures. *Journal of AOAC International.* 102 (5): 1397–1400. https://doi.org/10.1093/jaoac/102.5.1397

Tapia-Quirós, P., et al. 2024. Microwave-assisted extraction with natural deep eutectic solvents for polyphenol recovery from agrifood waste: mature for scaling up? *Science of the Total Environment.* 912: 168716. https://doi.org/10.1016/j.scitotenv.2023.168716

Tomšik, A., et al. 2017. Subcritical water extraction of wild garlic (*Allium ursinum* L.) and process optimization by response surface methodology. *The Journal of Supercritical Fluids.* 128: 79–88. https://doi.org/10.1016/j.supflu.2017.05.012

Tsvetov, N., et al. 2023. Application of natural deep eutectic solvents for extraction of bioactive components from *Rhodiola rosea* (L.). *Molecules.* 28 (2): 912. https://doi.org/10.3390/molecules28020912

Turan, O., et al. 2024. Microwave-assisted extraction of pectin from orange peel using deep eutectic solvents. *Sustainable Chemistry and Pharmacy.* 37: 101352. https://doi.org/10.1016/j.scp.2023.101352

Usmani, Z., et al. 2023. Biobased natural deep eutectic system as versatile solvents: structure, interaction and advanced applications. *Science of the Total Environment.* 881: 163002. https://doi.org/10.1016/j.scitotenv.2023.163002

Uwineza, P.A. and A. Waśkiewicz. 2020. Recent advances in supercritical fluid extraction of natural bioactive compounds from natural plant materials. *Molecules.* 25 (17): 3847. https://doi.org/10.3390/molecules25173847

Vannuchi, N., et al. 2022. High-performance extraction process of anthocyanins from jussara (*Euterpe edulis*) using deep eutectic solvents. *Processes*. 10 (3): 615. https://doi.org/10.3390/pr10030615

Vardanega, R., et al. 2023. Valorization of granadilla waste (*Passiflora ligularis*, Juss.) by sequential green extraction processes based on pressurized fluids to obtain bioactive compounds. *The Journal of Supercritical Fluids*. 194: 105833. https://doi.org/10.1016/j.supflu.2022.105833

Wang, L. and C.L. Weller. 2006. Recent advances in extraction of nutraceuticals from plants. *Trends in Food Science & Technology*. 17 (6): 300–312. https://doi.org/10.1016/j.tifs.2005.12.004

Wianowska, D. and M. Wiśniewski. 2015. Simplified procedure of silymarin extraction from *Silybum marianum* L. Gaertner. *Journal of Chromatographic Science*. 53 (2): 366–372. https://doi.org/10.1093/chromsci/bmu049

Zhang, J.H., et al. 2024. Oil control and pore refining effects of *Fomes officinalis* extract based on green deep eutectic solvent. *Journal of Molecular Liquids*. 394: 123623. https://doi.org/10.1016/j.molliq.2023.123623

Zhou, J.J., et al. 2022. Pulsed electric fields (PEF), pressurized liquid extraction (PLE) and combined PEF plus PLE process evaluation: effects on *Spirulina* microstructure, biomolecules recovery and triple TOF LC–MS/MS polyphenol composition. *Innovative Food Science & Emerging Technologies*. 77: 102989 https://doi.org/10.1016/j.ifset.2022.102989

5 Molecular Mechanism of Action of Polyphenol Nutraceuticals Targeting Human Disease

Rajesh Saini, Madhvendra Shukla, Awadhesh Kumar Mishra, Aparna Singh, and Kavindra Nath Tiwari

5.1 INTRODUCTION

Polyphenols are a diverse group of plant-derived secondary metabolites defined by the presence of multiple phenolic structures. They are abundantly found in fruits, vegetables, tea, coffee, wine, cocoa, and various herbs. These compounds exhibit a broad spectrum of biological activities, most notably antioxidant, anti-inflammatory, and anticarcinogenic effects. As such, polyphenols have garnered significant scientific attention for their protective roles against the development and progression of chronic diseases, including cancer, cardiovascular disorders, diabetes, and neurodegenerative conditions such as Alzheimer's and Parkinson's disease (Scalbert et al., 2005). The significance of polyphenols and other nutraceuticals in human health stems from their ability to modulate key cellular and molecular processes. These include neutralization of reactive oxygen species (ROS), inhibition of pro-inflammatory mediators (such as NF-κB, COX-2, and TNF-α), modulation of cell cycle regulators, and activation of detoxifying enzymes. Through these mechanisms, polyphenols contribute to the maintenance of cellular homeostasis and the prevention of pathological conditions (Pandey & Rizvi, 2009; Liu, 2004). Epidemiological studies and clinical trials have consistently demonstrated that diets rich in polyphenols, like those found in Mediterranean and Asian diets, are associated with reduced risk of chronic diseases and improved life expectancy. For example, flavonoids in green tea and red wine have been linked to cardiovascular health, while resveratrol, curcumin, and quercetin have shown anticancer and neuroprotective effects in both in vitro and in vivo models.

To harness the full therapeutic potential of these compounds, however, it is essential to understand their molecular mechanisms of action. Knowledge at this level allows researchers and clinicians to pinpoint specific molecular targets (such as enzymes, receptors, or transcription factors); signalling pathways (e.g., MAPK, PI3K/Akt, Nrf2); and gene expression profiles affected by polyphenols. This mechanistic

DOI: 10.1201/9781003588504-5

insight enables the rational design of nutraceutical formulations with enhanced efficacy, better bioavailability, and few side effects (Williamson & Manach, 2005). Furthermore, it supports the development of precision nutrition approaches where dietary recommendations or nutraceutical interventions can be tailored based on individual genetic profiles, lifestyle factors, and disease risk. A thorough understanding of their molecular interactions not only reinforces their role as health-promoting agents but also paves the way for innovative therapeutic applications in integrative and personalized medicine.

5.2 HEALTH BENEFITS OF POLYPHENOLS

The variety of polyphenols, such as flavonoids, flavonols (e.g., quercetin) and flavan-3-ols (e.g., catechins), and phenolics (caffeic acid and ferulic acid), possess several health benefits. Polyphenols represent a promising avenue in the prevention and management of chronic diseases (Figure 5.1). Quercetin, a flavonol found in onions, apples, and berries, exhibits antioxidant, anti-inflammatory, and antiviral activities. It scavenges ROS, inhibits lipid peroxidation, and modulates inflammatory signalling pathways like NF-κB, MAPK, and PI3K/Akt, which are crucial for regulating inflammation, apoptosis, and cell proliferation (Boots et al., 2008). Quercetin has therapeutic potential for chronic diseases, such as reducing blood pressure, improving endothelial function, alleviating neuroinflammation, and inhibiting cancer cell proliferation.

Another polyphenol, catechins, particularly epigallocatechin gallate (EGCG) from green tea, are powerful antioxidants. EGCG modulates oxidative stress,

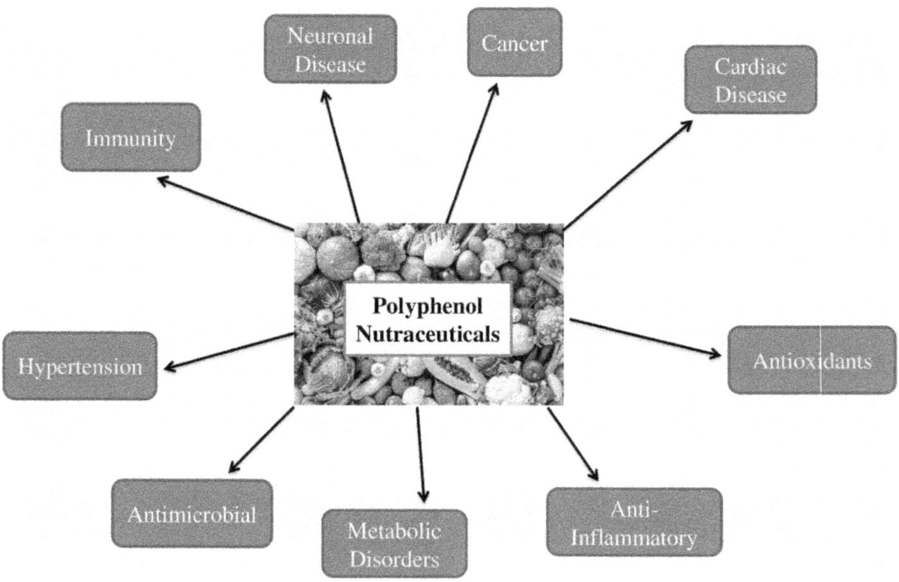

FIGURE 5.1 Health benefits of polyphenols.

inflammation, and detoxification by activating the Nrf2 pathway and inhibiting cyclooxygenase 2 (COX-2) and inducible nitric oxide synthesis (iNOS) (Yang et al., 2009). Catechins are beneficial for cardiovascular health by improving lipid profiles and vascular function, while also exhibiting anticancer properties and neuroprotection in diseases like Alzheimer's.

Phenolic acids are a group of polyphenolic compounds commonly found in fruits, vegetables, whole grains, and seeds. Two notable examples are caffeic acid and ferulic acid. Caffeic acid is found in coffee, fruits, and vegetables. It has antioxidant, anti-inflammatory, and anticancer properties by scavenging free radicals, modulating inflammatory pathways (e.g., NF-κB), and inhibiting cancer cell proliferation and promoting apoptosis (Choi et al., 2021). Ferulic acid, abundant in whole grains, oats, and rice, is known for its antioxidant activity. It protects against oxidative stress, enhances vascular function, and has shown potential in preventing ultraviolet (UV)-induced skin damage and promoting collagen synthesis (Kusuma et al., 2022). Stilbenes are a class of polyphenolic compounds found in plants, with resveratrol being the most well-known example. It is primarily found in grapes, red wine, peanuts, and berries. Resveratrol has antioxidant, anti-inflammatory, and anticancer properties, primarily by activating the SIRT1 pathway and modulating AMPK and NF-κB signalling. These actions contribute to its cardioprotective and neuroprotective effects. Recent studies also suggest its potential in promoting longevity and mitigating the effects of ageing (Zhang et al., 2020a). Lignans are a group of plant compounds that have antioxidant, anti-inflammatory, and anticancer effects. Secoisolariciresinol is one of the most-studied lignans, found in flaxseeds, sesame seeds, and whole grains. It has been shown to exert cardioprotective effects, improve lipid profiles, and reduce cancer cell proliferation. Lignans are converted into enterolignans by gut microbiota, which may further contribute to their health benefits (Teng et al., 2020).

5.3 MOLECULAR MECHANISMS OF ACTION

5.3.1 Antioxidant Activity

5.3.1.1 Free Radical Scavenging

Polyphenols exhibit strong antioxidant activity by scavenging free radicals, particularly ROS and reactive nitrogen species (RNS). These free radicals are highly reactive molecules that can damage cellular components like DNA, lipids, and proteins, leading to diseases such as cancer, cardiovascular diseases, and neurodegenerative disorders. Polyphenols neutralize these radicals by donating electrons, thus preventing oxidative stress and cellular damage (Gao et al., 2020).

5.3.1.2 Activation of Endogenous Antioxidant Enzymes via the Nrf2 Pathway

The Nrf2 signalling pathway, which controls the expression of genes encoding antioxidant enzymes such as glutathione peroxidase and superoxide dismutase, is activated by polyphenols. Nrf2 activation promotes cellular defence mechanisms against oxidative stress by inducing the transcription of cytoprotective genes. This process has therapeutic potential in managing oxidative stress–related diseases, such as Alzheimer's and Parkinson's disease (Zhao et al., 2021a).

5.3.2 Anti-Inflammatory Effects

5.3.2.1 Inhibition of NF-κB Signalling

One important regulator of inflammation, NF-κB (nuclear factor kappa-light-chain-enhancer of activated B cells) signalling, is inhibited by polyphenols. NF-κB activation leads to the expression of pro-inflammatory cytokines and adhesion molecules that contribute to chronic inflammation. Polyphenols reduce the phosphorylation and degradation of IκB proteins, which prevent NF-κB from translocating to the nucleus and promoting inflammation (Liu et al., 2020).

5.3.2.2 Suppression of Pro-Inflammatory Cytokines (IL-6, TNF-α)

Polyphenols suppress the expression of key pro-inflammatory cytokines like IL-6 and TNF-α, which are involved in inflammation and immune responses. Elevated levels of these cytokines are linked to several chronic conditions, including rheumatoid arthritis, cardiovascular diseases, and cancer. By down-regulating cytokine production, polyphenols help reduce inflammation and promote immune regulation (Santos et al., 2021a).

5.3.3 Epigenetic Modifications

DNA methylation, a crucial epigenetic alteration that controls gene expression, is influenced by polyphenols. A methyl group is added to the DNA molecule during DNA methylation, which usually silences gene expression. Polyphenols can alter the DNA methylation patterns of tumour-suppressor genes, reactivating their expression and inhibiting cancer cell growth (Cheng et al., 2020). This modification may also help in the prevention of other diseases associated with epigenetic alterations. It modulateshistone acetylation and deacetylation by inhibiting histone deacetylases (HDACs). This action enhances histone acetylation, which leads to an open chromatin structure and promotes gene transcription. By modulating histone modifications, polyphenols influence gene expression related to cell cycle regulation, apoptosis, and inflammation (Wu et al., 2021).

Polyphenols can regulate miRNA expression, impacting the post-transcriptional regulation of gene expression. This regulation influences several biological processes, including cancer progression, immune responses, and stress adaptations (Yang et al., 2020).

5.3.4 Signal Transduction Modulation

Polyphenols modulate various signal transduction pathways that are critical for regulating cell survival, growth, and metabolism. They have been shown to inhibit the mitogen-activated protein kinase (MAPK) signalling pathway, which plays a key role in inflammation and cancer cell proliferation (Li et al., 2021a) (Figure 5.2). Additionally, polyphenols can activate the phosphoinositide 3-kinase/protein kinase B (PI3K/

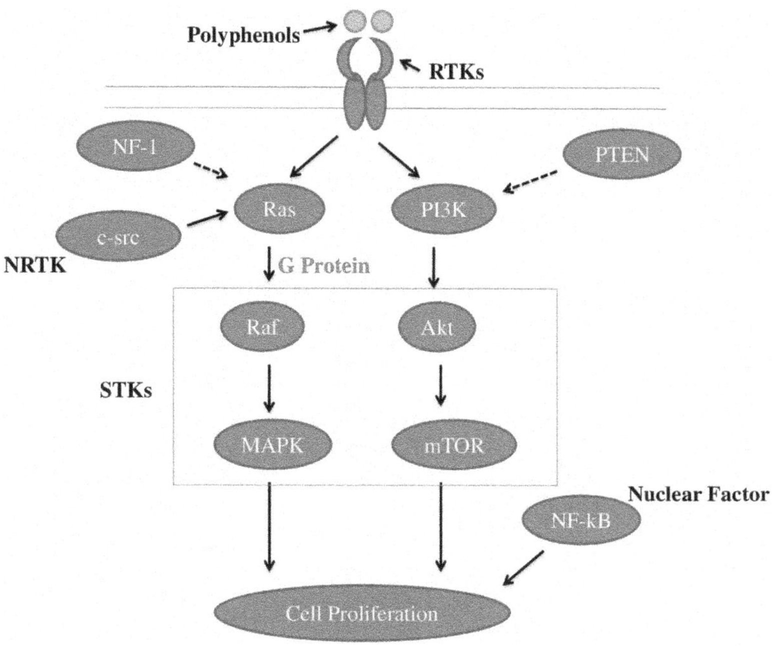

FIGURE 5.2 Key proliferation signalling pathways include the PI3K/Akt/mTOR and Ras/Raf/MAPK cascades. Upon binding of polyphenols to receptor tyrosine kinases (RTKs), these pathways are activated. Specifically, activation of RTKs leads to the stimulation of Ras/Raf/MAPK and/or PI3K/Akt/mTOR pathways. The kinases Raf, MAPK, Akt, and mTOR are all serine/threonine-specific protein kinases (STKs). Additionally, the intracellular tyrosine kinase c-Src can activate the Ras/Raf/MAPK pathway. The NF-κB also plays a significant role in regulating cell proliferation. Solid arrows represent activation, while dashed arrows indicate suppression.

Akt) pathway, thereby promoting cell survival and inhibiting apoptosis (Peng et al., 2020). They also stimulate AMP-activated protein kinase (AMPK) activity, a major regulator of energy homeostasis and metabolism (Li et al., 2021b). Furthermore, polyphenols inhibit the Wnt/β-catenin signalling pathway, which is involved in cell differentiation and cancer progression (Peng et al., 2020). Through these mechanisms, polyphenols exert significant protective effects against various pathological conditions.

Polyphenols interact with hormone receptors such as oestrogen (ER) and androgen (AR) receptors. By modulating these receptors, polyphenols influence gene expression related to reproductive health, cancer progression, and metabolic disorders (Peng et al., 2020).

5.3.5 Cell Cycle Arrest and Apoptosis

Polyphenols play a crucial role in inducing tumour suppressor pathways such as p53 and p21, leading to cell cycle arrest and inhibiting the proliferation of cancer cells.

Activation of p53 by polyphenols triggers DNA damage responses, apoptosis, and senescence, all of which are essential for cancer prevention (Cao et al., 2021). In addition, polyphenols activate caspases, key enzymes responsible for initiating apoptosis, and modulate the expression of Bcl-2 family proteins that regulate cell survival and death. By up-regulating pro-apoptotic proteins like Bax and down-regulating anti-apoptotic proteins like Bcl-2, polyphenols promote the apoptotic elimination of cancer cells (Liu et al., 2020).

5.3.6 Antimicrobial and Antiviral Actions

Polyphenols exhibit strong antimicrobial properties by inhibiting pathogen adhesion to host cells and disrupting replication processes, which are vital for preventing bacterial, viral, and fungal infections. They achieve this by interacting with microbial membranes and enzymes, thereby impairing pathogen replication and virulence (Zhang et al., 2020b). Additionally, polyphenols modulate host-pathogen interaction pathways by enhancing immune responses and interfering with pathogen virulence factors, ultimately strengthening the host's defence mechanisms against infections (Tang et al., 2021).

5.4 ROLE IN SPECIFIC HUMAN DISEASES

5.4.1 Cancer (Breast, Colon, Prostate, etc.)

Targeting several facets of cancer progression, including apoptosis induction, cell proliferation inhibition, and modulation of important signalling pathways implicated in carcinogenesis, polyphenols demonstrate strong anticancer effects. In breast cancer, polyphenols like EGCG from green tea inhibit ER signalling, reduce HER2 expression, and trigger apoptosis through caspase activation (Hosseini et al., 2020). In colon cancer, polyphenols from fruits such as berries and citrus effectively inhibit colorectal cancer cell proliferation, induce cell cycle arrest, and modulate gut microbiota, thereby influencing cancer development (Zhao et al., 2021b). Similarly, in prostate cancer, polyphenols like resveratrol suppress cell proliferation by modulating AR signalling and promoting apoptosis pathways (Valko et al., 2020).

5.4.2 Cardiovascular Diseases

Polyphenols improve endothelial function, lower inflammation, and decrease oxidative stress, all of which lower the risk of cardiovascular diseases (CVDs). Their antioxidant activity neutralizes ROS and decreases oxidative modification of low-density lipoprotein (LDL) cholesterol, a major contributor to atherosclerosis (Zhang et al., 2020c). Additionally, polyphenols exert anti-inflammatory effects by suppressing pro-inflammatory cytokines and inhibiting pathways like NF-κB, thereby reducing arterial inflammation (Santos et al., 2021b). They also enhance vascular function by improving endothelial performance and stimulating nitric oxide production, which helps maintain vascular health and prevent hypertension (González-Correa & Rodríguez-Rodríguez, 2021). In neurodegenerative disorders like Alzheimer's and Parkinson's disease, polyphenols provide neuroprotection through antioxidant, anti-inflammatory, and epigenetic mechanisms. In Alzheimer's disease, polyphenols such as resveratrol and flavonoids reduce amyloid-beta (Aβ) plaque formation and tau hyperphosphorylation while promoting neurogenesis and cognitive function (Sánchez-González et al., 2021). In Parkinson's disease, they protect dopaminergic neurons by mitigating oxidative stress and inflammation and by modulating neurotrophic factors like brain-derived neurotrophic factor (BDNF), crucial for neuronal survival and repair (Manczak et al., 2020).

Polyphenols help manage metabolic disorders by improving insulin sensitivity, reducing inflammation, and promoting lipid metabolism. In diabetes, they enhance glucose metabolism by activating AMPK, increasing insulin sensitivity, and modulating glucose absorption in the intestines, while also lowering oxidative stress and inflammation, which are major contributors to diabetic complications (Ganesan et al., 2021). In obesity, polyphenols from sources like berries and green tea inhibit adipogenesis, promote fat oxidation, and regulate the gut microbiota, which plays a critical role in the development of obesity and metabolic syndrome (Chandra &Wali, 2020).

5.4.3 Autoimmune and Inflammatory Diseases

Polyphenols are good at controlling immune responses and reducing chronic inflammation linked to autoimmune disorders. In rheumatoid arthritis, polyphenols like curcumin and resveratrol reduce the production of inflammatory cytokines such as TNF-α and IL-6, leading to decreased joint inflammation and pain (Kim & Lee, 2021). In inflammatory bowel disease (IBD), polyphenols help regulate gut inflammation by modulating the immune response and gut microbiota composition, while also protecting against oxidative damage in the gut (Zhao et al., 2020c).

5.4.4 Infectious Diseases and Immune Modulation

Polyphenols have antimicrobial and immunomodulatory effects, helping to combat infections and regulate immune responses. They exhibit antibacterial, antifungal, and antiviral activities, particularly those derived from tea, wine, and berries, by interfering with pathogen adhesion, biofilm formation, and replication (Zhang et al., 2020c). Additionally, polyphenols enhance immune function by boosting phagocytic activity, stimulating T-cell responses, and regulating pro-inflammatory cytokines, thereby strengthening the body's defence mechanisms against viral and bacterial infections (Li et al., 2021b).

5.5 CLINICAL EVIDENCE AND HUMAN TRIALS

Polyphenols have been extensively studied in clinical trials for their health benefits. Green tea catechins (EGCG) have shown positive effects in breast cancer patients by reducing cancer recurrence and enhancing immune responses (Liu et al., 2019). A flavonoid-rich diet, including citrus and dark chocolate, significantly reduced blood pressure and improved vascular endothelial function in hypertensive patients (Alesci et al., 2020). Resveratrol supplementation in Alzheimer's patients improved cognitive function and reduced oxidative stress, providing neuroprotective benefits

(Yiannopoulou et al., 2020). Additionally, polyphenols from cinnamon and berries improved insulin sensitivity and glycaemic control in patients with type 2 diabetes (Ganesan et al., 2021). However, translating in vitro and in vivo findings to human settings faces challenges such as poor bioavailability, which limits efficacy due to poor absorption and rapid metabolism. Nanoencapsulation and co-delivery systems are being explored to enhance bioavailability (Basu et al., 2021). Variability in polyphenol content in foods and supplements also affects outcomes, prompting the development of standardized extracts for consistent dosing (Zhang et al., 2022). Human variability in genetics, gut microbiota, and health conditions influences responses to polyphenol supplementation, leading to interest in personalized medicine approaches (Li & Van de Wiele, 2023). Moreover, most human trials are short-term, highlighting the need for longitudinal studies to better understand the long-term safety and efficacy of polyphenols (Zeng et al., 2020).

5.6 CONCLUSION

Polyphenols, with their potent antioxidant, anti-inflammatory, and anticancer properties, demonstrate significant therapeutic potential in the prevention and management of various diseases, including cancer, CVDs, neurodegenerative disorders, and diabetes. Clinical trials have shown promising outcomes, particularly in improving immune function, reducing blood pressure, and enhancing cognitive function. These polyphenols act by modulating the immune system, e.g., reducing the production of inflammatory cytokines such as TNF-α and IL-6 and inhibiting pathways like NF-κB, encouraging epigenetic modification and signal transduction modulation, and controlling the cell cycle and apoptosis. Polyphenols enhance immune function by boosting phagocytic activity and stimulating T-cell responses. Polyphenols also regulate gut inflammation by modulating the immune response and gut microbiota composition and controlling oxidative damage in the gut. Obesity is also regulated by polyphenols as they inhibit adipogenesis, enhance fat oxidation, and regulate the composition of the gut microbiota. Polyphenols enhance vascular function by improving endothelial performance and stimulating nitric oxide production.

REFERENCES

Alesci, S., et al. (2020). Effects of flavonoid-rich diet on hypertension. *American Journal of Hypertension*, 33(8), 715–722.

Basu, S. M., Yadava, S. K., Singh, R., & Giri, J. (2021). Lipid nanocapsules co-encapsulating paclitaxel and salinomycin for eradicating breast cancer and cancer stem cells. *Colloids and Surfaces B: Biointerfaces*, 204, 111775. https://doi.org/10.1016/j.colsurfb.2021.111775.

Boots, A. W., Haenen, G. R., & Bast, A. (2008). Health effects of quercetin: From antioxidant to nutraceutical. *European Journal of Pharmacology*, 585(2–3), 325–337. https://doi.org/10.1016/j.ejphar.2008.03.008

Cao, Z., Li, X., & Liu, Y. (2021). Role of polyphenols in tumor suppressor activation and apoptosis in cancer therapy. *Cancer Research*, 81(3), 537–548. https://doi.org/10.1158/0008-5472.CAN-20-2726

Chandra, A., & Wali, J. (2020). Role of polyphenols in the treatment of obesity. *Phytochemistry Reviews*, 19(4), 935–946. https://doi.org/10.1007/s11101-020-09689-5

Cheng, H., Zhang, W., & Li, X. (2020). Epigenetic regulation of gene expression by polyphenols: Implications in disease prevention. *Biochimica et Biophysica Acta (BBA) – Molecular Basis of Disease*, 1866(7), 165728. https://doi.org/10.1016/j.bbadis.2020.165728

Choi, S. Y., Jung, J. Y., & Kim, M. Y. (2021). Caffeic acid: A review of its biological activities and potential applications. *Molecules*, 26(2), 357. https://doi.org/10.3390/molecules26020357

Ganesan, K., et al. (2021). Effect of polyphenols on glucose metabolism in type 2 diabetes. *Clinical Nutrition*, 40(4), 1142–1149.

Gao, J., Zhang, X., & Zhou, Q. (2020). Polyphenols as antioxidants in health and disease. *Nutrients*, 12(5), 1408. https://doi.org/10.3390/nu12051408

González-Correa, J. A., & Rodríguez-Rodríguez, R. (2021). Polyphenols in the prevention of cardiovascular diseases: A review. *Antioxidants*, 10(6), 926. https://doi.org/10.3390/antiox10060926

Hosseini, A., Farkhondeh, T., & Pourbagher, R. (2020). Polyphenols and breast cancer: Mechanisms and potential applications. *Nutrition and Cancer*, 72(2), 288–307. https://doi.org/10.1080/01635581.2020.1711147

Kim, H. K., & Lee, C. (2021). The potential of polyphenols in the management of autoimmune diseases. *Current Rheumatology Reports*, 23(1), 10. https://doi.org/10.1007/s11926-021-00954-3

Kusuma, I. H., Cho, M. S., & Kim, H. J. (2022). Ferulic acid and its potential effects in the prevention of chronic diseases: A review. *Antioxidants*, 11(8), 1550. https://doi.org/10.3390/antiox11081550

Li, Q., & Van de Wiele, T. (2023). Gut microbiota as a driver of the interindividual variability of cardiometabolic effects from tea polyphenols. *Critical Reviews in Food Science and Nutrition*, 63(11), 1500–1526.

Li, S., Zhang, S., & Yang, Z. (2021a). Polyphenols and their regulatory effects on signal transduction pathways. *Critical Reviews in Food Science and Nutrition*, 61(8), 1330–1345. https://doi.org/10.1080/10408398.2020.1799119

Li, S., Zhang, X., & Li, C. (2021b). The role of polyphenols in immune system regulation and infectious diseases. *Frontiers in Pharmacology*, 12, 655768. https://doi.org/10.3389/fphar.2021.655768

Liu, R. H. (2004). Potential synergy of phytochemicals in cancer prevention: Mechanism of action. *The Journal of Nutrition*, 134(12 Suppl), 3479S–3485S. https://doi.org/10.1093/jn/134.12.3479S

Liu, X., Yang, S., & Zhang, Y. (2020). Polyphenols and their apoptotic effects in cancer therapy. *Journal of Cancer Research and Clinical Oncology*, 146(9), 2155–2167. https://doi.org/10.1007/s00432-020-03224-3

Liu, Y., Li, H., & Zhang, L. (2020). Polyphenols modulate NF-κB pathway and reduce inflammation in cancer and cardiovascular disease. *Critical Reviews in Food Science and Nutrition*, 60(6), 937–948. https://doi.org/10.1080/10408398.2019.1666220

Manczak, M., Ayyadevara, S., & Iavicoli, I. (2020). Polyphenols and neuroprotection in Parkinson's disease: Mechanisms and therapeutic implications. *Frontiers in Neuroscience*, 14, 276. https://doi.org/10.3389/fnins.2020.00276

Pandey, K. B., & Rizvi, S. I. (2009). Plant polyphenols as dietary antioxidants in human health and disease. *Oxidative Medicine and Cellular Longevity*, 2(5), 270–278. https://doi.org/10.4161/oxim.2.5.9498

Peng, L., Wang, Y., & Liu, X. (2020). Modulation of hormone receptor signaling by polyphenols: Mechanisms and therapeutic potential. *Frontiers in Pharmacology*, 11, 540313. https://doi.org/10.3389/fphar.2020.540313

Sánchez-González, C., García-García, F., & García-Morales, V. (2021). Polyphenols in Alzheimer's disease prevention: A review of their antioxidant and anti-inflammatory properties. *Molecules*, 26(9), 2761. https://doi.org/10.3390/molecules26092761

Santos, A. M., Almeida, A., & Ferreira, D. (2021a). Polyphenols and their anti-inflammatory effects: A molecular approach. *Phytomedicine*, 85, 153578. https://doi.org/10.1016/j.phymed.2021.153578
Santos, A. M., Almeida, A., & Ferreira, D. (2021b). Cardiovascular protective effects of polyphenols: Mechanisms and therapeutic potential. *Phytomedicine*, 72, 152827. https://doi.org/10.1016/j.phymed.2020.152827
Scalbert, A., Johnson, I. T., & Saltmarsh, M. (2005). Polyphenols: Antioxidants and beyond. *The American Journal of Clinical Nutrition*, 81(1), 215S–217S. https://doi.org/10.1093/ajcn/81.1.215S
Tang, J., Wang, Z., & Lin, M. (2021). Polyphenols as modulators of host-pathogen interactions: A review of their antiviral and antibacterial properties. *Critical Reviews in Food Science and Nutrition*, 61(17), 2941–2952. https://doi.org/10.1080/10408398.2020.1837749
Teng, X., Yin, J., & Li, H. (2020). Lignans and their health benefits: A review on secoisolariciresinol and its potential role in human health. *Molecules*, 25(12), 2843. https://doi.org/10.3390/molecules25122843
Valko, M., Rhodes, C., & Moncol, J. (2020). Polyphenols in prostate cancer: Mechanisms and therapeutic implications. *Cancer Letters*, 496, 1–10. https://doi.org/10.1016/j.canlet.2020.02.004
Williamson, G., & Manach, C. (2005). Bioavailability and bio efficacy of polyphenols in humans. II. Review of 93 intervention studies. *The American Journal of Clinical Nutrition*, 81(1), 243S–255S. https://doi.org/10.1093/ajcn/81.1.243S
Wu, X., Zhao, L., & Sun, C. (2021). Role of polyphenols in the inhibition of HDACs and their effects in cancer prevention. *Journal of Nutritional Biochemistry*, 88, 108579. https://doi.org/10.1016/j.jnutbio.2020.108579
Yang, C. S., Wang, X., Lu, G., & Picinich, S. C. (2009). Cancer prevention by tea: Animal studies, molecular mechanisms, and human relevance. *Nature Reviews Cancer*, 9(6), 429–439. https://doi.org/10.1038/nrc2641
Yang, L., Liu, Y., & Zhang, H. (2020). Regulation of miRNAs by polyphenols and their implications in health. *Frontiers in Nutrition*, 7, 583869. https://doi.org/10.3389/fnut.2020.583869
Yiannopoulou, K. et al. (2020). Resveratrol in Alzheimer's disease: A clinical trial. *Neurochemical Research*, 45(9), 2035–2044.
Zeng, Y., Pu, X., Du, J., Yang, X., Li, X., Mandal, M. S., Yang, T., & Yang, J. (2020). Molecular mechanism of functional ingredients in barley to combat human chronic diseases. *Oxidative Medicine and Cellular Longevity*, 2020(1), 3836172. https://doi.org/10.1155/2020/3836172.
Zhang, B., Zhang, Y., Xing, X., & Wang, S. (2022). Health benefits of dietary polyphenols: Insight into interindividual variability in absorption and metabolism. *Current Opinion in Food Science*, 48, 100941. https://doi.org/10.1016/j.cofs.2022.100941.
Zhang, Q., Zhang, J., & Liang, Y. (2020a). Resveratrol and its therapeutic effects: A review of recent developments. *Food & Function*, 11(9), 7872–7883. https://doi.org/10.1039/D0FO01377H
Zhang, Y., Liu, X., & Yu, L. (2020c). Role of polyphenols in cardiovascular diseases: A review of recent advances. *Nutrients*, 12(4), 1221. https://doi.org/10.3390/nu12041221
Zhang, Z., Song, X., & Liu, C. (2020b). Antimicrobial and antiviral activities of polyphenols: Mechanisms and therapeutic implications. *Frontiers in Microbiology*, 11, 1077. https://doi.org/10.3389/fmicb.2020.01077
Zhao, C., Luo, J., & Liu, Y. (2021a). Nrf2 signaling pathway in the regulation of antioxidant and anti-inflammatory effects of polyphenols. *Frontiers in Pharmacology*, 12, 643702. https://doi.org/10.3389/fphar.2021.643702
Zhao, Y., Wu, Z., & Yang, X. (2021b). The role of polyphenols in colorectal cancer prevention. *Molecules*, 26(8), 2312. https://doi.org/10.3390/molecules26082312
Zhao, Y., Wang, P., & Zhang, L. (2020c). Polyphenols in inflammatory bowel diseases: Mechanisms and therapeutic potential. *Phytomedicine*, 75, 153258. https://doi.org/10.1016/j.phymed.2020.153258

6 Method of Synthesis of Polyphenol Nutraceutical Formulation

Renisha Pinto, Nidhi Minz, Tripti Pal,
Aqsa Majgaonkar, and Dwaipayan Sinha

6.1 INTRODUCTION

Polyphenols are a diverse and structurally complex group of secondary metabolites found abundantly in the plant kingdom (Lattanzio et al. 2009; Zagoskina et al. 2023). They are characterised by multiple phenolic structural units and are primarily involved in plant defence mechanisms against ultraviolet radiation (Šamec et al. 2021), pathogens (Kumar et al. 2020), and environmental stressors (Tuladhar, Sasidharan, and Saudagar 2021) These compounds are widely distributed in dietary sources such as fruits, vegetables, tea, cocoa, coffee, wine, and various medicinal herbs (El Gharras 2009). Based on their chemical structure, polyphenols are broadly classified into flavonoids, phenolic acids, stilbenes, and lignans (Singla et al. 2019).

Recent advances in nutritional biochemistry have established polyphenols as significant bioactive compounds with considerable therapeutic potential (Cherian et al. 2023; Rudrapal et al. 2022; Choi et al. 2012). Their antioxidative (Quan et al. 2019; Turkmen et al. 2007; Gülçin et al. 2011), anti-inflammatory (Yahfoufi et al. 2018; Oliviero et al. 2018; Larrosa et al. 2009), and antiproliferative (Huang et al. 2014; Ferreira, Silva, and Nunes 2018; Choi et al. 2012) effects are attributed to their ability to scavenge free radicals, chelate transition metals, and regulate enzyme activity. These properties not only protect cellular structures from oxidative damage but also modulate key biochemical pathways involved in the pathophysiology of chronic diseases (Rudrapal et al. 2022; Sharifi-Rad et al. 2020). Epidemiological evidence consistently associates polyphenol-rich diets with a lower risk of cardiovascular diseases (Noad et al. 2016), type 2 diabetes (Guasch-Ferré et al. 2017), neurodegenerative conditions (Grabska-Kobyłecka et al. 2023; Ross et al. 2024), and certain types of cancer, such as prostate cancer and colorectal cancer (Almanza-Aguilera et al. 2023; Zamora-Ros et al. 2018; Bo' et al. 2019). However, their efficacy is often influenced by several variables, including the food matrix, metabolic conversion, interaction with gut microbiota, and bioavailability following ingestion (Kumar Singh et al. 2019). The development of polyphenol-based nutraceuticals requires a thorough

DOI: 10.1201/9781003588504-6

and interdisciplinary method involving extraction, purification, stabilisation, and the design of delivery systems. The intrinsic instability of polyphenols – attributable to their vulnerability to oxidation, pH fluctuations, and enzymatic degradation – necessitates the implementation of effective processing techniques that maintain their bioactivity during manufacturing, storage, and consumption.

This chapter explores the strategies for converting raw plant-derived polyphenols into clinically verified nutraceutical compositions. We delineate the technical considerations at each stage, encompassing plant selection, pre-processing, extraction, purification, formulation, and scale-up. Focus is directed on methodological advances that conform to regulatory standards and enhance industrial scalability, facilitating a practical and significant shift from laboratory to market. This chapter functions as a reference guide for academicians, formulation scientists, and industry stakeholders seeking to maximise the therapeutic potential of polyphenols through the integration of latest research trends and process optimisation methodologies in rational product design.

6.2 EXTRACTION-GRADE PLANT MATERIAL SELECTION CRITERIA

The selection of plant material from which polyphenols are to be extracted is a key factor in the production of high-quality extracts with optimum bioactivity. Of prime importance here is the concentration of polyphenols in the plant because the more concentrated the polyphenols, the more powerful the extract. Green tea, for instance, is selected based on its rich content of catechins, while grapes are selected based on the level of resveratrol content (Meng et al. 2021). The content of polyphenols is usually assayed using techniques like high-performance liquid chromatography (HPLC), which provides accuracy in content quantification (Motilva, Serra, and Macià 2013).

Purity and quality guarantee the success of the extraction process. Organic farming methods are normally preferred as they restrict the use of chemicals and synthetic pesticides, which may interfere with the purity of the extract (Prado and Rostagno 2022). Second, plants under low-stress conditions produce lower levels of polyphenols. However, plants cultivated in moderate stress conditions, e.g., scarce water or ultraviolet (UV) radiation, may yield higher quantities of polyphenolic compounds as part of their natural defence mechanisms (Sharma et al. 2019). Harvesting time also significantly affects polyphenol extraction. Polyphenol concentration changes throughout the growth period of a plant, and each compound attains a peak at a different stage. For example, grapes typically reach optimum levels of resveratrol at late ripening, and berries contain optimum polyphenol levels when fully ripe (Siriwoharn et al. 2004). Harvesting plants at the appropriate time maximises yield and ensures that the extracted polyphenols do not lose their desired bioactivity. Lastly, environmental conditions such as soil, sun exposure, and temperature affect the polyphenol content. Photosynthetically active plants that expose themselves to enough sun yield more polyphenolic compounds, which act as protection against UV damage (Petruk et al. 2018). This explains the need to select plant material carefully according to polyphenol content, quality, harvesting time, and environmental conditions to formulate effective and bioactive extracts.

6.3 PRE-PROCESSING AND STORAGE CONDITIONS

Polyphenols are frequently dependent on processing and environmental conditions, and are susceptible to degradation produced by heat, enzymes, oxygen, and light. . To keep them intact between the period of harvest and the extraction operation, pre-processing and storage conditions must be optimised (Corrêa et al. 2016; Bhatta, Janezic, and Ratti 2020).

Pre-treatment washing, cutting, and drying reduce water activity and inhibit microbial growth but must be controlled to avoid degradation of thermolabile compounds. Low-temperature drying methods like freeze-drying and vacuum drying are used ideally because they have the potential to regain antioxidant activity better than the traditional thermal drying methods (Bhatta, Janezic, and Ratti 2020).

Mechanical reduction processes like milling or blending enhance the surface area and lead to improved extraction yields. Yet excessive processing in the presence of oxygen can lead to polyphenol oxidation and bioactivity loss. Reduction in size under oxygen-limited conditions reduces these impacts. Moreover, polyphenol oxidase (PPO), which is a naturally occurring enzyme in many plant tissues, can catalyse the degradation of phenolics. For PPO activity inhibition, blanching or rapid drying techniques are generally applied (Corrêa et al. 2016).

The preservation of polyphenol stability is significantly influenced by storage conditions. Plant samples that are high in polyphenols are best stored in airtight, opaque containers under low temperatures, such as refrigeration (4 °C) or freezing (–20 °C), to avert the risk of oxidation and photodegradation. Inert gases such as nitrogen added to packaging atmospheres serve to safeguard further these susceptible molecules (Aktaş and Kurek 2024). Quality control measures like regular checking of moisture level, pH, and microbiological status are necessary to ensure long-term product stability and safety. Standardisation of pre-processing and storage conditions enables the reduction of polyphenol breakdown so that reproducibility and the design of functional products are feasible.

6.4 EXTRACTION OF POLYPHENOLS

Polyphenol extraction encompasses the numerous procedures used to remove these bioactive molecules from plant or synthetic sources. Natural resource extraction involves both age-old methods – e.g., maceration, decoction, and Soxhlet – and more recent techniques such as ultrasound-assisted, microwave-assisted, and supercritical fluid extraction, which yield increased efficiency and greater protection for the heat-sensitive phenolics. When low amounts or seasonally fluctuating supplies come from natural sources, chemical synthesis provides a scalable and reproducible method of preparing polyphenols via reactions like Friedel–Crafts acylation and Suzuki–Miyaura cross-coupling to replicate complex structures (Farhang et al. 2022; Keylor, Matsuura, and Stephenson 2015; Sartori and Maggi 2006). Enzymatic synthesis, on the other hand, employs cellulase and pectinase enzymes to hydrolyse cell wall components, facilitating the release and yield of the bound polyphenols under mild and friendly conditions (Calderón-Oliver and Ponce-Alquicira 2021; Bitwell et al. 2023). Each method has its benefits depending on the polyphenol extract's purity, scalability, and application.

6.4.1 Extraction of Polyphenols From Natural Resources

Extraction is the first step in isolating desired natural products from raw materials. Common techniques include solvent extraction, distillation, pressing, and sublimation, depending on the extraction principle. Among these, solvent extraction is most widely used. The extraction process generally involves (1) the solvent penetrating the solid matrix, (2) the solute dissolving in the solvent, (3) the solute diffusing out of the matrix, and (4) collection of the extract (Zhang, Lin, and Ye 2018).

The efficiency of extraction depends on factors such as the solvent's properties, particle size of the material, solvent-to-solid ratio, temperature, and duration (Feudjio et al. 2022). Solvent selection is crucial – it should have suitable solubility, selectivity, low cost, and a good safety profile. Solvents with polarity similar to the target compound typically work best, following the principle of "like dissolves like" (Lee, Jayakody, and Kim 2024). Ethanol and methanol are commonly used in phytochemical extractions (Michiels et al. 2012).

Reducing particle size increases the surface area, improving contact between the sample and solvent and enhancing yield (Prasedya, Frediansyah, and Martyasari 2021). Higher temperatures improve solubility, diffusion, and mass transfer, while lowering solvent viscosity and surface tension, thereby speeding up extraction (Sheng et al. 2014; Oroian, Ursachi, and Dranca 2020; Getachew et al. 2022). Generally, longer extraction times further improve efficiency (Hassanloofard et al. 2024).

6.4.1.1 Conventional Extraction Technologies

Polyphenolic compounds are usually present in small quantities in foods, drinks, or plants, so extraction techniques are essential. Pre-treatments like drying, crushing, or grinding may be required. After extraction, further isolation and purification yield the active compounds (Chuo, Nasir, and Mohd-Setapar 2022). Traditional extraction methods include percolation, decoction, heat reflux extraction, Soxhlet extraction, and maceration. These methods are simple and widely used but often require long extraction times, consume more energy, and may lead to solvent loss (Sridhar, Ponnuchamy, and Kumar 2021).

6.4.1.1.1 Percolation

Percolation involves packing finely powdered plant material into a percolator and gradually passing solvent through from top to bottom (Alara, Abdurahman, and Ukaegbu 2021). The material is first moistened, then left to stand for about 24 hours before percolation begins. Solvent is added until about 75% of the final volume is collected. The extract is obtained by filtration, and the marc is pressed to recover residual extract (Abubakar and Haque 2020).

6.4.1.1.2 Decoction

Decoction is a water-based method where plant parts are boiled in water for a set time (15 minutes to 2 hours), cooled, and filtered. It's suitable for heat-stable, water-soluble compounds. Softer plant parts like leaves and flowers are boiled briefly, while harder materials like bark may be boiled longer (Hlatshwayo et al. 2025).

6.4.1.1.3 Heat Reflux Extraction

Heat reflux extraction is widely used for herbal extracts. The solvent is heated, vaporised, condensed, and returned to the material, maintaining a constant temperature (typically 100 °C). This method enhances diffusion rates, speeds up extraction, and allows batch processing, making it suitable for industrial-scale production (Kim, Shin, and Seo 2014; Olalere, Gan, and Taiwo 2023; Lai et al. 2024).

6.4.1.1.4 Soxhlet Extraction

Soxhlet extraction continuously cycles fresh solvent through the sample via distillation and siphoning, improving analyte removal compared to simple soaking. It is faster and uses less solvent than maceration or percolation. The setup includes a distillation flask, thimble, siphon, and condenser. Solvent vapours extract compounds, condense them, and return them to the flask, repeating the cycle until extraction is complete (Sander 2017; Torres-Rodriguez, Darvishzadeh, and Skidmore 2024; Sinha, Mukherjee, and Chowdhury 2022).

6.4.1.1.5 Maceration

In maceration, coarsely ground material is soaked in solvent in a sealed container for at least 3 days with occasional stirring. The mixture is then filtered or decanted, and the extract is concentrated by evaporation. This method is especially suited for thermolabile (heat-sensitive) materials (Abubakar and Haque 2020).

6.4.1.2 Sustainable Extraction of Polyphenols

Recent years have seen a rise in environmental awareness, prompting research to develop sustainable extraction technologies that yield safe and high-quality extracts for use in the food, pharmaceutical, and cosmetic sectors. Considering these requirements, researchers coined the phrase "green extraction" grounded in six principles (Chemat, Vian, and Cravotto 2012):

- Utilisation of renewable and sustainable biological resources.
- Utilisation of eco-friendly solvents or water.
- Reduced energy input.
- Production of co-products from waste.
- Minimal quantity of operational units.

6.4.1.2.1 Microwave-Assisted Extraction

Microwave-assisted extraction (MAE) is widely used to extract bioactive compounds from plants. It works by rapidly heating a solvent–sample mixture using microwave energy, helping transfer analytes from the matrix to the solvent. A key advantage is its ability to quickly heat mixtures, making it suitable even for heat-sensitive compounds.

MAE efficiency depends on the solvent type, sample material, and dielectric properties of the compounds. Critical parameters to optimise include solvent polarity, solvent volume, sample size, extraction time, temperature, and microwave power (Kataoka, n.d.).

TABLE 6.1
Recent Examples of Microwave-Assisted Extraction

Matrix	Solvent Used	Extraction Condition	Total Polyphenol Content Recovered	Reference
Bitter orange industrial waste	Water–ethanol (50%)	Microwave 600 W, 80 °C, 10 min	~35 mg GAE/g dry weight	(García-Martín et al. 2023)
Blackcurrant by-products	Water–ethanol (50%)	Microwave 500 W, 60 °C, 15 min	~29 mg GAE/g dry weight	(Alchera, Ginepro, and Giacalone 2022)
Crataegus monogyna (hawthorn)	Water–ethanol (50%)	Microwave 450 W, 70 °C, 10 min	~48 mg GAE/g dry weight	(Gavrila, Damian, and Rosca 2025)
Sorghum bran	Water–ethanol (60%)	Microwave 600 W, 100 °C, 15 min	~41 mg GAE/g dry weight	(Duke et al. 2024)
Olive pomace (*Žutica* variety)	Water–ethanol (60%)	Microwave 500 W, 80 °C, 10 min	~39 mg GAE/g dry weight	(Mardokić, Maldonado, and Klosz 2023)

Microwave heating works mainly through **ionic conduction** and **dipole rotation**. Ionic conduction generates heat through friction as ions move under an electromagnetic field. Dipole rotation involves dipoles constantly realigning with the applied field, creating molecular motion and heat. At 2450 MHz (common in commercial systems), dipoles reorient nearly 5 billion times per second, producing rapid heating (Eskilsson and Björklund 2000). Table 6.1 illustrates selected examples of the extraction of polyphenolic substances from plants and plant products using MAE.

6.4.1.2.2 Ultrasonic-Assisted Extraction

Ultrasonic-assisted extraction (UAE) uses ultrasonic waves to improve extraction efficiency by disrupting cell walls and enhancing mass transfer (Zahari et al. 2020). Cavitation – created by sound waves – breaks cell structures and releases bioactive compounds, making UAE particularly effective for extracting phenolics and other phytochemicals. It is considered a sustainable method, requiring less solvent and energy, and can operate at room temperature and atmospheric pressure, making it safe, cost-effective, and reproducible (Medina-Torres et al. 2017).

The process generates intense local conditions (up to 4500 °C and 50 MPa) that break down cell membranes and release intracellular substances. UAE can be direct (with a sonotrode submerged in the solvent–sample mixture) or indirect (using an ultrasonic bath), both requiring cleaning afterward. Key parameters include temperature, sonication time, particle size, and extraction duration (Weggler et al. 2020). UAE is preferred over traditional methods because of its lower energy use, faster processing, reduced degradation of actives, and better yields (Sun et al. 2019). Table 6.2 illustrates selected examples of the extraction of polyphenolic substances from plants and plant products using UAE.

TABLE 6.2
Recent Applications of UAE to Polyphenol Extraction

Matrix	Solvent Used	Extraction Condition	Total Polyphenol Content Recovered	Reference
Pinus elliottii needles (pine needles)	50% ethanol–water	Ultrasound 40 kHz, 400 W, 40 °C, 30 min	~38.5 mg GAE/g dry weight	(Zhang et al. 2024)
Picea abies bark (Norway spruce)	50% ethanol–water	Ultrasound 35 kHz, 30 °C, 40 min	~76 mg GAE/g dry weight	(Chmelová et al. 2020)
Applewood	60% ethanol–water	Ultrasound 20 kHz, 300 W, 60 °C, 30 min (pilot and lab scale)	~27 mg GAE/g dry weight	(Withouck 2023)
Chinese propolis	70% ethanol–water	Ultrasound 40 kHz, 50 °C, 30 min	~138 mg GAE/g dry weight	(Peng 2023)
Phyllanthi Fructus (amla)	50% ethanol–water	Ultrasound 40 kHz, 400 W, 45 °C, 30 min	~92 mg GAE/g dry weight	(Che 2024)

6.4.1.2.3 Supercritical Fluid Extraction

Supercritical fluid extraction (SFE) is an eco-friendly method that extracts bioactive compounds from plants, herbs, and spices using supercritical fluids, typically CO_2, valued for its safety and non-oxidising environment (Uwineza and Waśkiewicz 2020). The process, involving extraction and separation steps, relies on supercritical CO_2 flowing through a solid sample to dissolve target compounds, which are then separated by adjusting pressure or temperature.

Systems can vary from simple to complex, with options for recycling CO_2 and collecting volatile fractions. SFE can operate in static (no flow) or dynamic (continuous flow) modes (Brunner 2005; Uwineza and Waśkiewicz 2020). This method reduces solvent use and protects sensitive compounds, making it ideal for high-purity extracts. Table 6.3 illustrates selected examples of the extraction of polyphenolic substances from plants and plant products using SFE.

6.4.1.2.4 Pressurized Liquid Extraction

Pressurized liquid extraction (PLE), also called accelerated or subcritical solvent extraction, uses solvents at elevated temperatures and pressures below their critical points, keeping them in liquid form (Alvarez-Rivera et al. 2020). Higher temperatures reduce viscosity and improve solute diffusion and matrix wetting, enhancing extraction efficiency. Water is often highlighted as a green solvent due to its changing properties at higher temperatures (Carabias-Martínez et al. 2005).

PLE equipment includes a solvent reservoir, high-pressure pump, heated extraction cell, and collection system. Dynamic PLE offers greater control over solvent flow and conditions. This method is fast, uses less solvent, and is suitable for both analytical and preparative work with a lower environmental impact (Osorio-Tobón and

TABLE 6.3
Recent Applications of SFE to Polyphenol Extraction

Matrix	Solvent Used	Extraction Condition	Total Polyphenol Content Recovered	Reference
Pouteria lucuma seeds	Supercritical CO_2 (no co-solvent)	35 MPa, 60 °C, 120 min	~12 mg GAE/g dry weight	(Chauca-Cerrutti 2024)
Apple peels	Supercritical CO_2 + 10% ethanol modifier	35 MPa, 50 °C, 90 min	~15 mg GAE/g dry weight	(Aresta, Vietro, and Cotugno 2024)
Hemp seeds	Supercritical CO_2 + ethanol modifier (15%)	35 MPa, 60 °C, 60 min	~9 mg GAE/g dry weight	(Allay et al. 2025)
Rheum tataricum roots	Supercritical CO_2	40 MPa, 50 °C, 60 min	~25 mg GAE/g dry weight	(Amangeldinova et al. 2024)
Mexican oregano (*Lippia graveolens*)	Supercritical CO_2	30 MPa, 50 °C, 60 min	~18 mg GAE/g dry weight	(Picos-Salas et al. 2024)

TABLE 6.4
Recent Applications of PLE to Polyphenol Extraction

Matrix (Plant Material/Powder/Extract, Etc.)	Solvent Used	Extraction Condition	Total Polyphenol Content Recovered	Reference
Oak leaves	Ethanol–water mixtures (50%)	PLE 100 °C, 10 min	~108 mg GAE/g dry weight	(Álvarez, Rocha-Guzmán, and Gallegos-Infante 2025)
Moringa oleifera leaves	Ethanol–water (50%)	PLE, 120 °C, 10 min, 1500 psi	~85 mg GAE/g dry weight	(Chatzimitakos 2024)
Olive leaves	Ethanol–water (50%)	PLE, 90 °C, 15 min	~62 mg GAE/g dry weight	(Huamán-Castilla 2024)
Red seaweed	Water–ethanol (60:40)	PLE, 120 °C, 10 min	~30 mg GAE/g dry weight	(Huamán-Castilla et al. 2024)
Eucalyptus marginata leaves	Ethanol–water–acetone (40:40:20)	PLE, 100 °C, 15 min, 1500 psi	~95 mg GAE/g dry weight	(Hasni et al. 2024)

Meireles 2013). Table 6.4 illustrates selected examples of the extraction of polyphenolic substances from plants and plant products using PLE.

6.4.1.2.5 High-Voltage Electrical Discharge

High-voltage electrical discharge (HVED) is a non-thermal technique that applies pulsed high-voltage discharges (20–80 kV/cm) to liquid suspensions, generating

TABLE 6.5
Recent Applications of HVED

Matrix (Plant Material/Powder/Extract, Etc.)	Solvent Used	Extraction Condition	Total Polyphenol Content Recovered	Reference
Grapefruit peel	Deep eutectic solvents (choline chloride:-lactic acid, choline chloride:glycerol), aqueous glycerol	HVED, room temperature, short-duration pulses	Up to ~4.4 mg GAE/g dry weight (DES + HVED)	(El Kantar et al. 2019)
Orange peel	Water + enzymes (pectinase/cellulase), no organic solvent	HVED + enzymatic hydrolysis at 45 °C, pH 5, 2h	~8.6 mg GAE/g dry weight	(El Kantar et al. 2018)
Dill seed	Subcritical water, water (HVED), ethanol–water (microwave)	HVED (15 kV/cm, 20 pulses), subcritical water (150 °C, 30 min), microwave-assisted (400 W, 10 min)	HVED: ~10.2 mg GAE/g; Subcritical: ~17.5 mg GAE/g; Microwave: ~13.1 mg GAE/g	(Nastić et al. 2025)
Oregano	Water (HVED)	HVED (20 kV/cm, 30 pulses), room temperature	~18.4 mg GAE/g dry weight	(Nutrizio et al. 2022)
Blueberry pomace	Water–ethanol mixtures (50% v/v), deep eutectic solvents	Ultrasound-assisted extraction (20 kHz, 400 W), MAE (400 W, 10 min)	~17–24 mg GAE/g dry weight depending on method	(Lončarić et al. 2020)

shock waves, cavitation, UV light, and radicals that break cell walls and release bioactive compounds (Xi, He, and Yan 2017; Abbaspour et al. 2024). The process includes a streamer discharge phase followed by an arc discharge, where effects are stronger (Li, Fan, and Xi 2019).

HVED offers advantages over traditional methods, including shorter extraction times, higher yields, lower energy use, and better preservation of bioactive compound structure and activity (Brianceau et al. 2016). Table 6.5 illustrates selected examples of the extraction of polyphenolic substances from plants and plant products using HVED.

6.4.2 Chemical Synthesis of Polyphenols

Chemical synthesis is a controllable, scalable process for plant polyphenol extraction that overcomes the restrictions of seasonal availability, limited availability, or minimal concentration of natural sources. Chemical synthesis ensures consistent molecular structure, purity, and reproducibility of bioactivity – issues of paramount concern in the pharmaceutical, nutraceutical, and cosmetic applications (Quideau et al. 2011; Ameer, Shahbaz, and Kwon 2017).

Polyphenols are structurally diverse and typically entail more than one hydroxyl group on an aromatic ring. They are prepared using conventional organic reactions to form an accurate assembly of these ring systems and functional groups. Some of the common methods used are discussed next.

6.4.2.1 Friedel–Crafts Acylation

Inserts acyl groups into aromatic rings with the help of Lewis acid catalysts like $AlCl_3$. The process is crucial for the synthesis of phenolic acid derivatives and flavonoid skeletons (Sartori and Maggi 2006).

6.4.2.2 Oxidative Coupling

Mimics natural biosynthesis by phenolic unit coupling by carbon–carbon or carbon–oxygen bonds. It is typically used on dimeric or oligomeric polyphenols (Quideau et al. 2011).

6.4.2.3 Aldol Condensation

Forms carbon–carbon bonds between ketones and aldehydes, which assists in the development of flavonoid or stilbene structures such as chalcones and resveratrol precursors (Keylor, Matsuura, and Stephenson 2015; Rajendran, Bhanu, and Aruchamy 2022).

6.4.2.4 Suzuki–Miyaura Cross-Coupling

A reaction catalysed by palladium that couples boronic acids and aryl halides. The reaction is highly selective and has seen widespread use in the synthesis of resveratrol analogues and complex flavonoids (Farhang et al. 2022; Keylor, Matsuura, and Stephenson 2015).

For instance, the synthetic route of resveratrol includes total synthesis with Heck or Wittig reactions for stilbene core assembly, followed by introduction of phenolic compounds through hydroxylation (Giacomini et al. 2016).

Although it is accurate, chemical synthesis is marred by issues such as stereoselectivity, regioselectivity, and yield restrictions. Green chemistry principles and chemoenzymatic processes using biocatalysts and eco-friendly solvents are increasingly preferred to address this (Rudroff et al. 2018).

Furthermore, synthetic analogues of polyphenols – i.e., quercetin's methylated or glycosylated derivatives – are in use with improved bioavailability, stability, and targeted action compared to the natural ones (Kawabata, Mukai, and Ishisaka 2015).

Therefore, chemical synthesis is necessary to produce rare, complex, or optimised polyphenols for industry-scale uses with controlled functionality and enhanced therapeutic capacity.

6.4.3 Enzymatic Synthesis

Phytochemical extraction from the plant matrix is often hindered by the presence of rigid polysaccharide–lignin complexes that are stabilised by hydrogen bonding and hydrophobic interactions, such as van der Waals forces. These structural barriers restrict solvent access, especially when bioactives are confined in the cytoplasm or firmly bound to the cell wall matrix. Enzyme-assisted extraction (EAE) is an innovative strategy that can potentially address these limitations by preferably breaking down structural polymers with the view to enhancing phytochemical release and yield (Calderón-Oliver and Ponce-Alquicira 2021).

Enzymes such as cellulase, pectinase, and amylase are commonly used to degrade cellulose, hemicellulose, pectin, and starch, making the cell wall soft and permitting the release of bound phytochemicals (Bitwell et al. 2023). Enzymes work under mild conditions, and it is a selective and sustainable process of extraction without harsh chemicals or high temperatures. Enzyme-assisted aqueous extraction (EAAE) and enzyme-assisted cold pressing (EACP) are among the most frequently employed techniques. EAAE enhances oil and phytochemical recovery via polysaccharide–protein colloid breaking and cell wall solubilisation at the expense of emulsion formation that would hinder further downstream processing (Concha et al. 2004). EACP, on the contrary, is under a non-aqueous condition under which enzymatic action is constrained to cell wall breakdown without leading to colloidal dissolution, thus barring emulsion problems.

Growing demand for green and efficient extraction procedures in the pharmaceutical and food industries has increased the demand for EAE. Its good specificity, regioselectivity, and performance under environmentally safe conditions are in perfect harmony with the concept of sustainable bioprocessing (Puri, Sharma, and Barrow 2012; Chen et al. 2011). Therefore, EAE not only enhances the efficiency of extraction but also supports cleaner and more functional bioactive formulation development. Table 6.6 illustrates selected examples of EAE of phenolic compounds from plant sources.

6.4.4 Factors Affecting Extraction Yield

Temperature, pH, particle size, concentration of the solvent, and speed of agitation play critical roles in the control of the yield and purity of polyphenols extracted. Ethanol–water solvents of up to 30% water content are applied to enhance the phenolic yield in grape marc extractions but lead to diminishing extract concentration at concentrations higher than 50% owing to the lower solubility of certain polyphenolic compounds. Although higher temperatures (e.g., 60 °C) enhance the efficiency of extraction, long treatment for over 20 hours is likely to lead to thermal degradation of bioactive compounds (Spigno, Tramelli, and De Faveri 2007). Temperature plays a double-duty function: it can enhance diffusion rates and solubility, but, conversely, excessive heat would also destroy vulnerable molecules like catechins and anthocyanins. pH conditions are no exception; low pH is favourable for anthocyanins' stability but could suppress the recovery of other polyphenols (Enaru et al. 2021).

Recent research indicates the green solvent potential of solvents such as deep eutectic solvents (DESs) as possessing high extraction efficiency with a reduced environmental and health hazard (Ruesgas-Ramón, Figueroa-Espinoza, and Durand 2017). Additionally, pre-treatments such as enzymatic hydrolysis and mechanical size reduction increase solvent penetration and compound availability. In general,

TABLE 6.6
Enzyme-Assisted Extraction of Polyphenols

Plant Species	Plant Part	Major Compound Detected/Extracted	Optimum Conditions of the Process	Reference
Padina gymnospora	Whole body	Alginate, fucoidan, and phenolic compound	Temperature: 65 °C Water-to-sample ratio: 61.31 mL/g Enzyme loading: 32% Extraction time: 1.95 h	(Nguyen et al. 2024)
Grape pomace	Whole sample	Phenolic compounds	Temperature: 37 °C–50 °C Extraction time: 1 h–3 h Solid-to-liquid ratio-1/80–1/20 g/mL	(Stanek-Wandzel, Krzyszowska, and Zarębska 2024)
Morus nigra	Fruits	Phenolic compounds	Temperature: 40 °C Enzyme units: 38.46 enzyme units per g of sample pH: 4 Time: 20 min	(Tizón Alba et al. 2023)
Olea europaea	Leaves	Phenolic compounds	Temperature: 25 °C–50 °C pH: 4.0 Solid to liquid ratio: 1:10 Time: 1h	(Vardakas et al. 2024)
Hippophae rhamnoides	Leaves	Phenolics	Temperature: 45 °C pH: 4.9 Time: 3.25h Enzyme concentration: 1% v/w	(Puzeryte, Martusevice, and Sousa 2023)
Camellia sinensis	Leaves	Polyphenols	pH: 4.4 Time: 36 minutes Solvent and material ratio: 5.9	(Oanh et al. 2023)
Prunus avium	Fruits		Temperature: 30 °C–70 °C Time: 30–300 min Enzyme con centration: 1–120 μL/g pH –3.0–8.0	(Domínguez-Rodríguez, Marina, and Plaza 2021)
Pinus koraiensis	Nut	Flavonoid	Extraction time: 2 h Material/liquid ratio: 1:20 (g/mL) Enzyme concentration: 90 U/g	(Zhang, Ma, and Wang 2021)
Ribes nigrum	Pressed cake	Anthocyanins and polyphenols	Temperature: 40 °C–50 °C Solid: solvent ratios: 1:10 and 1:4 w/v Enzyme concentration: 200 mg/L	(Granato, Fidelis, and Haapakoski 2022)
Olive	Pomace	Phenolic compounds	Temperature: 50 °C pH: 5.0 Enzyme concentration: 2% (w/w) Time: 120 min	(Macedo et al. 2021)

extraction conditions should be optimised to obtain maximum yield without compromising on functional and sensory attributes of polyphenol-enriched extracts.

6.5 PURIFICATION AND ISOLATION TECHNIQUES

Phenolic extracts may include various impurities and interfering compounds, which need to be eliminated before identification and quantification (Silvia et al. 2025).

6.5.1 Filtration and Centrifugation

One kind of membrane separation technique used to separate and purify polyphenols is ultrafiltration. The compound's particle size is one of the most important factors in ultrafiltration and the force that propels transport over the membrane. Using this approach, molecules with molecular weights ranging from 3000 Da to 100 KDa and higher are separated. Because polyphenolics have a low molecular weight, ultrafiltration using semi-permeable membranes makes separation simple and quick.

To recover bioactive components from purple sweet potatoes (PSPs), a study assessed a sustainable technique that combines UAE, centrifugation, and ultrafiltration. Centrifugation (2500 rpm, 6 min) was performed after ultrasonic extraction (840 W, 59 kHz) with water as a solvent. This process improved polyphenol recovery and successfully eliminated proteins, maintaining the majority of the anthocyanin content. After effective separation was made possible by ultrafiltration, 99% anthocyanin recovery was obtained from the permeate. A fouling coefficient of 29.5 m^{-1} and a polyphenol recovery of 34.5% were obtained under ideal centrifugation conditions (4000 rpm for 7.74 min). Crucially, centrifugation did not break down important anthocyanins, according to an HPLC-DAD-ESI-MS2 study, confirming its function in protein removal without anthocyanin loss. An efficient and environmentally responsible method for PSP valuation is provided by the integrated process (Zhu et al. 2016).

6.5.2 Chromatographic Methods

These techniques enable the isolation, purification, and identification of the components of a mixture for qualitative and quantitative analysis. The mixture's diverse components separate because they move at different rates. Differential partitioning between the stationary and mobile phases serves as the basis for the separation. The stationary phase is always made up of a solid phase or a liquid layer that has been adsorbed onto a solid support. The mobile phase is always made up of a liquid or gaseous component. Chemicals or phytochemicals have been separated on or by a stationary phase according to their molecular integrity or polarity. These unique methods are employed to separate chemicals from mixtures according to their charge, size, and form.

The separation method known as HPLC uses the adsorption principles to accomplish effective separation. It works effectively for separating inorganic and organic substances. In HPLC, the stationary phase is made up of solid particles that are tightly packed together, while the mobile phase is made up of an appropriate solvent. The interaction of the mixture's chemicals with the stationary phase's solid particles starts the separation process.

A solvent reservoir, sample injector, pressure pump, HPLC column, and diode detector are typical parts of an HPLC system. The mixture to be separated is injected at the bottom of the HPLC column to start the process. Concurrently, a suitable solvent is added to the solvent reservoir. When the tap is opened, a pressure pump forces the solvent downhill, where it combines with the injected sample. The mixture that is produced then goes into the diode detector, where compound separation takes place. After eliminating trash, the detector sends the finished product to processing facilities for additional examination or gathering (Ekbbal et al. 2024).

6.5.3 Membrane Filtration and Dialysis

Polyphenols are characterised by their poor water solubility. Possible intermolecular interactions between polyphenols and polysaccharides might account for the high recovery of polyphenols. The centrifuged wine lees include compounds with identical molecular weights that hinder efficient membrane fractionation. As a result, membrane processing works better for recovering and concentrating bioactive substances, particularly those that are sensitive to temperature, like polyphenols.

To extract polyphenols and organic acids from wine lees, a study investigated an environmentally friendly, two-step membrane technique that involves *ultrafiltration* and *nanofiltration*. Compounds were identified using HPLC, and ferric reducing and DPPH tests were utilised to measure antioxidant activity. A liquid concentrate rich in polyphenols (1351 ppm) was produced by the technique, and it was spray-dried into a powder that retained its antioxidant qualities. Thiobarbituric acid reactive substances (TBARSs) and oxymyoglobin testing verified that it successfully decreased lipid oxidation when added to a dietary system. According to the study, one helpful technique for obtaining useful bioactive chemicals is membrane separation (Filippou, Mitrouli, and Vareltzis 2022).

6.5.4 Drying and Concentration of the Extract

The polyphenol extract is frequently still liquid after purification. It is dried and condensed to make it more stable and simpler to use or store. To create novel pulse components, a study investigated the pilot-scale spray drying of plant-based microparticles derived from pea protein (PP) and blackcurrant polyphenols (BC). After being described, the resultant PP-BC microparticles were utilised in protein bars. They demonstrated good pH-dependent solubility and a high phenolic content (48.2 mg GAE/g). Over 8 weeks, encapsulated polyphenols in PP-BC exhibited noticeably greater stability (95.2% retention) than non-encapsulated ones (78.2% retention). Furthermore, bars produced using PP-BC exhibited better sensory qualities. This process offers a viable way to create pulse-based meals that are palatable, wholesome, and stable (Hoskin et al. 2025).

6.6 FORMULATION DEVELOPMENT STRATEGIES

6.6.1 Selection of Carriers and Excipients

Notwithstanding their health advantages, polyphenols have drawbacks, including limited absorption; an unpleasant taste; instability in light, oxygen, and heat environments; and interactions with dietary ingredients that lower nutritional value. In order to get around this, polyphenols are encapsulated in carriers, which shield them from deterioration throughout digestion, storage, and processing.

To guarantee structural integrity, increase solubility, improve absorption, and enable targeted distribution without compromising sensory qualities, polyphenols are encased in carrier systems in a promising technique known as *nanoencapsulation*. In addition to stabilising polyphenols, a suitable carrier formulation enables regulated and modulated release.

Because they enhance stability, water solubility, and oral bioavailability, lipid-based nanocarriers – such as solid lipid nanoparticles and nanostructured lipid carriers – are often employed, particularly for hydrophobic polyphenols like resveratrol and curcumin. These carriers are appropriate for oral usage since they are frequently composed of generally recognised as safe (GRAS) ingredients. Lipid carriers have demonstrated efficacy with hydrophilic polyphenols, despite being commonly employed for hydrophobic substances.

To put it straightforwardly, carriers are essential for improving polyphenols' functioning and distribution, which are what makes them suitable for use in nutraceutical applications (Pimentel-Moral et al. 2018).

6.6.2 Encapsulation Techniques (Nanoemulsion, Liposomes, Microencapsulation)

The majority of phenolic compounds have a disagreeable taste, which further restricts their use. These limitations can be efficiently mitigated by using encapsulated polyphenols rather than free components. The technologies of encapsulation of polyphenols include spray drying, coacervation, liposome entrapment, inclusion complexation, co-crystallisation, microencapsulation, freeze drying, yeast encapsulation, and nanoemulsion.

Microencapsulation. Small vesicles or particles, known as microcapsules, can have sizes ranging from sub-micron to several millimetres. Although a variety of shapes may be created for encapsulation, two main shapes are frequently observed: aggregates, which have several cores embedded in a matrix, and mononuclear capsules, which have a single core encased in a shell. The process methods and the wall materials used to make the capsules have an impact on their particular forms in various systems.

There are several methods for encapsulation. Encapsulating bioactive substances generally involves three steps: (1) creating a wall around the item to be encapsulated, (2) making sure that unwanted leaks don't happen, and (3) ensuring that unwanted materials are kept out.

Liposomes are colloidal particles with an aqueous gap or spaces encapsulated by a membrane system made of lipid bilayers. Because liposomes include both lipid and aqueous phases, they may be used to entrap, transport, and release molecules that are amphiphilic, lipid soluble, and water soluble.

Emulsion technology is often used to encapsulate bioactives in aqueous solutions that may either be employed immediately in their liquid condition or, following emulsification, can be dried to create powders (for example, by spraying, roller drying, or freeze drying). As a result, it truly is a step in the encapsulation process. In essence, an emulsion is made up of two or more immiscible liquids, often water and oil, with one of the liquids distributed as tiny, spherical droplets in the other (Fang and Bhandari 2010).

6.6.3 Stability Enhancement Approaches

Plant polyphenols provide important health benefits, such as hypoglycaemic, anti-inflammatory, and antioxidant activities. Chemical instability, however, frequently restricts their use in food items. This study examines innovative composite delivery methods intended to improve the stability and bioavailability of plant polyphenols.

High-value substances like polyphenols (flavonoids, stilbenoids, lignans, ellagic acids, and polymers of phenolic acids and flavonoids, specifically hydrolysable and condensed tannins) are in high demand. In this regard, the market's continuous expansion is facilitated by the creation of stability-friendly processing techniques, as well as the regulation of humidity, heating, pH, oxygen, light, interaction with metal ions and other food ingredients, enzymes, high pressure, and the in vitro digestive process. It may also determine the structure of degradation products and their rates of degradation with the use of mass spectrometry and nuclear magnetic resonance (NMR) analysis. *Chemical modification, nanotechnologies, encapsulation, spray drying, emulsification, lyophilisation*, and other processes can all be used to increase their stability. Redesigning flowsheets based on scientific evidence of non-covalent and covalent polyphenol–protein interactions; cold plasma use versus thermal methods; short-time treatment and low-temperature, long-time treatment; and storage conditions on their stability, bioaccessibility, and bioavailability are made possible by the potential to combine various operations for this stability purpose that offers intriguing new insights. Phenolic compound extraction, purification, and preservation can benefit greatly from the use of cutting-edge sustainable technology (Cao et al. 2021).

6.7 OPTIMISATION OF THE FORMULATION PROCESS

Optimisation of the extraction of polyphenols is necessary to achieve increased yield, preserve bioactivity, and allow economic scaling. Statistical methodologies such as factorial design and response surface methodology (RSM) are typically used to optimise the critical parameters of the solvent, temperature, time of extraction, and solid-to-solvent ratio.

Selection of the solvent is critical; polar solvents like ethanol, methanol, and acetone possess good ability in isolating a broad range of phenolic compounds (Galanakis et al. 2013). Ethanol–water mixtures are especially suitable as they possess maximum polarity. The concentration of the solvent also affects selectivity and efficiency of recovery. Temperature influences diffusion rates and solubility. Moderate temperatures (40–60 °C) are optimal, favouring extraction without degrading heat-sensitive compounds like anthocyanins and catechins (Antony and Farid 2022). Reducing particle size exposes more surface area for solvent interaction, though very fine powders may cause filtration problems. Agitation and pH also affect extraction, with low pH stabilising anthocyanins but possibly reducing flavanol solubility.

Cellulase- or pectinase-assisted extraction has been found to increase release by breaking down plant cell walls. Enzyme conditions, however, must be strictly maintained lest degradation of polyphenols take place. Co-extracted species like proteins, lipids, and sugars reduce purity, and therefore defatting and protein removal pre-treatment become mandatory (Gerde et al. 2013).

The formulation should take into account the light, oxygen, and temperature sensitivity of polyphenols. Stabilisers and encapsulating agents like spray drying, freeze drying, and nanoencapsulation promote stability and shelf life. These processes should be compatible with the target compounds' physicochemical properties. Lastly, efficient optimisation preserves bioactivity retention together with scaling, sensory quality, and economic viability.

6.7.1 Design of Experiments and Response Surface Methodology

Design of experiments (DOE) is a scientific approach to understanding how different variables influence critical responses. DOE finds application in polyphenol formulation to unveil the best combinations of levels of ingredients and process conditions, i.e., solvent concentration, mixing speed, pH, and temperature.

DOE is less time- and cost-consuming than sequential one-variable-at-a-time experimentation. DOE provides a global picture of factor interactions, which is critical when creating intricate functional products. RSM is a more advanced version of DOE that utilises mathematical and statistical techniques in modelling and solving problems with many variables that affect a response (Bezerra et al. 2008). For instance, RSM has been employed to simulate maximum encapsulation efficiency of polyphenols in gum arabic and maltodextrin mixtures. The method enables scientists to determine the ideal wall material ratio that can ensure maximum antioxidant retention in spray drying (Mahdavi et al. 2016).

Software packages, such as Design-Expert, Minitab, and JMP, are mostly used to implement DOE and RSM (Rampado and Peer 2023). The software comes with predictive models, contour plots, and optimisation parameters, which aid in decision-making during product development. Application of DOE and RSM not only optimises polyphenol yield but also enhances reproducibility and product uniformity, which are critical for scaling up and regulatory needs (Belwal et al. 2020). It is a data-driven, cutting-edge approach towards food and nutraceutical development.

6.7.2 Stability and Shelf-Life Testing

Polyphenols are susceptible to light-induced degradation, oxygen, moisture, temperature fluctuations, and enzymatic breakdown. Hence, determining their stability in various storage conditions becomes of utmost significance in guaranteeing the quality and activity of polyphenol-based products. Stability tests are conducted both in real and accelerated conditions to simulate long-term and stress storage conditions. Real-time testing simulates regular storage conditions to observe natural degradation patterns over time, and accelerated testing places the product under increased temperature and humidity to predict the shelf life of the product and identify possible stability problems earlier. These tests contribute to the assessment of the stability of polyphenolic compounds under various conditions (e.g., light, temperature, humidity) (Corradini 2018).

Key parameters measured during stability testing include:

1. Total polyphenol content (TPC) via Folin–Ciocalteu assay.
2. Antioxidant activity using DPPH or ABTS radical scavenging assays.
3. HPLC to monitor individual compound degradation.
4. Moisture content and water activity (aw).
5. Microbial stability.
6. Sensory evaluation at various intervals.

Encapsulation processes such as spray drying and coacervation are frequently used to stabilise polyphenols (Fang and Bhandari 2010). Wall materials, including maltodextrin, whey proteins, or starches with modification, serve as protecting agents for light and oxidation. Encapsulated products always have enhanced antioxidant activity retention over storage time (Deladino et al. 2008). Inert packaging atmospheres, especially nitrogen flushing, further protect polyphenols from oxidative degradation. Non-transparent and airtight packages are also essential (Cao et al. 2021). Shelf-life labelling, regulatory approval, and customer confidence are supported by stability testing. Stability testing also informs formulation, packing, and processing method changes to ensure that the final product continues to function and remain visually appealing within its intended storage life.

6.7.3 Scale-Up Considerations

Large-scale production involves numerous technical and practical challenges in scaling up from laboratory extraction. Whereas laboratory rooms allow precise control, commercial facilities need to ensure consistency, safety, and affordability without compromising the quality of the product. Among the main concerns is equipment compatibility. Industrial extractors, encapsulators, and mixers must mimic laboratory conditions to maintain phenolic integrity. An example is spray-drying scale-up, which requires control of inlet and outlet temperatures to prevent degradation of polyphenols (Verma and Singh 2015). Process variables like mixing time, shear force, and thermal exposure have to be optimised on a larger scale. Any change can lead to loss of antioxidant activity or sensory attributes.

Pilot-scale experiments are thus required for verification of the reproducibility of formulation outcomes (Pickardt et al. 2015). Batch-to-batch variation must be kept to a minimum using rigorous quality control measures. Real-time monitoring systems such as process analytical technology (PAT) are utilised to monitor important parameters and ensure product homogeneity (Simon et al. 2015). Regulatory compliance is also important. Additives to the formula, like process aids and additives and solvents, are pharmaceutical grade or food grade, as appropriate to the final application. Good labelling, traceability, and safety testing are important factors in commercial feasibility. The raw material supply chain also has an impact on scalability. Seasonality, price, and availability of plant materials must be considered to maintain constant production. Procurement of farmers or using food industry by-products (e.g., grape pomace or citrus peel) can provide sustainable sourcing opportunities (Ferdosh et al. 2025). Finally, economic sustainability must be considered. Technical performance, energy use, labour, and packaging must be balanced with the intended retail price and demand. Success in scale-up relies on matching technical performance, regulatory compliance, and commercial objectives.

6.8 CONCLUSION

The development of polyphenol-based nutraceutical formulations is a challenging but promising endeavour that harnesses improvements in plant biochemistry, extraction technology, formulation science, and regulatory compliance. Polyphenols, renowned for their antioxidant, anti-inflammatory, and disease-preventive attributes, have significant potential in enhancing human health via functional foods and supplements. Nonetheless, obstacles such as instability, inadequate solubility, and diminished bioavailability persist in constraining their clinical and economic feasibility. This chapter highlighted the significance of a methodical approach, commencing with the meticulous selection of plant materials shaped by genetic, environmental, and harvesting variables. The implementation of both traditional and environmentally friendly extraction technologies, such as ultrasound-assisted, microwave-assisted, supercritical fluid, and enzyme-assisted procedures, increases polyphenol yield while maintaining bioactivity. Downstream purification and concentration processes, together with encapsulation techniques such as nanoemulsions, liposomes, and spray drying, have demonstrated substantial enhancement in the shelf life, stability, and functional efficacy of formulations. Optimisation with statistical tools like DOE and RSM guarantees reproducibility and scalability, while sophisticated characterisation and quality control techniques facilitate standardisation between batches. Real-time and expedited stability testing, along with packaging advancements, further ensure product integrity during storage and distribution. Notwithstanding advancements, enduring challenges – such as variable bioavailability, formulation instability, and regulatory disunity – necessitate ongoing interdisciplinary cooperation. The future involves the integration of intelligent delivery systems, sustainable processing, and standardised protocols to provide evidence-based, plant-derived health treatments to a worldwide audience. This chapter is intended to function as a pragmatic guide and strategic resource for converting polyphenol knowledge into advanced nutraceutical breakthroughs.

REFERENCES

Abbaspour, Leila, Nazila Ghareaghajlou, Mohammad Reza Afshar Mogaddam, and Zahra Ghasempour. 2024. "An Innovative Technique for the Extraction and Stability of

Polyphenols Using High Voltage Electrical Discharge." *Current Research in Food Science* 9: 100928. https://doi.org/10.1016/j.crfs.2024.100928.

Abubakar, A.R., and M. Haque. 2020. "Preparation of Medicinal Plants: Basic Extraction and Fractionation Procedures for Experimental Purposes." *J Pharm Bioallied Sci* 12 (1): 1–10. https://doi.org/10.4103/jpbs.JPBS_175_19.

Aktaş, Havva, and Marcin A. Kurek. 2024. "Deep Eutectic Solvents for the Extraction of Polyphenols from Food Plants." *Food Chemistry* 444 (138629). https://doi.org/10.1016/j.foodchem.2024.138629.

Alara, O.R., N.H. Abdurahman, and C.I. Ukaegbu. 2021. "Extraction of Phenolic Compounds: A Review." *Curr Res Food Sci* 4: 200–214. https://doi.org/10.1016/j.crfs.2021.03.011.

Alchera, F., M. Ginepro, and G. Giacalone. 2022. "Microwave-Assisted Extraction of Polyphenols from Blackcurrant By-Products and Possible Uses of the Extracts in Active Packaging." *Foods* 11 (18). https://doi.org/10.3390/foods11182727.

Allay, Aymane, Chaymae Benkirane, Abdessamad Ben Moumen, Marie-Laure Fauconnier, Hamza Bouakline, Jacques Nkengurutse, Hana Serghini Caid, Ahmed Elamrani, and Farid Mansouri. 2025. "Optimizing Ethanol-Modified Supercritical CO_2 Extraction for Enhanced Bioactive Compound Recovery in Hemp Seed Oil." *Scientific Reports* 15 (1): 8551. https://doi.org/10.1038/s41598-025-91441-x.

Almanza-Aguilera, Enrique, Daniel Guiñón-Fort, Aurora Perez-Cornago, Miriam Martínez-Huélamo, Cristina Andrés-Lacueva, Anne Tjønneland, Anne Kirstine Eriksen, Verena Katzke, Rashmita Bajracharya, and Matthias B. Schulze. 2023. "Intake of the Total, Classes, and Subclasses of (Poly) Phenols and Risk of Prostate Cancer: A Prospective Analysis of the EPIC Study." *Cancers* 15 (16): 4067. https://doi.org/10.3390/cancers15164067.

Álvarez, S.A., N.E. Rocha-Guzmán, and J.A. Gallegos-Infante. 2025. "Pressurized Liquid Extraction of Oak Leaf Polyphenols: Solvent Selection via Hansen Parameters, Antioxidant Evaluation and Monoamine-Oxidase-a Inhibition Analysis." *Food Chem* 463 (Pt 2). https://doi.org/10.1016/j.foodchem.2024.141212.

Alvarez-Rivera, G., M. Bueno, D. Ballesteros-Vivas, J.A. Mendiola, and E. Ibañez. 2020. "Pressurized Liquid Extraction." In *Liquid-Phase Extraction*. Elsevier: 375–98. https://doi.org/10.1016/B978-0-12-816911-7.00013-X.

Amangeldinova, Madina, Mehmet Ersatır, Adem Necip, Mustafa Abdullah Yilmaz, Mehmet Cimentepe, Nataliya Kudrina, Nina V. Terletskaya, Ozge Ozturk Cimentepe, and Metin Yildirim. 2024. "Simultaneous Quantitative Screening of 53 Phytochemicals from Rheum Tataricum L. Roots: A Comparative Study of Supercritical CO2, Subcritical Ethanol, and Ultrasound-Assisted Extraction for Enhanced Antioxidant, Antibacterial Activities, and Molecular Docking Study." *Frontiers in Plant Science* 15 (December): 1513875. https://doi.org/10.3389/fpls.2024.1513875.

Ameer, K., H.M. Shahbaz, and J.H. Kwon. 2017. "Green Extraction Methods for Polyphenols from Plant Matrices and Their Byproducts: A Review." *Compr Rev Food Sci Food Saf* 16 (2): 295–315. https://doi.org/10.1111/1541-4337.12253.

Antony, Anila, and Mohammed Farid. 2022. "Effect of Temperatures on Polyphenols during Extraction." *Applied Sciences* 12 (4): 2107. https://doi.org/10.3390/app12042107.

Aresta, A., N. Vietro, and P. Cotugno. 2024. "Supercritical Fluid Extraction of Bioactive Components from Apple Peels and Their Modulation of Complex I Activity in Isolated Mitochondria." *Antioxidants* 13 (3). https://doi.org/10.3390/antiox13030307.

Belwal, Tarun, Farid Chemat, Petras Rimantas Venskutonis, Giancarlo Cravotto, Durgesh Kumar Jaiswal, Indra Dutt Bhatt, Hari Prasad Devkota, and Zisheng Luo. 2020. "Recent Advances in Scaling-up of Non-Conventional Extraction Techniques: Learning from Successes and Failures." *TrAC Trends in Analytical Chemistry* 127 (115895). https://doi.org/10.1016/j.trac.2020.115895.

Bezerra, Marcos Almeida, Ricardo Erthal Santelli, Eliane Padua Oliveira, Leonardo Silveira Villar, and Luciane Amélia Escaleira. 2008. "Response Surface Methodology (RSM) as a Tool for Optimization in Analytical Chemistry." *Talanta* 76 (5): 965–77. https://doi.org/10.1016/j.talanta.2008.05.019.

Bhatta, Sagar, Tatjana Stevanovic Janezic, and Cristina Ratti. 2020. "Freeze-Drying of Plant-Based Foods." *Foods* 9 (1): 87. https://doi.org/10.3390/foods9010087.

Bitwell, C., S.S. Indra, C. Luke, and M.K. Kakoma. 2023. "A Review of Modern and Conventional Extraction Techniques and Their Applications for Extracting Phytochemicals from Plants." *Sci Afr* 19: e01585. https://doi.org/10.1016/j.sciaf.2023.e01585.

Bo', Del, Stefano Bernardi Cristian, Mirko Marino, Marisa Porrini, Massimiliano Tucci, Simone Guglielmetti, Antonio Cherubini, Barbara Carrieri, Benjamin Kirkup, and Paul Kroon. 2019. "Systematic Review on Polyphenol Intake and Health Outcomes: Is There Sufficient Evidence to Define a Health-Promoting Polyphenol-Rich Dietary Pattern?" *Nutrients* 11 (6): 1355. https://doi.org/10.3390/nu11061355.

Brianceau, S., M. Turk, X. Vitrac, and E. Vorobiev. 2016. "High Voltage Electric Discharges Assisted Extraction of Phenolic Compounds from Grape Stems: Effect of Processing Parameters on Flavan-3-Ols, Flavonols and Stilbenes Recovery." *Innov Food Sci Emerg Technol* 35: 67–74. https://doi.org/10.1016/j.ifset.2016.04.006.

Brunner, G. 2005. "Supercritical Fluids: Technology and Application to Food Processing." *J Food Eng* 67 (1–2): 21–33. https://doi.org/10.1016/j.jfoodeng.2004.05.060.

Calderón-Oliver, M., and E. Ponce-Alquicira. 2021. "Environmentally Friendly Techniques and Their Comparison in the Extraction of Natural Antioxidants from Green Tea, Rosemary, Clove, and Oregano." *Molecules* 26 (7). https://doi.org/10.3390/molecules26071869.

Cao, Hui, Oznur Saroglu, Ayse Karadag, Zoriţa Diaconeasa, Gianni Zoccatelli, Carlos Adam Conte-Junior, and Gustavo A. Gonzalez-Aguilar. 2021. "Available Technologies on Improving the Stability of Polyphenols in Food Processing." *Food Frontiers* 2 (2): 109–39. https://doi.org/10.1002/fft2.65.

Carabias-Martínez, R., E. Rodríguez-Gonzalo, P. Revilla-Ruiz, and J. Hernández-Méndez. 2005. "Pressurized Liquid Extraction in the Analysis of Food and Biological Samples." *J Chromatogr A* 1089 (1–2): 1–17. https://doi.org/10.1016/j.chroma.2005.06.072.

Chatzimitakos, T. 2024. "Moringa oleifera PLE Optimization." *Int J Mol Sci* 25 (9).

Chauca-Cerrutti, A. 2024. "Supercritical Fluid Extraction of Pouteria Lucuma." *Front Chem* 12 (1491479).

Che, H. 2024. "Ultrasound-Assisted Extraction of Phyllanthi Fructus." *Ultrason Sonochem* 111 (107083).

Chemat, F., M.A. Vian, and G. Cravotto. 2012. "Green Extraction of Natural Products: Concept and Principles." *Int J Mol Sci* 13: 8615–27. https://doi.org/10.3390/ijms13078615.

Chen, S., X.H. Xing, J.J. Huang, and M.S. Xu. 2011. "Enzyme-Assisted Extraction of Flavonoids from Ginkgo Biloba Leaves: Improvement Effect of Flavonol Transglycosylation Catalyzed by Penicillium Decumbens Cellulase." *Enzyme Microb Technol* 48 (1): 100–5. https://doi.org/10.1016/j.enzmictec.2010.09.017.

Cherian, Sam, Kushvar Mammadova Hacisayidli, Renju Kurian, and Allan Mathews. 2023. "Therapeutically Important Bioactive Compounds of the Genus Polygonum L. and Their Possible Interventions in Clinical Medicine." *Journal of Pharmacy and Pharmacology* 75 (3): 301–27. https://doi.org/10.1093/jpp/rgac105.

Chmelová, D., D. Škulcová, B. Legerská, M. Horník, and M. Ondrejovič. 2020. "Ultrasonic-Assisted Extraction of Polyphenols and Antioxidants from Picea Abies Bark." *J Biotechnol* 315: 25–33. https://doi.org/10.1016/j.jbiotec.2020.04.003.

Choi, Dong-Young, Young-Jung Lee, Jin Tae Hong, and Hwa-Jeong Lee. 2012. "Antioxidant Properties of Natural Polyphenols and Their Therapeutic Potentials for Alzheimer's Disease." *Brain Research Bulletin* 87 (2): 144–53. https://doi.org/10.1016/j.brainresbull.2011.11.014.

Chuo, S.C., H.M. Nasir, and S.H. Mohd-Setapar. 2022. "A Glimpse into the Extraction Methods of Active Compounds from Plants." *Crit Rev Anal Chem* 52 (4): 667–96. https://doi.org/10.1080/10408347.2020.1820851.

Concha, J., C. Soto, R. Chamy, and M.E. Zuniga. 2004. "Enzymatic Pretreatment on Rose-Hip Oil Extraction: Hydrolysis and Pressing Conditions." *J Am Oil Chem Soc* 81: 549–52. https://doi.org/10.1007/s11746-004-0933-2.

Corradini, Maria G. 2018. "Shelf Life of Food Products: From Open Labeling to Real-Time Measurements." *Annual Review of Food Science and Technology* 9 (March): 251–69. https://doi.org/10.1146/annurev-food-030117-012433.

Corrêa, Rúbia C.G., Rosane M. Peralta, Charles W.I. Haminiuk, Giselle Maria Maciel, Adelar Bracht, and Isabel C.F.R. Ferreira. 2016. "The Past Decade Findings Related with Nutritional Composition, Bioactive Molecules and Biotechnological Applications of Passiflora Spp. (Passion Fruit)." *Trends in Food Science & Technology* 58: 79–95. https://doi.org/10.1016/j.tifs.2016.10.006.

Deladino, Lorena, Pablo S. Anbinder, Alba S. Navarro, and Miriam N. Martino. 2008. "Encapsulation of Natural Antioxidants Extracted from Ilex Paraguariensis." *Carbohydrate Polymers* 71 (1): 126–34. https://doi.org/10.1016/j.carbpol.2007.05.030.

Domínguez-Rodríguez, Gloria, María Luisa Marina, and Merichel Plaza. 2021. "Enzyme-Assisted Extraction of Bioactive Non-Extractable Polyphenols from Sweet Cherry (Prunus Avium L.) Pomace." *Food Chemistry* 339 (March): 128086. https://doi.org/10.1016/j.foodchem.2020.128086.

Duke, K., C. Syeunda, J.F. Brantsen, S. Nindawat, and J.M. Awika. 2024. "Polyphenol Recovery from Sorghum Bran Waste by Microwave Assisted Extraction: Structural Transformations as Affected by Grain Phenolic Profile." *Food Chem* 444 (138645). https://doi.org/10.1016/j.foodchem.2024.138645.

Ekbbal, Rustam, Mahaveer Singh, Dr Gaurav, and Sapna Salar. 2024. "Chromatography Techniques for Isolation of Phytoconstituents from Medicinal Plants." In *Futuristic Trends in Pharmacy & Nursing*, edited by Dr Dharmendra Singh, Dr Gaurav Gautam, Ms Ishita Nath, Mr Jagat Pal Yadav, and First, Vol. 3. Iterative International Publishers, Selfypage Developers Pvt Ltd: 257–76. https://www.iipseries.org/viewpaper.php?pid=4795&pt=chromatography-techniques-for-isolation-of-phytoconstituents-from-medicinal-plants.

El Gharras, Hasna. 2009. "Polyphenols: Food Sources, Properties and Applications–a Review." *International Journal of Food Science and Technology* 44 (12): 2512–18. https://doi.org/10.1111/j.1365-2621.2009.02077.x.

El Kantar, Sally, Nadia Boussetta, Hiba N. Rajha, Richard G. Maroun, Nicolas Louka, and Eugène Vorobiev. 2018. "High Voltage Electrical Discharges Combined with Enzymatic Hydrolysis for Extraction of Polyphenols and Fermentable Sugars from Orange Peels." *Food Research International* 107 (May): 755–62. https://doi.org/10.1016/j.foodres.2018.01.070.

El Kantar, Sally, Hiba N. Rajha, Nadia Boussetta, Eugène Vorobiev, Richard G. Maroun, and Nicolas Louka. 2019. "Green Extraction of Polyphenols from Grapefruit Peels Using High Voltage Electrical Discharges, Deep Eutectic Solvents and Aqueous Glycerol." *Food Chemistry* 295 (October): 165–71. https://doi.org/10.1016/j.foodchem.2019.05.111.

Enaru, Bianca, Georgiana Drețcanu, Teodora Daria Pop, Andreea Stănilă, and Zorița Diaconeasa. 2021. "Anthocyanins: Factors Affecting Their Stability and Degradation." *Antioxidants* 10 (12). https://doi.org/10.3390/antiox10121967.

Eskilsson, C.S., and E. Björklund. 2000. "Analytical-Scale Microwave-Assisted Extraction." *J Chromatogr A* 902 (1): 227–50. https://doi.org/10.1016/S0021-9673(00)00921-3.

Fang, Zhongxiang, and Bhesh Bhandari. 2010. "Encapsulation of Polyphenols – a Review." *Trends in Food Science & Technology* 21 (10): 510–23. https://doi.org/10.1016/j.tifs.2010.08.003.

Farhang, M., A.R. Akbarzadeh, M. Rabbani, and A.M. Ghadiri. 2022. "A Retrospective-Prospective Review of Suzuki-Miyaura Reaction: From Cross-Coupling Reaction to Pharmaceutical Industry Applications." *Polyhedron* 227 (116124). https://doi.org/10.1016/j.poly.2022.116124.

Ferdosh, Sahena, Nurul Ashikin Abdul Bari, Bulan Wu, and Md Zaidul Islam Sarker. 2025. "Supercritical Fluid Extraction of Phenolics from Anisophyllea Disticha (Jack) Baill. and Evaluation of Their Antioxidant Activities." *The Natural Products Journal* 15 (2): 070623217749. https://doi.org/10.2174/2210315513666230607123047.

Ferreira, Sandrine S., Amélia M. Silva, and Fernando M. Nunes. 2018. "Citrus Reticulata Blanco Peels as a Source of Antioxidant and Anti-Proliferative Phenolic Compounds." *Industrial Crops and Products* 111: 141–48. https://doi.org/10.1016/j.indcrop.2017.10.009.

Feudjio, Césaire, Guy Sedar Singor Njateng, Mathilde Julie Klang, Muhammad Arfat Yameen, Muhammad Ahsan Khan, and Jules-Roger Kuiate. 2022. "Response Surface Methodology Applied to the Optimization of the Preparation of Antioxidant and Antidiabetic Extracts from Phragmanthera Capitata (Spreng.) Balle: Effect of Particle Size, Powder-to-Solvent Ratio, and Temperature." Edited by Ihsan Ul Haq. *Evidence-Based Complementary and Alternative Medicine* 2022 (September): 1–18. https://doi.org/10.1155/2022/8397250.

Filippou, Polychronis, Soultana T. Mitrouli, and Patroklos Vareltzis. 2022. "Sequential Membrane Filtration to Recover Polyphenols and Organic Acids from Red Wine Lees: The Antioxidant Properties of the Spray-Dried Concentrate." *Membranes* 12 (4): 353. https://doi.org/10.3390/membranes12040353.

Galanakis, C.M., V. Goulas, S. Tsakona, G.A. Manganaris, V., and Gekas. 2013. "A Knowledge Base for The Recovery of Natural Phenols with Different Solvents." *International Journal of Food Properties* 16 (2): 382–96. https://doi.org/10.1080/10942912.2010.522750.

García-Martín, J.F., C.H. Feng, N.M. Domínguez-Fernández, and P. Álvarez-Mateos. 2023. "Microwave-Assisted Extraction of Polyphenols from Bitter Orange Industrial Waste and Identification of the Main Compounds." *Life (Basel)* 13 (9). https://doi.org/10.3390/life13091864.

Gavrila, A.I., E.J. Damian, and A. Rosca. 2025. "Optimization of Microwave-Assisted Extraction of Polyphenols from Crataegus Monogyna L." *Antioxidants (Basel)* 14 (3). https://doi.org/10.3390/antiox14030357.

Gerde, Jose A., Tong Wang, Linxing Yao, Stephanie Jung, Lawrence A. Johnson, and Buddhi Lamsal. 2013. "Optimizing Protein Isolation from Defatted and Non-Defatted Nannochloropsis Microalgae Biomass." *Algal Research* 2 (2): 145–53. https://doi.org/10.1016/j.algal.2013.02.001.

Getachew, A.T., S.L. Holdt, A.S. Meyer, and C. Jacobsen. 2022. "Effect of Extraction Temperature on Pressurized Liquid Extraction of Bioactive Compounds from Fucus Vesiculosus." *Mar Drugs* 20 (4). https://doi.org/10.3390/md20040263.

Giacomini, E., S. Rupiani, L. Guidotti, M. Recanatini, and M. Roberti. 2016. "The Use of Stilbene Scaffold in Medicinal Chemistry and Multi-Target Drug Design." *Curr Med Chem* 23 (23): 2439–89. https://doi.org/10.2174/0929867323666160510125648.

Grabska-Kobyłecka, Izabela, Piotr Szpakowski, Aleksandra Król, Dominika Książek-Winiarek, Andrzej Kobyłecki, Andrzej Głąbiński, and Dariusz Nowak. 2023. "Polyphenols and Their Impact on the Prevention of Neurodegenerative Diseases and Development." *Nutrients* 15 (15): 3454. https://doi.org/10.3390/nu15153454.

Granato, D., M. Fidelis, and M. Haapakoski. 2022. "Enzyme-Assisted Extraction of Anthocyanins and Other Phenolic Compounds from Blackcurrant (Ribes Nigrum L.) Press Cake: From Processing to Bioactivities." *Food Chem* 391 (133240). https://doi.org/10.1016/j.foodchem.2022.133240.

Guasch-Ferré, Marta, Jordi Merino, Qi Sun, Montse Fitó, and Jordi Salas-Salvadó. 2017. "Dietary Polyphenols, Mediterranean Diet, Prediabetes, and Type 2 Diabetes: A Narrative Review of the Evidence." *Oxidative Medicine and Cellular Longevity* 2017: 1–16. https://doi.org/10.1155/2017/6723931.

Gülçin, İlhami, S. Beyza Öztürk Sarikaya Fevzi Topal, Ercan Bursal, Gökhan Bilsel, and Ahmet C. Gören. 2011. "Polyphenol Contents and Antioxidant Properties of Medlar (Mespilus Germanica L." *Records of Natural Products* 5 (3): 158.

Hasni, Soumaya, Hajer Riguene, Jose A. Mendiola, Elena Ibáñez, Lidia Montero, Gloria Domínguez-Rodríguez, Hanene Ghazghazi, Ghayth Rigane, and Ridha Ben Salem. 2024. "Optimization of a Pressurized Extraction Process Based on a Ternary Solvent System for the Recovery of Neuroprotective Compounds from Eucalyptus Marginata Leaves." *International Journal of Molecular Sciences* 26 (1): 94. https://doi.org/10.3390/ijms26010094.

Hassanloofard, Z., M. Zandi, M. Gharekhani, A. Ganjloo, and L. Roufegarinejad. 2024. "Novel Green Method for Enhancing the Extraction Efficiency of Bioactive Compounds: Falcaria Vulgaris Extract." *J Food Biochem* 2024 (9974099). https://doi.org/10.1155/2024/9974099.

Hlatshwayo, S., N. Thembane, S.B.N. Krishna, N. Gqaleni, and M. Ngcobo. 2025. "Extraction and Processing of Bioactive Phytoconstituents from Widely Used South African Medicinal Plants for the Preparation of Effective Traditional Herbal Medicine Products: A Narrative Review." *Plants* 14 (2). https://doi.org/10.3390/plants14020206.

Hoskin, Roberta Targino, Fábio Gonçalves Macêdo Medeiros, Juan Ruano, Mary Grace, Abhishek Shetty, Keith B. Rodenhausen, Mary Ann Lila, and Marvin Moncada. 2025. "Pilot-Scale Production of Spray Dried Pulse Protein-Polyphenol Particles and Protein Bars." *Journal of Food Process Engineering* 48 (4): 70121. https://doi.org/10.1111/jfpe.70121.

Huamán-Castilla, N.L. 2024. "PLE from Olive Leaves." *Foods* 13 (2).

Huamán-Castilla, Nils Leander, Erik Edwin Allcca-Alca, Frank Hervas Nina, Nilton Cesar León-Calvo, Franz Zirena Vilca, and Yesica Luz Vilcanqui Chura. 2024. "Pressurized Liquid Extraction of Antioxidant and α-Amylase-Inhibitory Compounds from Red Seaweed Using Water–Ethanol Mixtures." *Molecules* 29 (21): 5018. https://doi.org/10.3390/molecules29215018.

Huang, Doudou, Yun Jiang, Wansheng Chen, Fengyan Yao, and Lianna Sun. 2014. "Polyphenols with Anti-Proliferative Activities from Penthorum Chinense Pursh." *Molecules* 19 (8): 11045–55. https://doi.org/10.3390/molecules190811045.

Kataoka, H. n.d. "Pharmaceutical Analysis | Sample Preparation." *Encyclopedia of Analytical Science* 2019: 231–55. https://doi.org/10.1016/B978-0-12-409547-2.14358-6.

Kawabata, K., R. Mukai, and A. Ishisaka. 2015. "Quercetin and Related Polyphenols: New Insights and Implications for Their Bioactivity and Bioavailability." *Food Funct* 6 (5): 1399–1417. https://doi.org/10.1039/C5FO00138E.

Keylor, M.H., B.S. Matsuura, and C.R. Stephenson. 2015. "Chemistry and Biology of Resveratrol-Derived Natural Products." *Chem Rev* 115 (17): 8976–9027. https://doi.org/10.1021/acs.chemrev.5b00250.

Kim, J.H., H.K. Shin, and C.S. Seo. 2014. "Optimization of the Extraction Process for the Seven Bioactive Compounds in Yukmijihwang-Tang, an Herbal Formula, Using Response Surface Methodology." *Pharmacogn Mag* 10 (Suppl 3). https://doi.org/10.4103/0973-1296.139798.

Kumar, Santosh, Md Minhajul Abedin, Ashish Kumar Singh, and Saurav Das. 2020. "Role of Phenolic Compounds in Plant-Defensive Mechanisms." In *Plant Phenolics in Sustainable Agriculture*, edited by Rafiq Lone, Razia Shuab, and Azra N, Kamili. Singapore: Springer: 517–32. http://link.springer.com/10.1007/978-981-15-4890-1_22.

Kumar Singh, Amit, Célia Cabral, Ramesh Kumar, Risha Ganguly, Harvesh Kumar Rana, Ashutosh Gupta, Maria Rosaria Lauro, Claudia Carbone, Flávio Reis, and Abhay K. Pandey. 2019. "Beneficial Effects of Dietary Polyphenols on Gut Microbiota and Strategies to Improve Delivery Efficiency." *Nutrients* 11 (9): 2216. https://doi.org/10.3390/nu11092216.

Lai, C., M. Huang, Q. Xiong, Y. Liang, Y. Jiang, and J. Zhang. 2024. "Green and Efficient Approach to Extract Bioactive Flavonoids with Antioxidant, Antibacterial, Antiglycation, and Enzyme Inhibitory Activities from Navel Orange Peel." *Sustain Chem Pharm* 38 (101479). https://doi.org/10.1016/j.scp.2024.101479.

Larrosa, Mar, Cristina Luceri, Elisa Vivoli, Chiara Pagliuca, Maura Lodovici, Gloriano Moneti, and Piero Dolara. 2009. "Polyphenol Metabolites from Colonic Microbiota Exert Anti-inflammatory Activity on Different Inflammation Models." *Molecular Nutrition & Food Research* 53 (8): 1044–54. https://doi.org/10.1002/mnfr.200800446.

Lattanzio, Vincenzo, Paul A. Kroon, Stéphane Quideau, and Dieter Treutter. 2009. "Plant Phenolics – Secondary Metabolites with Diverse Functions." *Recent Advances in Polyphenol Research* 1: 1–35. https://doi.org/10.1002/9781444302400.ch1.

Lee, J.-E., J.T.M. Jayakody, and J.-I. Kim. 2024. "The Influence of Solvent Choice on the Extraction of Bioactive Compounds from Asteraceae: A Comparative Review." *Foods* 13 (19). https://doi.org/10.3390/foods13193151.

Li, Z., Y. Fan, and J. Xi. 2019. "Recent Advances in High Voltage Electric Discharge Extraction of Bioactive Ingredients from Plant Materials." *Food Chem* 277: 246–60. https://doi.org/10.1016/j.foodchem.2018.10.119.

Lončarić, Ante, Maria Celeiro, Antun Jozinović, Josip Jelinić, Tihomir Kovač, Stela Jokić, Jurislav Babić, Tihomir Moslavac, Sandra Zavadlav, and Marta Lores. 2020. "Green Extraction Methods for Extraction of Polyphenolic Compounds from Blueberry Pomace." *Foods* 9 (11): 1521. https://doi.org/10.3390/foods9111521.

Macedo, G.A., Á.L. Santana, L.M. Crawford, S.C. Wang, F.F. Dias, and J.M. Moura Bell. 2021. "Integrated Microwave-and Enzyme-Assisted Extraction of Phenolic Compounds from Olive Pomace." *LWT* 138 (110621). https://doi.org/10.1016/j.lwt.2020.110621.

Mahdavi, Sahar Akhavan, Seid Mahdi Jafari, Elham Assadpoor, and Danial Dehnad. 2016. "Microencapsulation Optimization of Natural Anthocyanins with Maltodextrin, Gum Arabic and Gelatin." *International Journal of Biological Macromolecules* 85: 379–85. https://doi.org/10.1016/j.ijbiomac.2016.01.011.

Mardokić, A., A.E. Maldonado, and K. Klosz. 2023. "Optimization of Conditions for Microwave-Assisted Extraction of Polyphenols from Olive Pomace of Žutica Variety: Waste Valorization Approach." *Antioxidants (Basel)* 12 (6). https://doi.org/10.3390/antiox12061175.

Medina-Torres, N., T. Ayora-Talavera, H. Espinosa-Andrews, A. Sánchez-Contreras, and N. Pacheco. 2017. "Ultrasound Assisted Extraction for the Recovery of Phenolic Compounds from Vegetable Sources." *Agronomy (Basel)* 7 (3). https://doi.org/10.3390/agronomy7030047.

Meng, T., D. Xiao, A. Muhammed, J. Deng, L. Chen, and J. He. 2021. "Anti-Inflammatory Action and Mechanisms of Resveratrol." *Molecules* 26 (1): 229. https://doi.org/10.3390/molecules26010229

Michiels, J.A., C. Kevers, J. Pincemail, J.O. Defraigne, and J. Dommes. 2012. "Extraction Conditions Can Greatly Influence Antioxidant Capacity Assays in Plant Food Matrices." *Food Chem* 130: 986–93. https://doi.org/10.1016/j.foodchem.2011.07.117.

Motilva, Maria-José, Aida Serra, and Alba Macià. 2013. "Analysis of Food Polyphenols by Ultra High-Performance Liquid Chromatography Coupled to Mass Spectrometry: An Overview." *Journal of Chromatography A* 1292: 66–82. https://doi.org/10.1016/j.chroma.2013.01.012.

Nastić, Nataša, Ante Lončarić, Siniša Simić, Kristian Pastor, Marija Banožic, Stela Jokić, Krunoslav Aladić, and Jelena Vladić. 2025. "*Anethum Graveolens* L. (Dill) Seed Polyphenolic Fractions: Assessment of High-voltage Electrical Discharges, Subcritical Water, and Microwave-Assisted Extraction Systems." *Journal of Food Science* 90 (2): e17664. https://doi.org/10.1111/1750-3841.17664.

Nguyen, H.C., K.N. Ngo, H.K. Tran, and C.J. Barrow. 2024. "Enzyme-Assisted Coextraction of Phenolics and Polysaccharides from Padina Gymnospora." *Mar Drugs* 22 (1). https://doi.org/10.3390/md22010042.

Noad, Rebecca L., Ciara Rooney, Damian McCall, Ian S. Young, David McCance, Michelle C. McKinley, Jayne V. Woodside, and Pascal P. McKeown. 2016. "Beneficial Effect of a

Polyphenol-Rich Diet on Cardiovascular Risk: A Randomised Control Trial." *Heart* 102 (17): 1371–79. https://doi.org/10.1136/heartjnl-2015-309218.
Nutrizio, Marinela, Anet Režek Jambrak, Tonči Rezić, and Ilija Djekic. 2022. "Extraction of Phenolic Compounds from Oregano Using High Voltage Electrical Discharges–Sustainable Perspective." *International Journal of Food Science & Technology* 57 (2): 1104–13. https://doi.org/10.1111/ijfs.15476.
Oanh, Doan Thi Yen, Ly Thi Minh Hien, Nguyen Kha Duyen, Nguyen Huu Hieu, and Dong Thi Anh Dao. 2023. "Effect of Enzyme – Assisted Extraction on Total Polyphenol Content and Antioxidant Capacity from Fresh Tea Leaves (*Camellia Sinensis*)." *Vietnam Journal of Chemistry* 61 (5): 551–62. https://doi.org/10.1002/vjch.202300109.
Olalere, O.A., C.-Y. Gan, and A.E. Taiwo. 2023. "Heat-Reflux Processing of Black Peppercorn into Bioactive Antioxidant Oleoresins: A Three-Functioned Taguchi-Based Grey Relational Grading." *Food Measure* 17: 4140–48. https://doi.org/10.1007/s11694-023-01951-3.
Oliviero, Francesca, Anna Scanu, Yessica Zamudio-Cuevas, Leonardo Punzi, and Paolo Spinella. 2018. "Anti-inflammatory Effects of Polyphenols in Arthritis." *Journal of the Science of Food and Agriculture* 98 (5): 1653–59. https://doi.org/10.1002/jsfa.8664.
Oroian, M., F. Ursachi, and F. Dranca. 2020. "Ultrasound-Assisted Extraction of Polyphenols From Crude Pollen." *Antioxidants (Basel)* 9 (4). https://doi.org/10.3390/antiox9040322.
Osorio-Tobón, J.F., and M.A.A. Meireles. 2013. "Recent Applications of Pressurized Fluid Extraction: Curcuminoids Extraction with Pressurized Liquids." *Food Public Health* 3 (6): 289–303. https://doi.org/10.5923/j.fph.20130306.05.
Peng, S. 2023. "Ultrasound-Assisted Extraction of Chinese Propolis." *Front Sustain Food Syst* 7 (1131959).
Petruk, Ganna, Rita Del Giudice, Maria Manuela Rigano, and Daria Maria Monti. 2018. "Antioxidants from Plants Protect against Skin Photoaging." *Oxidative Medicine and Cellular Longevity* 1: 1454936. https://doi.org/10.1155/2018/1454936.
Pickardt, Claudia, Peter Eisner, Dietmar R. Kammerer, and Reinhold Carle. 2015. "Pilot Plant Preparation of Light-Coloured Protein Isolates from de-Oiled Sunflower (Helianthus Annuus L.) Press Cake by Mild-Acidic Protein Extraction and Polyphenol Adsorption." *Food Hydrocolloids* 44: 208–19. https://doi.org/10.1016/j.foodhyd.2014.09.020.
Picos-Salas, Manuel Adrian, Nayely Leyva-López, Pedro De Jesús Bastidas-Bastidas, Marilena Antunes-Ricardo, Luis Angel Cabanillas-Bojórquez, Miguel Angel Angulo-Escalante, J. Basilio Heredia, and Erick Paul Gutiérrez-Grijalva. 2024. "Supercritical CO2 Extraction of Naringenin from Mexican Oregano (Lippia Graveolens): Its Antioxidant Capacity under Simulated Gastrointestinal Digestion." *Scientific Reports* 14 (1): 1146. https://doi.org/10.1038/s41598-023-50997-2.
Pimentel-Moral, S., M.C. Teixeira, A.R. Fernandes, D. Arráez-Román, A. Martínez-Férez, A. Segura-Carretero, and E.B. Souto. 2018. "Lipid Nanocarriers for the Loading of Polyphenols – a Comprehensive Review." *Advances in Colloid and Interface Science* 260 (October): 85–94. https://doi.org/10.1016/j.cis.2018.08.007.
Prado, Juliana M., and Mauricio A. Rostagno. 2022. *Natural Product Extraction: Principles and Applications*. Vol. 71. Royal Society of Chemistry.
Prasedya, E.S., A. Frediansyah, and N.W.R. Martyasari. 2021. "Effect of Particle Size on Phytochemical Composition and Antioxidant Properties of Sargassum Cristaefolium Ethanol Extract." *Sci Rep* 11 (17876). https://doi.org/10.1038/s41598-021-95769-y.
Puri, M., D. Sharma, and C.J. Barrow. 2012. "Enzyme-Assisted Extraction of Bioactives from Plants." *Trends Biotechnol* 30 (1): 37–44. https://doi.org/10.1016/j.tibtech.2011.06.014.
Puzeryte, V., P. Martusevice, and S. Sousa. 2023. "Optimization of Enzyme-Assisted Extraction of Bioactive Compounds from Sea Buckthorn (Hippophae Rhamnoides L.) Leaves: Evaluation of Mixed-Culture Fermentation." *Microorganisms* 11 (9). https://doi.org/10.3390/microorganisms11092180.

Quan, Nguyen, Tran Xuan, Hoang-Dung Tran, Nguyen Thuy, Le Trang, Can Huong, Yusuf Andriana, and Phung Tuyen. 2019. "Antioxidant, α-Amylase and α-Glucosidase Inhibitory Activities and Potential Constituents of Canarium Tramdenum Bark." *Molecules* 24 (3): 605–605. https://doi.org/10.3390/molecules24030605.

Quideau, S., D. Deffieux, C. Douat-Casassus, and L. Pouységu. 2011. "Plant Polyphenols: Chemical Properties, Biological Activities, and Synthesis." *Angew Chem Int Ed Engl* 50 (3): 586–621. https://doi.org/10.1002/anie.201000044.

Rajendran, G., D. Bhanu, and B. Aruchamy. 2022. "Chalcone: A Promising Bioactive Scaffold in Medicinal Chemistry." *Pharmaceuticals* 15 (10). https://doi.org/10.3390/ph15101250.

Rampado, Riccardo, and Dan Peer. 2023. "Design of Experiments in the Optimization of Nanoparticle-Based Drug Delivery Systems." *Journal of Controlled Release* 358: 398–419. https://doi.org/10.1016/j.jconrel.2023.05.001.

Ross, F.C., D.E. Mayer, J. Horn, J.F. Cryan, D.Del Rio, E. Randolph, and C.I.R. Gill. 2024. "Potential of Dietary Polyphenols for Protection from Age-Related Decline and Neurodegeneration: A Role for Gut Microbiota?" *Nutritional Neuroscience* 27 (9): 1058–76. https://doi.org/10.1080/1028415X.2023.2298098.

Rudrapal, Mithun, Shubham J. Khairnar, Johra Khan, Abdulaziz Bin Dukhyil, Mohammad Azam Ansari, Mohammad N. Alomary, Fahad M. Alshabrmi, Santwana Palai, Prashanta Kumar Deb, and Rajlakshmi Devi. 2022. "Dietary Polyphenols and Their Role in Oxidative Stress-Induced Human Diseases: Insights Into Protective Effects, Antioxidant Potentials and Mechanism(s) of Action." *Frontiers in Pharmacology* 13 (February): 806470. https://doi.org/10.3389/fphar.2022.806470.

Rudroff, F., M.D. Mihovilovic, H. Gröger, R. Snajdrova, H. Iding, and U.T. Bornscheuer. 2018. "Opportunities and Challenges for Combining Chemo- and Biocatalysis." *Nat Catal* 1 (1): 12–22. https://doi.org/10.1038/s41929-017-0010-4.

Ruesgas-Ramón, Mariana, Maria Cruz Figueroa-Espinoza, and Erwann Durand. 2017. "Application of Deep Eutectic Solvents (DES) for Phenolic Compounds Extraction: Overview, Challenges, and Opportunities." *Journal of Agricultural and Food Chemistry* 65 (18): 3591–3601. https://doi.org/10.1021/acs.jafc.7b01054.

Šamec, Dunja, Erna Karalija, Ivana Šola, Valerija Vujčić Bok, and Branka Salopek-Sondi. 2021. "The Role of Polyphenols in Abiotic Stress Response: The Influence of Molecular Structure." *Plants* 10 (1): 118. https://doi.org/10.3390/plants10010118.

Sander, L.C. 2017. "Soxhlet Extractions." *J Res Natl Inst Stand Technol* 122 (1). https://doi.org/10.6028/jres.122.004.

Sartori, G., and R. Maggi. 2006. "Use of Solid Catalysts in Friedel–Crafts Acylation Reactions." *Chem Rev* 106 (3): 1077–104. https://doi.org/10.1021/cr040695c.

Sharifi-Rad, Mehdi, Nanjangud V. Anil Kumar, Paolo Zucca, Elena Maria Varoni, Luciana Dini, Elisa Panzarini, Jovana Rajkovic, et al. 2020. "Lifestyle, Oxidative Stress, and Antioxidants: Back and Forth in the Pathophysiology of Chronic Diseases." *Frontiers in Physiology* 11 (July). https://doi.org/10.3389/fphys.2020.00694.

Sharma, Anket, Babar Shahzad, Abdul Rehman, Renu Bhardwaj, Marco Landi, and Bingsong Zheng. 2019. "Response of Phenylpropanoid Pathway and the Role of Polyphenols in Plants under Abiotic Stress." *Molecules* 24 (13): 2452. https://doi.org/10.3390/molecules24132452.

Sheng, Z., Y. Wang, P. Wan, and Y. Li. 2014. "Ultrasound-Assisted Extraction of Total Flavonoids from Leaves of Syringa Oblata Lindl." *Latin Am Appl Res* 44 (2): 131–35.

Silvia, Netala, Atla Srinivasa Rao, Yalla Kranthi, Gangolu Yohan, Darisi Saketh, Gajula Meghamala, and Kolli Prabhanjan Kumar. 2025. "Innovative Approaches for the Extraction, Purification, and Quantification of Polyphenols: A Comprehensive Review." *International Journal of Pharmaceutical Investigation* 15 (3): 725–39. https://doi.org/10.5530/ijpi.20250192.

Simon, Levente L., Hajnalka Pataki, György Marosi, Fabian Meemken, Konrad Hungerbühler, Alfons Baiker, and Srinivas Tummala. 2015. "Assessment of Recent Process Analytical

Technology (PAT) Trends: A Multiauthor Review." *Organic Process Research and Development* 19 (1): 3–62. https://doi.org/10.1021/op500261y.

Singla, Rajeev K., Ashok K. Dubey, Arun Garg, Ramesh K. Sharma, Marco Fiorino, Sara M. Ameen, Moawiya A. Haddad, and Masnat Al-Hiary. 2019. "Natural Polyphenols: Chemical Classification, Definition of Classes, Subcategories, and Structures." *Journal of AOAC International* 102 (5): 1397–400. https://doi.org/10.5740/jaoacint.19-0133.

Sinha, Dwaipayan, Suchetana Mukherjee, and Shahana Chowdhury. 2022. "Methods of Extraction of Phytochemicals." 250–79. https://doi.org/10.4018/978-1-6684-7337-5.ch010.

Siriwoharn, Thanyaporn, Ronald E. Wrolstad, Chad E. Finn, Clifford B. Pereira, and Sampling Blackberry. 2004. "Influence of Cultivar, Maturity." *Journal of Agricultural and Food Chemistry* 52 (26): 8021–30. https://doi.org/10.1021/jf048619y.

Spigno, Giorgia, Lorenza Tramelli, and Dante Marco De Faveri. 2007. "Effects of Extraction Time, Temperature and Solvent on Concentration and Antioxidant Activity of Grape Marc Phenolics." *Journal of Food Engineering* 81 (1): 200–208. https://doi.org/10.1016/j.jfoodeng.2006.10.021.

Sridhar, A., M. Ponnuchamy, and P.S. Kumar. 2021. "Techniques and Modeling of Polyphenol Extraction from Food: A Review." *Environ Chem Lett* 19 (4): 3409–43. https://doi.org/10.1007/s10311-021-01217-8.

Stanek-Wandzel, N., A. Krzyszowska, and M. Zarębska. 2024. "Evaluation of Cellulase, Pectinase, and Hemicellulase Effectiveness in Extraction of Phenolic Compounds from Grape Pomace." *Int J Mol Sci* 25 (24). https://doi.org/10.3390/ijms252413538.

Sun, A., X. Chi, X. Yang, J. Feng, Y. Li, and J. Zhou. 2019. "Applications and Prospects of Ultrasound Assisted Extraction in Chinese Herbal Medicine." *Open Access J Biomed Sci* 1 (1): 5–15. https://doi.org/10.38125/OAJBS.000103.

Tizón Alba, Ainara, María José Aliaño-González, Miguel Palma, Gerardo Fernández Barbero, and Ceferino Carrera. 2023. "Enhancing Efficiency of Enzymatic-Assisted Extraction Method for Evaluating Bioactive Compound Analysis in Mulberry: An Optimization Approach." *Agronomy* 13 (10): 2548. https://doi.org/10.3390/agronomy13102548.

Torres-Rodriguez, A., R. Darvishzadeh, and A.K. Skidmore. 2024. "High-Throughput Soxhlet Extraction Method Applied for Analysis of Leaf Lignocellulose and Non-Structural Substances." *MethodsX* 12 (102644). https://doi.org/10.1016/j.mex.2024.102644.

Tuladhar, Palistha, Santanu Sasidharan, and Prakash Saudagar. 2021. "Role of Phenols and Polyphenols in Plant Defense Response to Biotic and Abiotic Stresses." In *Biocontrol Agents and Secondary Metabolites*. Elsevier: 419–41. https://www.sciencedirect.com/science/article/pii/B978012822919400017X.

Turkmen, Nihal, Y. Sedat Velioglu, Ferda Sari, and Gokce Polat. 2007. "Effect of Extraction Conditions on Measured Total Polyphenol Contents and Antioxidant and Antibacterial Activities of Black Tea." *Molecules* 12 (3): 484–96. https://doi.org/10.3390/12030484.

Uwineza, P.A., and A. Waśkiewicz. 2020. "Recent Advances in Supercritical Fluid Extraction of Natural Bioactive Compounds from Natural Plant Materials." *Molecules* 25 (17). https://doi.org/10.3390/molecules25173847.

Vardakas, A., A. Kechagias, N. Penov, and A.E. Giannakas. 2024. "Optimization of Enzymatic Assisted Extraction of Bioactive Compounds from Olea Europaea Leaves." *Biomass* 4 (3): 647–57. https://doi.org/10.3390/biomass4030035.

Verma, Anjali, and Satya Vir Singh. 2015. "Spray Drying of Fruit and Vegetable Juices–a Review." *Critical Reviews in Food Science and Nutrition* 55 (5): 701–19. https://doi.org/10.1080/10408398.2012.672939.

Weggler, B.A., B. Gruber, P. Teehan, R. Jaramillo, and F.L. Dorman. 2020. "Inlets and Sampling." *Separation Science and Technology* 12: 141–203. https://doi.org/10.1016/B978-0-12-813745-1.00005-2.

Withouck, H. 2023. "Applewood Polyphenol Extraction." *Foods* 12 (17).

Xi, J., L. He, and L.G. Yan. 2017. "Continuous Extraction of Phenolic Compounds from Pomegranate Peel Using High Voltage Electrical Discharge." *Food Chem* 230: 354–61. https://doi.org/10.1016/j.foodchem.2017.03.072.

Yahfoufi, Nour, Nawal Alsadi, Majed Jambi, and Chantal Matar. 2018. "The Immunomodulatory and Anti-Inflammatory Role of Polyphenols." *Nutrients* 10 (11): 1618. https://doi.org/10.3390/nu10111618.

Zagoskina, Natalia V., Maria Y. Zubova, Tatiana L. Nechaeva, Varvara V. Kazantseva, Evgenia A. Goncharuk, Vera M. Katanskaya, Ekaterina N. Baranova, and Maria A. Aksenova. 2023. "Polyphenols in Plants: Structure, Biosynthesis, Abiotic Stress Regulation, and Practical Applications (Review)." *International Journal of Molecular Sciences* 24 (18): 13874. https://doi.org/10.3390/ijms241813874.

Zahari, N.A.A., G.H. Chong, L.C. Abdullah, and B.L. Chua. 2020. "Ultrasonic-Assisted Extraction (UAE) Process on Thymol Concentration from Plectranthus Amboinicus Leaves: Kinetic Modeling and Optimization." *Processes* 8 (3). https://doi.org/10.3390/pr8030322.

Zamora-Ros, Raul, Valerie Cayssials, Mazda Jenab, Joseph A. Rothwell, Veronika Fedirko, Krasimira Aleksandrova, and Anne Tjønneland. 2018. "Dietary Intake of Total Polyphenol and Polyphenol Classes and the Risk of Colorectal Cancer in the European Prospective Investigation into Cancer and Nutrition (EPIC) Cohort." *European Journal of Epidemiology* 33 (11): 1063–75. https://doi.org/10.1007/s10654-018-0408-6.

Zhang, M., W. Ma, and C. Wang. 2021. "Optimization of Enzyme-Assisted Extraction and Purification of Flavonoids from Pinus Koraiensis Nut-Coated Film and Antioxidant Activity Evaluation." *Molecules* 26 (7). https://doi.org/10.3390/molecules26071950.

Zhang, Qing-Wen, Li-Gen Lin, and Wen-Cai Ye. 2018. "Techniques for Extraction and Isolation of Natural Products: A Comprehensive Review." *Chinese Medicine* 13 (1): 20. https://doi.org/10.1186/s13020-018-0177-x.

Zhang, Siheng, Hongzhao Xie, Jie Huang, Qiumei Chen, Xin Li, Xiaopeng Chen, Jiezhen Liang, and Linlin Wang. 2024. "Ultrasound-Assisted Extraction of Polyphenols from Pine Needles (Pinus Elliottii): Comprehensive Insights from RSM Optimization, Antioxidant Activity, UHPLC-Q-Exactive Orbitrap MS/MS Analysis and Kinetic Model." *Ultrasonics Sonochemistry* 102 (January): 106742. https://doi.org/10.1016/j.ultsonch.2023.106742.

Zhu, Zhenzhou, Tian Jiang, Jingren He, Francisco J. Barba, Giancarlo Cravotto, and Mohamed Koubaa. 2016. "Ultrasound-Assisted Extraction, Centrifugation and Ultrafiltration: Multistage Process for Polyphenol Recovery from Purple Sweet Potatoes." *Molecules* 21 (11): 1584. https://doi.org/10.3390/molecules21111584.

7 Challenges in the Design of Nutraceuticals

Rajesh Saini, Awadhesh Kumar Mishra, Praveen Kumar Shukla, Jyoti Vishwakarma, and Kavindra Nath Tiwari

7.1 INTRODUCTION

Nutraceuticals are dietary supplements present naturally in food and provide health or medical advantages (Chauhan et al., 2013). The name was invented in 1989 by Dr. Stephen Defelice, who combined the words "nutrition" and "pharmaceutical". Because industrialization has used a lot of different electromagnetic waves, heavy metals, chemicals, and other potentially dangerous synthetic materials, it has contaminated soil, water, and air. The prevalence of vascular illnesses, diabetes, physiological issues, various malignancies, obesity, and other degenerative diseases has increased because of these problems. The expense of medical treatment has skyrocketed because of the growing need for medical care. The investigation of new food items with therapeutic qualities has led to a rebirth in the study of nutrition and health. A new era of research has beenlaunched by nutraceuticals to enhance quality of life. Even if there is plenty of evidence that supports the superiority of one polyphenol over another, the entire extent of their efficacy is still somewhat unknown. In addition, little is known about the chemical variability of these molecules, and fresh information is emerging daily that indicates there is still much to learn about their chemistry and bioactivity, particularly those found in and derived from food plants (Piccolella et al., 2019).

Although several food and beverage ingredients can be included to form the nutraceuticals, there are various challenges in this path. A proper identification of the functional foods and beverages present in nature is necessary, andproper awareness is needed about their effectiveness and side effects. The process of separating, purifying, and identifying the various secondary metabolites found in plants is necessary for natural product chemists to uncover a variety of intricate chemical structures that may only be accessible for analysis in microgram quantities (Piccolella et al., 2019). A significant problem is the need to supply a nutraceutical dose that works for a particular health benefit, since this affects the final product's flavour and scent (Puri et al., 2022). These challenges should be addressed by the agency concerned to globalize the nutraceuticals and to make them trustworthy for human consumption.

DOI: 10.1201/9781003588504-7

7.2 DIFFICULTIES WITH NUTRACEUTICAL DISEASE CONTROL: CLINICAL ASSESSMENT

Clinical research on nutraceuticals is more challenging because there are no valid biomarkers and little information about how nutraceuticals interact with food and medicine. Regulators have realized that to conduct clinical studies effectively and with definite results, they must raise the bar for current nutraceutical standards and offer guidance tailored to nutraceuticals. Science-based understanding of nutraceuticals is also necessary, as is the assessment of their pharmacological and therapeutic effects on humans (Chaturvedi et al., 2022).

Aspirin, digoxin, and statins are just a few of the commonly utilized cardiovascular medications that were discovered in dietary food sources which were once employed as herbal remedies for a wide range of illnesses. While supporters of nutraceuticals frequently highlight their health benefits, there is a lack of solid proof from carefully conducted trials demonstrating improvement in clinically meaningful outcomes (Sharma and Felker, 2021).

The loss of vitally significant human anatomical and biochemical features in the in vitro approach is a major limitation for cell culture research employing nutraceuticals. Numerous nutraceuticals demonstrating a potential effect with in vitro cell cultures may not work well in people, according to pre-existing research, particularly from drug testing. This is because the complexity of the human biological network is absent from isolated cells. As a result, the scientific validity of nutraceuticals' health claims based on in vitro assays will be questioned in human studies (Mahabir, 2014).

7.2.1 Evaluation Instruments

Measuring the intake of nutraceuticals and their correlations with relevant diseases or health outcomes is crucial when addressing human populations. Most of the instruments in use today are not intended to fully capture the wide range of available nutraceuticals. For a more conventional dietary assessment, researchers would typically compare two dietary instruments, such as a food frequency questionnaire (FFQ) or diet history, to 24-hour food recalls or records as the standard technique. Either self-reporting or interviewing isused to collect data, and information is gathered over a period that is either less than or greater than a year. However, there is a significant associated error in the provided dietary data (Kipnis et al., 2003).

7.2.2 Biomarkers

Biomarkers are chemical compounds thatare crucial for a particular disease and can be administered using nutraceuticals. Given that biomarkers are not dependent on self-reports of dietary or nutritional consumption, there is little chance of mistakes, and it should be confirmed with a questionnaire approach and should be correlated (Subar et al., 2003). It would be highly beneficial to establish biomarkers for the consumption, metabolism, and excretion of nutraceutical goods. Biomarkers are helpfulfor use in epidemiological research for lipids, trace metals, and numerous vitamins and minerals. Using the folate biomarker as an example, research has shown that smokers have lower folate quantity in plasma and red blood cells than non-smokers do; however, it is unclear if this is due to decreased consumption or higher requirements (Ortega et al., 2004).

7.2.3 Validation Challenges

Vitamins, minerals, and vitamins with minerals are among the products for which test procedures are easily accessible and scientifically sound. Methodologies and unique selling point (USP) monographs are available for some of these goods, and the techniques employed are somewhat contemporary and include electrophoresis, high-performance liquid chromatography (HPLC), gas chromatography (GC), and others. Unfortunately, there are still laborious and imprecise microbiological procedures for some vitamins in the USP (such as those for cyanocobalamin, biotin, and vitamin D). Finding the active components in botanical goods is one example of a nutraceutical product where technique development and validation can be challenging (Mehta, 2010). Nutraceutical product formulations can be extremely complex, containing multiple chemicals in a single liquid, tablet, or capsule. The substitution of an active component source is a prime illustration of the difficulties in formulating vitamins and minerals.

7.3 THE DIFFICULTIES OF INCLUDING NUTRACEUTICALS IN FOOD

Nutraceutical fortification, assimilation, or addition to food is a complicated procedure. Taste, texture, and shelf life may be affectedwhen a product is fortified or has a new component (such as a nutraceutical) added to it. These changes have an impact on the physicochemical composition and sensory qualities of the original recipe of food and drink. The design of the formulation and processes utilized in the synthesis of the nutraceutical ingredient areinfluenced by thekind, format, and solubility. The desired qualities of the food with the added nutraceuticals are influenced by this. Generally, it is not possible to directly incorporate nutraceuticals into food products because many of them are likely to degrade and could interact with other substances in meals and decrease the nutraceutical ingredient's bioactivity and the quality of the food item. Instead, the nutraceuticals must be guarded by a well-thought-out delivery system that is designedfor the intended food application (Montes et al., 2019). The low bioavailability of nutraceuticals, which are typically eliminated from the body without offering any therapeutic benefit, is a significant problem in this industry (Zanella et al., 2015). The inability to dissolve gastrointestinal fluids and the development of insoluble complexes with other gastrointestinal tract constituents (GITs) (Rimbach et al., 2008), restricted liberation from the food matrix (Moelants et al., 2012), poor permeability through the epithelium cells, and/or molecular transformations in the GIT are some of the physicochemical and/or physiological processes that may contribute to the low bioavailability of nutraceuticals (Actis-Goretta et al., 2013).

7.4 ASSURANCE OF QUALITY IN NUTRACEUTICALS

Dedication to quality and safety has become a crucial topic of focus to meet the challenges and changes of this rapidly changing sector. Although it is a difficult task, quality assurance (QA) is a crucial element that should not be ignored. Upholding high standards requires teamwork, good manufacturing procedures (GMP), and other QA methods to assist in determining the product's quality benchmarks. Significant resource savings and performance gains can be achieved by following the rules and requirements of QA reviewers and regulatory agencies like the Food and Drug Administration (FDA) in the United States and the Food Safety and Standards Authority of India (FSSAI). Nutraceuticals are becoming more and more necessary, yet the worldwide regulatory definition and framework for these goods have not yet been established (Inbathamizh et al., 2022).

The optimization of the agriceutical crop is the first step in a systems approach that ensures the quality of nutraceuticals. Agronomic practice records (good agricultural practices), biodata records, farm site identification (via global positioning systems), and botanical identity of the crop are all necessary for anintelligent productdelivery systems philosophy. Monitoring and analyses of the bioactive ingredients in nutraceuticals and/or biomarkers are necessary to collect phytochemical fingerprinting data. Providing real-time consistency and standardization with documentation and change, observedduring dealing, harvesting, preparing, and manufacturing, necessitates the formation of quick analytical methods and instruments. It is necessary to identify and address the issues preventing full quality management before outdated techniques automatically become the norm (Lachance and Saba, 2002).

In terms of the global commercialization of this product category, this lack of coherence could lead to uncertainty and inconsistency. For example, several nations' present laws do not use the word "nutraceutical" (Chopra et al., 2022). On the other hand, according to the Food and Drug Administration Modernization Act of 1997 (FDAMA) and the Dietary Supplement Health and Education Act of U.S. regulations, makers of nutritional supplements are not required to obtain FDA approval before going on sale; hence, some of these items may be released onto the market without demonstrating clinical efficacy (Danielski, 2023).

7.5 GLOBAL REGULATORY COMPLIANCE IN NUTRACEUTICALS

7.5.1 International Regulations

The global nutraceutical industry faces increasing scrutiny regarding the safety and efficacy of bioactive compounds such as polyphenols. Regulatory frameworks differ significantly across regions, presenting both opportunities and challenges for manufacturers seeking global market access. Core regulatory concerns focus on three primary areas: product safety, claim substantiation, and manufacturing quality standards.

7.5.2 FDA and the Dietary Supplement Health and Education Act

The Dietary Supplement Health and Education Act (DSHEA) of 1994 controls polyphenol-based dietary supplements in the United States. This set of laws makes it easier to get them to market than pharmaceuticals because it doesn't require pre-market approval. The FDA does, however, have strong post-market surveillance powers, especially when it comes to products that make false claims about diseases, like those that say they help with heart or brain health (FDA, 2024).

The New Dietary Ingredient (NDI) notification requirement is a crucial regulatory factor under the DSHEA. An NDI notification is usually required if a business uses a new extraction method or introduces a novel component, such as one that increases the bioavailability of polyphenols derived from blueberries by 40%. To support the safety of the new ingredient under its suggested conditions of use, this process necessitates the submission of extensive safety data, frequently including 90-day sub-chronic toxicity studies, as well as a careful analysis of the body of scientific literature already in existence (Manzoor et al., 2024).

7.5.3 EFSA's Evidence-Based Approach

The European Food Safety Authority (EFSA) has established a stringent framework for health claims on polyphenols (EFSA Pannel, 2025). Since 2008, fewer than a dozen of over 500 submitted claims have been approved. EFSA demands high-quality human clinical evidence and mechanistic data, as shown by its 2012 rejection of heart health claims for olive polyphenols. More recent approvals have incorporated data on gene-nutrient interactions and gut microbiome modulation (Santini et al., 2018).

The Novel Food Regulation (EU) 2015/2283 also imposes substantial barriers for new polyphenol sources. For example, market approval for Chilean maqui berry extracts required over €250,000 and 18–24 months of review. As a result, many companies focus on traditional herbs with documented EU use prior to 1997 (European Commission, 2015).

7.5.4 Asia-Pacific Regulatory Landscape

Japan's Foods for Specified Health Uses (FOSHU) system is regarded as a global benchmark for scientifically validated functional foods. Its approval of persimmon leaf polyphenols for blood glucose control in 2021 highlights its responsiveness to emerging science. However, clinical trials with ≥100 participants over 12 weeks make certification costly for small producers.

China's evolving Health Food Registration has moved toward recognizing individual compounds, as seen with the 2020 "Blue Hat" certification of grape seed polyphenols. Conversely, India's FSSAI continues to face challenges in standardization, exemplified by heavy metal contamination issues in pomegranate polyphenol supplements in 2022 (Manzoor et al., 2024).

7.6 EMERGING CHALLENGES AND REGULATORY SOLUTIONS

The absence of global regulatory harmonization hampers cross-border market entry. A product approved in the United States might require reformulation for EU markets due to differing solvent residue limits. To address this, blockchain-enabled smart labelling systems allow dynamic, country-specific compliance (Stephens and

Bigelow, 2019). Widespread adulteration is another concern; in 2023, 32% of commercial curcumin supplements were found to contain synthetic dyes. Advanced tools such as DNA meta-barcoding and isotope ratio mass spectrometry are increasingly used for authentication and quality control (Franchetti, 2024).

7.7 STANDARDIZATION OF POLYPHENOL CONTENT

The chemical complexity of polyphenols presents unique challenges for standardization, in contrast to single-compound pharmaceuticals. Effective standardization requires a multistep analytical strategy using diverse complementary machines such as HPLC systems with photodiode array (PDA) detectors, spectroscopy, etc. HPLC with PDA can now quantify over 15 major polyphenols within a 20-minute run. Ultra-high-performance supercritical fluid chromatography (UHPSFC) offers superior separation of highly polar compounds like anthocyanins. However, liquid chromatography-tandem mass spectrometry (LC-MS/MS) remains suboptimal for structural confirmation and detecting synthetic adulterants (Zhang et al., 2022). Two-dimensional liquid chromatography (2D-LC) provides exceptional resolution. For example, a recent study identified 87 distinct polyphenols in green tea using 2D-LC, compared to only 32 with standard HPLC (Rosales and Fabi, 2023). Conventional Folin-Ciocalteu assays are increasingly being replaced. Modified singleton methods with gallic acid standards align better with bioactive content. Near-infrared spectroscopy (NIRS) enables non-destructive raw material screening, with calibration models achieving R^2 values above 0.95 (Cao et al., 2021). Advanced antioxidant assessments now include the cellular antioxidant activity (CAA) assay using HepG2 cells, offering greater physiological relevance than DPPH. Electron paramagnetic resonance (EPR) spectroscopy also provides direct detection of free radicals (Girdhar et al., 2017).

7.8 VALIDATION OF REFERENCE MATERIALS AND CONTROLLING PROCESS VARIABILITY

The lack of certified reference standards for minor polyphenols remains a bottleneck. In 2023, the first validated reference materials for oligomeric proanthocyanidins were released through collaboration between the National Institute of Standards and Technology (NIST) and industry. Analytical method validation typically follows ICH Q2(R2) standards, requiring specificity, 90–110% accuracy, and <5% relative standard deviation (RSD) (Zhang et al., 2022). Processing can cause 30–80% polyphenol loss. Optimizing spray drying at 160–180°C with maltodextrin carriers can retain up to 95% of polyphenols. Pulsed electric field (PEF) extraction techniques have reduced thermal degradation while improving yields by 15–25% (Cao et al., 2021).

7.9 GOOD MANUFACTURING PRACTICES AND QUALITY ASSURANCE

Modern facilities utilize modular cleanrooms (ISO Class 7 or better), with separate zones for each processing stage. Equipment is typically stainless steel with electropolished surfaces (Ra <0.5 μm) to prevent contamination. Closed-loop systems with nitrogen blanketing protect polyphenols from oxidative degradation (Manzoor et al., 2024). Traditional sensory evaluations have been replaced with hyperspectral imaging for real-time adulteration detection. DNA authentication techniques now include single-nucleotide polymorphism (SNP) profiling for accurate cultivar identification. Blockchain-enabled smart contracts automatically reject materials not meeting Council on Accreditation (COA) standards (Franchetti, 2024). Process analytical technology (PAT) tools have revolutionized quality control. In-line ultraviolet–visible light (UV-Vis) probes monitor extraction in real time, while Raman spectroscopy ensures structural integrity. Multivariate statistical models can predict final product quality within the first 30% of production (Girdhar et al., 2017).

7.10 STRATEGIES FOR QUALITY ASSURANCE

7.10.1 Predictive Stability Modelling

Beyond traditional International Conference on Harmonisation (ICH) protocols, predictive tools using the Arrhenius equation and moisture sorption isotherms allow accurate shelf-life forecasts within 8 weeks. Molecular dynamics simulations help select optimal excipients to stabilize polyphenols (Zhang et al., 2022).

7.10.2 Advanced Contaminant Detection

High-resolution mass spectrometry (HRMS) enables contaminant screening down to partspertrillion. Modern environmental monitoring includes adenosine triphosphate (ATP) bioluminescence and polymerase chain reaction (PCR)–based microbial assays. Supplier audits now often integrate environmental and sustainability metrics (Santini et al., 2018).

7.10.3 Digital Quality Management

Cloud-based quality systems offer real-time deviation tracking and predictive analytics. Artificial intelligence tools anticipate quality issues based on historical data. Digital twins of production environments allow virtual quality assessments before manufacturing begins (Stephens and Bigelow, 2019).

7.11 MARKETING OF NUTRACEUTICALS

For both present and potential consumers, marketing is crucial to the success of functional and nutraceutical foods. Before launching the finished products onto the market, a few prerequisites must be recognized and thoroughly examined. Having access to patents, trademarks, and intellectual property is essential for the success of functional products. It's crucial to realize that budget plays a significant role in marketing and that both upstream and downstream marketing are essential to the success of the items. It is crucial to allocate funds for marketing at the appropriate moment (Datta, 2017). To solve these glaring shortcomings, special information and communication initiatives are desperately needed because consumers are not wellinformed regarding the potential health risks of recently created functional constituents. This is especially important for pioneering companies that are entering a particular market sector for which targeted information activities to consumers and opinion leaders (such as doctors and nutritional advisers) are essential success factors (Vergari et al., 2010).

A key research challenge lies in identifying consumer segments and understanding their sociodemographic and behavioural preferences in an era dominated by advanced predictive analytics. In such a landscape, digital marketers risk reinforcing discriminatory practices in the absence of clear ethical or legal boundaries. There is a pressing need to further explore consumer comfort and digital marketing ethics. Expanding our understanding of consumer perceptions, anxieties, and perceived risks is essential to identifying emerging segments and aligning online advertising practices with ethical standards, behavioural trends, and sociodemographic profiles (Zoupos and Spais, 2023).

In parallel, regulatory and market barriers significantly complicate the commercialization of polyphenol-based products. The EFSA's rigorous requirements have led to the rejection of approximately 95% of submitted health claims related to polyphenols (Santini et al., 2018), creating formidable obstacles to product differentiation. These challenges are exacerbated for polyphenols derived from novel or exotic botanical sources, which often require extensive safety dossiers to obtain novel food approval. Moreover, widespread adulteration further undermines market integrity – up to 30% of commercial grape seed extracts are estimated to be adulterated with cheaper fillers or synthetic substitutes (Rossi et al., 2012). Collectively, these factors contribute to a highly complex and demanding environment for product formulators and marketers.

7.12 CRITICAL COMMERCIALIZATION LESSONS IN THE POLYPHENOL MARKET

Valuable insights into successful polyphenol product commercialization have emerged from both market triumphs and setbacks. A strong intellectual property (IP) strategy has proven essential, exemplified by Chromadex's effective patenting of pTeroPure pterostilbene, which secured significant market exclusivity and a competitive advantage. Vertical integration has also shown clear benefits; companies such as Asahi Group have enhanced both quality assurance and cost efficiency by overseeing the entire apple polyphenol supply chain – from orchard cultivation to extract production. Market dynamics increasingly prioritize clinical validation, with human trials becoming central to premium product positioning. High-end formulations, such as nano-encapsulated polyphenols, command three to five times the price of conventional extracts, largely due to their enhanced bioavailability and consumer-perceived value.

Perhaps most critically, the regulatory pathway selected significantly influences market access. Japan's FOSHU approval system, for instance, provides a powerful gateway to high-value Asian markets. In contrast, the U.S. DSHEA framework, while offering easier entry, can limit the credibility and market positioning of products targeting health claims (Rambaran, 2022).

7.12.1 Turning Science into Solutions: Lab-to-Market Models in Nutraceutical Innovation

A crucial stage in the development of polyphenol nutraceuticals is the shift from laboratory discovery to products that are ready for the market. Technical optimization, legal compliance, and market strategy all need careful consideration at this point. The case studies that follow provide examples of effective models that can be used for new developments in the future. Chromadex'spTeroPure pterostilbene, a synthetic form of a polyphenol that occurs naturally in blueberries, is one prominent example. Pterostilbene, a dimethylated analogue of resveratrol, has been shown to have better oral bioavailability and superior metabolic stability, with pre-clinical models reporting a bioavailability of about 80% as opposed to less than 1% for resveratrol (Nagarajan et al., 2022; Smoliga et al., 2012). Chromadex started with thorough pharmacological characterization and efficacy profiling as part of an integrated development approach. Extensive pharmacokinetic and toxicological analyses ensued, which were crucial in obtaining self-affirmed generally recognized as safe (GRAS) status – a crucial regulatory milestone that permits the launch of a product (Stephens and Bigelow, 2019).

To create a strong competitive barrier, Chromadex also pursued a strong IP strategy, obtaining patents pertaining to extraction processes, purification methods, and therapeutic applications (Rambaran, 2022). The development of proprietary purification technologies that can reliably produce ≥98% pure pterostilbene at commercial scale helped to overcome the technological challenges associated with the transition from lab-scale synthesis to industrial production (Dutta et al., 2023). To further enhance the compound's already advantageous bioavailability, the company developed lipid-based delivery systems in parallel, demonstrating innovation in formulation (Zhang et al., 2022).

The collaboration between Well Science and BGG, which aims to develop supplements based on polyphenols to support pollen season, is another example. Their product is a clinically optimized formulation of vitamin C and quercetin from *Sophora japonica*. The collaboration guarantees strict quality control from the procurement of raw materials to the manufacturing of final products by promoting BGG's vertically integrated supply chain. The use of blockchain technology for real-time quality traceability in this instance is a notable innovation that addresses growing consumer demand for nutraceutical transparency.

Clinical validation through actual user feedback is another innovative approach to product development and marketing that this partnership exemplifies. The business enables quick, data-driven product iterations, reducing risk and time to market by incorporating efficacy testing directly into consumer channels. This model serves as an example of how consumer-centric development, technological transparency, and strategic alliances can expedite the lab-to-market transition and enhance the credibility of the product (BGG World, 2024).

7.12.2 Nutraceuticals from Local to Global

The global burden of malnutrition has significant socioeconomic implications that affect people individually, in families, and across nations. India has consistently scored low on the Global Hunger Index, even though it is among the world's top producers of a variety of agricultural goods. Funds are provided by the World Food Programme to help malnourished children in low- and medium-income countries (LMICs) receive Plumpy Nut, a nutraceutical product. These products are expensive to buy on the global market, and shipping goods across borders adds to pollution. Budgetary restrictions have prevented enough food from being provided to feed almost 40% youngsters of India whosuffer from malnutrition despite these efforts (Chaurasia et al., 2024).

The global nutraceutical business is presently growing quickly; however, there are several obstacles to this expansion, such as poor manufacturing and fragmented laws that create trade hurdles internationally. While various countries have rules and regulations governing nutraceutical products, these policies are not uniform and vary from one nation to the next. Certain nations have strict restrictions in place, while others do not have organized, strict regulations regarding nutraceuticals (Bansal and Dhiman, 2020).

7.12.3 Technical and Scientific Obstacles in the Development of Polyphenol Products

The creation of successful polyphenol-based products still faces several scientific and technical obstacles, despite growing commercial interest. Poor bioavailability is a major worry. In human studies, for example, early resveratrol formulations showed less than 1% bioavailability (Smoliga et al., 2012). To increase absorption and therapeutic efficacy, this limitation has spurred extensive research into sophisticated delivery systems, such as lipid-based carriers and nanoencapsulation technologies. Another significant barrier is stability. Formulation strategies are complicated by the rapid degradation of some polyphenols, like curcumin, in aqueous environments (Tabanelli et al., 2021). To counteract this, strategies like cyclodextrin inclusion complexes and nanoemulsions have been investigated to improve chemical stability and extend shelf life.

7.13 CONCLUSION

There is great potential for improving health and wellness through the creation and marketing of products based on polyphenols. However, enduring technical, scientific, and regulatory obstacles limit this potential. Significant difficulties in formulation and standardization arise from problems like low bioavailability, chemical instability, and raw material variability. Market entry and scalability are further hampered by complicated regulations, high production costs, and scant IP protections. A multidisciplinary strategy combining developments in formulation science, reliable analytical techniques, open regulatory frameworks, and strategic consumer education will be needed to overcome these challenges. The full therapeutic potential of polyphenols can be more successfully realized in the global health marketplace with sustained innovation and cross-sector collaboration.

REFERENCES

Actis-Goretta L, Leveques A, Rein M, Teml A, Schäfer C, Hofmann U, Li H, Schwab M, Eichelbaum M, Williamson G. Intestinal absorption, metabolism, and excretion of

(–)-epicatechin in healthy humans assessed by using an intestinal perfusion technique. *The American Journal of Clinical Nutrition*. 2013 Oct 1;98(4):924–33.

Bansal R, Dhiman A. Nutraceuticals: A comparative analysis of regulatory framework in different countries of the world. *Endocrine, Metabolic & Immune Disorders-Drug Targets (Formerly Current Drug Targets-Immune, Endocrine & Metabolic Disorders)*. 2020 Dec 1;20(10):1654–63.

BGG World. "Well Science Partners with BGG to Deliver Innovative Nutrition for the Pollen Season | BGG World." 2024. https://bggworld.com/well-science-partners-with-bgg-to-deliver-innovative-nutrition-for-the-pollen-season/.

Cao H, Saroglu O, Karadag A, Diaconeasa Z, Zoccatelli G, Conte-Junior CA, Gonzalez-Aguilar GA, Ou J, Bai W, Zamarioli CM, de Freitas LA. Available technologies on improving the stability of polyphenols in food processing. *Food Frontiers*. 2021;2(2):109–39. https://doi.org/10.1002/fft2.65.

Chaturvedi K, Shah H, Gajera B. Challenges in oral delivery of nutraceuticals: From formulation scale-up to clinical assessment. In *Nutraceutical Delivery Systems* (pp 241–78). Apple Academic Press, 2022 Aug 18.

Chauhan B, Kumar G, Kalam N, Ansari SH. Current concepts and prospects of herbal nutraceutical: A review. *Journal of Advanced Pharmaceutical Technology & Research*. 2013 Jan 1;4(1):4–8.

Chaurasia S, Pati RK, Padhi SS, Gavirneni S. Is localization better than globalization for sustainability? Evidence from the nutraceuticals industry for managing malnutrition in India. *International Journal of Production Economics*. 2024 Feb 1;268:109106.

Chopra AS, Lordan R, Horbańczuk OK, Atanasov AG, Chopra I, Horbańczuk JO, Arkells N. The current use and evolving landscape of nutraceuticals. *Pharmacological Research*. 2022;175:106001.

Danielski R. Challenges hindering the commercialization of nutraceuticals derived from agri-foodby-products. *Journal of Food Bioactives*. 2023;24. https://doi.org/10.31665/JFB.2023.18358.

Datta S. Marketing challenges and strategies. In *Developing New Functional Food and Nutraceutical Products* (pp 47–62). Academic Press, 2017 Jan 1.

Dutta BJ, Rakshe PS, Maurya N, Chib S, Singh S. Unlocking the therapeutic potential of natural stilbene: Exploring pterostilbene as a powerful ally against aging and cognitive decline. *Ageing Research Reviews*. 2023;92:102125. https://doi.org/10.1016/j.arr.2023.102125.

European Commission. "Regulation – 2015/2283 – EN – EUR-Lex." European Union. 2015 Nov 25. https://eur-lex.europa.eu/eli/reg/2015/2283/oj/eng.

EFSA Panel on Nutrition, Novel Foods and Food allergens (NDA), et al. Phenolic compounds naturally present in olive oil and lowering of blood LDL-cholesterol and systolic blood pressure, therefore reducing the risk of coronary heart disease: Evaluation of a health claim pursuant to Article 14 of Regulation (EC) No 1924/2006. *EFSA Journal*. 2025;23(5):e9470.

FDA. "Dietary Supplements Guidance Documents & Regulatory Information." FDA. 2024 Sept 9. https://www.fda.gov/food/guidance-documents-regulatory-information-topic-food-and-dietary-supplements/dietary-supplements-guidance-documents-regulatory-information.

Franchetti J. "The Role of Quality Assurance in Nutraceutical Production." JAF Consulting, Inc. (blog). 2024. https://jafconsulting.com/blog/the-role-of-quality-assurance-in-nutraceutical-production/.

Girdhar S, Pandita D, Girdhar A, Lather V. Safety, quality and regulatory aspects of nutraceuticals. *Applied Clinical Research, Clinical Trials and Regulatory Affairs*. 2017;4(1):36–42. https://doi.org/10.2174/2213476X03666160801150607.

Inbathamizh L, Prabavathy D, Sudha S. Quality assurance of nutraceuticals and their approval, registration, marketing. *Handbook of Nutraceuticals and Natural Products: Biological, Medicinal, and Nutritional Properties and Applications*. 2022 May 13;2:337–60.

Kipnis V, Subar AF, Midthune D, Freedman LS, Ballard-Barbash R, Troiano RP, Bingham S, Schoeller DA, Schatzkin A, Carroll RJ. Structure of dietary measurement error: Results of the OPEN biomarker study. *American Journal of Epidemiology*. 2003 Jul 1;158(1):14–21.

Lachance PA, Saba RG. Quality management of nutraceuticals: Intelligent product-delivery systems and safety through traceability. *ACS Symposium Series*. 2002;803:2–9.

Mahabir S. Methodological challenges conducting epidemiological research on nutraceuticals in health and disease. *PharmaNutrition*. 2014 Jul 1;2(3):120–5.

Mehta J. Practical challenges of stability testing of nutraceutical formulations. In Huynh-Ba K (ed), *Pharmaceutical Stability Testing to Support Global Markets. Biotechnology: Pharmaceutical Aspects*, vol XII. New York, NY: Springer, 2010. https://doi.org/10.1007/978-1-4419-0889-6_12.

Manzoor MF, Riaz S, Verma DK, Waseem M, Goksen G, Ali A, Zeng XA. Nutraceutical tablets: Manufacturing processes, quality assurance, and effects on human health. *Food Research International*. 2024;115197. https://doi.org/10.1016/j.foodres.2024.115197.

Moelants KR, Lemmens L, Vandebroeck M, Van Buggenhout S, Van Loey AM, Hendrickx ME. Relation between particle size and carotenoid bioaccessibility in carrot-and tomato-derived suspensions. *Journal of Agricultural and Food Chemistry*. 2012 Dec 5;60(48):11995–2003.

Montes C, Villaseñor MJ, Rios A. Analytical control of nano delivery lipid-based systems for encapsulation of nutraceuticals: Achievements and challenges. *Trends in Food Science & Technology*. 2019 Aug 1; 90:47–62.

Nagarajan S, Mohandas S, Ganesan K, Xu B, Ramkumar KM. New insights into dietary pterostilbene: Sources, metabolism, and health promotion effects. *Molecules*. 2022;27(19):6316. https://doi.org/10.3390/molecules27196316.

Ortega RM, et al. Smoking and passive smoking as conditioners of folate status in young women. *Journal of the American College of Nutrition*.2004 Aug;23(4):365–71.

Piccolella S, Crescente G, Candela L, Pacifico S. Nutraceutical polyphenols: New analytical challenges and opportunities. *Journal of Pharmaceutical and Biomedical Analysis*. 2019 Oct 25; 175:112774.

Puri V, Nagpal M, Singh I, Singh M, Dhingra GA, Huanbutta K, Dheer D, Sharma A, Sangnim T. A comprehensive review on nutraceuticals: Therapy support and formulation challenges. *Nutrients*. 2022 Nov 3;14(21):4637.

Rambaran TF. A patent review of polyphenol nano-formulations and their commercialization. *Trends in Food Science & Technology*. 2022;120:111–22. https://doi.org/10.1016/j.tifs.2022.01.011.

Rimbach G, Pallauf J, Moehring J, Kraemer K, Minihane AM. Effect of dietary phytate and microbial phytase on mineral and trace element bioavailability literature review. *Current Topics in Nutraceutical Research*. 2008 Aug 1;6(3).

Rosales TK, Fabi JP. Valorization of polyphenolic compounds from food industry by-products for application in polysaccharide-based nanoparticles. *Frontiers in Nutrition*. 2023;10:1144677. https://doi.org/10.3389/fnut.2023.1144677.

Rossi D, Guerrini A, Bruni R, Brognara E, Borgatti M, Gambari R, Maietti S, Sacchetti G. Trans-resveratrol in nutraceuticals: Issues in retail quality and effectiveness. *Molecules*. 2012;17(10):12393–405.

Santini A, Cammarata SM, Capone G, Ianaro A, Tenore GC, Pani L, Novellino E. Nutraceuticals: Opening the debate for a regulatory framework. *British Journal of Clinical Pharmacology*. 2018;84(4):659–72. https://doi.org/10.1111/bcp.13496.

Sharma A, Felker GM. Challenges and opportunities in the evaluation of nutraceuticals in cardiovascular diseases. *Basic to Translational Science*. 2021 Jan 1;6(1):22–4.

Smoliga JM, Vang O, Baur JA. Challenges of translating basic research into therapeutics: Resveratrol as an example. *Journals of Gerontology Series A: Biomedical Sciences and Medical Sciences*. 2012;67(2):158–67. https://doi.org/10.1093/gerona/glr062.

Stephens G, Bigelow S. Emerging regulatory pathways for nutraceuticals: Each product class offers unique options for making claims and communicating benefits to consumers and healthcare professionals. *Nutraceuticals World*. 2019;22(3):28–31.

Subar A, et al. Using intake biomarkers to evaluate the extent of dietary misreporting in a large sample of adults: The OPEN study. *American Journal of Epidemiology*.2003;158:1–13.

Tabanelli R, Brogi S, Calderone V. Improving curcumin bioavailability: Current strategies and future perspectives. *Pharmaceutics*. 2021;13(10):1715. https://doi.org/10.3390/pharmaceutics13101715.

Vergari F, Tibuzzi A, Basile G. An overview of the functional food market: From marketing issues and commercial players to future demand from life in space. In *Bio-farms for Nutraceuticals: Functional Food and Safety Control by Biosensors* (pp 308–21), Advances in Experimental Medicine and Biology, 2010 Jan 1.

Zanella M, Ciappellano SG, Venturini M, Tedesco E, Manodori L, Benetti F. Nutraceuticals and nanotechnology. *Diet, Ingredients, Supply*. 2015 Jul 1; 26:26–31.

Zhang Y, Cai P, Cheng G, Zhang Y. A brief review of phenolic compounds identified from plants: Their extraction, analysis, and biological activity. *Natural Product Communications*. 2022;17(1):1–14. https://doi.org/10.1177/1934578X211069721.

Zoupos D, Spais G. Digital marketing of nutraceutical and pharmaceutical supplements: Marketing ethics and consumer comfort. *Journal of Marketing Analytics*. 2023 Sep;11(3):331–51.

8 Formulation Strategies and Challenges for Polyphenol Nutraceuticals

Romila Rawat Bisht and Ishwar Singh

8.1 INTRODUCTION

Polyphenol nutraceuticals are phenolic compounds occurring in a wide range of edible and non-edible plants that provide various health benefits (Singh, 2017). Stephen DeFelice coined and defined the term 'nutraceutical' in 1989 as "a food or part of a food that has a medical or health benefit, including the prevention and treatment of disease" (DeFelice, 2010; Keservani et al., 2017). There is a fine distinction between food supplements and nutraceuticals. Nutraceuticals differ from food supplements in the sense that these are exploited more for their use in disease prevention and cure than their role in nutrition. Moreover, a nutraceutical can be categorized as either a drug or dietary/food supplement depending on its ingredients, which are usually polyphenols.

Bioactives, including polyphenolic nutraceuticals, target specific molecular sites at the cellular level, which makes them very useful in the management of life-threatening disorders. Thus, polyphenols possessing nutraceutical properties have attained special regard due to their potential applications in food and pharmaceutical industries (Mehta et al., 2018). There has been exponential growth in the nutraceuticals market value in recent years, and it is estimated that the global nutraceutical market will surpass 700 billion USD with a compound annual growth rate (CAGR) of 8.3% by 2027 (Jha et al., 2021; Puri et al., 2022).

The effectiveness and pharmaceutical utility of polyphenols depend on their stability and bioactivity. Polyphenolics are generally heat-labile and sensitive to environmental conditions. They become unstable and lose their efficacy in terms of bioactivity and bioavailability (Larrauri et al., 1997). Many nutraceuticals have had limited success because of several drawbacks including low stability, poor bioavailability, and rapid clearance (Mohanty and Sahoo, 2012; Yashaswini et al., 2017; Di Costanzo and Angelico, 2019; Nair et al., 2023). Therefore, methods must be specially designed and prepared for nutraceutical formulations considering specific requirements. Encapsulation is an old and time-tested method developed to preserve the bioactivity of the active compounds. In this process, bioactive compounds are coated with an inert material that maintains the desired properties of the bioactive compounds and ensures their delivery (Fang and Bhandari, 2010). New findings can offer

DOI: 10.1201/9781003588504-8

alternatives for enhancing the stability of bioactive compounds. Innovative methods are being designed for novel delivery carriers. Studies are focused on experimenting with novel carrier systems such as polymeric nanoparticles, solid-lipid nanoparticles, lipid vesicles, emulsions, inclusion complexes, and nanosponges (Akombaetwa et al., 2023; Mehta et al., 2018). Many innovative ways such as spray drying, spray cooling, liposome entrapment, coacervation, centrifugal extrusion, extrusion coating, and rotational suspension separation have been incorporated into encapsulation (Desai and Park, 2005). This chapter focuses on methods of incorporation of formulations of polyphenols into functional food. The important encapsulation methods, namely, pan coating, fluidized bed coating, spray drying, centrifugal extrusion, solvent evaporation, polymerization, phase coacervation, vibrational nozzle, supercritical fluid anti-solvent method, and thermal gelation, are also discussed.

8.2 ENCAPSULATION

Encapsulation is an old and tested method developed around 80 years ago. It preserves the bioactivity and bioavailability of active compounds including nutraceuticals. In encapsulation, gelatin coats are synthesized around oil liquids using colloidal solutions (Fang and Bhandari, 2010).

This involves entrapping a functionally active biomolecule inside an inert material called a coating material or capsule (Figure 8.1). Multiple names such as payload,

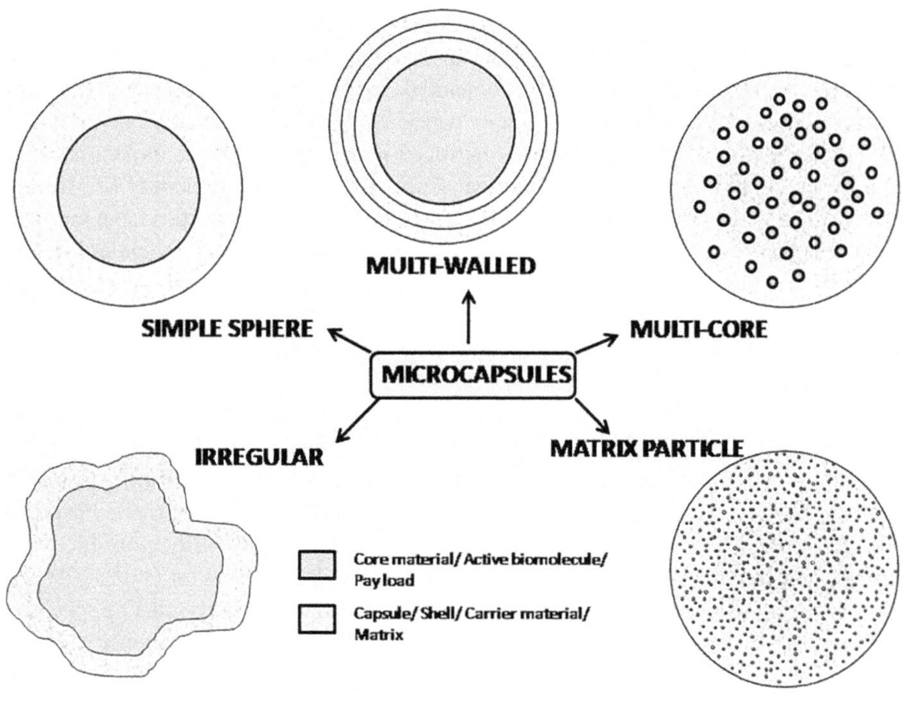

FIGURE 8.1 Types of microcapsules entrapping a functionally active biomolecule inside an inert material called a coating material or capsule.

internal phase fill, or core material are given to active biomolecules; similarly, the capsule is also named a wall, shell, matrix, membrane, or carrier material (Timilsena et al., 2020).

Encapsulation technology provides many benefits related to stability, acceptability, and targeted nutrient delivery. Encapsulation maintains the active molecular form until consumption (Bartosz and Irene, 2015). The shell protects the core material from adverse effects of temperature, light, oxygen, and moisture (Palanikumar et al., 2018). It ensures time-bound and controlled release of the core material at the desired site as and when required. It also preserves the flavor and aroma and protects sensitive material from nutritional loss. Moreover, it enhances the palatability of phenolic compounds by masking their unpleasant taste and smell. The method suitable for encapsulation depends on many factors, such as the molecular weight of the polymer, the solubility of the polymer in solution, emulsifying properties, diffusibility, and film-forming properties (Gharsallaoui et al., 2007).

Capsules are suitable for drug delivery because of their high surface area–to–volume ratio, which helps with quicker diffusion of the core material. Capsules can be designed in variable forms as per the requirement (Figure 8.1) (Gouin, 2004; Desai and Park, 2005; Gharsallaoui et al., 2007). The large-sized capsules, usually millimeter-sized, having more core capacity are called macrocapsules. In 1942, Green developed microencapsulation, a modified process in which tiny capsules, called microspheres or microcapsules, are formed (Figure 8.1) (Ahmad et al., 2021). Compared to macrocapsules, microcapsules are of smaller size, 1 μm to 1000 μm in diameter, and hold tiny amounts of core material (Hamad et al., 2011). It is safe to handle and store microcapsules. Microcapsules can also change the apparent properties of the core material; for example, it can change liquid into fine-grade powder. The delivery of a payload to various sites depends on its particle size (Kawashima, 2001; Hughes, 2005). Therefore, efforts have been made in recent years on nanoencapsulation, which reduces the size a step down at the nanoscale. Nanoencapsulation is similar to microencapsulation except for the size. The size of nanocapsules ranges from 1 to 100 nm (Suganya and Anuradha, 2017). The nanocapsule is a vesicular system in which a bioactive substance is surrounded by a unique polymer membrane. In nanospheres, the bioactive substance is uniformly dispersed, forming a matrix system (Couvreur et al., 1995).

8.3 ENCAPSULATION METHODS

The encapsulation process involves embedding the active particles in a matrix by dispersing, mixing, drying, grinding, or sieving. This ensures the immobilization of active particles and controls their release rate. The type of particles produced depends on factors such as the physical and chemical properties of the core, the chemical composition of the wall, the chosen microencapsulation technique, and preparation cost. The core material can be solid, liquid, or a blend of active ingredients, stabilizers, diluents, excipients, and agents that modify the release rate. Coating materials vary widely, including polymers, waxes, resins, and polysaccharides, and are categorized as water-soluble or water-insoluble resins, lipids, or proteins (Ngwuluka et al., 2021). Water-soluble resins include gelatin, gum arabic, arabinogalactan,

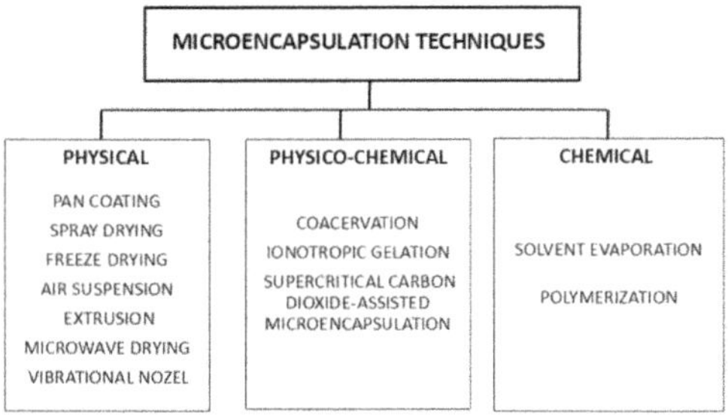

FIGURE 8.2 Type of microencapsulation methods.

PVP/CMC, methylcellulose, polyvinyl acrylate, and polyacrylic acid; water-insoluble options include ethyl cellulose, polyethylene, polymethacrylate, cellulose nitrate, and silicon-based compounds. Common lipids for coatings are paraffin, carnauba wax, beeswax, stearic acid, and stearyl alcohol, while protein options include skim milk protein, milk protein isolates, soy proteins, and pea proteins. The chosen coating material should be economic food-grade substances that must exhibit excellent properties related to strength, flexibility, impermeability, optical characteristics, non-hygroscopic nature, tastelessness, and stability. Furthermore, the coating material must be compatible and non-reactive with the core material. Encapsulation is a multistep process. The available encapsulation methods can be classified into physical, chemical, and physicochemical processes (Figure 8.2) (Poncelet, 2006).

Examples of various types of encapsulation methods are shown in Figure 8.2 and described next.

8.3.1 Pan Coating

The process of tumbling solid particles larger than 600 μm in a pan while applying a coating material is called pan coating. It is one of the oldest industrial procedures for producing small coated particles (Singh et al., 2010; Anusha et al., 2023). The coating material is sprayed using atomizers, and then the coated material is dried by hot air. Consequently, a rigid layer is formed that provides the strength (Cole et al., 1995; Bhowmik et al., 2019; Anusha et al, 2023). Pan coating not only provides chemical and physical protection to the active principle but also helps mask the bitter taste of active principles (Bhowmik et al., 2019; Kulkarni et al., 2015). Because of the frequent application in the pharmaceutical industry, pan coating is often called a tablet coating method. However, it is equally applied in the food and confectionery industry. Ayurvedic dietary supplement (ADS) granules are commonly coated by a pan-coating process. To overcome the bitter taste of ADS, Kulkarni et al. (2015) used an acid-soluble coating material called Eurdagit E100. Since it is insoluble in saliva, the contact between plant extract and taste buds is inhibited (Kulkarni et al., 2015).

The pan coating machine consists of a circular metal pan mounted at 40 degrees and rotated horizontally by a motor (Figure 8.3). This method results in uniform coating. However, many factors such as pan variables, air quality, temperature, flow

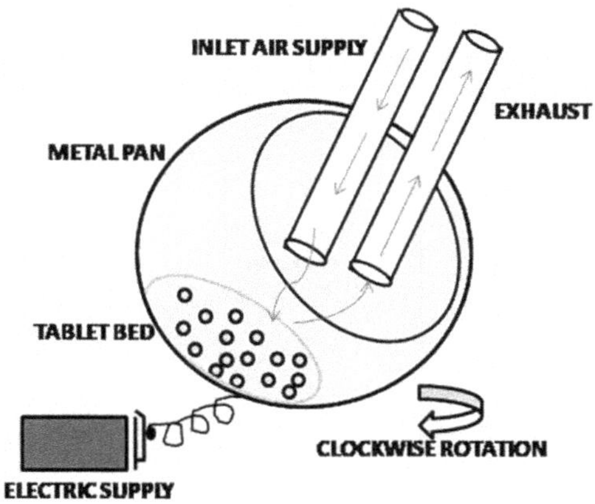

FIGURE 8.3 Sketch of a coating pan consisting of a metal pan raised at 45 degrees and rotated at the horizontal axis with the help of a motor using an electric supply. Hot air is supplied with the help of an inlet air supply, and the moist air is removed from the exhaust. The metal pan contains a tablet bed at the base where the solvent is sprayed on the active particles.

rate, and spray variables can affect the coating process. Among these, pan speed was found to be the most significant factor related to coating uniformity (Rege et al., 2002; Pandey et al., 2006). Pan coating is a highly skilled and time-consuming process, and its scalability is still challenging (Pandey et al., 2006). It requires a large coating material for the desired thickness and uniformity.

8.3.2 Spray Drying

Spray drying is a quick and economical method of encapsulation that gives reproducible results. Unlike pan coating, the scalability of the process is not a problem in this method. During the process, an atomizer sprays a fluid in the drying chamber containing hot gas to change it into the powdered form (Gohel et al., 2008). Spray-dried ingredients have smaller particle sizes (<100 um) and high solubility. Thus spray drying not only produces microcapsules of good solubility but has also been used for nanocapsule production (Okuyama et al., 2006; Arpagaus et al., 2017). This technique has been efficiently applied in the food and pharmaceutical industry for years (Gouin, 2004). The cost of the spray drying method is 30 to 50 times lower than the freeze-drying method (referred to in Section 8.3.3) (Desobry et al., 1997).

Spray drying comprises three basic steps: (i) preparation of the emulsion or dispersion, (ii) homogenization of the dispersion, and (iii) atomization by fluid spraying in the drying chamber (Fernández-Quiroz et al., 2023). The atomized droplets are converted into a powder after exposure to a hot gas current (Figure 8.4). The liquid in the atomizer can be a solution or suspension or an emulsion, while the drying medium used generally is air or some inert gas. The solid component is separated from the liquid through evaporation by a spray drier. As the moisture leaves the liquid, solid powder is obtained, which is collected through the outlet. Spray drying is largely considered a dehydration process.

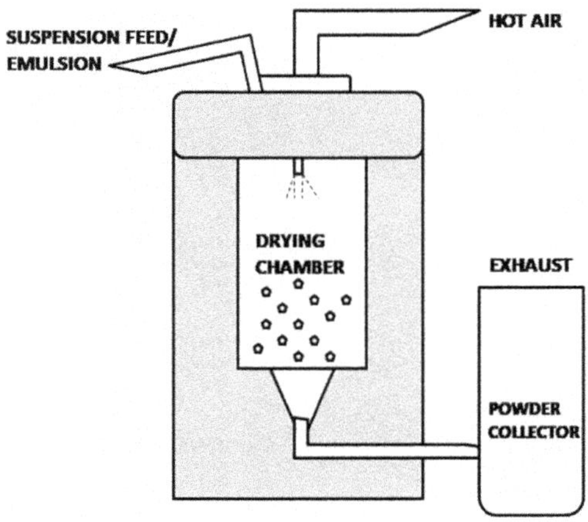

FIGURE 8.4 A diagrammatic representation of a spray dryer consisting of a drying chamber, spray nozzle/atomizer, inlet for suspension feed and hot air, outlet for exhaust, and powder separation system.

Depending on the feeding material and operating conditions of the spray drying, very fine (10–50 μm) or coarse (2–3 mm) particles are synthesized (Gharsallaoui et al., 2007). Furthermore, the choice of the feeding medium affects the texture of the final product. For instance, when encapsulation of the aqueous acidified blackberry extract was done using gum arabic, brighter spherical particles having smooth surfaces were obtained, whereas particles had spherical rough surfaces when polydextrose was the feeding material (Rigon and Noreña, 2016). Microcapsules of cardamom oil were generated by the spray-drying method using different gum ratios (Beristain et al., 2001). Similarly, gelatin microparticles containing propolis extractive solution have also been prepared using this technique with mannitol (Bruschi et al., 2003). In a study, spray drying improved the viability of *Lactobacillus paracasei* subsp. *paracasei* CRL431 in whole milk powder due to a lower moisture content and molecular mobility in the matrix (Poddar et al., 2014). Similarly, the viability of *Bifidobacterium animalis* subsp. *lactis* KMP H9–01 was maintained for more than 6 months at room temperature by encapsulating it in skim milk or trehalose using spray drying (Boontun et al., 2024).

Spray drying can be equally applied to water-soluble and -insoluble active compounds with insignificant thermal degradation. Because of the fast drying in this method, the active compounds have limited exposure to heat (Sosnik and Seremata, 2015). Furthermore, the particle size obtained remains relatively uniform in the submicron-to-micron scale (Sosnik and Seremata, 2015). Although recent advances have been made in spray-drying techniques, the process is still not fully controllable. Factors such as temperature and feed rate continue to affect the outcome of spray drying. In cinnamon, the total phenolic content and antioxidant capacity morphology were affected by temperature (Santiago-Adame et al., 2015). Unlike the fluidized bed methods (where a shell or matrix is built around a large-sized particle by spraying drying the solution over a bed of inert solids), spray drying does not produce full microencapsulation but makes particles with a much more stable shelf life.

8.3.3 Freeze Drying

Freeze drying is also known as lyophilization or cryodesiccation. In this encapsulation method, an emulsion of the desired compound and the coating materials is made, which is later converted into microcapsules (Rodríguez et al., 2021). Freeze drying removes water without affecting nutrients and leaves behind a rich aroma, flavor, and color. Therefore, it is primarily used in the food and cosmetic industry. Fruits, seeds, rinds, meat, fish, and dairy products can be freeze-dried. To get a desired product that has a smaller size and aggregation, it is important to optimize effective factors in formulations (Mohammady et al., 2020). Before drying, the oil is mixed in water and frozen at –90 to –40°C. The frozen material is then heated at reduced pressure so that the frozen water within the material can sublimate directly into gas. Unlike spray-dried products, freeze-dried products retain a higher amount of volatile compounds. The freeze-drying method also increases the durability of nanoparticles, which have applications in drug delivery (Arora et al., 2024).

8.3.4 Air Suspension

The air suspension or Wurster process is similar to pan coating but has better control and flexibility. In this process, the fine core particles of size less than 100 μm are suspended in a vertical air current along with the spray of coating material (Figure 8.5). Subsequently, evaporation takes place, which causes the deposition of a layer of coating material on core particles (Farzi and Gheysipour, 2023). Compared

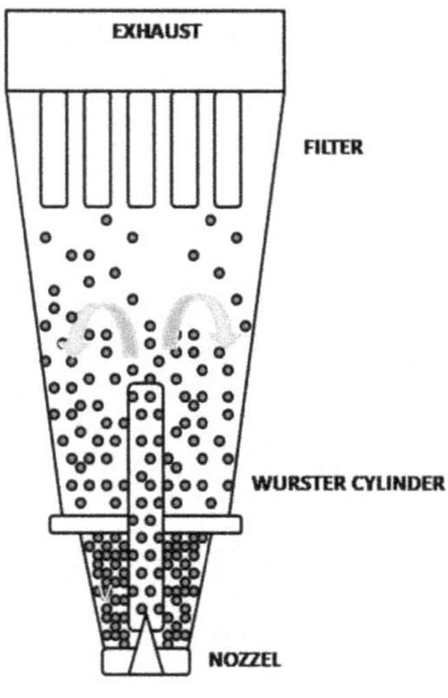

FIGURE 8.5 Outline diagram of a Wurster coating process in a fluidized bed chamber. An air stream is used to coat particles that come from the spray nozzle.

to spray-drying encapsulation, Wurster processes layers with a more distinct film coating over the core (Werner et al., 2007). Among the major advantages, it is a continuous process with minimal manual interruption and has reproducible assurance (Sonar and Rawat, 2015). The core particle size in this process remains relatively large, which increases further due to agglomeration if the process continues unregulated. In this process, powdered particles are recycled from the spray zone until the coating of the requisite thickness is obtained, which makes the entire process time consuming and expensive. However, these successive coatings are useful for controlled release. Since it produces oral liquid suspensions, it is an innovative technology that benefits children and people with swallowing difficulties (Mohylyuk et al., 2020).

8.3.5 Extrusion

Extrusion is the process of obtaining granules with active liquid ingredients (Tackenberg and Kleinebudde, 2015). Products obtained by extrusion have less moisture, have a longer shelf life, and are safe from microbial activities. There are many procedures to make value-added and fortified extruded products with different raw materials (Pathak and Kochhar, 2018). Generally, core material is mixed in a molten matrix and poured into a hot barrel containing a rotating screw (Figure 8.6). When heated to an appropriate temperature, the core forms a homogenous melt, which is subsequently moved into a compression zone. Finally, the material is extruded through an orifice to achieve a defined shape (Figure 8.6). The final product comprises a solid wall of material that covers the core material entrapped in a matrix. The granules are recovered by filtration or centrifugation, and the residual liquid/solvent is removed by air/vacuum drying (Mittala et al., 2023; Singh et al., 2017). Extrusion is extensively used in flavor encapsulation such as orange terpenes flavor in maltodextrin matrix (Tackenberg et al., 2015). Food components like starch, lipids, proteins, oligosaccharides, and fibers are frequently required in the extrusion process (Pathak and Kochhar, 2018). Extrusion technology has been used to develop millet-based contemporary foods (Devi et al., 2014). In malt extrusion, surface-washed particles are used in a vertical screwless process. The centrifugal extrusion process, which consists of a concentric feed tube, is currently used in the manufacture of encapsulated vitamin A acetate (Mujica-Álvarez et al., 2020). In spray drying of sweetened yogurt, the mixed culture of *Streptococcus thermophilus* and *Lactobacillus bulgaricus* is encapsulated using sodium alginate as a coating material (Seth et al., 2017). The extrusion

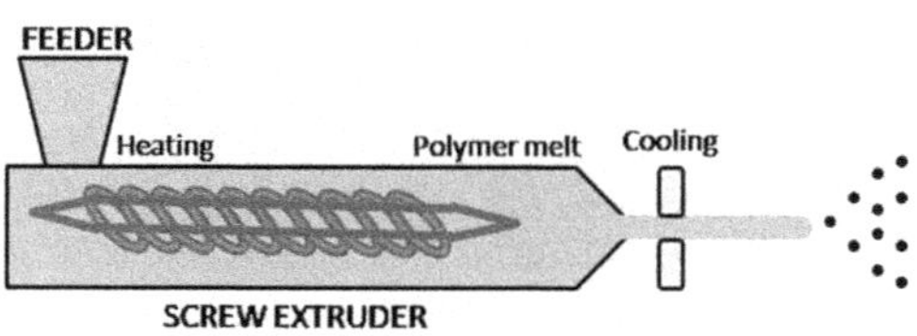

FIGURE 8.6 Diagram of a single screw extruder (SSE) consisting of a screw, drive mechanism, barrel, feeder and cooling unit, and orifice.

technique also has applications in tissue engineering because it maintains the structure and properties of bio-ingredients such as stem cells (Arabpour et al., 2023).

8.3.6 Microwave Drying

This technique is based on a unique volumetric heating mode and internal vapor generation facilitated by electromagnetic radiation of 2450 MHz (Feng et al., 2012). The heat generated within a food product creates internal pressure that pushes moisture out, significantly reducing drying time while enhancing product quality. This process is especially advantageous for encapsulating temperature-sensitive products. A study conducted on the processing of tea flowers showed that the combined effect of microwave-assisted drying and air drying (MAD-AD) had higher amounts of active nutraceutical compounds and possessed higher antioxidant activities in contrast to freeze drying (Shi et al., 2021). The microwave-drying technique has been successfully applied in many industries, but its acceptance rate is slow, especially in the primary manufacturing sector (Mcloughlin et al., 2000).

8.3.7 Vibrational Nozzle Technology

Vibrational nozzle technology is a co-extrusion process where liquid core and wall materials are separately expelled from two concentric nozzles (Ngwuluka et al., 2021). The feed materials exit the nozzle as a single fluid stream that forms droplets of core covered by the shell under vibration (Figure 8.7). This is based on Rayleigh

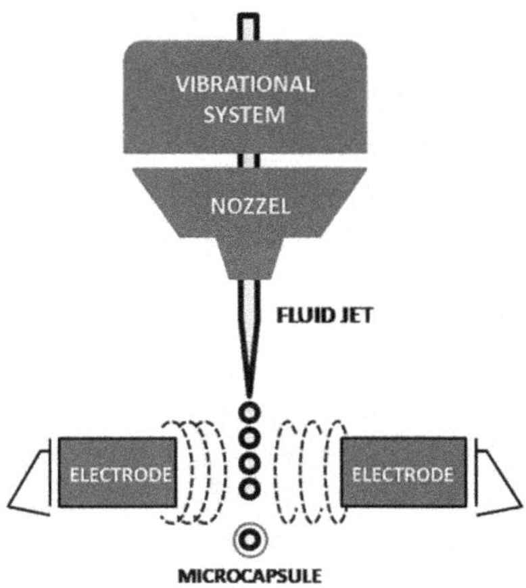

FIGURE 8.7 Vibrational nozzle microencapsulation method. It consists of a vibrational system, a nozzle, a fluid jet, and a voltage system. The voltage system applies an electric potential between the nozzle and an electrode positioned directly beneath it. As droplets move through the electrode, an induced surface charge causes them to repel one another. This repulsion diverts the droplets during separation, preventing mid-air collisions and allowing them to enter the hardening solution over a wider area. This process facilitates the formation of uniform, consistent, and spherical microspheres

instability, where a liquid of low density is pushed through another liquid medium of higher density resulting in the instability of two interfaces (Bennacef et al., 2023).

The vibration of the nozzle can affect the droplets' volume and shape. The droplets are passed through an electrostatic field to ensure uniform beads and to avoid large size distributions due to coalescence effects (Figure 8.7) (Whelehan and Marison, 2011). With this technology beads do not hit each other during the flight and separate as individual microcapsules. High-efficiency vibrational technology (HEVT), has also been used to obtain live cell–loaded polymethyl-methacrylate (PMMA) microcapsules (Pisani et al., 2020). The microcapsules obtained by HEVT showed 50% yield and 27% cell encapsulation efficiency (Pisani et al., 2020).

8.3.8 Coacervation

The coacervation technique of microencapsulation is based on liquid-liquid phase separation. The term 'coacervate' typically refers to spherical clusters of colloidal droplets bound by hydrophobic forces. The two liquid phases consist of a polymer-rich phase (coacervate) and a polymer-depleted phase (equilibrium solution) (Timilsena et al., 2020). It involves the interaction of oppositely charged macro-ions.

Based on the phase separation process, coacervation methods can be divided into simple and complex types. In simple coacervation, a desolation agent such as alcohol or natural salt is added for phase separation, whereas complex coacervation involves complexation between the two polymers of opposite charges, such as gelatin, gum arabic, and silk fibroin chitosan (Deveci and Basal, 2009).

The general process of coacervation includes three steps, which are conducted under continuous agitation (Dubey et al., 2016). The process begins with the preparation of three non-miscible chemical phases, namely, the liquid phase, the core material phase, and the coating material phase. Next comes the coating step, in which a layer of coating material phase is laid around the core material phase, causing the separation of core and liquid phases. Finally, the coating is solidified through crosslinking using the desolvation method or thermal treatment (Azarmi et al., 2006). Consequently, the structural changes begin in biomolecules like proteins, causing their clumping at some level. Eventually, crosslinking results in nanoparticle formation.

In this method, particle size and uniformity can be controlled by modifying the amount of polymer used, its molecular weight, the viscosity of the non-solvent, and other related factors (Dubey et al., 2016). The major limitations of producing food ingredients using coacervation are complicated processing, the high cost of production, and potential health concerns due to the chemical crosslinking agents used in the process (Gouin, 2004). This technique has been used for both polar (hydrophilic) and non-polar (hydrophobic) compounds (Moulik et al., 2022). For example, the hydrophobic vanillin complex used in dairy products was encapsulated in the coacervate of rice bran protein and flaxseed gum to enhance its shelf life (Hasanv and Rafe, 2019). Furthermore, algal oil that prevents oxygen penetration was encapsulated in soy protein and chitosan (Yuan et al., 2017). Similarly, hydrophilic molecules such as sugar, starch, cellulose, and proteins have also been encapsulated by this technique.

8.3.9 Ionotropic Gelation

This method, also called polyelectrolyte complexation, relies on the interaction between cations or anions and an ionic polymer to form a densely crosslinked structure. This technique does not use harsh chemicals or high temperatures (Ahirrao et al., 2014). The method involves dissolving polysaccharides (such as alginate, gellan, and pectin) in water, or in a mildly acidic medium, followed by gradually adding this mixture dropwise, under constant stirring, into a solution containing oppositely charged ions. The polysaccharides then undergo ionic gelation, resulting in the synthesis of spherical, bead-like particles (Racoviţă et al., 2009). The ionotropic gelation technique can formulate hydrogel beads using various polymers by a similar method. Generally, sodium alginate is first dissolved in distilled water, followed by the addition of polymers with continuous stirring, dispersion of drugs, and drop-wise addition of ionic solution through a needle placed on a magnetic stirrer, resulting in the formation of microspheres (Chan et al., 2006). The encapsulation efficiency was reported to increase by more than 60% in a study when the encapsulation of *Stevia* crude extract was done in different alginate blends using ionic gelation (Nathalia et al., 2019). Both micro- and nano-particles are easily obtained by ionotropic gelation (Racoviţă et al., 2009).

8.3.10 Supercritical Fluid Technology

Carbon dioxide is a very common supercritical fluid that acts as a drying medium to remove moisture. This extrusion technology is a flourishing method used in producing various food products and nutraceuticals (Jelena et al., 2016). The system comprises a high-pressure steel container, a hot bath with a temperature controller, a magnetic stirrer, a pressure sensor, and a high-pressure CO_2 liquid pump (Da Costa, 2013). Essential oil components are extracted using supercritical carbon dioxide. In this process, the oil and wall material are submerged in a water bath maintained below 80°C, after which CO_2 is introduced into a cell until the target pressure is reached. It ensures the solubilization of the core material in the supercritical carbon dioxide. And then finally, the microcapsules are obtained (Kamali et al., 2015). The nutrient loss remains minimal in this process as it completes quickly. High temperature also destroys trypsin inhibitors and improves the digestibility of proteins (Jelena et al., 2016).

8.3.11 Solvent Evaporation

The solvent evaporation method has been widely employed for preparing microspheres, nanospheres, and microcapsules (O'Donnell and McGinity, 1997). This process consists of three steps: first, a solution or dispersion of the core material is made in an aqueous medium. In the second step, an emulsion of the core material (oil-in-water emulsion) is made using an appropriate stabilizer. The third step involves the removal of solvent from the dispersed phase, either through evaporation or extraction, which leads to precipitation and the formation of microspheres (O'Donnell and McGinity, 1997; Freitas et al., 2005; Li et al., 2008). Floating microspheres, which is a novel technique for the prolonged release of drugs, are effectively prepared by this approach (Solanki et al., 2016). Hence, it finds maximum use in pharmaceuticals. Polyphenols in many plants, such as *Moringa oleifera*, *Citrus sinensis*, and *Euphorbia helioscopia*, have also been extracted by the solvent evaporation method (Bennour et al., 2020; Gómez-Urios et al., 2023; Bourgou et al., 2016).

8.3.12 Polymerization

Polymerization is a chemical process to synthesize large molecules – polymers – from monomers (smaller building blocks). Polymerization microencapsulation involves adding a thin layer of polymers around the core particle. It includes steps of dispersion, monomer initiation, polymerization, stabilization, and termination. Polymerization can be classified as discussed in the following sections.

8.3.12.1 Interfacial Polymerization

In this process, two reactive monomers or prepolymers are dissolved separately in two immiscible phases – one in an aqueous phase and the other in an organic phase. When the monomers meet at the liquid-liquid interface, they react rapidly to form microcapsules (Nguon et al., 2018).

8.3.12.2 In Situ Polymerization

This is one of the most extensively used methods in the cosmetic industry in which monomer polymerization occurs directly on the core particle's surface (Lu and Kinan, 2012; Nguon et al., 2018). For this, a solution containing monomeric or oligomeric wall material is added to the dispersed core phase and adjusted to the desired size. For instance, *in situ* polymerization was employed for the synthesis of melamine-formaldehyde (MF) resin-walled microcapsules containing styrene in an oil-in-water emulsion (Wang et al., 2009). The microcapsule synthesis begins by adding oil to the MF solution and its sonication to make an oil-in-water emulsion. This emulsion is continuously stirred, and a resin is added in the next step. The pH of the final emulsion is set to an acidic range. Shell formation of the microcapsules occurs as melamine reacts with formaldehyde at the oil droplet interface, forming a crosslinked MF polymer film (Wang et al., 2009).

8.3.12.3 Matrix Polymerization

The core material is embedded within a polymeric matrix during particle formation. This polymer matrix is favored for its ease of use, along with properties such as abrasion and corrosion resistance, light weight, high strength, and stiffness (Rajak et al., 2019).

8.3.12.4 Single Emulsion Method

Oil and polymer are mixed with proper sonication, resulting in an emulsion of water and oil. It is combined with crosslinking polymers, heated to cause denaturation, and microspores are formed in an organic solvent. The microspores so formed are then separated and isolated by centrifugation, washing, and a separation process (Verma et al., 2019).

8.3.12.5 Double Emulsion Method

In this process, the inner and dispersed polymer phases are mixed to form a water-in-oil emulsion. Subsequently, surfactants and other chemicals are added to this emulsion and stirred to make another water-in-oil emulsion. Finally, by heating and evaporation, rigid microcapsules are formed (Verma et al., 2019).

8.4 INCORPORATION OF BIOACTIVE COMPOUNDS INTO FUNCTIONAL FOODS

Active ingredients, when directly added to functional foods, improve nutritional value but they may slowly degrade or lose activity over time (Augustin and Sanguansri, 2014). Microencapsulation offers promising solutions to the problems of change in taste, color, texture, and aroma of functional food. For example, directly mixing fish oil and milk changes milk's flavor. However, the level of eicosapentaenoic acid (EPA) and docosahexaenoic acid (DHA) in milk is increased without any adverse effect on its flavor by incorporating encapsulated fish oil (Nelson and Martini, 2009). Similarly, milk can be fortified with vitamin D entrapped in liposomes (Polzonetti et al., 2020). Microencapsulated iron ingredients can maintain the bioavailability of iron compounds and prevent an undesirable flavor (Muñoz-More et al., 2023). Similarly, probiotic bacterial cells encapsulated in calcium alginate are produced in fermented frozen dairy products (Krasaekoopt et al., 2003). Microencapsulation of beta-galactosidase in liposomes can be used to act in vivo but protected from acting on lactose during storage (Rodriguez-Nogales and Delgadillo, 2005). Interestingly, ascorbic acid is added while preparing bread to improve its quality. However, this acid degrades in the presence of water and oxygen, which can be prevented by encapsulation (Vargas and Simsek, 2021). Thus microencapsulation can be an effective tool for preserving desired qualities and avoiding undesired ones like unpleasant taste and smell of different nutritional ingredients such as soy extract, bitter herbs, and omega-3 oils. Moreover, hygroscopic properties of materials like sodium chloride may be reduced by microencapsulations (Singh et al., 2010).

8.5 DIETARY SUPPLEMENTS AND CAPSULE FORMULATIONS

Dietary supplements, also called food or nutritional supplements, are formulations designed to top up nutrients like essential amino acids, fatty acids, vitamins, minerals, or other bioactive substances in the body. The quantity of such nutrients in a person's diet is generally insufficient, which affects the normal physiological functioning and causes various health-related problems. Nutritional supplements support health and wellness by addressing nutritional deficiencies and sometimes through their therapeutic properties (Augustin and Sanguansri, 2014).

Dietary supplement formulations often contain nutrients such as vitamins, minerals, amino acids, and metabolites or ingredients like herbs, concentrates, and extracts or their combinations. However, there is no global consensus on the definition and categorization of dietary supplements, and therefore, different countries treat them as either foods or drugs (Thakkar et al., 2020). These products are always meant for oral consumption, and these are never indicated for any other use like topical application, injection, or inhaling purposes (Blaze, 2021). The growing awareness of health and wellness, along with increasing scientific validation of the benefits of certain nutraceuticals, has led to significant growth in the dietary supplement industry. Furthermore, over-the-counter availability of these in the majority of countries also adds to their demand and popularity. These products are consumed to improve or maintain overall health, prevent diseases, or enhance physical and mental stamina. Dietary supplements are also used to increase the performance of athletes and sports players (Maughan et al., 2018).

8.5.1 Capsule Formulations

Dietary supplements are available in different dosage forms such as tablets, capsules, powders, liquids, and energy bars. Among these, capsule formulations, followed by tablet form, are the most popular dosage forms for dietary supplements (Singh et al., 2018). In contrast to other dosage forms, capsule formulations provide certain advantages in terms of versatility, ease of manufacturing and consumption, and ability to obscure unwanted properties of the ingredients and to protect sensitive ingredients. Unlike tablets, which are solid compressed forms of drugs or active ingredients, capsules are a solid dosage form in which the active ingredients are covered with a shell. The shell of the capsules is made up of either gelatin or non-gelatin compounds such as hydroxypropyl methylcellulose (HPMC), polyvinyl alcohol, or carrageenan (Aliyu et al., 2020).

Capsules are basically carriers that are taken mostly via the oral route to deliver a pre-determined amount of active compounds in various forms such as powders, granules, semisolids, and liquids (Darji et al., 2018). The capsules used for stuffing different ingredients, including dietary supplements, are primarily of two types: hard capsules and soft gelatin capsules (softgels).

Hard capsules are typically composed of two separate pieces – a body and a cap (Cole et al., 2008). The body is a cylindrical tube closed at one end which remains filled with powdered or granular ingredients, whereas the cap is a short tube used to seal the body. Hard capsules are preferred as these offer flexibility in dosage and are commonly used for dry formulations, such as powdered herbs or vitamins. They are easy to produce and made from animal-derived gelatin, but vegetarian alternatives made from HPMC, starch, carrageenan, and other compounds are also available (Darji et al., 2018).

Unlike hard capsules, softgels are single-piece capsules of variable shapes filled with liquid or semi-solid ingredients (Benza and Munyendo, 2011). The ingredients of the softgels are also the same as that of hard capsules. However, for requisite flexibility, different plasticizers such as sorbitol, propylene glycol, and glycerin are added. Softgels are an ideal choice for encapsulating oils, fat-soluble vitamins, and other liquid formulations. These are preferred for their smooth texture, ease of swallowing, and ability to mask unpleasant tastes or odors. Furthermore, they better protect sensitive ingredients from oxidation, moisture, and light.

Innovations in capsule formulation technology have made available some additional forms of capsules, namely, modified-release capsules, liquid-filled hard capsules, and combination capsules (Hadi and Rao, 2012). Modified-release capsules release active ingredients at the desired site in a regulated manner. Further modifications in capsule formulations such as delayed-release or enteric-coated capsules ensure that ingredients bypass the stomach and are released in the intestines, thus protecting sensitive components like probiotics or enzymes from the enzymatic action of digestive juices. Liquid-filled hard capsules combine the advantages of hard capsules with the ability to encapsulate liquid ingredients, providing extra versatility in formulation options. Combination capsules contain multiple compartments, allowing different ingredients to be separated within a single dose, reducing the risk of ingredient interactions and ensuring more stable formulations (Hadi et al., 2013).

Capsules intended for dietary supplements must comply with regulatory standards established by organizations like the FDA and EFSA (Maheshwari et al., 2018). Manufacturers must ensure that the capsules are safe for consumption, the dosage is accurate, and the ingredients are stable throughout the product's shelf life (Sieniawska, 2017). Regulations also mandate proper labeling, including disclosing all ingredients, usage instructions, and potential allergen information.

8.5.1.1 Advantages and Challenges of Capsule Formulations

Capsule formulations offer several advantages, which make them suitable for a wide range of dietary supplements (Benza and Munyendo, 2011):

- **Ease of consumption**: Capsules are easier to swallow than tablets, especially for people who find swallowing solid dosage forms difficult.
- **Taste and odor masking**: Capsules effectively mask the taste and odor of ingredients, particularly in the case of bitter or unpleasant-tasting compounds.
- **Protection of ingredients**: Ingredients like omega-3 fatty acids, probiotics, and fat-soluble vitamins are sensitive to environmental conditions like light, oxygen, and moisture (Hadi et al., 2013). Capsules, especially softgels, protect such sensitive compounds, prolong shelf life, and maintain the potency of the active ingredients.
- **Customization and flexibility**: Capsules offer flexibility in terms of formulation design. For instance, various forms of active ingredients such as powders, beads, and even mini tablets can be filled in hard capsules, allowing the possibility of a combination of multiple ingredients (Hadi and Rao, 2012). Similarly, softgels are suitable for delivering liquid formulations, including oils, essential fatty acids, and lipid-based ingredients.

While capsule formulations offer numerous benefits, they also present several formulation challenges, including:

- **Moisture sensitivity**: The shell of capsules, especially when it contains gelatin or HPMC, may absorb moisture from the environment, which may cause the capsules to soften or harden. This affects the stability of the active ingredients and makes the proper packaging and storage of capsules essential (Hadi et al., 2013).

- **Ingredient compatibility**: Some active ingredients or excipients may interact with the capsule shell, leading to degradation or reduced efficacy. For instance, liquid fills for softgels may require specific compatibility testing to ensure they do not compromise the shell's integrity.
- **Release characteristics**: Capsules are generally designed to release their contents in the stomach. However, release of certain ingredients (e.g., probiotics or enteric-coated substances) in the stomach make them inactive as they are prone to being destroyed by gastric juices. Therefore, capsules must be modified to ensure their release in the intestines.

Thus, capsule formulations have become a cornerstone in the dietary supplement industry due to their versatility, convenience, and ability to protect sensitive ingredients. However, challenges related to moisture sensitivity, ingredient compatibility, and maintaining stability must be addressed during product development. As technology advances, capsule formulations will further improve, offering more innovative and effective ways to deliver nutraceutical ingredients.

8.6 CHALLENGES IN STABILITY AND SHELF LIFE

The stability and shelf life of nutraceutical formulations are critical aspects of product development. These parameters directly affect the efficacy, safety, and consumer acceptance of the final product. The shelf life of any product is the time since its manufacturing after which the product loses its desired properties due to decomposition or structural changes and becomes unfit for consumption (Calvino et al., 2021). Unlike conventional pharmaceuticals, nutraceutical formulations are often complex and contain a wider range of active ingredients, each having unique stability profiles. Therefore, nutraceuticals are more prone to stability issues and a shorter shelf life. Understanding the factors that influence the stability and shelf life of these formulations is essential for ensuring product quality throughout the intended shelf life. Several factors, including hydration, dehydration, oxidation, and microbial spoilage, may affect the desired constitution of nutraceutical products and hence their marketability (Calvino et al., 2021).

8.6.1 Ingredient Characteristics

The physicochemical properties of active ingredients play a pivotal role in defining the stability of nutraceutical formulations. Many bioactive compounds are inherently unstable due to their sensitivity to chemical processes such as oxidation, hydrolysis, and photodegradation (Maheshwari et al., 2018). For instance, vitamin C and polyunsaturated fatty acids are highly susceptible to oxidation and may degrade under a range of radiation and temperature (Yin et al., 2022; Carvalho et al., 2019). Additionally, some ingredients, such as amino acids and herbal extracts, due to their hygroscopic nature have a higher tendency to absorb moisture. Certain biomolecules like vitamin C remain active in a narrow range of pH, which necessitates careful formulation to maintain their stability (Yin et al., 2022).

8.6.2 Formulation Factors

Besides active ingredients, nutraceutical formulations, similar to pharmaceutical preparations, contain other ingredients, which are collectively termed excipients. Excipients are also known by various terms like diluents, disintegrants, binders, glidants, lubricants, stabilizers, solvents and co-solvents, and coating agents. These are inactive substances that help preserve the requisite physical and chemical properties of the dosage forms (Darji et al., 2018). The choice of excipients and the overall formulation design significantly influence the stability of active ingredients in various formulations, including nutraceutical products (Calvino et al., 2021). To avoid adverse interactions between excipients and active ingredients, the excipients must be compatible with the active ingredients of the formulation (Darji et al., 2018). The dosage form, which may be a tablet, capsule, powder, or liquid, also affects stability. For example, liquid formulations are more susceptible to microbial growth and chemical degradation than solid dosage forms.

Additionally, encapsulation techniques, such as microencapsulation or liposomal delivery, can protect sensitive ingredients from environmental factors like air, moisture, and temperature, thereby enhancing stability and prolonging the shelf life of active ingredients like vitamins and oils (Delaporte et al., 2024; Bakry et al., 2016). In capsules, the shell composition is important in increasing its shelf life by providing stability to the core material. Delaporte and colleagues (2024) recently investigated how varying the wall material–to–active (WM:A) ratio influences the stability of vitamin C encapsulated in a blend of maltodextrin and gum arabic. They found that a higher WM:A ratio (7:1) significantly enhanced both the stability and antioxidant activity of the active ingredient during storage (Delaporte et al., 2024). In another similar study, encapsulation increased the oxidative stability of fish oil, which is full of omega-3 fatty acids and lipid-soluble vitamins, compared to untreated oil. Furthermore, whey protein was more effective than maltodextrin and gum arabic (Selim et al., 2021). Thus, the choice of wall material is an important factor for longer stability and shelf life of the nutraceuticals.

8.6.3 Packaging Considerations

Packaging is important for protecting nutraceutical formulations from environmental factors that can enhance degradation, microbial spoilage, and contamination and provides information regarding the product and dosage (Das et al., 2018). The material used for packaging – whether glass, plastic, or metal – must provide adequate barrier properties against moisture, oxygen, and light. For instance, amber glass bottles are often used to protect light-sensitive ingredients from ultraviolet (UV) exposure, while high-barrier plastics can minimize oxygen permeability (Ibrahim et al., 2022). The integrity of the packaging seal is also vital, as poor sealing can allow the ingress of moisture and oxygen, leading to product spoilage. Moreover, the headspace oxygen in the packaging can influence the oxidation of sensitive components, necessitating the use of oxygen absorbers or nitrogen flushing in some cases (Gupta, 2023).

8.6.4 Environmental Conditions

Storage conditions are a critical determinant of the shelf life of nutraceutical products. Elevated temperatures generally promote chemical reactions and microbial growth, leading to faster degradation of active ingredients (Palanikumar et al., 2018). Similarly, high humidity can promote moisture uptake in hygroscopic materials, resulting in clumping, degradation, or microbial contamination. Light exposure, particularly UV light, can degrade photosensitive ingredients, while oxygen exposure can lead to the oxidation of fatty acids and vitamins (Yin et al., 2022). For example, alpha-lipoic acid (ALA) decomposes under light and heat, causing an unpleasant odor (Jovicic-Bata et al., 2024). Controlling these environmental factors through proper storage conditions, such as cool, dry, and dark environments, is essential for maintaining the product's stability.

8.6.5 Microbial Contamination

Contamination of microorganisms, mycotoxins, pesticides, heavy metals, etc., is a substantial concern for the herbal health industry (White, 2020). The occurrence of contaminants in health products not only decreases their quality but also exposes consumers to serious health problems. Nutraceutical formulations, particularly those containing water or organic materials, provide a conducive environment for microbial growth. The initial bioburden of the product, which refers to the number of viable microorganisms present, must be minimized through stringent Good Manufacturing Practices (GMP) (White, 2020). The effectiveness of preservatives in inhibiting microbial growth is also critical for formulations with water or other susceptible components. Ensuring microbial stability through proper formulation, packaging, and storage conditions is essential to prevent spoilage and ensure product safety (Maheshwari et al., 2018).

8.6.6 Manufacturing Process

The conditions under which nutraceutical products are manufactured can have a profound impact on their stability. Processing conditions, such as heat, pressure, and exposure to oxygen during manufacturing, can affect the stability of sensitive ingredients (Maheshwari et al., 2018). For example, excessive heat during drying or tableting processes can degrade heat-labile compounds. Additionally, adherence to GMP (good manufacturing practices) is essential to ensure that the product is consistently produced to meet quality standards, thereby reducing variability in stability and shelf life.

8.6.7 Regulatory and Compliance Factors

Regulatory guidelines, such as those provided by the International Conference on Harmonisation (ICH), Dietary Supplement Health and Education Act (DSHEA), and the World Health Organization (WHO), emphasize the importance of quality assurance and stability testing for the proper stability and shelf life of nutraceutical products (White, 2020). Similarly, the nutraceutical industry in India is regulated under the Food Safety and Standard Act (2006) by the Food Safety and Standard Authority of India (FSSAI). It covers various aspects such as manufacturing, packaging and labeling, selling, and restrictions on the advertisement of food products including nutraceuticals (Sahoo et al., 2020). Comprehensive stability studies, conducted under accelerated and real-time conditions, are necessary to identify potential degradation pathways and to determine the product's shelf life (Maheshwari et al., 2018). These studies help ensure that the product remains safe and effective throughout its intended shelf life, in compliance with regulatory requirements. However, in most countries, such tests are not mandatory for dietary supplements as the latter are often classified as foods instead of pharmaceutical products (Binns et al., 2018). Thus, nutritional supplements get registered for marketing without any strict legal requirement of quality control tests, leading to substandard products. Lalić-Popović et al. (2023) compared the seven different available ALA formulations (five dietary supplements and two drugs) for uniformity of content, disintegration time, and dissolution rates; significant variations were reported in the collected samples, suggesting a need for stringent regulations on the formulation of dietary supplements (Lalić-Popović et al., 2023)

8.6.8 Consumer Usage and Handling

How consumers handle and store nutraceutical products can also affect their stability and shelf life. Exposure to air, frequent opening and closing of packaging, and improper sealing after use can introduce contaminants or environmental factors that accelerate degradation. Educating consumers about good storage practices, such as keeping products in a cool and dry place under dark conditions, can help maintain the product's integrity over time.

8.7 CONCLUSION

The arena of polyphenol nutraceutical formulations has significantly advanced due to the development efforts and ongoing research focusing on enhancing the stability, bioavailability, and therapeutic efficacy of polyphenols. Innovative encapsulation techniques, like microencapsulation and nanoencapsulation, along with controlled-release technologies, have addressed some of the core challenges of polyphenol stability and targeted delivery in functional foods and dietary supplements. Despite these advancements, polyphenol formulations still face obstacles, particularly regarding environmental sensitivities and cost-effectiveness in production.

Each encapsulation method, from microencapsulation to advanced techniques like nanoencapsulation, spray drying, freeze drying, and coacervation, offers unique advantages and challenges based on the peculiar characteristics of the active ingredients and the desired applications. For example, spray drying and freeze drying are cost-effective methods that create stable, dry powders suitable for food and supplement formulations. At the same time, nanoencapsulation and coacervation offer precise control over release rates and protection for sensitive ingredients.

As the demand for nutraceuticals continues to rise, particularly for products with natural and functional benefits, there is a growing imperative for industry and researchers to collaborate in refining these formulations. Future studies should prioritize optimizing formulation techniques and developing solutions to stability challenges to ensure that polyphenol-based nutraceuticals meet high standards for efficacy and safety. By doing so, they can better fulfill their potential as accessible, therapeutic dietary supplements that contribute meaningfully to preventative health and wellness.

REFERENCES

Ahmad SU, Li B, Sun J, Arbab S, Dong Z, Cheng F, Zhou X, Mahfuz S, Zhang J. (2021). Recent advances in microencapsulation of drugs for veterinary applications. *J Vet Pharmacol Ther* 44(3): 298–312. https://doi.org/10.1111/jvp.12946.

Ahirrao SP, Gide PS, Shrivastav B, Sharma P. (2014). Ionotropic gelation: a promising cross linking technique for hydrogels. *Res Rev J Pharm Nanotech* 2(1): 1–13.

Akombaetwa N, Iiangala AB, Thom L, Mrmvanga PB, Witika BA, Aristotle B. (2023). Current advances in lipid nanosystems intended for topical and transdermal drug delivery applications. *Pharmaceutics* 15(2): 656. https://doi.org/10.3390/pharmaceutics15020656.

Aliyu RS, Lawal AM, Chasta P, Sharma GK. (2020). Capsules: types, manufacturing, formulation, quality control tests and, packaging and storage – a comprehensive review. *World J Pharm Life Sci* 6(8): 93–104.

Anusha L, Sharma GS, Kiran S, Rao TR. (2023). Review on pan coating. *Int J Modern Pharma Res* 7(11): 8–14.

Arabpour Z, Fath-Bayati L, Sefat F, Navaei-Nigjeh M. (2023). Microencapsulation: extrusion principles of biomaterials encapsulation. In: *Biomaterials*. Woodhead Publishing Series, 1(19): 393–409.

Arora S, Dash SK, Dhawan D, Sahoo PK, Jindal A, Gugulothu D. (2024). Freeze-drying revolution: unleashing the potential of lyophilization in advancing drug delivery systems. *Drug Deliv Transl Res* 14(5): 1111–1153. https://doi.org/10.1007/s13346-023-01477-7.

Arpagaus C, John P, Collenberg A, Ruetti DP. (2017). Nanocapsules formation by nano spray drying. In: *Nanoencapsulation Technologies for the Food and Nutraceutical Industries*. Elsevier Inc. A, 10: 346–401.

Augustin MA, Sanguansri L. (2014). Challenges and solutions to incorporation of nutraceuticals in foods. *Ann Rev Food Sci Technol* 6(1). https://doi.org/10.1146/annurev-food-022814-015507.

Azarmi S, Huang Y, Chen H, McQuarrie S, Abrams D, Roa W, Finlay WH, Miller GG, Löbenberg R. (2006). Optimization of a two-step desolvation method for preparing gelatin nanoparticles and cell uptake studies in 143 Bosteosarcoma cancer cells. *J Pharm Sci* 9: 124–132.

Bakry AM, Abbas S, Ali B, Majeed H, Abouelwafa MY, Mousa A, Liang L. (2016). Microencapsulation of oils: a comprehensive review of benefits, techniques, and applications. *Comprehens Rev Food Sci Food Saf* 15(1): 143–182. https://doi.org/10.1111/1541-4337.12179.

Bartosz T, Irene T. (2015). Polyphenols encapsulation – application of innovation technologies to improve stability of natural products. *Phy Sci Rev* 1(2): 20150005. https://doi.org/10.1515/psr-2015-0005.

Bennacef C, Desobry-Banon S, Probst L, Desobry S. (2023). Alginate core-shell capsules production through coextrusion methods: principles and technologies. *Marine Drugs* 21(4): 235.

Bennour N, Mighri H, Eljani H, Zammouri T, Akrout A. (2020). Effect of solvent evaporation method on phenolic compounds and the antioxidant activity of *Moringa oleifera* cultivated in Southern Tunisia. *S African J Bot* 129: 181–190.

Benza HI, Munyendo WLL. (2011). A review of progress and challenges in soft gelatin capsules formulations for oral administration. *Int J Pharma Sci Rev Res* 10: 20–24.

Beristain CI, Garcíab HS, Vernon-Carter EJ. (2001). Spray-dried encapsulation of cardamom (*Elettaria cardamomum*) essential oil with mesquite (*Prosopis juliflora*) gum. *Lebensmittel-Wissenschaft und-Technologie J Food Sci Technol* 34 (6): 398–401.

Bhowmik D, Bhanot R, Kumar KPS. (2019). Tablet coating technique. *Res Pharma Dosage Forms Tech* 11(1): 1–9. https://doi.org/10.5958/0975-4377.2019.00001.6.

Binns CW, Lee MK, Lee AH. (2018). Problems and prospects: public health regulation of dietary supplements. *Annu Rev Public Health* 39: 403–420.

Blaze J. (2021). A comparison of current regulatory frameworks for nutraceuticals in Australia, Canada, Japan, and the United States. *Innov Pharm* 12(2): 8.

Boontun C, Vatanyoopaisarn S, Phalakornkule C, Domrongpokkaphan V, Thitisak P, Thaveetheptaikul P, Bamrungchue, N. (2024). Influence of protectant for encapsulation by freeze-drying and spray-drying techniques, and packaging environments on the stability of the probiotic *Bifidobacterium animalis* subsp. *lactis* strain KMP-H9–01 during storage. *Drying Technol* 42: 762–774.

Bourgou S, Sairi Beji R, Medini F, Ksouri R. (2016). Effect of the solvent and the extraction method on the phenol content and the antioxidant potential of *Euphorbia helioscopia. J New Sci* 28: 1649–1655.

Bruschi ML, Cardoso MLC, Lucchesi MB, Gremião MPD. (2003). Gelatin microparticles containing propolis obtained by spray-drying technique: preparation and characterization. *Int J Pharma* 264(1–2): 45–55.

Calvino MM, Lisuzzo L, Cavallaro G, Lazzara G, Milioto S. (2021). Non-isothermal thermogravimetry as an accelerated tool for the shelf-life prediction of paracetamol formulations. *Thermo Chimica Acta* 70: 178940. https://doi.org/10.1016/j.tca.2021.178940.

Carvalho JDD, Oriani VB, deOliveira GM, Hubinger MD. (2019). Characterization of ascorbic acid microencapsulated by the spray chilling technique using palm oil and fully hydrogenated palm oil. *LWT Food Sci Technol* 101: 306–314. https://doi.org/10.1016/j.lwt.2018.11.043.

Chan LW, Lee HY, Heng PWS. (2006). Mechanisms of external and internal gelation and their impact on the functions of alginate as a coat and delivery system. *Carbohydrate Polym* 63(2): 176–187.

Cole ET, Cadéb D, Benameur H. (2008). Challenges and opportunities in the encapsulation of liquid and semi-solid formulations into capsules for oral administration. *Adv Drug Deliv Rev* 60: 747–756.

Cole G, Hogan J, Aulton M. (1995). *Pharmaceutical Coating Technology*. London, Taylor and Francis, 1–5.

Couvreur P, Dubernet C, Puisieux F. (1995). Controlled drug delivery with nanoparticles: current possibilities and future trends. *Euro J Pharma Biopharma* 41: 2–13.

Da Costa L. (2013). Microencapsulation of oregano essential oil in starch-based materials using supercritical fluid technology. *Innov Food Sci Emerg Technol* 20: 140–145.

Darji MA, Lalge RM, Marathe SP, Tarul M, Tasnim F, Alia A, Hyung L, Michael R, Narasimha M. (2018). Excipient stability in oral solid dosage forms: a review. *AAPS Pharm Sci Tech* 19: 12–26. https://doi.org/10.1208/s12249-017-0864-4.

Das PS, Saha P, Das R. (2018). Pharmaceutical packaging technology: a brief outline. *Res J Pharm Dos Technol* 10: 23–28.

DeFelice S. (2010). Nutraceuticals: reflections. In: Pathak Y (ed), *Handbook of Nutraceuticals: Ingredients, Formulations, and Applications*. CRC Press Taylor & Francis Group, 357–365.

Delaporte A, Duchemin B, Grisel M, Gore E. (2024). Impact of wall material-to-active ratio in the stability of spray-dried ascorbic acid using maltodextrin and gum arabic. *Molecules* 29(15): 3587. https://doi.org/10.3390/molecules29153587.

Desai KGH, Park HJ. (2005). Recent developments in microencapsulation of food ingredients, *Dry Technol* 23: 1361–1394.

Desobry S, Netto FM, Labuza TP. (1997). Comparison of spray-drying, drum-drying and freeze-drying for beta-carotene encapsulation and preservation. *J Food Sci* 62(6): 1158–1162.

Deveci SS, Basal G. (2009). Preparation of PCM microcapsules by complex coacervation of silk fibroin and chitosan. *Coll Polymer Sci* 287: 1455–1467.

Devi NL, Shobha S, Alavi S, Kalpana K, Soumya M. (2014). Extrusion technology for the development of millet based complementary foods. *J Food Sci Technol* 51(10): 2845–2850. https://doi.org/10.1007/s13197-012-0789-6.

Di Costanzo A, Angelico R. (2019). Formulation strategies for enhancing the bioavailability of silymarin: the state of the art. *Molecules* 24: 2155. https://doi.org/10.24926/iip.v12i2.3694.

Dubey S, Mody N, Sharma R, Agrawal U, Vyas SP. (2016). Nanobiomaterials: novel nanoplatforms for protein and peptide delivery In Grumezescu AM (ed), *Nanobiomaterials in Drug Delivery*. William Andrew Publishing, 111–146. https://doi.org/10.1016/B978-0-323-42866-8.00004-6.

Fang Z, Bhandari B. (2010). Encapsulation of polyphenols – a review. *Trends Food Sci Technol* 21(10): 510–523.

Farzi G, Gheysipour M. (2023). Microencapsulation: air suspension technique. In Sefat F, Farzi G, Mozafari M (eds), *Principles of Biomaterials Encapsulation*. Elsevier, 297–303.

Feng H, Yin Y, Tang J. (2012). Microwave drying of food and agricultural materials: basics and heat and mass transfer modeling. *Food Eng Rev* 4(2): 1–18. https://doi.org/10.1007/s12393-012-9048-x.

Fernández-Quiroz D, Tohidi MM, Paymard B, Lucero-Acuña A. (2023). Immobilization of essential oils in biopolymeric matrices: recent approaches for controlled delivery systems. In Rahman A (ed), *Studies in Natural Products Chemistry*. Elsevier, 78(9): 365–401. https://doi.org/10.1016/B978-0-323-91253-2.00005.

Freitas S, Merkle HP, Gander B. (2005). Microencapsulation by solvent extraction/evaporation: reviewing the state of the art of microsphere preparation process technology. *J Controlled Release* 102: 313–332.

Gharsallaoui A, Roudaut G, Chambin O, Voilley A, Saurel R. (2007). Advances in colloid and interface. *Science* 223: 40–54. https://doi.org/10.1016/j.cis.2015.05.003.

Gohel M, Parikh RK, Nagori SA Gandhi AV, Shroff MS, Patel PK, Gandhi CS, Patel V, Bhagat NY, Poptani SD, Kharadi SR, Pandya R, Patel TC. (2008). Spray drying: a review. *Pharma Rev* 7(5).

Gómez-Urios C, Viñas-Ospino A, Puchades-Colera P, Blesa J, López-Malo D, Frígola A, Esteve MJ. (2023). Choline chloride-based natural deep eutectic solvents for the extraction and stability of phenolic compounds, ascorbic acid, and antioxidant capacity from *Citrus sinensis* peel. *LWT* 177(1): 114595.

Gouin S. (2004). Microencapsulation: industrial appraisal of existing technologies and trends. *Trends Food Sci Tech* 15: 330–347. https://doi.org/10.1016/j.tifs.2003.10.005.

Gupta P. (2023). Role of oxygen absorbers in food as packaging material, their characterization and applications. *J Food Sci Technol*. https://doi.org/10.1007/s13197-023-05681-8.

Hadi MA, Rao NGR. (2012). Novel techniques in formulations: an overview. *World J Pharma Res* 1: 1–17.

Hadi MA, Rao NGR, Rao AS, Shiva G, Akram JW. (2013). Impact of capsules as a carrier for multiple unit drug delivery and the importance of HPMC capsules. *Res J Pharm Tech* 6(1): 34–43.

Hamad SA, Dyab AFK, Stoyanov SD, Paunov VA. (2011). Encapsulation of living cells into sporopollenin microcapsules. *J Mater Chem* 21: 18018–18023.

Hasanv E, Rafe A. (2019). Development of vanillin/β-cyclodexterin inclusion microcapsules using flax seed gum-rice bran protein complex coacervates. *Int J Biol Macromol* 131: 60–66.

Hughes GA. (2005). Nanostructure-mediated drug delivery, nanomedicine: nanotechnology. *Disease Month* 51(6): 342–361.

Ibrahim ID, Hamam Y, Sadiku ER, Ndambuki JM, Kupolati WK, Jamiru T, Eze AA, Snyman J. (2022). Need for sustainable packaging: an overview. *Polymers* 14(20): 4430. https://doi.org/10.3390/polym14204430.

Jelena BP, Djurdjica A, Antun J, Jurislav B, Borislav M, Stela J, Biljana P, Drago S. (2016). Application of supercritical carbon dioxide extrusion in food processing technology. *Hemijska Industrija* 71: 24–24. https://doi.org/10.2298/HEMIND150629024P.

Jha SK, Roy P, Chakrabarty S. (2021). Nutraceuticals and pharmaceutics: its importance and their applications. *Int J Drug Dev Res* 13(3): 2.

Jovicic-Bata J, Todorović N, Krstonošic V, Ristic I, Kovačević Z, Vukovic M, Lalic-Popovic M. (2024). Liquid- and semisolid-filled hard gelatin capsules containing alpha-lipoic acid as a suitable dosage form for compounding medicines and dietary supplements. *Pharmaceutics* 16: 892.

Kamali H, Aminimoghadamfarouj N, Golmakani E, Nematollahi A. (2015). The optimization of essential oils supercritical CO_2 extraction from *Lavandula hybrida* through static-dynamic steps procedure and semi-continuous technique using response surface method. *Pharmacognosy Res* 7(1): 57–65. https://doi.org/10.4103/0974-8490.147209.

Kawashima Y. (2001). Nanoparticulate system for improved drug delivery. *Adv Drug Deliv Rev* 47: 1–2.

Keservani RK, Kesharwani RK, Sharma AK, Gautam SP, Verma SK. (2017). Nutraceutical formulations and challenges. *Dev New Func Food Nutraceutical Prod* 161–177. https://doi.org/10.1016/b978-0-12-802780-6.00009-2.

Krasaekoopt W, Bhandari B, Deeth H. (2003). Evaluation of encapsulation techniques of probiotics for yoghurt. *Int Dairy J* 13(1): 3–13. https://doi.org/10.1016/S0958-6946(02)00155-3.

Kulkarni A, Rumalla S, Andhale S, Kolatkar A, Kuchekar B. (2015). Taste masking of ayurvedic nutraceutical formulation by pan coating process. *Phcog J* 7(4): 215–220.

Lalić-Popović MN, Vuković MM, Jovičić-Bata JN, Čanji-Panić JM, Todorović NB. (2023). Comparison of formulation characteristics of drugs and dietary supplements containing alpha-lipoic acid relevant to therapeutic efficacy. *Eu Rev Med Pharmacolog Sci* 27: 3159–3170.

Larrauri JA, Rupérez P, Saura-Calixto F. (1997). Effect of drying temperature on the stability of polyphenols and antioxidant activity of red grape pomace peels. *J Agric Food Chem* 45(4): 1390–1393. https://doi.org/10.1021/jf960282f.

Li M, Rouaud O, Poncelet D. (2008). Microencapsulation by solvent evaporation: state of the art for process engineering approaches. *Int J Pharma* 363: 26–39.

Lu Y, Kinan P. (2012). Microencapsulation: methods and pharmaceutical applications. In: *Encyclopedia of Pharmaceutical Science and Technology*, 4(1): 1–13. https://doi.org/10.3109/9781841848204.000.

Maheshwari R, Todke P, Soni S, Raval N, Deb PK, Amarji B, Ravipati NVAK, Tekade RK. (2018). Stability and degradation studies for drug and drug product In: Tekade RK (ed), *Advances in Pharmaceutical Product Development and Research, Dosage Form Design Consideration*. Academic Press, 225–257.

Maughan RJ, Burke LM, Dvorak J, Larson-Meyer DE, Peeling P, Phillips SM, Engebretsen, L. (2018). IOC consensus statement: dietary supplements and the high-performance athlete. *Br J Sports Med* 52(7): 439–455. https://doi.org/10.1136/bjsports-2018.

Mcloughlin CM, McMinn WAM, Magee TRA. (2000). Microwave drying of pharmaceutical powders. *Food Bioprod Proces* 78(2): 90–96. https://doi.org/10.1205/096030800532798.

Mehta P, Pawar A, Mahadik K, Bothiraja C. (2018). Emerging novel drug delivery strategies for bioactive flavonol Fisetin in biomedicine. *Biomed Pharmacother* 106: 1282–1291.

Mittala YG, Kamblea P, Gotea G, Patilb Y, Mehtaa AK, Karunakarana KP. (2023). A novel analytical model for screw extrusion of thermoplastic ABS with emphasis on additive manufacturing. *Manuf Lett* 35: 652–657.

Mohammady M, Mohammadi Y, Yousefi G. (2020) Freeze-drying of pharmaceutical and nutraceutical nanoparticles: the effects of formulation and technique parameters on nanoparticles characteristics. *J Pharm Sci* 109(11): 3235–3247. https://doi.org/10.1016/j.xphs.2020.07.015.

Mohanty C, Das M, Sahoo SK. (2012). Emerging role of nanocarriers to increase the solubility and bioavailability of curcumin. *Expert Opin Drug Deliv* 9: 1347–1364.

Mohylyuk V, Patel K, Scott N, Richardson C, Murnane D, Liu F. (2020). Wurster Fluidised bed coating of microparticles: towards scalable production of oral sustained-release liquid medicines for patients with swallowing difficulties. *AAPS Pharm Sci Tech* 21: 3. https://doi.org/10.1208/s12249-019-1534-5.

Moulik SP, Rakshit AK, Pan A, Naskar B. (2022). An overview of coacervates: the special disperse state of amphiphilic and polymeric materials in solution. *Colloids Interfaces* 6(3): 45. https://doi.org/10.3390/colloids6030045.

Mujica-Álvarez J, Castell O, Barra P, Amparo RG, Bustos R, Faccini M, Matiacevich S. (2020). Encapsulation of vitamins A and E as spray-dried additives for the feed industry. *Molecules* 25(6): 1357. https://doi.org/10.3390/molecules25061357.

Muñoz-More HD, Nole-Jaramillo JM, Valdiviezo-Marcelo J, Espinoza-Delgado MP, Socola-Juarez ZM, Ruiz-Flores LA, Espinoza-Espinoza LA. (2023). Microencapsulated iron in food, techniques, coating material, efficiency, and sensory analysis: a review. *Front Sustain Food Syst* 7: 1–18. https://doi.org/10.3389/fsufs.2023.1146873.

Nair AB, Dalal P, Kadian V, Kumar S, Garg M, Rao R, Almuqbil RM, Alnaim AS, Aldhubiab B, Alqattan F. (2023). Formulation strategies for enhancing pharmaceutical and nutraceutical potential of sesamol: a natural phenolic bioactive. *Plants* 12: 1168. https://doi.org/10.3390/plants12051168.

Nathalia D, Arriola A, Chater I, Wilcox M, Lucini L, Rochetti G, Dalmina M, Pearson JP, Melo RD, Amboni C. (2019). Encapsulation of *Stevia rebaudiana* Bertoni aqueous crude extracts by ionic gelation – effects of alginate blends and gelling solutions on the polyphenolic profile. *Food Chem* 275(1): 123–134.

Nelson KAS, Martini S. (2009). Increasing omega fatty acid content in cow's milk through diet manipulation: effect on milk flavor. *J Dairy Sci* 92(4): 1378–86. https://doi.org/10.3168/jds.2008-1780.

Nguon, O, Lagugne-Labarthet F, 3M Company Canada, Gillies ER. (2018). Microencapsulation by in situ polymerization of amino resins. *Chem Publ* 110. https://ir.lib.uwo.ca/chempub/110.

Ngwuluka NC, Abu-Thabit NY, Uwaezuoke OJ, Erebor JO, Ilomuanya MO, Mohamed RR, Soliman SMA, Abu Elella MH, Ebrahim NAA. (2021). Natural polymers in micro- and nanoencapsulation for therapeutic and diagnostic applications: part I: lipids and fabrication techniques. *IntechOpen*. https://doi.org/10.5772/intechopen.94856.

O'Donnell PB, McGinity JW. (1997). Preparation of microspheres by the solvent evaporation technique. *Adv Drug Del Rev* 28: 25–42.

Okuyama K, Abdullah M, Lenggoro W, Iskandar F. (2006). Preparation of functional nanostructured particles by spray drying. *Adv Powder Technol* 17(6): 587–611.

Palanikumar L, Choi ES, Oh JY, Park SA, Choi H, Kim K, Kim C, Ryu JH. (2018). Importance of encapsulation stability of nanocarriers with high drug loading capacity for increasing in vivo therapeutic efficacy. *Biomacromolecules* 19(7): 3030–3039. https://doi.org/10.1021/acs.biomac.8b00589.

Pandey P, Turton R, Joshi N, Hammerman E, Ergun J. (2006). Scale-up of a pan-coating process. *AAPS Pharm Sci Tech* 7(4): 102. http://www.aapspharmscitech.org.

Pathak N, Kochhar A. (2018). Extrusion technology: solution to develop quality snacks for malnourished generation. *Int J Cur Microbio App Sci* 7(1): 1293–1307.

Pisani S, Dorati R, Genta I, Chiesa E, Modena T, Conti B. (2020). High efficiency vibrational technology (HEVT) for cell encapsulation in polymeric microcapsules. *Pharmaceutics* 212(5): 469. https://doi.org/10.3390/pharmaceutics12050469.

Poddar D, Das S, Jones G, Palmer JS, Jameson GB, Haverkamp R, Singh H. (2014). Stability of probiotic *Lactobacillus paracasei* during storage as affected by the drying method. *Int Dairy J* 39(1): https://doi.org/10.1016/j.idairyj.2014.04.007.

Polzonetti V, Pucciarelli S, Vincenzetti S, Polidori P. (2020). Dietary intake of vitamin d from dairy products reduces the risk of osteoporosis. *Nutrients* 12(6): 1743–1757. https://doi.org/10.3390/nu12061743.

Poncelet D. (2006). Microencapsulation: fundamental, methods and application. In: Blitz P, Gun'ko VM (eds), *Surface Chemistry in Biomedical and Environment Science*. Springer, 23–34. https://doi.org/10.1007/1-4020-4741-X_3.

Puri V, Nagpal M, Singh I, Singh M, Dhingra AG, Huanbutta K, Dheer D, Sharma A, Sangnim T. (2022). A comprehensive review on nutraceuticals: therapy support and formulation challenges. *Nutrients* 14(21): 4637. https://doi.org/10.3390/nu14214637.

Racoviţă S, Silvia AB, Vasiliu B, Popaa M, Lucab C. (2009). Polysaccharides based on micro- and nanoparticles obtained by ionic gelation and their applications as drug delivery systems. *Revue Roumaine Chimie* 54(9): 709–718.

Rajak DK, Pagar DD, Menezes PL, Linul E. (2019). Fiber-reinforced polymer composites: manufacturing, properties, and applications. *Polymers* 11(10): 1667. https://doi.org/10.3390/polym11101667.

Rege BD, Gawel J, Kou JH. (2002). Identification of critical process variables for coating actives onto tablets via statistically designed experiments. *Int J Pharma* 237(1–2): 87–94.

Rigon RT, Noreña CPZ. (2016). Microencapsulation by spray-drying of bioactive compounds extracted from blackberry (*Rubus fruticosus*). *J Food Sci Technol* 53(3): 1515–1524. https://doi.org/10.1007/s13197-015-2111-x.

Rodríguez MC, Oliveira PG, Lourenço-Lopes C, Simal-Gandara J, Corral MF, Preito MA, Gandara JS. (2021). Freeze-drying encapsulation as a mechanism of choice in oils: methods and mechanism. In Gomez-Javaglia (ed), *A Basic Protocols in Encapsulation of Food Ingredients*. Humana Press, Springer Protocols, 91–101. https://doi.org/10.1007/978-1-0716-1649-9_9.

Rodriguez-Nogales JM, Delgadillo A. (2005). Stability and catalytic kinetics of microencapsulated β-galactosidase in liposomes prepared by the dehydration–rehydration method. *J Mol Catal B Enzymatic* 33(1–2): 15–21. https://doi.org/10.1016/j.molcatb.2005.01.003.

Sahoo J, Kamalaja T, Rajeshwari K. (2020). Nutraceuticals monitoring safety systems in India. In: Arya P (ed), *Advances in Nutrition*. Delhi: AkiNik Publication. https://doi.org/10.22271/ed.book.792.

Santiago-Adame R, Medina-Torres L, Gallegos-Infante JA, Calderas F, González-Laredo RF, Rocha-Guzmán NE, Ochoa-Martínez LA, Bernad-Bernad MJ. (2015). Spray drying-microencapsulation of cinnamon infusions (*Cinnamomum zeylanicum*) with maltodextrin. *LWT Food Sci Technol* 64(2): 571–577. https://doi.org/10.1016/j.lwt.2015.06.020.

Selim KA, Alharthi SS, Abu El-Hassan AM, Elneairy NA, Rabee LA, Abdel-Razek AG. (2021). The effect of wall material type on the encapsulation efficiency and oxidative stability of fish oils. *Molecules* 26(20): 6109.

Seth D, Mishra HN, Deka SC. (2017). Effect of microencapsulation using extrusion technique on viability of bacterial cells during spray drying of sweetened yoghurt. *Int J Biol Macromol* 103: 802–807. https://doi.org/10.1016/j.ijbiomac.2017.05.099.

Shi L, Kim E, Yang L, Huan Y, Ren N, Li N, He P, Tu Y, Wu Y. (2021). Effect of a combined microwave-assisted drying and air drying on improving active nutraceutical compounds, flavor quality, and antioxidant properties of *Camellia sinensis* L. (cv. Longjing 43) flowers. *Food Qual Saf* 5(4): 1–7. https://doi.org/10.1093/fqsafe/fyaa040.

Sieniawska E. (2017). Concerns about quality and safety of dietary supplements. *EC Pharma Toxicol* 3(1): 13–14.

Singh B, Sharma C, Sharma S. (2017). Fundamentals of Extrusion processing In Nanda V, Sharma S (eds), *Novel Food Processing Technologies*. New India Publishing Agency-Nipa, 1–10.

Singh I. (2017). Antimicrobials in higher plants: classification, mode of action and bioactivities. *Chem Bio Lett* 4(1): 48–62.

Singh MN, Hemant KSY, Ram M, Shivakumar HG. (2010). Microencapsulation: a promising technique for controlled drug delivery. *Res Pharma Sci* 5(2): 65–77.

Singh S, Desai I, Noronha S. (2018) Capsules: the go-to dietary supplement dosage form. *Tablets and Capsules*. https://www.tabletscapsules.com/3641-Technical-Articles/592540-Capsules-The-Go-To-Dietary-Supplement-Dosage-Form/.

Solanki N, Mehta M, Satija S, Pandet P, Dahiya M. (2016). Solvent evaporation technique: an innovative approach to increase gastric retention. *Int J Adv Sci Res* 1(4): 60–67.

Sonar G, Rawat S. (2015). Optimization of pantoprazole enteric pellets coating process by QBD: effect of coating process variables on the intermediate quality of the product and scale up. *Intl J Pharmacy Pharm Sci* 7(12): 80–87.

Sosnik A, Seremeta K P. (2015). Advantages and challenges of the spray-drying technology for the production of pure drug particles and drug-loaded polymeric carriers. *Adv Colloid Interface Sci* 223: 40–54. https://doi.org/10.1016/j.cis.2015.05.003.

Suganya V, Anuradha V. (2017). Microencapsulation and nanoencapsulation: a review. *Int J Pharma Clinical Res* 9(3): 233–239. https://doi.org/10.25258/ijpcr.v9i3.8324.

Tackenberg MW, Kleinebudde P. (2015). Encapsulation of liquids via extrusion – a review. *Curr Pharm Des* 21(40): 5815–5828. https://doi.org/10.2174/1381612821666151008150142.

Tackenberg MW, Krauss R, Schuchmann HP, Kleinebudde P. (2015). Encapsulation of orange terpenes investigating a plasticisation extrusion process. *J Microencap* 32(4): 408–17. https://doi.org/10.3109/02652048.2015.1035686.

Thakkar S, Anklam E, Xu A, Ulberth F, Li J, Li B, Tong W. (2020). Regulatory landscape of dietary supplements and herbal medicines from a global perspective. *Reg Toxicol Pharma* 114: 104647. https://doi.org/10.1016/j.yrtph.2020.104647.

Timilsena Y, Haque MA, Adhikari B. (2020). Encapsulation in the food industry: a brief historical overview to recent developments. *Food Nutr Sci* 11(6): 481–508. https://doi.org/10.4236/fns.2020.116035.

Vargas MCA, Simsek S. (2021). Clean label in bread. *Foods* 10(9): 2054. https://doi.org/10.3390/foods10092054.

Verma R, Verma S, Kumar S. (2019). Microsphere-a novel drug delivery system. *Res Chron Health Sci* 5(1): 5–14.

Wang H, Yuan Y, Rong M, Zhang M. (2009). Microencapsulation of styrene with melamine-formaldehyde resin. *Colloid Polym Sci* 287(9): 1089–1097. https://doi.org/10.1007/s00396-009-2072-6.

Werner RLS, Jones JR, Paterson A, Archer R, Pearce DL. (2007). Air-suspension particle coating in the food industry: part I – state of the art. *Powder Technol* 171: 25–33. https://doi.org/10.1016/j.powtec.2006.08.014.

Whelehan M, Marison IW. (2011). Microencapsulation using vibrating technology. *J Microencapsul* 28: 669–688. https://doi.org/10.3109/02652048.2011.586068.

White CM. (2020). Dietary supplements pose real dangers to patients. *Ann Pharmacother* 106002801990050. https://doi.org/10.1177/1060028019900504.

Yashaswini PS, Kurrey NK, Singh SA. (2017). Encapsulation of sesamol in phosphatidyl choline micelles: enhanced bioavailability and anti-inflammatory activity. *Food Chem* 228: 330–337.

Yin X, Chen K, Cheng H, Chen X, Feng S, Song Y, Liang L. (2022). Chemical stability of ascorbic acid integrated into commercial products: a review on bioactivity and delivery technology. *Antioxidants* 11(1): 153. https://doi.org/10.3390/antiox11010153.

Yuan Y, Kong ZY, Sun YE, Zeng QZ, Yang XQ. (2017). Complex coacervation of soy protein with chitosan: constructing antioxidant microcapsule for algal oil delivery. *LWT* 75: 171–179.

9 Standardization of the Nutraceutical Products

Shivam Kumar, Akhil Singh, Rajeeva Gaur, Satyendra Pratap Singh, and Reena Yadav

9.1 INTRODUCTION

The nutraceutical industry has rapidly grown in recent years, driven by increasing consumer demand for health-promoting products that lie at the intersection of nutrition and medicine (Ariyo & Akinyemi, 2024). Nutraceuticals, a term derived from "nutrition" and "pharmaceutical," encompass a broad range of products, including dietary supplements, herbal extracts, functional foods, and other bioactive compounds that provide medical or health benefits beyond basic nutritional functions (Teli et al., 2024). These products are acclaimed for their potential in preventing and managing chronic diseases, improving immunity, enhancing physical performance, and promoting general well-being (Wang & Li, 2014). However, the diversity of ingredients, sources, and methods of preparation poses significant challenges to maintaining consistency and quality across nutraceutical products. Unlike conventional pharmaceuticals, nutraceuticals often lack comprehensive regulations or standardized methods for assessment, leading to considerable variability in efficacy, safety, and quality. Standardization in nutraceuticals is essential to ensure product consistency, bioactivity, and safety, as well as to build consumer confidence (Komala et al., 2023). The process of standardizing nutraceuticals involves several critical steps, including the identification and quantification of active ingredients, validation of product formulations, and development of quality control measures. In addition, standardized testing protocols, regulatory compliance, and validation of health claims are necessary to authenticate and optimize the functional properties of nutraceuticals (Sadgrove, 2022). This chapter delves into the fundamentals and importance of nutraceutical standardization. It covers the methodologies employed to assess purity, potency, and bioavailability of nutraceutical products and discusses the role of standardization in addressing regulatory challenges, protecting public health, and fostering market growth summarized in Figure 9.1 and Table 9.1. Through standardized approaches, the nutraceutical industry can strive to provide safe, effective, and high-quality products that meet the needs of modern consumers seeking preventive and therapeutic health options.

 DOI: 10.1201/9781003588504-9

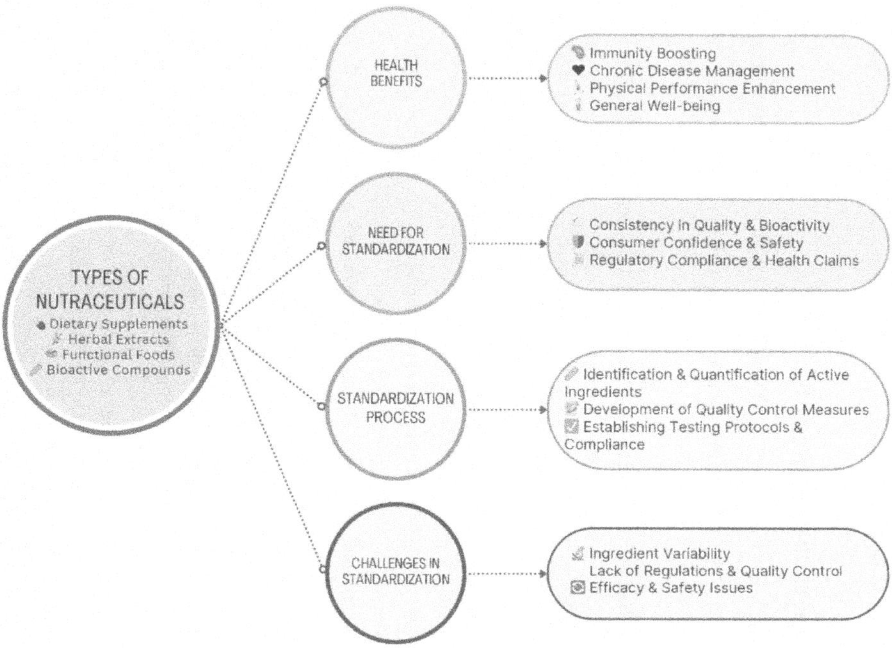

FIGURE 9.1 Overview of the standardization process in the nutraceutical industry.

9.2 CHALLENGES IN STANDARDIZING NUTRACEUTICAL PRODUCTS

The standardization of nutraceutical products poses novel and complicated issues due to their heterogeneous sources, complex mixtures of compounds, and differing biological efficacies. These include complex mixtures of compounds and different efficacies in biological activities. Such products are sourced from different natural sources including plants, animals, and microorganisms, all of which are sensitive to variability due to environmental factors like climate, soil quality, and harvesting methods. Furthermore, nutraceuticals often comprise intricate combinations of bioactive compounds, whose amounts and efficacies may vary according to extraction, processing, and storage methods (Bertola et al., 2021). The biological effect of such compounds may also vary considerably with individual genetic and metabolic differences. Some of the major challenges are discussed in the following sections.

9.2.1 Ingredient Variety and Complexity

Nutraceuticals are composed of a variety of components that include vitamins, minerals, amino acids, plant extracts, probiotics, and biochemicals (Rajat et al., 2012). These components may be present in various plant sections, from animal sources,

TABLE 9.1
Overview of Key Aspects of Active Ingredients and Descriptions of Polyphenolic Compounds

Aspect	Description	Parameters/Methods	Examples	References
Identification of active ingredients	Ensures the presence of bioactive compounds crucial for therapeutic effects	Spectroscopy (UV, IR, NMR, MS), chromatography (HPLC, GC), DNA barcoding	Curcumin in turmeric, resveratrol in grapes	Kotra et al., 2019
Quantification of bioactive components	Measures the exact concentration of active compounds	HPLC, LC-MS/MS, GC-MS, and ELISA	Quantifying polyphenols in green tea extract	Lee et al., 2014
Purity assessment	Assures absence of contaminants and unwanted materials	Heavy metal analysis (AAS, ICP-MS), pesticide residue tests, and microbial testing	Ensuring arsenic-free spirulina	Shariful et al., 2021
Stability testing	Evaluates the shelf life and stability of active ingredients	Accelerated stability tests under varied temperature, humidity, and light conditions	Stability of lycopene in tomato-based nutraceuticals	Kumar et al., 2022
Bioavailability assessment	Determines the extent and rate of absorption of bioactives in the body	In vitro digestion models, Caco-2 cell models, pharmacokinetic studies	Nanoformulations to enhance curcumin bioavailability	Liu et al., 2020
Standardization of dosage forms	Ensures consistency in dosage for therapeutic efficacy	Uniformity of weight, dissolution tests, and particle size analysis	Standardized capsules of omega-3 fatty acids	Vestland et al., 2016
Validation of manufacturing processes	Ensures reproducibility and quality during production	Process analytical technology (PAT), GMPs	Fermentation process for probiotics	Fenster et al., 2019
Toxicological assessment	Evaluates safety and potential adverse effects	Acute, sub-acute, and chronic toxicity tests; in vitro cytotoxicity assays	Assessment of toxicity levels in herbal supplements	Woo et al., 2012
Regulatory compliance	Aligns with global or regional standards for nutraceuticals	Compliance with FDA, EFSA, FSSAI, or WHO guidelines	Approved probiotics in the U.S. and EU markets	Arora & Baldi, 2017
Functional testing	Confirms the claimed health benefits through experimental validation	Pre-clinical and clinical trials, biomarkers for efficacy (e.g., cholesterol levels, antioxidant activity assays)	Anti-inflammatory effects of resveratrol in randomized trials	Samsami et al., 2015

or created genetically, and all are classified as having bioactive properties. The high degree of variability makes the establishment of uniform criteria for quality, strength, or efficacy across products very difficult.

9.2.2 Variability in Raw Materials

The ingredients that primarily comprise nutraceuticals, such as plants, are subject to a lot of fluctuation from one region to another owing to aspects such as geographical location, soil type, weather, and harvesting time, among others (Khan et al., 2021). Such variables contribute to different levels of active compounds, and thus consistency and quality of starting materials used in the production of various batches can be challenging.

9.2.3 Lack of Regulatory Harmonization

The regulation of nutraceuticals varies from region to region and even from country to country, resulting in variations in the quality and safety standards observed. For instance, in some nations, particular nutraceuticals may be classified as food supplements; in others, such products may be viewed as drugs or food preservatives, bearing different rules (Santini et al., 2018). The absence of similar laws makes it difficult to ensure standardization, which, in turn, makes it difficult to maintain a uniform quality all over the world.

9.2.4 Problems with Analytically Measuring Active Substances

The precise determination of active principles of nutraceuticals is possible on a quantitative basis, but this is rather challenging since the compositions are variable and complex (Piccolella et al., 2019). Simultaneously, quality control operations call for standardized analytical techniques, although these may not always be suitable for all types of nutraceuticals. Additionally, low concentration levels of the active principals, which are usually the basis of therapeutic activity, can also be cumbersome to analyse, hence requiring sophisticated techniques and experienced professionals.

9.2.5 Quality Control and Consistency

One of the main difficulties faced by manufacturers is ensuring batch-to-batch uniformity. In the case of chemical-based drugs where it is possible to reproduce the same chemical composition every time, nutraceuticals are often characterized by variations between batches (Chander et al., 2021). In order to prevent this, there is need for standardization of processes of extraction, purification, and formulation. However, these often prove difficult to enforce as the processes would need to vary from one raw material's quality to the other.

9.2.6 Stability and Shelf Life

As with many consumables, nutraceuticals may deteriorate due to factors such as temperature, humidity, light, and the passage of time (Dima et al., 2020). One of the important considerations is how to stabilize the active constituents of the products and prolong shelf life without losing the functional effectiveness of the formulations. Lastly, the interaction of different bioactive components may also compromise the stability of the formulation, which necessitates proper design and packaging of the product.

9.2.7 Scientific Validation of Efficacy

Nutraceuticals are promoted for their positive effects on health, but the scientific data confirming such benefits may differ in terms of conclusiveness (Espín et al., 2007). Often, there is a dearth of well-designed clinical studies akin to those performed on medicinal compounds, and responses to nutraceuticals differ based on genetic and health profiles. Acceptable grounds for such claims would also involve the need to develop and agree on efficacy testing standards across the nutraceutical industry.

9.2.8 Consumer Misinformation and Labelling

The nutraceutical industry is riddled with problems of mislabelling as well as provision of misinformation regarding products (Finan, 2023). These labels reveal little or no information about the dosage, where the risks of side effects may lie, or how the substance may interfere with other drugs, thus increasing the chances of misuse. It is of utmost importance to introduce and implement uniform labelling practices as this will help reduce confusion among consumers with regard to the information being correct or otherwise being presented to them.

9.2.9 Economic Constraints

For the lesser nutraceutical firms establishing processes, running clinical studies and having quality control may be prohibitively expensive (Bagchi & Nair, 2016). This is a deterrent for the investments made towards standardization, particularly in markets where there are consumers who value cheap affordability over certainty in quality.

9.2.10 Environmental and Sustainability Concerns

Another issue arises with regard to the challenge of sourcing raw materials in a sustainable manner, more so with plant-based and especially botanicals (Cardellina, 2002). The problem is exacerbated by overexploitation and unsustainable management of resources, which in turn causes loss of biodiversity and environmental destruction, making it difficult to source raw materials of high quality and a consistent type. It is, however, very difficult to strike a balance between standardization and sustainable practices in business.

9.3 NEED FOR STANDARDIZATION IN NUTRACEUTICAL PRODUCTS

Such rapid growth and popularity of nutraceutical products require standardization in the process of evaluation to ascertain the quality, safety, and effectiveness of products (Sadgrove, 2022). Standardize nutraceuticals is necessary to boost consumer confidence and ensure consistency across products for health benefits. Here are some of the major reasons why nutraceuticals should be standardized.

9.3.1 Ensuring Product Quality and Consistency

Standardization also facilitates a product's uniformity regarding strength, purity, and composition (Nafiu et al., 2017). Since raw materials derived from natural sources – particularly in the case of botanicals – vary significantly in terms of their levels of active compounds, products can vary widely in the amount of active constituents present. This variability is naturally mitigated by standardization, which provides consumers with a consistent profile of active constituents – a critical aspect for maintaining batch-to-batch efficacy.

9.3.2 Enhancing Safety and Reducing Risks

Standardization of nutraceuticals will prevent contamination or adulteration against laws. Through standardized testing and quality control measures, heavy metals or microbial and pesticide contaminants may be ruled out; hence, products might become safer for consumption. This is also very significant for defence, especially for vulnerable population segments such as children, older adults, and patients with chronic diseases (Sadgrove, 2022).

9.3.3 Building Consumer Confidence and Trust

Many people have nutraceuticals as the cornerstone of support for health and wellness. However, inconsistency in quality and false information will adversely affect consumer confidence. Clarity and reliability through standardization empower consumers with the ability to make informed choices and healthier decisions (Moloughney, 2023). A strong quality culture is thus necessary, especially for a global market, which is characterized by malicious scams and high-stakes economic competition.

9.3.4 Supporting Health Claims with Scientific Validation

Considering the fact that nutraceuticals are now established as health-enabling products, their benefits must be scientifically validated through research and testing. This shall entail the setting of protocols in relation to their safety, efficacy, and quality. Standardization offers a key role since consistent results between batches are possible because the same type of protocol can be used repeatedly for studies (Phillips & Castegnaro, 1999). These clinical studies are ultimately crucial in helping to authenticate health claims and providing clearly evidenced real-world information that can be confidently communicated to consumers and healthcare professionals. It becomes increasingly challenging to distinguish between truly beneficial nutraceuticals and those that do not offer demonstrated benefits, further eroding consumer confidence and the integrity of the nutraceutical industry in the absence of such authentication. Ultimately, the validation and standardization process supports public health, ensuring effective products in conjunction with enhancing the scientific credibility and market growth of the nutraceutical industry (Sadgrove, 2022).

9.3.5 Meeting Regulatory Requirements and Global Market Access

The regulations regarding the sale and marketing of nutraceutical products vary by country. Standardization can be very beneficial to companies when they have to operate within these regulations and engage in international trade to expand markets (Yeung et al., 2006). Even compliance with standardized protocols reduces the likelihood of recalls, fines, or bans, which could seriously damage a company's reputation and bottom line.

9.3.6 Improving Manufacturing Efficiency and Cost-Effectiveness

All manufacturing processes are standardized to maintain optimized production, reduced waste, and cost-effectiveness through fewer formulation adjustments and greater quality control (Mehta et al., 2024). These standard protocols thus help streamline ways for production in a complete and uniform manner, thus enabling each of the manufacturers to readily scale and maintain such high-quality products.

9.3.7 Fostering Research and Development

Standard regulatory norms create a framework for research and development by having clear-cut guidelines and methods through which researchers study bioactive compounds, optimize extraction methods, and develop innovative formulations (Krishnaswamy, 2024). This will lead to the advancement of science and new possible discoveries of nutraceuticals with improved benefits.

9.3.8 Promoting Ethical Marketing and Reducing Misinformation

Ethical labelling and advertising will be standardized through confirmed scientific evidence of health benefits. These health claims will make sure that there will be no misleading or exaggerated claims and will allow the consumer to make informed choices regarding products (Buchanan, 2000). Such ethical marketing would protect consumers from misleading products and promote responsible growth of the industry.

9.3.9 Addressing Sustainability Concerns

The increasing demand for natural ingredients has made the sustainable sourcing of raw materials a growing concern among consumers. Standardization can involve guidelines on ethical sourcing to minimize the environmental impacts of the manufacture of nutraceuticals (Coppens et al., 2006). Sustainable incorporation into the standardization process would enhance the contribution of the nutraceutical industry's efforts toward the conservation of biodiversity and responsible resource management.

9.3.10 Ensuring Long-Term Consumer Health and Wellness

To optimize the health outcomes in effective and safe nutraceutical products, all of which aim to promote health and well-being (Vlaicu et al., 2023), standardization is necessary. Standardization would ensure that these products not only define the product in terms of long-term health but are also genuine as far as constituents are concerned (Cochoy, 2005). Ensuring consistency of quality and efficacy across products directly contributes to positive health outcomes that empower the consumer to incorporate nutraceuticals as reliable constituents of their wellness regimen.

9.4 KEY ELEMENTS OF NUTRACEUTICAL STANDARDIZATION

The standardization of nutraceutical products requires a multifaceted approach to ensure product quality, safety, efficacy, and compliance with regulatory standards (Usmani & Almoselhy). Key elements involved in the standardization process are discussed in the following sections.

9.4.1 Identification and Authentication of Raw Materials

The quality of nutraceutical products is traced right from the source. True identification and authentication of raw materials are vital factors in ensuring the inclusion of only the finest quality and authentic materials in manufacturing (Sarkisyan et al., 2017). Techniques such as DNA barcoding, microscopy, and chromatography can offer authentication of botanicals and other natural ingredients, thus helping to avoid misplacing or bringing in wrong or contaminated raw materials (Seethapathy, 2019).

9.4.2 Quantification of Active Compounds

The effectiveness of any nutraceutical has much to do with the concentration of its bioactive components. Standardization relates to the identification and quantification of active compounds, usually through very precise analytical methods such as high-performance liquid chromatography (HPLC), has chromatography (GC), and mass spectrometry (MS) (Nikam et al., 2012). Accuracy in quantification thus ensures that each product batch has the specified level of active ingredients, which is required for effectiveness.

9.4.3 Setting Quality and Purity Standards

Nutraceuticals are also required to be safe enough for consumption, meeting very strict quality and purity parameters. This includes specifying limits on impurities, contaminants, and residual solvents. Heavy metals, pesticides, microbial contamination, and other toxic ingredients must be tested to ensure the safety standards of the products. The reason for this necessary definition of quality parameters is protection for consumers, as it ensures that products are free from toxic materials.

9.4.4 Ensuring Bioavailability and Stability

Bioavailability, which is the degree in which the active ingredient is absorbed and utilized by the body, is a critical factor in nutraceutical efficacy (Gonçalves et al., 2018). Standardization should result in addressing the bioavailability issue through optimized formulations, such as encapsulation or controlled-release technologies. Stability testing will be required to determine shelf life and potency of the product over time to ensure that it will remain effective up to its expiration date.

9.4.5 Development of Standardized Extraction and Processing Protocols

Different extraction and processing methods can also have different final products or results. Standardized protocols for extraction, purification, and subsequent processing maintain uniformity in the concentrations of active ingredients (Nafiu et al., 2017). These protocols specify the kind of solvents used and temperature and time used in processing to optimize yield while preserving bioactive compounds.

9.4.6 Adherence to Good Manufacturing Practices (GMPs)

GMPs are regulatory guidelines that ensure constant observation and control in the manufacturing process of products to maintain their quality (He et al., 2015). GMPs cover all aspects of production, from raw materials to packaging of the finished product. Following GMPs optimizes risk due to contamination and extremes of variability, hence assuring quality and safety criteria of products.

9.4.7 Implementation of Quality Control and Testing Protocols

Quality control (QC) within the process of nutraceutical standardization involves different levels of testing, ranging from in-process testing of ingredient purity and concentration to post-production contaminants, potency, and efficacy (Miller, 2016). In any case, QC protocols ensure that any batch of a product meets pre-determined standards before getting to consumers.

9.4.8 Validation of Health Claims through Scientific Research

Standardization of nutraceuticals also means that a number of health-related claims are verified through scientific studies and clinical trials (Borchers et al., 2016). Products are basically going to be marketed for a range of health benefits, which need to have due evidence supporting their efficacy and safety. The evidence-based approach will help in building trust among consumers and also ensure that regulatory requirements about what could be claimed regarding health benefits are complied with.

9.4.9 Standardized Labelling and Packaging

Precise and standard labelling gives the consumer important information about ingredient composition, dosage, allergens that could be present, conditions of storage, and dates of expiration (Bensa et al., 2023). Transparency in labelling ensures that there will be a sense of safety for the consumer, while building trust in the product. Standardized packaging protects the product from environmental factors that could affect its quality and stability.

9.4.10 Compliance with Regulatory Standards and Documentation

Each country has set out regulations that specifically control nutraceutical products. There needs to be confirmation of such regulatory standards to avoid any legal problems in terms of product acceptance in various markets (Santini et al., 2018). Proper documentation will be important for traceability, such that batches may be traced easily by tracking the production process for purposes of recalls or quality audits.

9.4.11 Sustainability and Ethical Sourcing

Guidelines for nutraceuticals concentrate their focus on the ethical and sustainable sourcing of ingredients, given the increased consumer concern about the environment (Sadgrove, 2022). This therefore increases biodiversity by mitigating negative impacts on the environment, devising fair labour practices, and conserving rare plant species. The subsequent processes ensure a continuous supply of high-quality input in the present and future.

9.4.12 Continuous Monitoring and Improvement

This is a process of standardization that will never be complete, as constant control, testing, and reworking are necessary. Established methodologies change based on consumer feedback, new technological developments, and even changes in regulatory requirements. Continuous improvement will help make sure the product quality stays high and the newest scientific understandings are folded into production.

9.5 ANALYTICAL TECHNIQUES IN NUTRACEUTICAL STANDARDIZATION

Nutraceutical standardization and the development of quality products involve engagement by way of applying analytical techniques that determine their quality, safety, and efficacy. This helps in the identification of active ingredients present in them, establishment of purity, detection of contaminants, and measurement of bioavailability, among others. The key analytical techniques most commonly used in nutraceutical standardization are discussed in the following sections.

9.5.1 High-Performance Liquid Chromatography

HPLC has become one of the most used techniques in the nutraceuticals area for the identification and quantification of bioactive compounds. In this regard, HPLC allows for the estimation of active ingredient concentrations, such as vitamins, flavonoids, and polyphenols, with great precision by separating components in a liquid sample (Bernal et al., 2011). It follows that this is highly sensitive and very precise; thus, it is important to ensure QC and batch-to-batch consistency of nutraceutical products.

9.5.2 Gas Chromatography

GC is more applied to volatile analysis: essential oils, acids, and other organic compounds present in nutraceuticals. The principle of separation in GC is according to volatility – the components are separated according to this characteristic (Wadood et al., 2020). This explains the fact that it is considerably detailed not only in terms of identity but also the concentration of the compounds. GC-MS has greatly increased sensitivity and selectivity, thereby allowing this detailed profiling to be performed even on very complex mixtures.

9.5.3 Mass Spectrometry

Generally, MS is coupled to techniques like chromatography, such as GC-MS or liquid chromatography (LC-MS), which will provide exact identification and quantification at the molecular level. Making use of circle ions' mass-to-charge ratio that MS measures assists in identifying the chemical structure and trace-level detection of bioactive compounds or contaminants (Chen et al., 2007). As observed, MS has high versatility and can assess purity and detect adulterants in bioactive compounds.

9.5.4 Fourier Transform Infrared Spectroscopy (FTIR)

FTIR analyses the molecular vibrations of compounds, which build up a sort of "molecular fingerprint" that allows for the identification of functional groups and structural features (Shukla, 2024). Because of this, FTIR is very useful in establishing the authenticity of raw materials; it is most important in plant-based nutraceuticals because it will identify contaminants or adulterants by comparing the spectra to reference compounds right away.

9.5.5 Nuclear Magnetic Resonance (NMR) Spectroscopy

This is an investigative technique in chemistry that is mainly concerned with providing a great deal of information on the molecular structure, relying on the magnetic properties of atomic nuclei (Svanberg & Svanberg, 2001). Further applications include structural elucidation of complex molecules and confirmation of compounds and impurity testing. Being such an important technique, authentication of natural products by means of NMR supports QC in plant extracts, essential oils, and amino acid profiles.

9.5.6 Atomic Absorption Spectroscopy (AAS) and Inductively Coupled Plasma (ICP) Analysis

AAS and ICP techniques have been essential in the testing of nutraceuticals to determine metal and mineral content (Smichowski & Londonio, 2018). There has also been some modification in ICP through the development of ICP-OES (Inductively Coupled Plasma-Optical Emission Spectroscopy) and ICP-MS techniques used in trace element detection, including heavy metals like lead, cadmium, and mercury, which uphold safety by confirming that the concentration of metals falls within accepted limits.

9.5.7 Thin Layer Chromatography (TLC)

TLC is one of the simplest and cheapest techniques that can be used as a starting point in identifying herbal extracts and establishing their purity (Kowalska & Sajewicz, 2022). TLC separates compounds in the stationary phase and enables visual comparison based on the retention factor. Though not as accurate, TLC finds application in the screening method and gives an overview regarding the presence of certain compounds.

9.5.8 Enzyme-Linked Immunosorbent Assay (ELISA)

ELISA is a biochemical technique for detecting and quantifying particular proteins or antigens in nutraceutical products (Shinkaruk et al., 2008). It becomes especially useful in the analysis of allergens, toxins, or certain bioactive proteins. A large number of ELISA kits have been customized for different applications, thus providing a very sensitive, high-throughput, and cost-effective way of performing quality control.

9.5.9 UV-Visible Spectroscopy

The spectroscopy of ultraviolet light–visible spectroscopy (UV-Vis) is well suited for the analysis of compounds that exhibit characteristic absorption either in the UV or visible region. Conventionally, it finds routine applications in analyses such as those of fat-soluble vitamins, phenolic compounds, and other bioactives of plant origin because of its ready availability and simplicity (Joshi & Joshi, 2012). This technique can enable qualitative and semi-quantitative data, though it is usually applied in combination with other analysis techniques.

9.5.10 Microbial and Stability Testing

Tests for microbial contamination, such as total aerobic plate count and pathogen detection testing, are a common activity in nutraceuticals (Som Chaudhury & Sinha, 2023). Stability testing involves accelerated stability studies under conditions of change in temperature and humidity that ensure the potency of the active ingredients over a given period of time, guaranteeing shelf life. Such tests are therefore crucial in ensuring a product remains safe and effective up to its expiration date.

9.5.11 Differential Scanning Calorimetry (DSC)

DSC is a technique that evaluates the thermal properties of compounds, including melting point, crystallinity, and glass transition temperature. This is useful in determining stability and purity assessments of nutraceuticals, especially solid or semi-solid ones. This way, DSC may be helpful for a formulation in order to optimize it based on the knowledge of how different components interact under changes in temperature (Gharanjig et al., 2020).

9.5.12 Microscopy Techniques

Microscopy, including optical, electron, and scanning electron microscopy, is used for structural and morphological analysis. It can verify the identity and quality of botanical materials, detect adulteration, and assess particle size and homogeneity in formulations. Microscopy techniques are especially useful for QC of raw materials and for detecting physical contaminants (Dery & Zaixiang, 2023).

9.5.13 Flow Cytometry and Particle Size Analysis

Flow cytometry and particle size analysis are used to assess the size, distribution, and concentration of particles in formulations such as emulsions, capsules, and powders (Helal et al., 2019). These techniques are essential for ensuring uniformity in dosage forms and improving bioavailability, particularly in nano- and micro-encapsulated nutraceutical products.

9.5.14 Polymerase Chain Reaction (PCR) for DNA Analysis

PCR and related DNA-based techniques are employed for species identification and authentication, especially in botanicals and probiotics (Newmaster et al., 2019). DNA fingerprinting or barcoding confirms the identity of plant species, preventing adulteration and ensuring compliance with botanical source claims.

9.6 QUALITY CONTROL AND ASSURANCE IN NUTRACEUTICAL STANDARDIZATION

QC and quality assurance (QA) have been two very vital processes in the standardization of nutraceuticals. These are two very important steps involved in nutraceutical production that assure safety, efficacy, and consistency of the final products (Kunle et al., 2012). While QC involves product quality testing and verification at different levels of production, QA takes into consideration the whole process, assuring that proper protocols, standards of operation, and practices are done to achieve consistent quality with time. Regarding this aspect, some of the very important issues representing QC and QA in nutraceutical standardization are discussed in the following sections.

9.6.1 Raw Material Sourcing and Testing

Nutraceutical companies should meticulously screen suppliers to guarantee that they supply top-quality, genuine, and sustainable raw materials (Chandra et al., 2019). Audits and certifications support in confirming that suppliers comply with industry standards.

9.6.2 Raw Material Testing

Extensive testing of the raw materials is done before incorporation to confirm their identity, potency, and purity. The tests include screening for contaminants like heavy metals, pesticides, and microbial agents to make sure the materials are safe from the very beginning (Salgueiro et al., 2010).

9.6.3 Authentication and Traceability

Authenticity and traceability in raw materials needs to be assured with a view to avoiding acts of substitution, adulteration, and contamination. Documentation helps trace the materials right from sourcing to the final product (Bose & Padmavati, 2024).

9.6.4 In-Process Quality Control

Monitoring of production processes at each stage of production, such as temperature, pH, and humidity, is continuously done to maintain uniformity. Real-time adjustments minimize variability to make every batch of products meet quality specifications (Kim et al., 2021).

9.6.5 Sampling and Testing at Critical Points

QC involves sampling and testing along the chain of manufacturing immediately after extraction, drying, or blending to confirm consistency and detect any deviation at the earliest point possible.

9.6.6 Standard Operating Procedures

Detailed standard operating procedures (SOPs) for each step of the production process reduce errors, maintain uniformity, and provide guidance for employees on handling raw materials, equipment, and processing methods.

9.6.7 Finished Product Testing

Potency and composition analysis ensures the concentration of active compounds in the final product, meeting the label claims; techniques used for accurate measurement include HPLC, GC, and MS. In addition, testing finished products for purity and screening against contaminants like heavy metals, pesticides, mycotoxins, and pathogens are done to assess whether or not safety and regulatory standards are met.

9.6.8 Physical Testing

Tests for physical properties such as colour, odour, taste, solubility, and hardness of tablets ensure that the sensory properties and the stability of the product remain consistent.

9.6.9 Stability Testing

The shelf-life stability test measures the effects of different temperatures, light levels, and humidity levels on a product's potency and purity during its shelf life. Therefore, stability testing guarantees that the products will be effective and safe until the end of the validity period.

9.6.10 Documentation and Traceability

Traceability and compliance with regulation are made easier with habitually comprehensive documents for every manufacturing batch. These include specification of raw materials, in-process checks, and testing of the final product. Batch records are essential for quality audit procedures in the company as well as investigation procedures in case of a product recall.

9.6.11 Quality Certificates and Compliance Documents

Certificates of analysis, together with other compliance documents, accompany each batch as proof of conformance to the required regulations. These records include any deviations from SOPs, the corrective actions taken to address them, and the preventive measures that were put in place to ensure that such instances will not happen again in the future (Gajjar et al., 2023).

9.6.12 Adherence to Good Manufacturing Practices

GMPs give tight control over manufacturing, testing, and QA. The nutraceutical companies have to follow GMP practices to ensure uniformity of their products, reduce the risk of contamination, and ensure safety in the manufacturing process. These include facility design, equipment maintenance, sanitation, and personnel training – all factors that combine to provide a contamination-free environment for high-quality manufacturing (Nath & Yeevani, 2014).

9.6.13 Quality Assurance Systems and Audits

Regular internal and external audits ensure that quality processes are adhered to and achieve multiregulatory and industry-recognized standards, thus highlighting gaps for corrective action and continuous improvement. Supplier audits are performed to validate the quality standards of raw materials with the view of ensuring they meet the required criteria on quality practices, environmental responsibility, and ethical sourcing to ensure the sustainability of supplies. QA teams identify the risk assessments for all stages of production and take necessary measures in advance to reduce the potential for failure (Inbathamizh et al., 2022).

9.7 CONSUMER SAFETY AND COMPLAINT MANAGEMENT

Adverse event monitoring refers to the monitoring and investigating of the flow of reported adverse events by the consumer, ensuring safety and providing an insight into possible improvements. Complaint handling and corrective actions are to be done with due urgency by nutraceutical companies through a review of consumer feedback, identification of the root cause of complaints, and implementation of corrective measures in order to avoid recurring issues (Shao, 2019).

9.7.1 Labelling and Regulatory Compliance

Product labels show the appropriate content, dosage, method of storage, and sell-by date, which help consumers make proper choices and reduce the risk of misuse of the product. Additionally, labels adhere to regulatory standards from the U.S. Food and Drug Administration (FDA) or the European Food Safety Authority (EFSA). QA teams monitor the change in regulations and take steps to ensure that products meet the legal standard regarding ingredients, claims, and manufacturing practices (Inbathamizh et al., 2022).

9.7.2 Continuous Improvement

Data-driven enhancements in quality make use of information from QC and QA processes to realize areas for further improvement, optimize production efficiency, and improve the quality of various products. In addition, regular training and development according to updated quality standards, regulatory changes, and best practices ensure that the quality culture keeps flowing within the organization.

9.8 REGULATORY FRAMEWORKS FOR NUTRACEUTICAL STANDARDIZATION

The regulatory environments for nutraceuticals are highly varied across regions and reflect the differences in definition, permissible health claims, labelling requirements, and standards of quality. The regulatory framework lays down the criteria on the safety of public health, efficacy, and quality of the products to inform consumers about the product they are consuming. Herein is an overview of some major regulatory frameworks that shape the standardization of nutraceuticals

9.8.1 U.S. Dietary Supplement Health and Education Act

In the United States nutraceuticals are considered a category of dietary supplements according to the Dietary Supplement Health and Education Act (DSHEA) of 1994. DSHEA classifies dietary supplements into categories such as vitamins, minerals, herbs, amino acids, and some other substances. The FDA says that the manufacturers need to adhere to GMP standards, which will help in maintaining uniformity regarding quality and purity. GMPs address manufacturing processes, recordkeeping, and QC. Under DSHEA, nutraceuticals can include "structure-function" claims, such as "supports immune health," but cannot claim to diagnose, treat, or cure diseases. Claims must be substantiated, and the FDA monitors product labelling for compliance. While pre-market approval is not generally required, manufacturers are responsible for ensuring the product is safe, and all new dietary ingredients must be submitted to the FDA for review 75 days before release to market (McNamara, 1997).

9.8.2 European Union – European Food Safety Authority

EFSA regulates nutraceuticals under the general food law framework and classifies them as "food supplements." It can be said that every EU country has its different requirements, but EFSA gives central oversight. Regulation EC No. 1924/2006 provides for health and nutrition claims. EFSA asks for scientific substantiation of all the health claims that are considered with thorough evaluation before approval is granted for such claims. Claims approved may only be on the labels of the product. The Novel Food Regulation (EU) 2015/2283 states that the marketing of foods containing such new ingredients or ingredients not commonly consumed in the EU before 1997 shall be subject to a pre-market authorization, with an overall safety assessment. Labelling shall truthfully reflect the composition of ingredients, dosages, conditions of storage, and allergens. These rules ensure that food labels are clear and easy to understand and accurate information is given about the content (Tedesco & Cagnardi, 2019).

9.8.3 Canada's Natural Health Product Regulations

Health Canada classifies nutraceuticals as "natural health products" (NHPs) and includes within this category vitamins, minerals, herbal remedies, homeopathic medicines, and probiotics. All must be approved by Health Canada before sale. To be licensed, manufacturers must submit data regarding safety, efficacy, and quality that Health Canada then reviews to ensure that the safety and health standards are met. NHPs sold in Canada can contain evidence-based claims, but the claims must not be misleading, and they definitely cannot purport to treat or cure serious diseases. Labels are required to list the active ingredients and provide usage and caution information, including possible side effects. Health Canada also mandates strict GMP standards to ensure that products are manufactured and stored in such a way as to not become adulterated and to maintain their quality (Nestmann et al., 2006).

9.8.4 Food Safety and Standards Authority of India (FSSAI)

Nutraceuticals do fall under the ambit of the FSSAI through specific legislation known as the Food Safety and Standards Act. Guidelines provided by FSSAI classify nutraceuticals, functional foods, and dietary supplements, each having its own set of criteria to fulfil. Every manufacturer will necessarily have to register their products with FSSAI and follow standards for safety, quality, and even labelling. New ingredients or formulations might require pre-market approval. Strict guidance on health claims, with proper substantiation and inability to mislead consumers as per FSSAI rules; ingredient listing; suggested dose; and caution against over-consumption must appear on the labelling and packaging. Furthermore, to ensure that all products guarantee a consistent level of quality and are not adulterated with any contaminants, FSSAI enforces GMP standards in their manufacturing (Keservani et al., 2014).

9.8.5 Japan Foods for Specified Health Use (FOSHU) and Foods with Function Claims (FFC)

Nutraceuticals in Japan are regulated as a category of food called Foods for Specified Health Use by the Ministry of Health, Labour and Welfare (MHLW). For FOSHU approval, products must undergo a strict procedure that requires the presentation of a health claim based on clinical data. To expedite approval, manufacturers may use a shortened route under the classification of Foods with Function Claims in which some health claims for the product may be asserted with scientific substantiation but clinical trials need not be as rigorous as those under FOSHU. Nevertheless, the proof has to be presented to the MHLW, with the aim of validation. Above all, FOSHU and FFC have to be labelled comprehensively. Labelling would enlighten consumers with regard to function and scientific evidence of the product that has to be put into use. Claims would simply be on functions alone – for example, "healthy digestion" – and not treatment (Bailey, 2005).

9.8.6 Australia and New Zealand – Therapeutic Goods Administration (TGA) and Food Standards Australia New Zealand (FSANZ)

In Australia and New Zealand, nutraceuticals fall into the purview of both the TGA and that FSANZ. Products categorized as complementary medicines are under the auspices of the TGA, while functional foods come under the FSANZ. The latter requires proof of safety, efficacy, and quality of the product. A registry of such approved products is kept by the TGA and is available publicly. Dietary supplements, on the other hand, might be less stringent under FSANZ. TGA and FSANZ have specific provisions on the kind of health claims allowed, with scientific evidence to support such claims. Labels should indicate active ingredients, usage directions, and possible side effects. GMPs by TGA ensure the safety, potency, and consistency of products. FSANZ also requires the quality standards concerning dietary supplements, focusing on the prevention of contamination and ensuring correct labelling (Komala et al., 2023).

9.8.7 China State Administration for Market Regulation (SAMR)

In the People's Republic of China, the SAMR is responsible for health food products, including nutraceuticals. Health foods have to be approved regarding certain health claims and have to be produced following the quality and safety standards laid down by the SAMR. There are two categories of health foods: functional foods, which can contain some health claims, and nutritional supplements. Only certain, specifically government-approved claims are permitted by the SAMR; these are supported by scientific data. Nutraceuticals purporting to have health values shall be registered in SAMR, which also necessitates providing data related to their safety and efficiency and the purity of ingredients. Unapproved claims are banned, and their misleading labelling is rigidly controlled (Yang, 2021).

9.8.8 Harmonization Efforts and Global Standards

The Codex Alimentarius, established by the World Health Organization (WHO) and the Food and Agriculture Organization of the United Nations (FAO), develops international food standards to protect the health of consumers and ensure fair trade. Guidelines set by Codex influence national standards for food supplements in areas such as labelling, safety, and hygiene. The International Standards Organization (ISO) has developed standards on the quality and safety of food supplements, including manufacturing practices, testing methods, and traceability. ISO standards are well recognized in most countries worldwide, representing the universalization of product quality. While the International Conference on Harmonisation (ICH) provides guidelines principally on pharmaceutical products, some of its guidelines spill over into and inform the quality, safety, and regulatory processes concerning nutraceuticals for countries embracing the harmonization of standards (Chopra et al., 2022).

9.9 WAY FORWARD

The standardization of nutraceutical products has to be addressed in an integrated approach, and it needs to be regulatory, technological, and sustainable. Well-defined regulatory frameworks need to be provided, which should be on par with the standards of the WHO and ISO for safety and quality. Research and development are needed to focus on evidence-based studies for efficacy and safety, backed by high-order analytical techniques like HPLC and GC-MS for proper quality assessment. GMP and HACCP (hazard analysis and critical control points) shall be compulsorily followed along with regular audits for enhanced QC. The consumer has to be educated on the correct usage, clear labelling, and hazards involved. The integration of blockchain and internet of things (IoT) will provide full traceability and transparency in the supply chain. Sustainability should be guaranteed through the use of eco-friendly ingredients and packaging. International collaboration will result in harmonization of standards, increase global trust, and provide consistency. In particular, such a holistic approach will be the driver of safety, quality, and consumer confidence for sustainable growth within the nutraceutical industry.

9.10 CONCLUSION

Standardization of nutraceutical products is the call of the day in terms of safety, efficacy, and quality. The need of the hour is a holistic approach to standardization: clear regulations, advanced QC, robust research, consumer education, and sustainability. By building trust, enhancing transparency, and supporting global growth, the nutraceutical industry will be able to achieve international collaboration and leverage modern technologies. Such measures will not only protect consumer interests but also pave the way for the sustainable development of the sector.

REFERENCES

Ariyo, O., & Akinyemi, T. (2024). Maximizing biotic food as nutraceuticals: A comprehensive exploration. *ABUAD International Journal of Natural and Applied Sciences*, 4(1), 33–38.

Arora, M., & Baldi, A. (2017). Comparative study of regulatory framework for probiotics: Current status and future recommendations. *Applied Clinical Research, Clinical Trials and Regulatory Affairs*, 4(2), 140–156.

Bagchi, D., & Nair, S. (2016). *Developing New Functional Food and Nutraceutical Products*. Academic Press.

Bailey, R. (2005). Functional foods in Japan: FOSHU ("foods for specified health uses") and "foods with nutrient function claims". In *Regulation of Functional Foods and Nutraceuticals: A Global Perspective* (pp. 247–261). Wiley Online Library.

Bensa, M., Vovk, I., & Glavnik, V. (2023). Resveratrol food supplement products and the challenges of accurate label information to ensure food safety for consumers. *Nutrients*, 15(2), 474.

Bernal, J., Mendiola, J., Ibáñez, E., & Cifuentes, A. (2011). Advanced analysis of nutraceuticals. *Journal of Pharmaceutical and Biomedical Analysis*, 55(4), 758–774.

Bertola, M., Ferrarini, A., & Visioli, G. (2021). Improvement of soil microbial diversity through sustainable agricultural practices and its evaluation by-omics approaches: A perspective for the environment, food quality and human safety. *Microorganisms*, 9(7), 1400.

Borchers, A. T., Keen, C. L., & Gershwin, M. E. (2016). The basis of structure/function claims of nutraceuticals. *Clinical Reviews in Allergy & Immunology*, 51, 370–382.

Bose, D., & Padmavati, M. (2024). Honey authentication: A review of the issues and challenges associated with honey adulteration. *Food Bioscience*, 105004.

Buchanan, D. R. (2000). *An Ethic for Health Promotion: Rethinking the Sources of Human Well-being*. Oxford University Press.

Cardellina, J. H. (2002). Challenges and opportunities confronting the botanical dietary supplement industry. *Journal of Natural Products*, 65(7), 1073–1084.

Chander, S., Kulkarni, G. T., Dhiman, N., & Kharkwal, H. (2021). Protein-based nanohydrogels for bioactive delivery. *Frontiers in Chemistry*, 9, 573748.

Chandra, S., Stanford, D., Fletcher, E., & Walker, L. A. (2019). Raw materials production and manufacturing process control strategies. In *The Science and Regulations of Naturally Derived Complex Drugs* (pp. 175–190). Springer Nature.

Chen, G., Zhang, L. K., & Pramanik, B. N. (2007). LC/MS: Theory, instrumentation, and applications to small molecules. In *HPLC for Pharmaceutical Scientist*. (pp. 281–346). Wiley Online Library.

Chopra, A. S., Lordan, R., Horbańczuk, O. K., Atanasov, A. G., Chopra, I., Horbańczuk, J. O., Jóźwik, A., Huang, L., Pirgozliev, V., & Banach, M. (2022). The current use and evolving landscape of nutraceuticals. *Pharmacological Research*, 175, 106001.

Cochoy, F. (2005). A brief history of 'customers,' or the gradual standardization of markets and organizations. *Sociologie du Travail*, 47, e36–e56.

Coppens, P., Da Silva, M. F., & Pettman, S. (2006). European regulations on nutraceuticals, dietary supplements and functional foods: A framework based on safety. *Toxicology*, 221(1), 59–74.

Dery, B., & Zaixiang, L. (2023). Scanning electron microscopy (SEM) as an effective tool for determining the morphology and mechanism of action of functional ingredients. *Food Reviews International*, 39(4), 2007–2026.

Dima, C., Assadpour, E., Dima, S., & Jafari, S. M. (2020). Bioavailability of nutraceuticals: Role of the food matrix, processing conditions, the gastrointestinal tract, and nanodelivery systems. *Comprehensive Reviews in Food Science and Food Safety*, 19(3), 954–994.

Espín, J. C., García-Conesa, M. T., & Tomás-Barberán, F. A. (2007). Nutraceuticals: Facts and fiction. *Phytochemistry*, 68(22–24), 2986–3008.

Fenster, K., Freeburg, B., Hollard, C., Wong, C., Rønhave Laursen, R., & Ouwehand, A. C. (2019). The production and delivery of probiotics: A review of a practical approach. *Microorganisms*, 7(3), 83.

Finan, J. (2023). *The Presence of Adulterants and Mislabeling of Caffeine Content in Australian Sports Supplements*. Murdoch University.

Gajjar, K., Patel, Z., Movaliya, V., Kanaki, N., Deshpande, S., & Zaveri, M. (2023). A comparative evaluation of Regulatory requirements for registration of dietary supplements in Brazil, Russia, India, China and South Africa. *International Journal of Drug Regulatory Affairs*, 11(3), 53–61.

Gharanjig, H., Gharanjig, K., Hosseinnezhad, M., & Jafari, S. M. (2020). Differential scanning calorimetry (DSC) of nanoencapsulated food ingredients. In *Characterization of Nanoencapsulated Food Ingredients* (pp. 295–346). Elsevier.

Gonçalves, R. F., Martins, J. T., Duarte, C. M., Vicente, A. A., & Pinheiro, A. C. (2018). Advances in nutraceutical delivery systems: From formulation design for bioavailability enhancement to efficacy and safety evaluation. *Trends in Food Science & Technology*, 78, 270–291.

He, T.-T., Ung, C. O. L., Hu, H., & Wang, Y.-T. (2015). Good manufacturing practice (GMP) regulation of herbal medicine in comparative research: China GMP, cGMP, WHO-GMP, PIC/S and EU-GMP. *European Journal of Integrative Medicine*, 7(1), 55–66.

Helal, N. A., Eassa, H. A., Amer, A. M., Eltokhy, M. A., Edafiogho, I., & Nounou, M. I. (2019). Nutraceuticals' novel formulations: The good, the bad, the unknown and patents involved. *Recent Patents on Drug Delivery & Formulation*, 13(2), 105–156.

Inbathamizh, L., Prabavathy, D., & Sudha, S. (2022). Quality assurance of nutraceuticals and their approval, registration, marketing. In *Handbook of Nutraceuticals and Natural Products: Biological, Medicinal, and Nutritional Properties and Applications* (Vol. 2, pp. 337–360). John Wiley & Sons, Inc.

Joshi, D. D., & Joshi, D. D. (2012). UV–Vis. Spectroscopy: Herbal drugs and fingerprints. In *Herbal Drugs and Fingerprints: Evidence Based Herbal Drugs* (pp. 101–120). Springer Nature.

Keservani, R. K., Sharma, A. K., Ahmad, F., & Baig, M. E. (2014). Nutraceutical and functional food regulations in India. In *Nutraceutical and Functional Food Regulations in the United States and around the World* (pp. 327–342). Elsevier.

Khan, M. M. H., Rafii, M. Y., Ramlee, S. I., Jusoh, M., & Al-Mamun, M. (2021). Bambara groundnut (*Vigna subterranea* L. Verdc): A crop for the new millennium, its genetic diversity, and improvements to mitigate future food and nutritional challenges. *Sustainability*, 13(10), 5530.

Kim, E. J., Kim, J. H., Kim, M.-S., Jeong, S. H., & Choi, D. H. (2021). Process analytical technology tools for monitoring pharmaceutical unit operations: A control strategy for continuous process verification. *Pharmaceutics*, 13(6), 919.

Komala, M. G., Ong, S. G., Qadri, M. U., Elshafie, L. M., Pollock, C. A., & Saad, S. (2023). Investigating the regulatory process, safety, efficacy and product transparency for nutraceuticals in the USA, Europe and Australia. *Foods*, 12(2), 427.

Kotra, V. S. R., Satyabanta, L., & Goswami, T. K. (2019). A critical review of analytical methods for determination of curcuminoids in turmeric. *Journal of Food Science and Technology*, 56, 5153–5166.

Kowalska, T., & Sajewicz, M. (2022). Thin-layer chromatography (TLC) in the screening of botanicals–its versatile potential and selected applications. *Molecules*, 27(19), 6607.

Kumar, M., Chandran, D., Tomar, M., Bhuyan, D. J., Grasso, S., Sá, A. G. A., . . .& Mekhemar, M. (2022). Valorization potential of tomato (*Solanum lycopersicum* L.) seed: Nutraceutical quality, food properties, safety aspects, and application as a health-promoting ingredient in foods. *Horticulturae*, 8(3), 265.

Krishnaswamy, S. (2024). Phytopharmaceutical biotechnology: Integration of botany, pharmacology and plant biotechnology to deliver the best therapeutic potential of herbs. In *Concepts in Pharmaceutical Biotechnology and Drug Development* (pp. 437–464). Springer.

Kunle, O. F., Egharevba, H. O., & Ahmadu, P. O. (2012). Standardization of herbal medicines-A review. *International Journal of Biodiversity and Conservation*, 4(3), 101–112.

Lee, L. S., Kim, S. H., Kim, Y. B., & Kim, Y. C. (2014). Quantitative analysis of major constituents in green tea with different plucking periods and their antioxidant activity. *Molecules*, 19(7), 9173–9186.

Liu, Z., Smart, J. D., & Pannala, A. S. (2020). Recent developments in formulation design for improving oral bioavailability of curcumin: A review. *Journal of Drug Delivery Science and Technology*, 60, 102082.

McNamara, S. H. (1997). Dietary supplement legislation enhances opportunities to market "nutraceutical"-type products. *Journal of Nutraceuticals, Functional & Medical Foods*, 1(1), 47–59.

Mehta, S., Sangwan, S., & Dhawan, V. (2024). Challenges in developing delivery systems of nutraceuticals and nanonutraceuticals. In *Handbook of Nutraceuticals: Science, Technology and Engineering* (pp. 1–25). Springer.

Miller, N. (2016). *Green Rooibos Neutraceutical: Optimisation of Hot Water Extraction and Spray-drying by Quality-by-design Methodology*. Stellenbosch: Stellenbosch University.

Moloughney, S. (2023). State of the industry: Experts examine top issues, trends, and challenges: With so many angles of the dynamic nutraceuticals world to cover, we consulted industry thought leaders and subject matter experts to present a robust view of today's marketplace. *Nutraceuticals World*, 26(10), 18–36.

Nafiu, M., Hamid, A., Muritala, H., & Adeyemi, S. (2017). Preparation, standardization, and quality control of medicinal plants in Africa. In *Medicinal Spices and Vegetables from Africa* (pp. 171–204). Academic Press.

Nath, D. K., & Yeevani, C. S. (2014). Current good manufacturing practices for nutraceuticals. In *Nutraceutical and Functional Food Regulations in the United States and around the World* (pp. 63–71). Elsevier.

Nestmann, E. R., Harwood, M., & Martyres, S. (2006). An innovative model for regulating supplement products: Natural health products in Canada. *Toxicology*, 221(1), 50–58.

Newmaster, S. G., Shanmughanandhan, D., Kesanakurti, P., Shehata, H., Faller, A., Noce, I. D., Lee, J. Y., Rudzinski, P., Lu, Z., & Zhang, Y. (2019). Recommendations for validation of real-time PCR methods for molecular diagnostic identification of botanicals. *Journal of AOAC International*, 102(6), 1767–1773.

Nikam, P. H., Kareparamban, J., Jadhav, A., & Kadam, V. (2012). Future trends in standardization of herbal drugs. *Journal of Applied Pharmaceutical Science*, 38–44.

Phillips, D., & Castegnaro, M. (1999). Standardization and validation of DNA adduct postlabelling methods: Report of interlaboratory trials and production of recommended protocols. *Mutagenesis*, 14(3), 301–315.

Piccolella, S., Crescente, G., Candela, L., & Pacifico, S. (2019). Nutraceutical polyphenols: New analytical challenges and opportunities. *Journal of Pharmaceutical and Biomedical Analysis*, 175, 112774.

Rajat, S., Manisha, S., Robin, S., & Sunil, K. (2012). Nutraceuticals: A review. *International Research Journal of Pharmacy*, 3(4), 95–99.

Sadgrove, N. J. (2022). Honest nutraceuticals, cosmetics, therapies, and foods (NCTFs): Standardization and safety of natural products. *Critical Reviews in Food Science and Nutrition*, 62(16), 4326–4341.

Salgueiro, L., Martins, A., & Correia, H. (2010). Raw materials: The importance of quality and safety. A review. *Flavour and Fragrance Journal*, 25(5), 253–271.

Samsami-Kor, M., Daryani, N. E., Asl, P. R., & Hekmatdoost, A. (2015). Anti-inflammatory effects of resveratrol in patients with ulcerative colitis: A randomized, double-blind, placebo-controlled pilot study. *Archives of Medical Research*, 46(4), 280–285.

Santini, A., Cammarata, S. M., Capone, G., Ianaro, A., Tenore, G. C., Pani, L., & Novellino, E. (2018). Nutraceuticals: Opening the debate for a regulatory framework. *British Journal of Clinical Pharmacology*, 84(4), 659–672.

Sarkisyan, V., Bessonov, V., & Kochetkova, A. (2017). Raw materials analysis and quality control. In *Developing New Functional Food and Nutraceutical Products* (pp. 195–211). Elsevier.
Seethapathy, G. S. (2019). *Ethnobotany, phytochemistry and DNA metabarcoding studies on Indian traditional medicine*. Thesis. https://www.duo.uio.no/handle/10852/68510.
Shao, A. (2019). Global adverse event reporting regulations for nutraceuticals, functional foods, and dietary/food/health supplements. In *Nutraceutical and Functional Food Regulations in the United States and around the WORLD* (pp. 619–624). Elsevier.
Shariful, I. M., Maamoun, I., Falyouna, O., Belal, E. B., Eljamal, O., & Saha, B. B. (2021). Biosorption of arsenic from contaminated water: Composite of algae and Fe-nano particles. In *Proceeding of International Exchange and Innovation Conference on Engineering & Sciences (IEICES)* (p. 7).
Shinkaruk, S., Lamothe, V., Schmitter, J.-M., Fructus, A., Sauvant, P., Vergne, S. b., Degueil, M., Babin, P., Bennetau, B., & Bennetau-Pelissero, C. (2008). Synthesis of haptens and conjugates for ELISA of glycitein: Development and validation of an immunological test. *Journal of Agricultural and Food Chemistry*, 56(16), 6809–6817.
Shukla, U. (2024). Fourier transform infrared spectroscopy: A power full method for creating fingerprint of molecules of nanomaterials. *Journal of Molecular Structure*, 140454.
Smichowski, P., & Londonio, A. (2018). The role of analytical techniques in the determination of metals and metalloids in dietary supplements: A review. *Microchemical Journal*, 136, 113–120.
Som Chaudhury, S., & Sinha, K. (2023). Microbiological evaluation of the nutraceuticals and nanonutraceuticals formulations. In *Handbook of Nutraceuticals: Science, Technology and Engineering* (pp. 1–38). Springer.
Svanberg, S., & Svanberg, S. (2001). *Atomic and Molecular Spectroscopy*. Springer.
Tedesco, D. E. A., & Cagnardi, P. (2019). Regulatory guidelines for nutraceuticals in the European union. In *Nutraceuticals in Veterinary Medicine*, 793–805. Springer Nature.
Teli, V., Tripathi, M. C., Tiwari, H., & Chandrul, K. K. (2024). Nutraceuticals: A review on. *European Journal of Modern Medicine and Practice*, 4(5), 417–426.
Usmani, A., & Almoselhy, R. I. (2024). Current trends in *Nigella sativa* L. (Black seed) from traditional to modern medicine with advances in extraction, formulation, quality control, regulatory status, and pharmacology. *International Journal of Pharmaceutical Chemistry and Analysis*, 11, 11–23. 10.18231/j.ijpca.2024.002.
Vestland, T. L., Jacobsen, Ø., Sande, S. A., Myrset, A. H., & Klaveness, J. (2016). Characterization of omega-3 tablets. *Food Chemistry*, 15(197), 496–502.
Vlaicu, P. A., Untea, A. E., Varzaru, I., Saracila, M., & Oancea, A. G. (2023). Designing nutrition for health – Incorporating dietary by-products into poultry feeds to create functional foods with insights into health benefits, risks, bioactive compounds, food component functionality and safety regulations. *Foods*, 12(21), 4001.
Wadood, S. A., Boli, G., Xiaowen, Z., Hussain, I., & Yimin, W. (2020). Recent development in the application of analytical techniques for the traceability and authenticity of food of plant origin. *Microchemical Journal*, 152, 104295.
Wang, C., & Li, S. (2014). Functional foods and nutraceuticals: Potential Role in Human Health. In *Clinical Aspects of Functional Foods and Nutraceuticals* (pp. 51–76). Taylor & Francis Group.
Woo, C. S. J., Lau, J. S. H., & El-Nezami, H. (2012). Herbal medicine: Toxicity and recent trends in assessing their potential toxic effects. In *Advances in Botanical Research* (Vol. 62, pp. 365–384). Academic Press.
Yang, M. S. (2021). Regulatory aspects of nutraceuticals: Chinese perspective. In *Nutraceuticals* (pp. 1281–1291). Elsevier.
Yeung, M. T., Hobbs, J. E., & Kerr, W. A. (2006). Accessing US and EU markets for nutraceuticals and functional foods: Legislative and technical barriers facing agribusiness marketers. *Journal of International Food & Agribusiness Marketing*, 19(1), 53–79.

10 Treatment with Polyphenol Nutraceuticals

Case Studies

Dwaipayan Sinha, Moumita Datta, Arindam Ganguly, Aqsa Majgaonkar, and Pratyusha De

10.1 INTRODUCTION

Polyphenols are naturally occurring molecules produced solely by plants, characterized by phenolic structures that help manage oxidative and inflammatory imbalances while supporting gut health through prebiotic-like actions (Bertelli et al. 2021). Polyphenols are widespread throughout the plant kingdom, typically playing roles in pollinator attraction (Zhang et al. 2016), structural functions (Bertelli et al. 2021), defense against ultraviolet radiation (Del Valle et al. 2020), and protection against microbial invasion (Slobodníková et al. 2016) and herbivores (Dini and Grumetto 2022). Polyphenols are among the most plentiful and pervasive categories of plant metabolites and are essential to human and animal nutrition. This diverse collection of chemicals in plant diets varies significantly, encompassing basic phenolic molecules to highly polymerized compounds with molecular weights exceeding 30,000 Da (Bravo 2009). The primary dietary sources of polyphenols consist of various fruits, vegetables, and drinks (Williamson 2017). Phenolic acids comprise approximately one-third of the total intake, whereas flavonoids constitute two-thirds (Rahman et al. 2021). The predominant flavonoids in the diet include flavanols (including catechins and proanthocyanidins), anthocyanins, and their oxidation derivatives. The primary dietary sources of polyphenols are fruits and beverages (such as fruit juice, wine, tea, coffee, chocolate, and beer), with vegetables, dry legumes, and cereals being secondary sources (Scalbert and Williamson 2000). Polyphenols are regarded as effective preventive agents for chronic and degenerative diseases owing to their complex and unique chemical structure that confers numerous biological roles (Grabska-Kobyłecka et al. 2023). Food polyphenols are prevalent antioxidants in dietary sources that can effectively inhibit atherosclerosis (Ziółkiewicz et al. 2023) and mitigate aging (Al-Regaiey 2024). Numerous research studies have corroborated the advantageous benefits of polyphenols in the prevention of cardiovascular (Guillermo Gormaz et al. 2016), neurological (Grabska-Kobyłecka et al. 2023), hepatic

DOI: 10.1201/9781003588504-10

(Ranneh et al. 2024), diabetic (Raina et al. 2024), and oncological disorders (Gao et al. 2018). Polyphenols have thus become an integral part of human diets due to their cheap and abundant presence in fruits, vegetables, and beverages. As mentioned, polyphenols confer protection against a wide array of disorders, thereby establishing their relevance in diets and as 'to be consumed' natural molecules. Thus, polyphenols highly satisfy the criteria of a nutraceutical due to their immense health benefits. This chapter is an attempt to highlight the beneficial role of polyphenols. Efforts have been made to highlight selected studies as 'cases' to highlight the potential of polyphenol nutraceuticals as a curative agent.

10.2 POLYPHENOLS: PATHWAYS OF ACTION IN DISEASE PREVENTION

Polyphenols are naturally occurring micronutrients that are vital in maintaining physiological homeostasis. Their remarkable ability to neutralize excessive oxidative species in normal cells has led to their recognition as nutraceuticals with potential therapeutic applications (Muntaha et al. 2025). Both isolated polyphenols and their complexes with herbs or foods exhibit dual antioxidant and pro-oxidant properties, suggesting their potential utility in chemoprevention. The scientific evidence supports the capacity of polyphenols to induce apoptosis, inhibit cell proliferation, suppress DNA synthesis, and modulate signal transduction pathways. Consequently, polyphenols have been implicated in the prevention and treatment of various conditions associated with inflammation and oxidative stress, including cancer, cardiovascular diseases, and neurodegenerative disorders.

10.2.1 Antioxidant Effect

Chronic oxidative stress can induce cellular damage and genetic modifications, triggering the inflammatory signaling cascade that underlies various diseases. Prolonged generation of free radicals during inflammation can intensify oxidative stress, contributing to chronic inflammation (Singh et al. 2020). Oxidants are crucial in activating Toll-like receptors (TLRs), which initiate inflammatory responses. Research has demonstrated that oxidant exposure up-regulates TLR4 translocation to the cell membrane, enhancing cellular responsiveness to danger signals and promoting pro-inflammatory signaling pathways. The nuclear factor kappa–light-chain-enhancer of activated B cells (NF-κB) pathway, a key signaling cascade responsive to free radicals, is essential for regulating inflammation. NF-κB activation occurs via canonical and alternative pathways, ultimately leading to its movement to the nucleus, triggering the activation of target gene expression. Under oxidative stress, antioxidant gene expression is up-regulated through nuclear factor erythroid 2-related factor 2 (Nrf2). Reactive oxygen species (ROS) activate Nrf2 by dissociating its inhibitor, Keap1, enabling Nrf2 to move to the nucleus and initiate the expression of antioxidant response genes (Zhang et al. 2022).

Polyphenols, well known for their potent antioxidant properties, are abundant in various foods, including fruits, vegetables, and cereals. Their robust antioxidant activity can be attributed to their unique structural characteristics. These compounds neutralize free radicals through multiple mechanisms, including hydrogen atom transfer, single-electron transfer, and transition metal chelation. During hydrogen atom transfer, polyphenols cleave O-H bonds to donate hydrogen atoms to free radicals (Muntaha et al. 2025). In contrast, single-electron transfer involves polyphenols acting as electron donors, transferring electrons to neutralize free radicals. Polyphenols also enhance antioxidant efficacy by scavenging ROS, including superoxide anions, hydroxyl radicals, and hydrogen peroxide (Singh et al. 2020). This is achieved through interactions with intracellular kinases, suppression of pro-oxidant gene activation, and up-regulation of antioxidant gene expression, including superoxide dismutases and catalases. Emerging evidence suggests that polyphenols may reverse epigenetic alterations, highlighting their potential in antioxidant prophylaxis and therapy (Muntaha et al. 2025). Plant-derived polyphenolic compounds, such as catechins, exhibit exceptional antioxidant properties, making them ideal candidates for antioxidant applications. Studies demonstrate that polyphenol-protein interactions can enhance antioxidant activity, improve functional properties, and modify protein structures (Grzesik et al. 2018).

10.2.2 Anti-Inflammatory Effect

Inflammation is a complex process involving immune cells, inflammatory signaling pathways, and cellular processes. Polyphenols and protein-polyphenol complexes can effectively inhibit the production of ROS and nitric oxide in lipopolysaccharide-activated cells. Studies on various protein-polyphenol combinations, including buckwheat protein, chia seed, pea proteins, blueberry polyphenols, black mulberry, and whey protein, have demonstrated significant anti-inflammatory effects. Plant polyphenols exhibit anti-inflammatory effects by inhibiting inflammatory cells, affecting cytokines and their secretion processes (Singh et al. 2020). Studies have demonstrated the anti-inflammatory activity of polyphenols, such as rutin and hesperidin, which reduce nitric oxide, interleukin-6, and tumor necrosis factor-alpha production. Polyphenols act as free radical scavengers, providing health benefits, including anti-inflammatory, antioxidant, and antiviral effects. Low-molecular-weight proteins and peptides also offer protection against inflammation.

10.2.3 Modulation of Gut Microbiota

The human gastrointestinal (GI) tract is teeming with a large and varied consortium of microorganisms, predominantly anaerobic bacteria, that have learned to thrive and survive on mucosal surfaces within the lumen of the gut. The composition and dynamics of gut microbiota are determined by multifactorial host variables such as genetics, immune system interactions, antibiotic exposures, and dietary habits (Buddington and Sangild 2011; Zhu et al. 2010). The human GI microbiota consists primarily of species from the phyla Firmicutes, Bacteroidetes, Actinobacteria, and Proteobacteria (Gerritsen et al. 2011). Dietary polyphenols are bioactive phytochemicals that are highly present in almost all fruits, vegetables, seeds, and herbs, as well as beverages like wine, beer, fruit juice, coffee, tea, and chocolate, with lower concentrations in dry legumes and cereals (Vinson et al. 2001). Thus, a healthy diet rich in polyphenols promotes gut health, favoring beneficial bacteria while inhibiting their harmful counterparts (Pistollato et al. 2016). Recent research has indicated a blood pressure–, triglyceride-, and uric acid–lowering action of red wine polyphenols, most probably as a consequence of increased growth of Bacteroides and Proteobacteria in the gut (Queipo-Ortuño et al. 2012). In a study, it was observed that blood glucose positively correlates with the Bacteroidetes: Firmicutes ratio in adults of Chinese descent with type 2 diabetes, coupled with increased abundance in Betaproteobacteria (Larsen et al. 2010). A study on European women with type 2 diabetes found a high colonization of *Clostridium clostridioforme* and a reduced presence of *Roseburia*, a butyrate-producing bacterium linked to improved insulin sensitivity (Karlsson et al. 2013). Polyphenol supplements can influence gut microbiota composition by modifying the Firmicutes: Bacteroides ratio. For instance, cranberry extract, rich in phenolic acids, flavonoids, and anthocyanins, might help to improve gut-related health by decreasing said ratio in mice subjected to a high-fat/high-sugar diet (Anhê et al. 2015). Dietary polyphenols such as rutin, quercetin, chlorogenic acid, and caffeic acid can modulate the Firmicutes: Bacteroides ratio, thus establishing the prospect for their use in the modulation of gut microbial balance. Combination treatment of grape pomace polyphenols and a probiotic strain, *Streptobacterium plantarum*, exhibited antimicrobial activity against pathogens such as *Escherichia coli*, *Bacillus megaterium*, and *Listeria monocytogenes* (Caponio et al. 2022).

10.2.4 Role in Cellular Signaling Pathway

Non-alcoholic fatty liver disease (NAFLD) represents a range of liver disorders, varying from a healthy liver to advanced cirrhosis, classified under genetic, environmental, and metabolic stress-related conditions (Zeng et al. 2014). It's been estimated that about 10% to 24% of the total population has NAFLD, and its prevalence has even been reported to be up to 75% among obese individuals. Due to the involvement of multiple pathways in the development of NAFLD, treatment for this condition has proved to be difficult (Van De Wier et al. 2017). Among numerous signaling mechanisms linked to insulin resistance, oxidative stress, and inflammation are NF-κB, AMP-activated protein kinase (AMPK), Janus kinase/signal transducers and activators of transcription (JAK/STAT), peroxisome proliferator–activated receptors (PPARs), SREBP-1c, phosphatidylinositol 3-kinase/protein kinase B (PI3K/Akt), and TLR (Michelotti et al. 2013). Polyphenols may thus act through at least four signaling pathways in preventing hepatocyte damage by NAFLD: (1) inhibition of lipogenesis through SREBP-1c down-regulation by AMPK activation, (2) augmentation of antioxidant defense via the Nrf2 pathway, (3) inhibition of inflammation by blockade of NF-κB pathway activation, and (4) up-regulation of PPARα activation enhancing β-fatty acid oxidation (Li et al. 2018). This SREBP, which is expressed mainly in the liver, regulates de novo lipogenesis through mediating the action of lipogenic enzymes and genes and thus is the target for metabolic modulation in the treatment of NAFLD (Zeng et al. 2014). Increased expression of hepatic SREBP-1c helps progress steatosis. Genistein (Shin et al. 2007), luteolin (Liu et al. 2011), rutin (Wu et al. 2011), and prunetin (Ahn et al. 2013) are some examples of polyphenols that have been shown to inhibit SREBP-1c, mainly through the reduced expression of SREBP-1c protein and its corresponding gene and activating AMPK that controls SREBP-1c transcription (Ferré and Foufelle 2010). AMPK is a vital regulator of cellular energy homeostasis (Zeng et al. 2014). It combines α, β, and γ subunits into a heterologous trimeric protein kinase. AMPK modulates fatty acid metabolism by activating the fatty acid biosynthetic pathway. AMPK activation inhibits the expression of acetyl-CoA carboxylase (ACC) by down-regulating SREBP-1c, reducing fatty acid, cholesterol, and triglyceride synthesis and promoting their uptake and β-oxidation (Zeng et al. 2014). Oxidative stress induces the expression of antioxidant genes mediated by Nrf2 (Ramos-Tovar and Muriel 2020). Nrf2 activation occurs when ROS releases their inhibitor, Keap1, thereby allowing Nrf2 accumulation in the nucleus and expressing genes pertinent to the antioxidant responses (Ramos-Tovar and Muriel 2020). The NF-κB pathway is involved in a multitude of cytokines that regulate inflammation. The inhibition of this pathway by polyphenols has been regarded as the possible mechanism for their anti-inflammatory activities. NF-κB signaling is activated by pro-inflammatory stimuli, which incite the degradation of the IκB kinase (IKK) complex to liberate NF-κB into the nucleus with a subsequent inflammatory response (Zeng et al. 2014). Kaempferol was found to inhibit the phosphorylation of insulin receptor substrate 1 (IRS-1), IKKα, and IKKβ, associated with inhibition of nuclear and cytoplasmic translocation of NF-κB, and lowering levels of TNF-α and IL-6 in a mouse model of insulin resistance and type 2 diabetes mellitus (Luo et al. 2015). Other polyphenols such as curcumin (Jiménez-Flores et al. 2014), morin (Wang et al. 2013), oligonol (Noh et al. 2011), and quercetin (Porras et al. 2017) have also been reported to inhibit the NF-κB pathway in NAFLD. The TNF receptor interacts with JNK, activating downstream signaling and activating AP-1 (Song and Lee 2012). PPARs, a group of nuclear receptors known to mediate various metabolic processes involving lipid and glucose, are among the probable targets for metabolic process regulation (Zeng et al. 2014). Among the three classes of PPARs described, PPARα and PPARγ are singled out for their role in the pathogenesis of NAFLD (Van De Wier et al. 2017). PPARα activation in the liver, which is highly expressed, regulates free fatty acid (FFA) transport and stimulates all relevant enzymes associated with β-oxidation. Therefore, stimulation of PPARα should help in resolving steatosis and hepatic inflammation. Many polyphenols have been extensively shown to stimulate PPARα. Some of these compounds can occupy the PPARα binding site as ligands or agonists (Medjakovic et al. 2010; Jia et al. 2013), whereas kaempferol (Chang et al. 2011), naringenin (Goldwasser et al. 2010), tiliroside (Goto et al. 2012), and glabridin (Lee et al. 2012) may up-regulate PPARα gene and/or protein expression.

10.3 CASE REPORTS ON THE HEALTH BENEFITS OF POLYPHENOL-BASED NUTRACEUTICALS

Polyphenols are secondary metabolites of plants, recognized as essential functional foods and nutraceuticals. These compounds, including phenolic acids, flavonoids, catechins, tannins, lignans, stilbenes, and anthocyanidins, are abundant in grains, cereals, pulses, vegetables, spices, fruits, chocolates, and beverages such as fruit juices, tea, coffee, and wine.

The global prevalence of cardiovascular disease, diabetes, cancer, neurodegenerative disorders, oxidative stress–related complications, and chronic age-related illnesses is steadily rising. This increase is partly due to the limitations of conventional medicinal treatments, including their toxicity, ineffectiveness in certain conditions, and the necessity of multiple drug therapies for some chronic diseases. As a result, there is a growing need to explore alternative treatment approaches.

Polyphenols have demonstrated antioxidant, anti-inflammatory, anticancer, and antidiabetic properties, along with protective effects on cardiovascular and nervous system health. Additionally, with increasing life expectancy and heightened media coverage of healthcare topics, public awareness of the benefits of nutritional support in disease prevention and management is expanding. Extensive scientific literature highlights the role of these natural compounds in reducing oxidative stress, managing diabetes and neurodegenerative diseases, and lowering the risk of cancer. Selected cases and studies on beneficial activity of polyphenols and flavonoid nutraceuticals are illustrated in Table 10.1.

10.4 CONCLUSION

Polyphenols represent a diverse and functionally significant class of phytochemicals crucial for health preservation and chronic disease prevention. Found abundantly in plant-based foods such as berries, leafy greens, whole grains, pulses, herbal beverages, fermented drinks, and chocolate products, these compounds are distinguished by their phenolic structures and notable biological effects. Their classification into primary groups – phenolic acids, flavonoids, stilbenes, and lignans – illustrates the structural complexity and chemical diversity that support their health-promoting attributes.

An increasing amount of scientific evidence underscores the positive impact of polyphenols on human well-being. Their antioxidant properties are extensively studied, including their ability to neutralize ROS and reduce oxidative strain, a key factor in the development of various non-communicable diseases. Furthermore, polyphenols have anti-inflammatory, anticarcinogenic, cardioprotective, neuroprotective, hepatoprotective, and antidiabetic characteristics. Many molecular processes, including the modulation of enzyme activity, inhibition of inflammatory signaling pathways, regulation of gene expression, and interactions with cellular receptors, mediate these effects.

Despite their therapeutic potential, several obstacles remain in fully harnessing the capabilities of dietary polyphenols. This includes bioavailability, metabolic stability, and individual variability in absorption and metabolism. The bioaccessibility of polyphenols is greatly influenced by their chemical structure, as well as factors like the food matrix, preparation techniques, and interactions with other dietary elements. Moreover, following consumption, polyphenols undergo considerable degradation in the GI tract and liver, potentially modifying their biological effects. Thus, understanding the pharmacokinetics and biotransformation of polyphenols is essential for improving their application in clinical and functional dietary settings.

TABLE 10.1
Selected Cases on the Beneficial Activity of Polyphenols and Flavonoid Nutraceuticals as Evaluated Through Pharmacological Studies

Sl no.	Name of the Nutraceutical/ Polyphenol	Source	Target Organism	Mode/ Type of Experiment	Dose	Outcome of the Study	Reference
1	Anthocyanins and soybean protein isolate (SPI)	Black rice extract powder	RAW264.7 cell line	In vitro	200 μg/mL	Significantly decreased the secretion of TNF-α and NO	(Z. Wang et al. 2022)
2	Anthocyanins, proanthocyanidins, catechins, and stilbenes	*Morus nigra* L. (black mulberry [BM])	Murine RAW264.7 macrophage cell line	In vitro	10 and 20 μg/mL	The optimized BM polyphenol-enriched whey matrix suppressed pro-inflammatory mediators and enhanced Nrf-2-dependent cytoprotective enzyme expression in LPS-induced macrophages, even at low doses.	(Ozleyen et al. 2022)
3	Astaxanthin (AST)	SML0982; Sigma-Alrich	STZ-induced DR (rats)	In vivo	10 and 20 mg/kg for 21 consecutive days	The administration of AST effectively reduced STZ-induced elevations in blood glucose levels. AST demonstrated protective effects against STZ-induced diabetic retinopathy (DR) by regulating oxidative stress markers (catalase) and NSE levels in retinal tissues, thereby mitigating DR progression.	(Subramanian et al. 2022)
4	Bacoside I and Bacoside A	*Bacopa monnieri*	Wistar rats	In vivo	80 mg/kg	Reduced neuroinflammation and oxidative stress	(Abhishek et al. 2022)

(Continued)

TABLE 10.1 (*Continued*)
Selected Cases on the Beneficial Activity of Polyphenols and Flavonoid Nutraceuticals as Evaluated Through Pharmacological Studies

Sl no.	Name of the Nutraceutical/ Polyphenol	Source	Target Organism	Mode/ Type of Experiment	Dose	Outcome of the Study	Reference
5	Caffeic acid (CA)	Fruits, olives, coffee beans, grains, propolis, and many dietary supplements	Atherogenic diet (Ath)–induced rat model	In vivo	25 mg/kg 50 mg/kg	Avoided gain in body weight. Prevented Ath-triggered disturbances in serum biomarkers. Escalated hemodynamic parameters. Enhanced oxidative stress induced by Ath.	(Y. Wang et al. 2022)
6	Calebin A	*Curcuma* species (*Curcuma longa/ Curcuma caesia*)	Obese individuals	In vivo	25 mg of Calebin A for 90 days	Levels of total cholesterol (−13.65 mg/dL), low-density lipoprotein cholesterol (LDL-C) (−9.53 mg/dL), very low-density lipoprotein cholesterol (VLDL-C) (−1.75 mg/dL), and triglycerides (−11.51 mg/dL) significantly decreased. Calebin A reduced leptin and hs-CRP levels, indicating its beneficial effects in managing MetS-induced inflammation.	(Majeed et al. 2023)
7	Catechin 7-O-b-dapiofuranoside (C7A)	*Ulmus davidiana* (U60E)	Human placental pericytes and human retinal microvascular endothelial cells	In vitro	(40 μg/mL) for 72 h	C7A in U60E did not directly affect endothelial cell permeability but prevented pericyte apoptosis caused by high glucose and TNF-α levels, thereby maintaining endothelial barrier integrity.	(Kim et al. 2023)

8	Catechins Green tea extract (GTE)	Green tea	Persons with metabolic syndrome (MetS) and healthy persons	In vivo	890 mg/d total catechins	Plasma triglycerides, total cholesterol, HDL-C, and uric acid decreased over time, regardless of treatment group or health status. Endotoxin levels were reduced, and glucose control was enhanced.	(Zeng et al. 2024)
9	Chlorogenic acid (CA) and soy protein isolate (SPI)	Beijing Solarbio Company (Beijing, China)	ABTS (2,20-azino-bis [3-ethylbenzothiazo-line-6-sulphonic acid]) and DPPH (1,1-di-phe-nyl-2-picrylhydrazyl) assay	In vitro	0–100 μmol/g	Foaming, emulsification, and antioxidant properties were enhanced.	(Ma et al. 2024)
10	A combination of two radioprotectors, pterostilbene (PT) and silibinin (SIL), with two radiomitigators, nicotinamide riboside (NR) and fibroblast-stimulating lipoprotein 1(FSL-1)	Merck Chemicals GmbH, Darmstadt, Germany and Abmole Bioscience Inc., Houston, TX	Swiss albino mice	In vivo	185 mg/kg/day + 100 mg PT/kg and 100 mg SIL/kg	The combination of pterostilbene and silibinin, along with two radiomitigators – nicotinamide riboside and fibroblast-stimulating lipoprotein 1 – provided long-term protection to normal tissues against a lethal dose of gamma radiation in mice.	(Obrador et al. 2023)
11	Covalently linked soy protein isolate (SPI)- (–)-epigallo-catechin-3-gallat (EGCG)	Cool Chemistry Co., Ltd. (Beijing, China)	ABTS (2,20-azino-bis [3-ethylbenzothiazo-line-6-sulphonic acid]) and DPPH (1,1-di-phe-nyl-2-picrylhydrazyl) assay	In vitro		DPPH and ABTS radical scavenging rates of the adducts increased by 74.21% and 82.23%, respectively. Antioxidant activity and foaming properties were significantly improved.	(Zhao et al. 2020)

(Continued)

TABLE 10.1 (*Continued*)
Selected Cases on the Beneficial Activity of Polyphenols and Flavonoid Nutraceuticals as Evaluated Through Pharmacological Studies

Sl no.	Name of the Nutraceutical/ Polyphenol	Source	Target Organism	Mode/ Type of Experiment	Dose	Outcome of the Study	Reference
12	Curcumin	MCE (MedChem-Exp-ress) Company, USA	BALB/C-NU nude mice	In vivo	20, 40, and 60 μM	The growth of subcutaneously grafted tumors was significantly suppressed. The invasion and metastasis of liver cancer cells were prevented.	(Zhao et al. 2022)
13	EGCG-based keratinocyte growth factor loaded in situ hydrogel	Aladdin biochemical technology (Shanghai, China)	In vitro (mouse fibroblast cell line derived from NIH/ Swiss mice embryo [NIH3T3 cells]) In vivo (TNBS [trinitrobenzenesul-fonic] acid–induced colitis in rats)	In vivo and in vitro	In vitro EGCG concentration in hydrogels (0.01 %, 0.1%, and 1%) In vivo (400 μg/ kg, rectal administration)	Boosted viscosity and adhesive force. Increased stability. Helped manage body weight loss induced by TNBS in rats. Rebuilt the morphology of the colon and functions of goblet cells.	(Lin et al. 2023)
14	Ellagic acid, gallic acid, quercetin, panicalin, cyanidin, panicalagin Pomegranate peel extract	Pomegranate (*Punica granatum L.*)	Persons with non-alco-holic fatty liver disease (NAFLD)	In vivo	1500 mg/day for 8 weeks	Metabolic syndrome risk factors were improved, including lipid profile, fasting blood sugar, systolic and diastolic blood pressure, body weight, fat mass, waist circumference, and fatty liver status.	(Barghchi et al. 2023)

15	Epicatechin (EC) and chlorogenic acid (CA) non-covalently bound with ovotransferrin (OVT)	TCI (Shanghai) Development Co., Ltd. (Shanghai, China)	RAW 264.7 cells	In vitro	500 µg/mL	Cell viabilities escalated, indicating that these complexes had no toxic effects on macrophages. The ovotransferrin-CGA (10:2) complex revealed the steadiest structure and the strongest immunomodulatory activity.	(Tang et al. 2024)
16	Epigallocatechin gallate (EGCG)	Green tea	Wistar albino rats	In vivo	100mg/kg	Regulated the degeneration and hyperchromatism of hepatic cells.	(Neelamegam et al. 2022)
17	Epigallocatechin-3-gallate (EGCG) with doxorubicin (Dox)		Lung adenocarcinoma (LAC) cell line, A549	In vitro	Dox (10 µM)-EGCG (0.5 µM)	Mitigated Dox-induced toxicity and enhanced antineoplastic efficacy through oxidative damage mediation.	(Datta and Sinha 2022)
18	Gallic acid (GA), caffeic acid (CA), and tannic acid (TA)	Sigma-Aldrich Corporation (St. Louis, MO, USA)	Soybean protein (SPI) upon covalent conjugation with GA, CA, and TA under alkali treatment	In vitro		Western blot and ELISA showed that the SPI-polyphenol complex significantly reduced IgE binding capacity. The SPI-polyphenol complex also improved antioxidant capacity and emulsifying properties.	(Pi et al. 2023)
19	Ginsenoside-Rg1 (GRg1)	Ginseng	Human retinal microvascular endothelial cells (hRMECs) and rats	In vitro and in vivo	In vitro 10 µM 48 h In vivo 0.5 mL (5 g/mL)	GRg1 suppresses the TLR4/NF-κB signaling pathway by up-regulating miR-216a-5p, inhibiting cell proliferation and cell cycle progression, reducing growth factors and inflammatory cytokines, limiting angiogenesis in DR, and preventing retinal damage.	(Xue et al. 2023)
20	Hydroxytyrosol (HT) and soy protein isolate (SPI)	Shanghai Aladdin Biochemical Technology Co., Ltd. (Shanghai, China).	ABTS (2,20-azino-bis [3-ethylbenzothiazoline-6-sulphonic acid]) and DPPH (1,1-diphenyl-2-picrylhydrazyl) assay	In vitro	0–70 µmol/g	The adducts showed a 245.03% and 26.35% increase in DPPH and ABTS radical scavenging rates, respectively. Antioxidant activity and foaming properties were enhanced.	(Gu et al. 2024)

(Continued)

TABLE 10.1 (*Continued*)
Selected Cases on the Beneficial Activity of Polyphenols and Flavonoid Nutraceuticals as Evaluated Through Pharmacological Studies

Sl no.	Name of the Nutraceutical/ Polyphenol	Source	Target Organism	Mode/ Type of Experiment	Dose	Outcome of the Study	Reference
21	Lutein (LU) and *Trapa bispinosa* roxb. extract (TBE)	*Trapa bispinosa*	Male C57BL/ KsJ-db/ db mice	In vivo	LU (13.3 mg/kg) and TBE (133.3 mg/kg)	Improved retinal blood flow regulation in response to systemic hyperoxia and flicker stimulation, preventing glial dysfunction and vascular endothelial growth factor (VEGF) expression in type 2 diabetic mice.	(Hanaguri et al. 2022)
22	Lutein-rich purple sweet potato leaf extract	Sweet potato	Diabetic retinopathy (DR) of streptozotocin (STZ)-induced male Sprague-Dawley (SD) rats	In vivo	200 or 400 mg/ kg/day for 12 weeks	Lutein-rich PSPL supplementation normalized FBG, sodium, chloride, ALP, and insulin levels in diabetes-induced rats. Lutein provided protective effects by preserving the structural integrity of the kidney, liver, retina, and pancreas in the 400 mg/kg treated group. As a potent antioxidant, lutein restored ferric-reducing ability of plasma (FRAP), glutathione (GSH), and total antioxidant capacity (TAC) levels, while reducing IL-17A levels in the serum and retina of diabetic rats.	(Hisamud-din et al. 2023)
23	Lychee seed Polyphenol (LSP)	*Lycopene*	Male db/db mice and hRECs (human retinal endothelial cells) (BNC C340909)	In vivo and in vitro	In vivo: 50/100/200 mg/ kg In vitro: 8, 16, or 32 μg/mL	LSP alleviated DR by inhibiting the NLRP3 (nucleotide-binding domain, leucine-rich repeat-containing family, pyrin domain-containing 3) inflammasome, reducing cell apoptosis, and preventing abnormal retinal capillary neovascularization in db/db mice and HG-induced hRECs.	(Xiang et al. 2023)

24	Metallothionein (MT)–kidney bean polyphenol complex	*Phaseolus vulgaris* L.	Rats with type 2 diabetes mellitus (T2DM)	In vivo	2.4 mg/kg body weight	Body weight increased by 13.36%. Fasting blood glucose levels dropped by 27.86%. Serum blood glucose levels declined by 28.18%. Serum insulin levels rose by 24.60%. LDL-C significantly decreased by 42.95%. HDL-C significantly decreased by 256.94%.	(Zuo et al. 2023)
25	Myofibrillar protein–chlorogenic acid	Green leaf Biotech Firm, Jiangsu, China	Rats with type 2 diabetes mellitus (T2DM)	In vivo	2.65 mg CGA/10 mg myofibrillar protein	Restored gut microbiota in T2DM rats. Enhanced intestinal barrier function and reduced intestinal inflammation.	(Zhou et al. 2023)
26	Naringenin	Sigma-Aldrich, St. Louis, MO, USA	HG-induced DR (zebrafish)	In vivo	25, 50, or 75 µM 3 h post-fertilization to 5 days post-fertilization	NG reduced glucose levels and prevented cellular damage, inhibiting retina macular degeneration. NG also suppressed the overexpression of VEGF.	(Sudhakaran et al. 2023)
27	Nicotinamide mononucleotide (NMN) with resveratrol or ginsenoside Rh2 and Rg3	Resveratrol was obtained from Baolong Zhongcheng Biotechnology Co. Ltd. Ginsenosides Rg3 and Rh2 (including 35.7% (20R)-ginsenoside Rg3 and 43.6% (20R)-ginsenoside Rh2) were purchased from Daosifu Biotechnology Co. Ltd.	Male C57/BL6 mice	In vivo	Resveratrol (50 mg/kg), ginsenoside Rh2 and Rg3 (50 mg/kg)	The combination of NMN and resveratrol increased NAD^+ levels by 1.59-fold in the heart and 1.72-fold in skeletal muscle ($p < .001$ for both). Additionally, NAD^+ levels in lung tissue were significantly elevated by 1.97-fold when NMN was combined with Rg3 and Rh2 ($p < .001$).	(Bai et al. 2022)

(Continued)

TABLE 10.1 (*Continued*)
Selected Cases on the Beneficial Activity of Polyphenols and Flavonoid Nutraceuticals as Evaluated Through Pharmacological Studies

Sl no.	Name of the Nutraceutical/ Polyphenol	Source	Target Organism	Mode/ Type of Experiment	Dose	Outcome of the Study	Reference
28	Palm oil mill effluent–derived beta-carotene (PBC)	Palm Oil Mill Sdn. Bhd. in Penang, Malaysia	STZ-induced DR rats	In vivo	50 and 100 mg/kg for 21 consecutive days	PBC improved DR-associated visual acuity responses in the optomotor response (OMR) and visual-cue function (VCFT) tests while modulating STZ-induced biomarker changes, including increased GSH (reduced glutathione) and catalase activity levels, along with decreased TBARS levels in retinal tissue.	(Para-maswaran et al. 2023)
29	Phycocyanin Spirulina liquid extract	*Arthrospira*	Healthy persons	In vivo	20 mg per day for 12 weeks	Significantly altered the lipid profile. Exhibited beneficial effects on oxidative stress.	(Koite et al. 2022)
30	Pinoresi-nol-4-O-β-D-gluco-pyranoside (PG)	European plum	Rat model	In vivo and in vitro	50 mg/kg	Fall in serum glucose and lipid peroxides level. Indicated hepatoprotective and antihyperglyce-mic activity.	(Youssef et al. 2020)
31	Pinostrobin	*Boesenbergia rotunda*	Sprague Dawley (SPD) male rats	In vivo	40 mg/kg	Significantly reduced CFTR-dependent chloride secretion and slowed the development of kidney cysts.	(Tonum et al. 2022)
32	Quercetin		Rat model	In vivo	10 mg/kg	Expression of β-catenin and Bcl-2 proteins decreased, whereas caspase-3 expression significantly increased.	(Tezerji et al. 2022)
33	Quercetin	MedChemExpress, New Jersey, HY-18085, China	male BALB/c mice	In vivo	25 mg/kg, 50 mg/kg, 100 mg/kg	Tumor volume and weight were significantly reduced in the quercetin treatment group.	(Wu et al. 2022)

34	Quercetin-sericin hydrolysate and rutin-sericin hydrolysate		(RAW 264.7 macrophage cells)	In vitro		The anti-inflammatory activities were evaluated, revealing that quercetin-sericin hydrolysate and rutin-sericin hydrolysate exhibited significantly enhanced ($p < .05$) anti-inflammatory and emulsifying properties compared to unmodified sericin hydrolysate.	(Omar et al. 2022)
35	Resveratrol	Absin, abs815905	C57BL/6J mouse lineage	In vivo	40 mg/kg/day for 4 weeks	Elevated hippocampal concentrations of inflammatory mediators (IL-1β, IL-6, and TNF-α) were observed, alongside the inhibition of the Sirt1/NF-κB signaling pathway. Additionally, the compound demonstrated anti-inflammatory, anxiolytic, and antidepressant effects.	(Wei et al. 2023)
36	MR-SNC, a sericin-based nanocarrier, co-delivering both resveratrol and melatonin	Sigma-Aldrich (St. Louis, USA)	MCF-7 breast carcinoma cells	In vitro	500 and 1000 μg/mL	Sericin, melatonin, and resveratrol combined effects demonstrated synergistic antioxidant activity. Significant cell toxicity was exhibited at pH 6 across all tested concentrations.	(Aghaz et al. 2023)
37	Total flavonoids (quercetin-3-O-rutinoside, hyperoside, isoquercitrin, kaempferol-3-O-rutinoside, luteolin-5-O-glucoside, astragaloside, 3,5,7,3’,4’-pentahydroxyflavone, 3,4’,5,7-tetrahydroxyflavone) from *Chimonanthus nitens*	*Chimonanthus nitens*	Male Kunming mouse, aged 6 weeks	In vivo	200 mg/kg	Reduced triglyceride (TG), total fat (TF), and free fatty acid (FFA) levels by preventing fat accumulation, thereby alleviating lipid metabolism disorders and oxidative stress.	(Meng et al. 2022)

Future studies should prioritize enhancing the bioavailability of polyphenols through advanced delivery systems like nanoformulations, encapsulation, and conjugation with biocompatible carriers. Additionally, integrative approaches, including metabolomics, nutrigenomics, and microbiome analysis, are essential for understanding personalized responses to polyphenol consumption. Furthermore, additional studies are needed to confirm the effectiveness of specific polyphenols or polyphenol-rich formulations in disease prevention and treatment.

In conclusion, dietary polyphenols represent promising natural chemicals for health improvement and disease prevention. Their multifaceted roles and prevalence in the human diet render them indispensable for developing functional foods and nutraceuticals. Continually exploring their mechanisms of action, bioefficacy, and clinical applications will enable their integration into preventive and therapeutic strategies for many chronic illnesses.

REFERENCES

Abhishek, Mishra, Singla Rubal, Kumar Rohit, Joshi Rupa, Sarma Phulen, Kaur Gurjeet, Sharma Amit Raj, et al. 2022. "Neuroprotective Effect of the Standardised Extract of Bacopa Monnieri (BacoMind) in Valproic Acid Model of Autism Spectrum Disorder in Rats." *Journal of Ethnopharmacology* 293 (July): 115199. https://doi.org/10.1016/j.jep.2022.115199.

Aghaz, Faranak, Zahra Asadi, Soraya Sajadimajd, Khosrow Kashfi, Elham Arkan, and Zohreh Rahimi. 2023. "Codelivery of Resveratrol Melatonin Utilizing pH Responsive Sericin Based Nanocarriers Inhibits the Proliferation of Breast Cancer Cell Line at the Different pH." *Scientific Reports* 13 (1): 11090. https://doi.org/10.1038/s41598-023-37668-y.

Ahn, Tae-Gue, Gabsik Yang, Heon-Myung Lee, Myung-Dong Kim, Ho-Young Choi, Kyoung-Sik Park, Sun-Dong Lee, Yoon-Bum Kook, and Hyo-Jin An. 2013. "Molecular Mechanisms Underlying the Anti-Obesity Potential of Prunetin, an O-Methylated Isoflavone." *Biochemical Pharmacology* 85 (10): 1525–33. https://doi.org/10.1016/j.bcp.2013.02.020.

Al-Regaiey, Khalid. 2024. "Crosstalk between Adipogenesis and Aging: Role of Polyphenols in Combating Adipogenic-Associated Aging." *Immunity & Ageing* 21 (1): 76. https://doi.org/10.1186/s12979-024-00481-w.

Anhê, Fernando F., Denis Roy, Geneviève Pilon, Stéphanie Dudonné, Sébastien Matamoros, Thibault V. Varin, Carole Garofalo, et al. 2015. "A Polyphenol-Rich Cranberry Extract Protects from Diet-Induced Obesity, Insulin Resistance and Intestinal Inflammation in Association with Increased *Akkermansia* Spp. Population in the Gut Microbiota of Mice." *Gut* 64 (6): 872–83. https://doi.org/10.1136/gutjnl-2014-307142.

Bai, Long-Bo, Lee-Fong Yau, Tian-Tian Tong, Wai-Him Chan, Wei Zhang, and Zhi-Hong Jiang. 2022. "Improvement of Tissue-specific Distribution and Biotransformation Potential of Nicotinamide Mononucleotide in Combination with Ginsenosides or Resveratrol." *Pharmacology Research & Perspectives* 10 (4): e00986. https://doi.org/10.1002/prp2.986.

Barghchi, Hanieh, Narges Milkarizi, Saba Belyani, Andisheh Norouzian Ostad, Vahid Reza Askari, Farnood Rajabzadeh, Ladan Goshayeshi, et al. 2023. "Pomegranate (*Punica Granatum* L.) Peel Extract Ameliorates Metabolic Syndrome Risk Factors in Patients with Non-Alcoholic Fatty Liver Disease: A Randomized Double-Blind Clinical Trial." *Nutrition Journal* 22 (1): 40. https://doi.org/10.1186/s12937-023-00869-2.

Bertelli, Alberto, Marco Biagi, Maddalena Corsini, Giulia Baini, Giorgio Cappellucci, and Elisabetta Miraldi. 2021. "Polyphenols: From Theory to Practice." *Foods* 10 (11): 2595. https://doi.org/10.3390/foods10112595.

Bravo, Laura. 2009. "Polyphenols: Chemistry, Dietary Sources, Metabolism, and Nutritional Significance." *Nutrition Reviews* 56 (11): 317–33. https://doi.org/10.1111/j.1753-4887.1998.tb01670.x.

Buddington, R. K., and P. T. Sangild. 2011. "Companion Animals Symposium: Development of the Mammalian Gastrointestinal Tract, the Resident Microbiota, and the Role of Diet in Early Life1." *Journal of Animal Science* 89 (5): 1506–19. https://doi.org/10.2527/jas.2010-3705.

Caponio, Giusy, Mirella Noviello, Francesco Calabrese, Giuseppe Gambacorta, Gianluigi Giannelli, and Maria De Angelis. 2022. "Effects of Grape Pomace Polyphenols and In Vitro Gastrointestinal Digestion on Antimicrobial Activity: Recovery of Bioactive Compounds." *Antioxidants* 11 (3): 567. https://doi.org/10.3390/antiox11030567.

Chang, Chia, Thing-Fong Tzeng, Shorong-Shii Liou, Yuan-Shiun Chang, and I-Min Liu. 2011. "Kaempferol Regulates the Lipid-Profile in High-Fat Diet-Fed Rats through an Increase in Hepatic PPAR α Levels." *Planta Medica* 77 (17): 1876–82. https://doi.org/10.1055/s-0031-1279992.

Datta, Suchisnigdha, and Dona Sinha. 2022. "Low Dose Epigallocatechin-3-gallate Revives Doxorubicin Responsiveness by a Redox-sensitive Pathway in A549 Lung Adenocarcinoma Cells." *Journal of Biochemical and Molecular Toxicology* 36 (4): e22999. https://doi.org/10.1002/jbt.22999.

Del Valle, José Carlos, Ma Luisa Buide, Justen B. Whittall, Fernando Valladares, and Eduardo Narbona. 2020. "UV Radiation Increases Phenolic Compound Protection but Decreases Reproduction in Silene Littorea." Edited by Mayank Gururani. *PLOS ONE* 15 (6): e0231611. https://doi.org/10.1371/journal.pone.0231611.

Dini, Irene, and Lucia Grumetto. 2022. "Recent Advances in Natural Polyphenol Research." *Molecules* 27 (24): 8777. https://doi.org/10.3390/molecules27248777.

Ferré, P., and F. Foufelle. 2010. "Hepatic Steatosis: A Role for *de Novo* Lipogenesis and the Transcription Factor SREBP-1c." *Diabetes, Obesity and Metabolism* 12 (s2): 83–92. https://doi.org/10.1111/j.1463-1326.2010.01275.x.

Gao, Jinhua, Hui Yu, Weikang Guo, Ying Kong, Lina Gu, Qi Li, Shanshan Yang, Yunyan Zhang, and Yaoxian Wang. 2018. "The Anticancer Effects of Ferulic Acid is Associated with Induction of Cell Cycle Arrest and Autophagy in Cervical Cancer Cells." *Cancer Cell International* 18 (1): 102. https://doi.org/10.1186/s12935-018-0595-y.

Gerritsen, Jacoline, Hauke Smidt, Ger T. Rijkers, and Willem M. De Vos. 2011. "Intestinal Microbiota in Human Health and Disease: The Impact of Probiotics." *Genes & Nutrition* 6 (3): 209–40. https://doi.org/10.1007/s12263-011-0229-7.

Goldwasser, Jonathan, Pazit Y. Cohen, Eric Yang, Patrick Balaguer, Martin L. Yarmush, and Yaakov Nahmias. 2010. "Transcriptional Regulation of Human and Rat Hepatic Lipid Metabolism by the Grapefruit Flavonoid Naringenin: Role of PPARα, PPARγ and LXRα." Edited by Catherine A. Wolkow. *PLoS ONE* 5 (8): e12399. https://doi.org/10.1371/journal.pone.0012399.

Goto, Tsuyoshi, Aki Teraminami, Joo-Young Lee, Kana Ohyama, Kozue Funakoshi, Young-Il Kim, Shizuka Hirai, et al. 2012. "Tiliroside, a Glycosidic Flavonoid, Ameliorates Obesity-Induced Metabolic Disorders via Activation of Adiponectin Signaling Followed by Enhancement of Fatty Acid Oxidation in Liver and Skeletal Muscle in Obese–Diabetic Mice." *The Journal of Nutritional Biochemistry* 23 (7): 768–76. https://doi.org/10.1016/j.jnutbio.2011.04.001.

Grabska-Kobyłecka, Izabela, Piotr Szpakowski, Aleksandra Król, Dominika Książek-Winiarek, Andrzej Kobyłecki, Andrzej Głąbiński, and Dariusz Nowak. 2023. "Polyphenols and their Impact on the Prevention of Neurodegenerative Diseases and Development." *Nutrients* 15 (15): 3454. https://doi.org/10.3390/nu15153454.

Grzesik, Michalina, Katarzyna Naparło, Grzegorz Bartosz, and Izabela Sadowska-Bartosz. 2018. "Antioxidant Properties of Catechins: Comparison with Other Antioxidants." *Food Chemistry* 241 (February): 480–92. https://doi.org/10.1016/j.foodchem.2017.08.117.

Gu, Meiyu, Jiahui Shi, Boya Zhang, Xibo Wang, and Xu Wang. 2024. "Covalent Modification of Soy Protein Isolates by Hydroxytyrosol: Effects on Structural and Functional Properties of Adducts." *LWT* 198 (April): 116041. https://doi.org/10.1016/j.lwt.2024.116041.

Guillermo Gormaz, Juan, Nicolas Valls, Camilo Sotomayor, Thomas Turner, and Ramón Rodrigo. 2016. "Potential Role of Polyphenols in the Prevention of Cardiovascular Diseases: Molecular Bases." *Current Medicinal Chemistry* 23 (2): 115–28. https://doi.org/10.2174/0929867323666151127201732.

Hanaguri, Junya, Harumasa Yokota, Akifumi Kushiyama, Sakura Kushiyama, Masahisa Watanabe, Satoru Yamagami, and Taiji Nagaoka. 2022. "Beneficial Effect of Long-Term Administration of Supplement with Trapa Bispinosa Roxb. and Lutein on Retinal Neurovascular Coupling in Type 2 Diabetic Mice." *Frontiers in Physiology* 13 (February): 788034. https://doi.org/10.3389/fphys.2022.788034.

Hisamuddin, Ahmad Safiyyu'd-din Bin, Ruth Naomi, Khairul Aiman Bin Manan, Hasnah Bahari, Fezah Othman, Hashim Embong, Amin Ismail, et al. 2023. "The Role of Lutein-Rich Purple Sweet Potato Leaf Extract on the Amelioration of Diabetic Retinopathy in Streptozotocin-Induced Sprague–Dawley Rats." *Frontiers in Pharmacology* 14 (May): 1175907. https://doi.org/10.3389/fphar.2023.1175907.

Jia, Yaoyao, Jin-Young Kim, Hee-jin Jun, Sun-Joong Kim, Ji-Hae Lee, Minh Hien Hoang, Hyun Sook Kim, et al. 2013. "Cyanidin is an Agonistic Ligand for Peroxisome Proliferator-Activated Receptor-Alpha Reducing Hepatic Lipid." *Biochimica et Biophysica Acta (BBA) – Molecular and Cell Biology of Lipids* 1831 (4): 698–708. https://doi.org/10.1016/j.bbalip.2012.11.012.

Jiménez-Flores, Lizbeth, Sergio López-Briones, Maciste Macías-Cervantes, Joel Ramírez-Emiliano, and Victoriano Pérez-Vázquez. 2014. "A PPARγ, NF-κB and AMPK-Dependent Mechanism May Be Involved in the Beneficial Effects of Curcumin in the Diabetic Db/Db Mice Liver." *Molecules* 19 (6): 8289–302. https://doi.org/10.3390/molecules19068289.

Karlsson, Fredrik H., Valentina Tremaroli, Intawat Nookaew, Göran Bergström, Carl Johan Behre, Björn Fagerberg, Jens Nielsen, and Fredrik Bäckhed. 2013. "Gut Metagenome in European Women with Normal, Impaired and Diabetic Glucose Control." *Nature* 498 (7452): 99–103. https://doi.org/10.1038/nature12198.

Kim, Iljin, Jieun Seo, Dong Hyun Lee, Yo-Han Kim, Jun-Hyung Kim, Myung-Bok Wie, Jun-Kyu Byun, and Jang-Hyuk Yun. 2023. "Ulmus Davidiana 60% Edible Ethanolic Extract for Prevention of Pericyte Apoptosis in Diabetic Retinopathy." *Frontiers in Endocrinology* 14 (May): 1138676. https://doi.org/10.3389/fendo.2023.1138676.

Koite, N'Deye Lallah Nina, N'gouro Issa Sanogo, Olivier Lépine, Jean-Marie Bard, and Khadija Ouguerram. 2022. "Antioxidant Efficacy of a Spirulina Liquid Extract on Oxidative Stress Status and Metabolic Disturbances in Subjects with Metabolic Syndrome." *Marine Drugs* 20 (7): 441. https://doi.org/10.3390/md20070441.

Larsen, Nadja, Finn K. Vogensen, Frans W. J. Van Den Berg, Dennis Sandris Nielsen, Anne Sofie Andreasen, Bente K. Pedersen, Waleed Abu Al-Soud, Søren J. Sørensen, Lars H. Hansen, and Mogens Jakobsen. 2010. "Gut Microbiota in Human Adults with Type 2 Diabetes Differs from Non-diabetic Adults." Edited by Stefan Bereswill. *PLoS ONE* 5 (2): e9085. https://doi.org/10.1371/journal.pone.0009085.

Lee, Joo-Won, Sung Sik Choe, Hagoon Jang, Jiyeong Kim, Hyun Woo Jeong, Hyunsun Jo, Kyeong-Hoon Jeong, et al. 2012. "AMPK Activation with Glabridin Ameliorates Adiposity and Lipid Dysregulation in Obesity." *Journal of Lipid Research* 53 (7): 1277–86. https://doi.org/10.1194/jlr.M022897.

Li, Sha, Hor Yue Tan, Ning Wang, Fan Cheung, Ming Hong, and Yibin Feng. 2018. "The Potential and Action Mechanism of Polyphenols in the Treatment of Liver Diseases." Edited by Sharad S. Singhal. *Oxidative Medicine and Cellular Longevity* 2018 (1): 8394818. https://doi.org/10.1155/2018/8394818.

Lin, Gaolong, Jiaojiao Yang, Jiayi Liu, Jianxun Shangguan, Hanxiao Pan, Yingying Zhang, Kunjie Ran, Dingwei Li, Fengnan Yu, and Helin Xu. 2023. "In Situ Polyphenol-Adhesive Hydrogel Enhanced the Noncarcinogenic Repairing of KGF on the Gut Epithelial Barrier on TNBS-Induced Colitis Rats." *International Journal of Biological Macromolecules* 231 (March): 123323. https://doi.org/10.1016/j.ijbiomac.2023.123323.

Liu, Jin-Feng, Ying Ma, Ying Wang, Zhi-Yan Du, Jing-Kang Shen, and Hong-Li Peng. 2011. "Reduction of Lipid Accumulation in HepG2 Cells by Luteolin is Associated with Activation of AMPK and Mitigation of Oxidative Stress." *Phytotherapy Research* 25 (4): 588–96. https://doi.org/10.1002/ptr.3305.

Luo, Cheng, Hui Yang, Chengyong Tang, Gaoqiong Yao, Lingxi Kong, Haixia He, and Yuanda Zhou. 2015. "Kaempferol Alleviates Insulin Resistance via Hepatic IKK/NF-κB Signal in Type 2 Diabetic Rats." *International Immunopharmacology* 28 (1): 744–50. https://doi.org/10.1016/j.intimp.2015.07.018.

Ma, Nan, Jiahui Duan, Guowei Zhou, and Xibo Wang. 2024. "Study of the Mechanism of Non-Covalent Interactions between Chlorogenic Acid and Soy Protein Isolate: Multi-Spectroscopic, in Vitro, and Computational Docking Analyses." *Food Chemistry* 457 (November): 140084. https://doi.org/10.1016/j.foodchem.2024.140084.

Majeed, Muhammed, Kalyanam Nagabhushanam, T. V. Devarajan, Santhosh Saklecha, S. Venkata Krishna Reddy, and Lakshmi Mundkur. 2023. "A Minor Metabolite from *Curcuma Longa* Effective against Metabolic Syndrome: Results from a Randomized, Double-Blind, Placebo-Controlled Clinical Study." *Food & Function* 14 (10): 4722–33. https://doi.org/10.1039/D2FO03627D.

Medjakovic, Svjetlana, Monika Mueller, and Alois Jungbauer. 2010. "Potential Health-Modulating Effects of Isoflavones and Metabolites via Activation of PPAR and AhR." *Nutrients* 2 (3): 241–79. https://doi.org/10.3390/nu2030241.

Meng, Wenya, Zitong Zhao, Lingli Chen, Suyun Lin, Yang Zhang, Jing He, Kehui Ouyang, and Wenjun Wang. 2022. "Total Flavonoids from Chimonanthus Nitens Oliv. Leaves Ameliorate HFD-Induced NAFLD by Regulating the Gut–Liver Axis in Mice." *Foods* 11 (14): 2169. https://doi.org/10.3390/foods11142169.

Michelotti, Gregory A., Mariana V. Machado, and Anna Mae Diehl. 2013. "NAFLD, NASH and Liver Cancer." *Nature Reviews Gastroenterology & Hepatology* 10 (11): 656–65. https://doi.org/10.1038/nrgastro.2013.183.

Muntaha, Sidra Tul, Allah Rakha, Hina Rasheed, Iraj Fatima, Masood Sadiq Butt, Gholam-Reza Abdi, and Rana Muhammad Aadil. 2025. "Polyphenol-Protein Particles: A Nutraceutical Breakthrough in Nutrition and Food Science." *Journal of Agriculture and Food Research* 19 (March): 101641. https://doi.org/10.1016/j.jafr.2025.101641.

Neelamegam, U., R. Muthuvel, and V. Suganthi. 2022. "Hepatoprotective Effect of Epigallocatechin-Gallate (Egcg) and Sorafenib against Den Induced Hepato Cellular Carcinoma in Experimental Animals." *Journal of Pharmaceutical Negative Results* 13: 921–9.

Noh, Jeong Sook, Chan Hum Park, and Takako Yokozawa. 2011. "Treatment with Oligonol, a Low-Molecular Polyphenol Derived from Lychee Fruit, Attenuates Diabetes-Induced Hepatic Damage through Regulation of Oxidative Stress and Lipid Metabolism." *British Journal of Nutrition* 106 (7): 1013–22. https://doi.org/10.1017/S0007114511001322.

Obrador, Elena, Rosario Salvador-Palmer, Blanca Pellicer, Rafael López-Blanch, J. Antoni Sirerol, Juan I. Villaescusa, Alegría Montoro, Ryan W. Dellinger, and José M. Estrela. 2023. "Combination of Natural Polyphenols with a Precursor of NAD+ and a TLR2/6 Ligand Lipopeptide Protects Mice against Lethal γ Radiation." *Journal of Advanced Research* 45 (March): 73–86. https://doi.org/10.1016/j.jare.2022.05.005.

Omar, Adil, Ahmidin Wali, Amina Arken, Yanhua Gao, Haji Akber Aisa, and Abulimiti Yili. 2022. "Covalent Binding of Flavonoids with Silk Sericin Hydrolysate: Anti-inflammatory, Antioxidant, and Physicochemical Properties of Flavonoid–Sericin Hydrolysate Conjugates." *Journal of Food Biochemistry* 46 (7). https://doi.org/10.1111/jfbc.14125.

Ozleyen, Adem, Zeynep Ozlem Cinar, Sercan Karav, Ayse Bayraktar, Aysenur Arslan, H. Mehmet Kayili, Bekir Salih, and Tugba Boyunegmez Tumer. 2022. "Biofortified Whey/Deglycosylated Whey and Chickpea Protein Matrices: Functional Enrichment by Black Mulberry Polyphenols." *Plant Foods for Human Nutrition* 77 (1): 51–61. https://doi.org/10.1007/s11130-021-00943-2.

Paramaswaran, Yamunna, Aswinprakash Subramanian, Nallupillai Paramakrishnan, Muthusamy Ramesh, and Arunachalam Muthuraman. 2023. "Therapeutic Investigation of Palm Oil Mill Effluent-Derived Beta-Carotene in Streptozotocin-Induced Diabetic Retinopathy via the Regulation of Blood–Retina Barrier Functions." *Pharmaceuticals* 16 (5): 647. https://doi.org/10.3390/ph16050647.

Pi, Xiaowen, Jiafei Liu, Yuxue Sun, Qingfeng Ban, Jianjun Cheng, and Mingruo Guo. 2023. "Protein Modification, IgE Binding Capacity, and Functional Properties of Soybean Protein upon Conjugation with Polyphenols." *Food Chemistry* 405 (March): 134820. https://doi.org/10.1016/j.foodchem.2022.134820.

Pistollato, Francesca, Sandra Sumalla Cano, Iñaki Elio, Manuel Masias Vergara, Francesca Giampieri, and Maurizio Battino. 2016. "Role of Gut Microbiota and Nutrients in Amyloid Formation and Pathogenesis of Alzheimer Disease." *Nutrition Reviews* 74 (10): 624–34. https://doi.org/10.1093/nutrit/nuw023.

Porras, David, Esther Nistal, Susana Martínez-Flórez, Sandra Pisonero-Vaquero, José Luis Olcoz, Ramiro Jover, Javier González-Gallego, María Victoria García-Mediavilla, and Sonia Sánchez-Campos. 2017. "Protective Effect of Quercetin on High-Fat Diet-Induced Non-Alcoholic Fatty Liver Disease in Mice is Mediated by Modulating Intestinal Microbiota Imbalance and Related Gut-Liver Axis Activation." *Free Radical Biology and Medicine* 102 (January): 188–202. https://doi.org/10.1016/j.freeradbiomed.2016.11.037.

Queipo-Ortuño, María Isabel, María Boto-Ordóñez, Mora Murri, Juan Miguel Gomez-Zumaquero, Mercedes Clemente-Postigo, Ramon Estruch, Fernando Cardona Diaz, Cristina Andrés-Lacueva, and Francisco J Tinahones. 2012. "Influence of Red Wine Polyphenols and Ethanol on the Gut Microbiota Ecology and Biochemical Biomarkers." *The American Journal of Clinical Nutrition* 95 (6): 1323–34. https://doi.org/10.3945/ajcn.111.027847.

Rahman, Md. Mominur, Md. Saidur Rahaman, Md. Rezaul Islam, Firoza Rahman, Faria Mannan Mithi, Taha Alqahtani, Mohannad A. Almikhlafi, et al. 2021. "Role of Phenolic Compounds in Human Disease: Current Knowledge and Future Prospects." *Molecules* 27 (1): 233. https://doi.org/10.3390/molecules27010233.

Raina, Jeevika, Atika Firdous, Gurvinder Singh, Rajesh Kumar, and Charanjit Kaur. 2024. "Role of Polyphenols in the Management of Diabetic Complications." *Phytomedicine* 122 (January): 155155. https://doi.org/10.1016/j.phymed.2023.155155.

Ramos-Tovar, Erika, and Pablo Muriel. 2020. "Free Radicals, Antioxidants, Nuclear Factor-E2-related Factor-2 and Liver Damage." *Journal of Applied Toxicology* 40 (1): 151–68. https://doi.org/10.1002/jat.3880.

Ranneh, Yazan, Alaa S. Bedir, Abdelghafar M. Abu-Elsaoud, and Seham Al Raish. 2024. "Polyphenol Intervention Ameliorates Non-alcoholic Fatty Liver Disease: An Updated Comprehensive Systematic Review." *Nutrients* 16 (23): 4150. https://doi.org/10.3390/nu16234150.

Scalbert, Augustin, and Gary Williamson. 2000. "Dietary Intake and Bioavailability of Polyphenols." *The Journal of Nutrition* 130 (8): 2073S–85S. https://doi.org/10.1093/jn/130.8.2073S.

Shin, Eui Seok, Hyoung Ho Lee, Si Young Cho, Hyun Woo Park, Sang Jun Lee, and Tae Ryong Lee. 2007. "Genistein Downregulates SREBP-1 Regulated Gene Expression by Inhibiting Site-1 Protease Expression in HepG2 Cells." *The Journal of Nutrition* 137 (5): 1127–31. https://doi.org/10.1093/jn/137.5.1127.

Singh, Amritpal, Yu Fung Yau, Kin Sum Leung, Hani El-Nezami, and Jetty Chung-Yung Lee. 2020. "Interaction of Polyphenols as Antioxidant and Anti-inflammatory Compounds in Brain–Liver–Gut Axis." *Antioxidants* 9 (8): 669. https://doi.org/10.3390/antiox9080669.

Slobodníková, Lívia, Silvia Fialová, Katarína Rendeková, Ján Kováč, and Pavel Mučaji. 2016. "Antibiofilm Activity of Plant Polyphenols." *Molecules* 21 (12): 1717. https://doi.org/10.3390/molecules21121717.

Song, Dong Hyun, and Jie-Oh Lee. 2012. "Sensing of Microbial Molecular Patterns by Toll-like Receptors." *Immunological Reviews* 250 (1): 216–29. https://doi.org/10.1111/j.1600-065X.2012.01167.x.

Subramanian, Aswinprakash, Jayaraman Thirunavukkarasu, and Arunachalam Muthuraman. 2022. "Astaxanthin Ameliorates Diabetic Retinopathy in Swiss Albino Mice via Inhibitory Processes of Neuron-Specific Enolase Activity." *Processes* 10 (7): 1318. https://doi.org/10.3390/pr10071318.

Sudhakaran, Gokul, Abhirami Chandran, A. R. Sreekutty, S. Madesh, Raman Pachaiappan, Bader O. Almutairi, Selvaraj Arokiyaraj, et al. 2023. "Ophthalmic Intervention of Naringenin Decreases Vascular Endothelial Growth Factor by Counteracting Oxidative Stress and Cellular Damage in In Vivo Zebrafish." *Molecules* 28 (14): 5350. https://doi.org/10.3390/molecules28145350.

Tang, Tingting, Yuanqi Lv, Yujie Su, Junhua Li, Luping Gu, Yanjun Yang, and Cuihua Chang. 2024. "The Differential Non-Covalent Binding of Epicatechin and Chlorogenic Acid to Ovotransferrin and the Enhancing Efficiency of Immunomodulatory Activity." *International Journal of Biological Macromolecules* 259 (February): 129298. https://doi.org/10.1016/j.ijbiomac.2024.129298.

Tezerji, Sajjad, Fatemeh Nazari Robati, Hamid Abdolazimi, Azadeh Fallah, and Behrouz Talaei. 2022. "Quercetin's Effects on Colon Cancer Cells Apoptosis and Proliferation in a Rat Model of Disease." *Clinical Nutrition ESPEN* 48 (April): 441–5. https://doi.org/10.1016/j.clnesp.2022.01.004.

Tonum, Kanlayanee, Napason Chabang, Somsak Fongsupa, Suphat Chantawarin, Chutima Jiarpinitnun, Patoomrattana Tuchinda, and Sunhapas Soodvilai. 2022. "Pinostrobin Inhibits Renal CFTR-Mediated Cl– Secretion and Retards Cyst Growth in Cell-Derived Cyst and Polycystic Kidney Disease Rats." *Journal of Pharmacological Sciences* 148 (4): 369–76. https://doi.org/10.1016/j.jphs.2022.02.003.

Van De Wier, Bregje, Ger H. Koek, Aalt Bast, and Guido R. M. M. Haenen. 2017. "The Potential of Flavonoids in the Treatment of Non-Alcoholic Fatty Liver Disease." *Critical Reviews in Food Science and Nutrition* 57 (4): 834–55. https://doi.org/10.1080/10408398.2014.952399.

Vinson, Joe A., Xuehui Su, Ligia Zubik, and Pratima Bose. 2001. "Phenol Antioxidant Quantity and Quality in Foods: Fruits." *Journal of Agricultural and Food Chemistry* 49 (11): 5315–21. https://doi.org/10.1021/jf0009293.

Wang, Xing, Dong-Mei Zhang, Ting-Ting Gu, Xiao-Qin Ding, Chen-Yu Fan, Qin Zhu, Yun-Wei Shi, Ye Hong, and Ling-Dong Kong. 2013. "Morin Reduces Hepatic Inflammation-Associated Lipid Accumulation in High Fructose-Fed Rats via Inhibiting Sphingosine Kinase 1/Sphingosine 1-Phosphate Signaling Pathway." *Biochemical Pharmacology* 86 (12): 1791–804. https://doi.org/10.1016/j.bcp.2013.10.005.

Wang, Ying, Gurpreet Kaur, Manish Kumar, Ajay Singh Kushwah, Atul Kabra, and Ritu Kainth. 2022. "Caffeic Acid Prevents Vascular Oxidative Stress and Atherosclerosis against Atherosclerogenic Diet in Rats." Edited by Ruchika Garg. *Evidence-Based Complementary and Alternative Medicine* 2022 (January): 1–8. https://doi.org/10.1155/2022/8913926.

Wang, Zhongjiang, Bailiang Li, Zengbo Wang, Jian Li, Lei Xing, Linyi Zhou, Xiaoshuang Zou, and Hua Yin. 2022. "Analysis of Structural Changes and Anti-inflammatory Capacity in Soybean Protein Isolates Conjugated with Anthocyanins." *Food Science and Technology* 42: e07922. https://doi.org/10.1590/fst.07922.

Wei, Ru-Meng, Yue-Ming Zhang, Yi-Zhou Feng, Kai-Xuan Zhang, Jing-Ya Zhang, Jing Chen, Bao-Ling Luo, Xue-Yan Li, and Gui-Hai Chen. 2023. "Resveratrol Ameliorates Maternal Separation-induced Anxiety- and Depression-like Behaviors and Reduces Sirt1-NF-kB Signaling-mediated Neuroinflammation." *Frontiers in Behavioral Neuroscience* 17 (May): 1172091. https://doi.org/10.3389/fnbeh.2023.1172091.

Williamson, G. 2017. "The Role of Polyphenols in Modern Nutrition." *Nutrition Bulletin* 42 (3): 226–35. https://doi.org/10.1111/nbu.12278.

Wu, Cheng-Hsun, Ming-Cheng Lin, Hsueh-Chun Wang, Mon-Yuan Yang, Ming-Jia Jou, and Chau-Jong Wang. 2011. "Rutin Inhibits Oleic Acid Induced Lipid Accumulation *via* Reducing Lipogenesis and Oxidative Stress in Hepatocarcinoma Cells." *Journal of Food Science* 76 (2). https://doi.org/10.1111/j.1750-3841.2010.02033.x.

Wu, Ruoxia, Ting Zhou, Jiaqing Xiong, Zhen Zhang, Sha Tian, Yongli Wang, Jindong Chen, and Xuefei Tian. 2022. "Quercetin, the Ingredient of Xihuang Pills, Inhibits Hepatocellular Carcinoma by Regulating Autophagy and Macrophage Polarization." *Frontiers in Bioscience-Landmark* 27 (12): 323. https://doi.org/10.31083/j.fbl2712323.

Xiang, Xiao-Hong, Jing Wei, Xiao-Fang Wang, Qin Xu, Chong-Lin Yu, Chang-Long He, Tao Long, et al. 2023. "Lychee Seed Polyphenol Ameliorates DR via Inhibiting Inflammasome/Apoptosis and Angiogenesis in hRECs and Db/Db Mice." *Biomedicine & Pharmacotherapy* 167 (November): 115478. https://doi.org/10.1016/j.biopha.2023.115478.

Xue, Liping, Min Hu, Qin Zhu, Yadi Li, Guanglong Zhou, Xiaofan Zhang, Yuan Zhou, Jieying Zhang, and Peng Ding. 2023. "GRg1 Inhibits the TLR4/NF-kB Signaling Pathway by Upregulating miR-216a-5p to Reduce Growth Factors and Inflammatory Cytokines in DR." *Molecular Biology Reports* 50 (11): 9379–94. https://doi.org/10.1007/s11033-023-08895-3.

Youssef, Fadia S., Mohamed L. Ashour, Hesham A. El-Beshbishy, Alaaeldin Ahmed Hamza, Abdel Nasser B. Singab, and Michael Wink. 2020. "Pinoresinol-4-*O* – *β*-D-Glucopyranoside: A Lignan from Prunes (*Prunus Domestica*) Attenuates Oxidative Stress, Hyperglycaemia and Hepatic Toxicity *in Vitro* and *in Vivo*." *Journal of Pharmacy and Pharmacology* 72 (12): 1830–9. https://doi.org/10.1111/jphp.13358.

Zeng, Lu, Wai J. Tang, Jin J. Yin, and Bei J. Zhou. 2014. "Signal Transductions and Nonalcoholic Fatty Liver: A Mini-Review." *International Journal of Clinical and Experimental Medicine* 7 (7): 1624–31.

Zeng, Min, Joanna K. Hodges, Avinash Pokala, Mona Khalafi, Geoffrey Y. Sasaki, Jillian Pierson, Sisi Cao, et al. 2024. "A Green Tea Extract Confection Decreases Circulating Endotoxin and Fasting Glucose by Improving Gut Barrier Function but without Affecting Systemic Inflammation: A Double-Blind, Placebo-Controlled Randomized Trial in Healthy Adults and Adults with Metabolic Syndrome." *Nutrition Research* 124 (April): 94–110. https://doi.org/10.1016/j.nutres.2024.02.001.

Zhang, F.-P., Q.-Y. Yang, and S.-B. Zhang. 2016. Dual Effect of Phenolic Nectar on Three Floral Visitors of Elsholtzia rugulosa (Lamiaceae) in SW China. *PLoS ONE* 11 (4): e0154381. https://doi.org/10.1371/journal.pone.0154381

Zhang, Zhiheng, Xiaojing Li, Shangyuan Sang, David Julian McClements, Long Chen, Jie Long, Aiquan Jiao, Zhengyu Jin, and Chao Qiu. 2022. "Polyphenols as Plant-Based Nutraceuticals: Health Effects, Encapsulation, Nano-Delivery, and Application." *Foods* 11 (15): 2189. https://doi.org/10.3390/foods11152189.

Zhao, Yue, Xu Wang, Dan Li, Honglin Tang, Dianyu Yu, Liqi Wang, and Lianzhou Jiang. 2020. "Effect of Anionic Polysaccharides on Conformational Changes and Antioxidant Properties of Protein-Polyphenol Binary Covalently-Linked Complexes." *Process Biochemistry* 89 (February): 89–97. https://doi.org/10.1016/j.procbio.2019.10.021.

Zhao, Zhongwei, Jielin Su, Jiaqi Zhao, Jiaxin Chen, Xinmu Cui, Manqing Sun, and Xuewu Zhang. 2022. "Curcumin Inhibits Invasion and Metastasis of Human Hepatoma Cells through Bclaf1-Mediated Wnt/β-Catenin Signalling." *Food and Agricultural Immunology* 33 (1): 664–76. https://doi.org/10.1080/09540105.2022.2113864.

Zhou, Zhiwei, Dan Wang, Xinyi Xu, Jin Dai, Guangjie Lao, Senlin Zhang, Xiaofang Xu, András Dinnyés, Youling Xiong, and Qun Sun. 2023. "Myofibrillar Protein-Chlorogenic Acid Complexes Ameliorate Glucose Metabolism via Modulating Gut Microbiota in a Type 2 Diabetic Rat Model." *Food Chemistry* 409 (May): 135195. https://doi.org/10.1016/j.foodchem.2022.135195.

Zhu, Baoli, Xin Wang, and Lanjuan Li. 2010. "Human Gut Microbiome: The Second Genome of Human Body." *Protein & Cell* 1 (8): 718–25. https://doi.org/10.1007/s13238-010-0093-z.

Ziółkiewicz, Agnieszka, Kamila Kasprzak-Drozd, Robert Rusinek, Ewa Markut-Miotła, and Anna Oniszczuk. 2023. "The Influence of Polyphenols on Atherosclerosis Development." *International Journal of Molecular Sciences* 24 (8): 7146. https://doi.org/10.3390/ijms24087146.

Zuo, Zhaohang, Weiqiao Pang, Wei Sun, Baoxin Lu, Liang Zou, Dongjie Zhang, and Ying Wang. 2023. "Metallothionein–Kidney Bean Polyphenol Complexes Showed Antidiabetic Activity in Type 2 Diabetic Rats by Improving Insulin Resistance and Regulating Gut Microbiota." *Foods* 12 (16): 3139. https://doi.org/10.3390/foods12163139.

11 Advantages, Limitations, Challenges, and Future of Polyphenol Nutraceuticals

Smaranika Pattnaik

11.1 INTRODUCTION

11.1.1 Polyphenol Nutraceuticals

Plant-derived polyphenols (Zhang et al., 2022) are of immense importance as nutraceuticals. Polyphenolics are produced via the shikimic acid pathway as secondary metabolites from the precursor phenylalanine (Pandey and Rizvi, 2009). These phytochemicals are considered chemical weapons of plants as they are used as defensive phytoconstituents against predators and grazing activities and to restrict microbial invasions. Polyphenols are produced with the onset of stress, like pH imbalance, drought, ultraviolet (UV) exposure or a nutrition deficit. Therefore, these biomolecules have enough potential to combat adverse effects. Keeping this view in mind, humans have long considered these phytoconstituents as nutraceuticals. This chapter discusses vegetables, fruits and cereals rich in polyphenols and their beneficial effects.

11.1.2 Secretion of Polyphenols

The polyphenols are secreted by a number of Angiosperm families. Several medicinal plants, specifically medicinal and aromatic plant clusters, are enriched with polyphenols. Among them are clove, peppermint, and cocoa; all types of berries including blackberries, high bush blueberries, raspberries, and strawberries; fruits like plums, black currants, sweet cherries, and apples; beans; nuts, including hazelnuts, walnuts, almonds, and pecans; vegetables, namely, artichokes, chicory, red onions, and spinach; soy like soy tempeh, soy flour, and tofu; black and green tea; and red wine are reported to have substantial amounts of polyphenolic compounds (www.https://www.healthline.co). The plant cells/tissue associated with polyphenol secretion are known as secretory glands. Variation in the structure of cavities containing the secreted products of polyphenols is reported in Castro and Demarco (2008). The sheath enclosing vascular bundles with cavities, idioblasts, laticifers, collers, and nuptial nectaries are prime locations of polyphenol reserves. Figure 11.1 depicts the source of polyphenols at a glance. Figure 11.2 is a three-dimensional

DOI: 10.1201/9781003588504-11

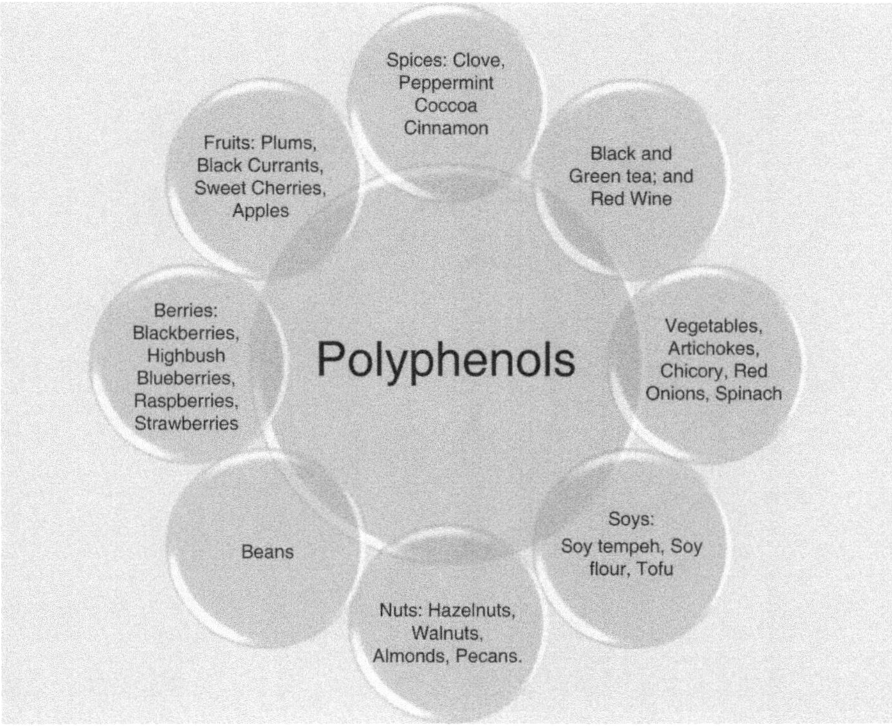

FIGURE 11.1 The distribution of polyphenols.

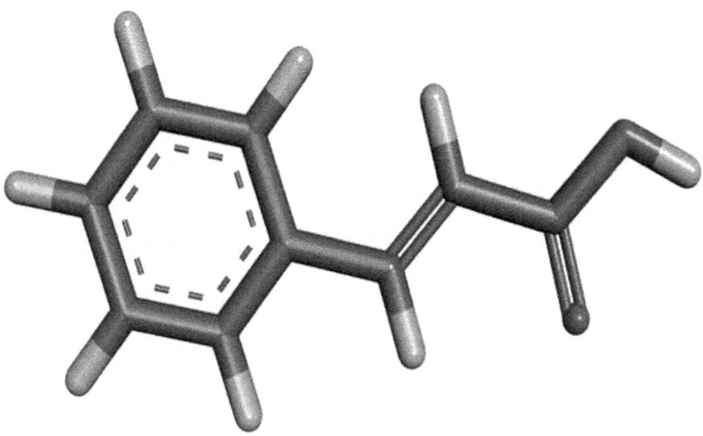

FIGURE 11.2 Molecular skeletal structure of cinnamic acid, a polyphenol found in cinnamon.

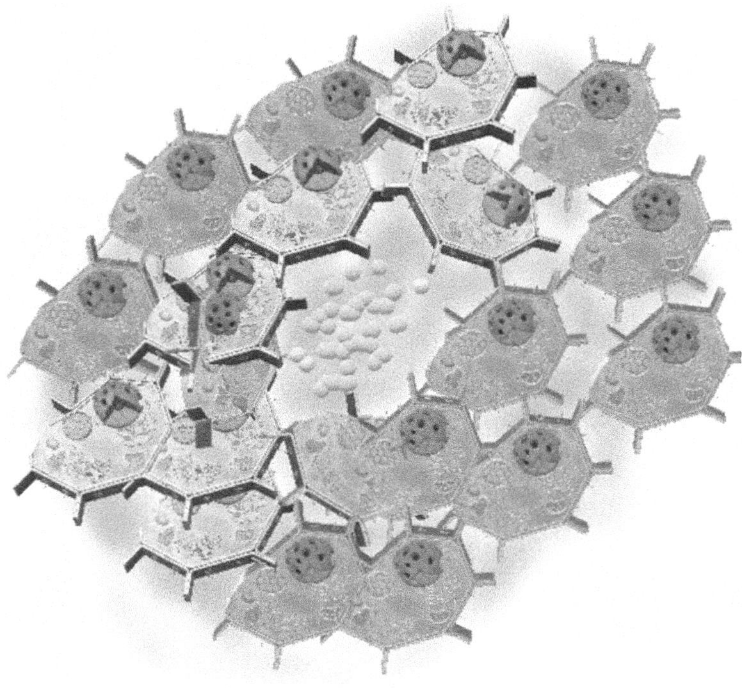

FIGURE 11.3 A typical plant tissue–based idioblast containing secreted polyphenols in its cavity.

structure of a typical polyphenol, namely cinnamic acid. Idioblasts are also reservoirs for secreted polyphenols (Figure 11.3).

11.2 ADVANTAGES OF POLYPHENOLIC NUTRACEUTICALS

11.2.1 Protection of Pancreatic Beta Cells From Oxidation

The pancreatic beta cells are well protected (Cory et al., 2018) from reactive oxygen species (ROS), as polyphenols act as scavengers of ROS produced by dysfunctional mitochondria. In addition, polyphenols can cause increased phosphorylation of tyrosine, leading to phosphorylation of insulin receptor substrate, which activates protein kinase B, which enters into the GluT4 storage vesicle (GSV) and leads to GluT4 translocation acceleration by polyphenols. Entry of glucose through the portal may lead to glycogenesis and/or lipogenesis. Moreover, polyphenols trigger the insulin signalling pathway as demonstrated in Figure 11.4.

11.2.2 Suppression of Advanced Glycation End Products

Polyphenols can suppress the production of advanced glycation end products (AGEPs). The glycation leads to the generation of free radicals, carbonyl species,

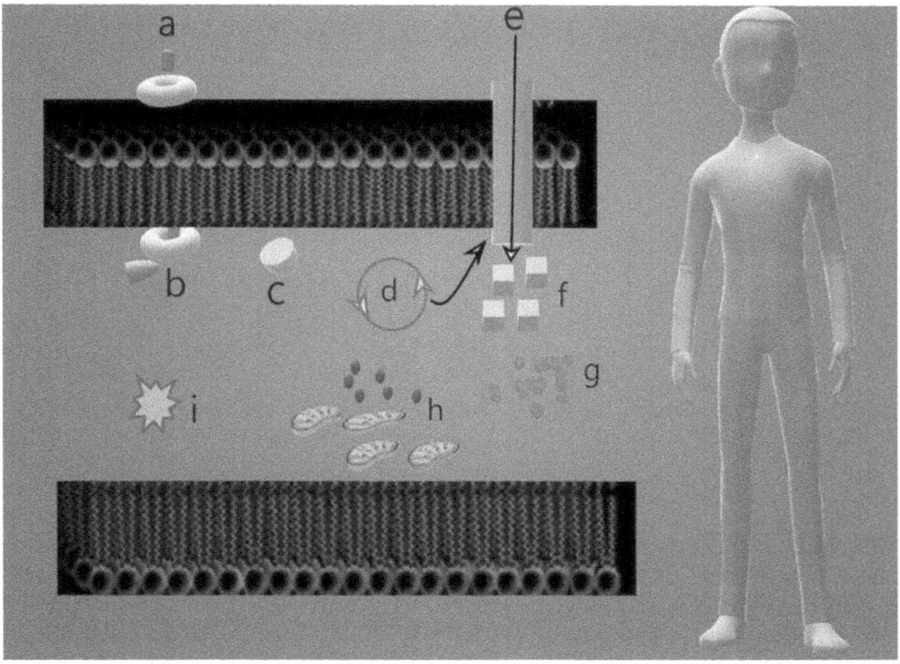

FIGURE 11.4 Mode of action of polyphenols. (a) In the event of facilitation of insulin through the human lipid bilayer membrane, (b) increased phosphorylation of tyrosine leads to phosphorylation of the insulin receptor substrate. (c) This leads to activation of protein kinase B, which enters into (d) GSV (GluT4 storage vesicle) and leads to GluT4 translocation acceleration by polyphenols. (f) When glucose enters the portal, it may lead to glycogenesis and/or lipogenesis. (h) Dysfunctional mitochondria generate ROS, which are scavenged by polyphenols. (i) The insulin signalling pathway is triggered by polyphenols.

and reactive carbonyl species. The instant post-translational modification of amino acids (Maillard reaction) is the root cause of the synthesis of AGEPs.

Methylglyoxal is a reactive aldehyde, produced endogenously by cells suffering from auto-oxidation. At this point, a physiological imbalance is induced, like protein glycation, oxidative stress, and even the onset of inflammation, thus exerting a toxic effect on insulin secretion from pancreatic Beta cells. As a consequence, the metabolic pathway suffers. Polyphenols are antiglycative (Yeh et al., 2017), which can prevent oxidative stress and ROS-induced physical injury leading to inflammation. Figure 11.5 depicts the putative role of polyphenols in consecutive events such as association with the (a) Schiff base, (b) Amadori products, (c) AGEs, and (d) ROS; (e) dissociating AGE from RAGE (receptor of RGE [receptor for advanced glycation end-products]); and (f) inhibiting the action of chemokines responsible for physical injury and inflammation, (g), thus protecting humans against chronic diseases

Polyphenols have an impact upon endothelial as well as vascular health, allowing smooth conduction of respective events, thus preventing health disorders. However, this activity lacks experimental proof and is derived from metadata Grosso et al., 2022). In this debate, a review report mentioned that kaempferol, quercetin, and

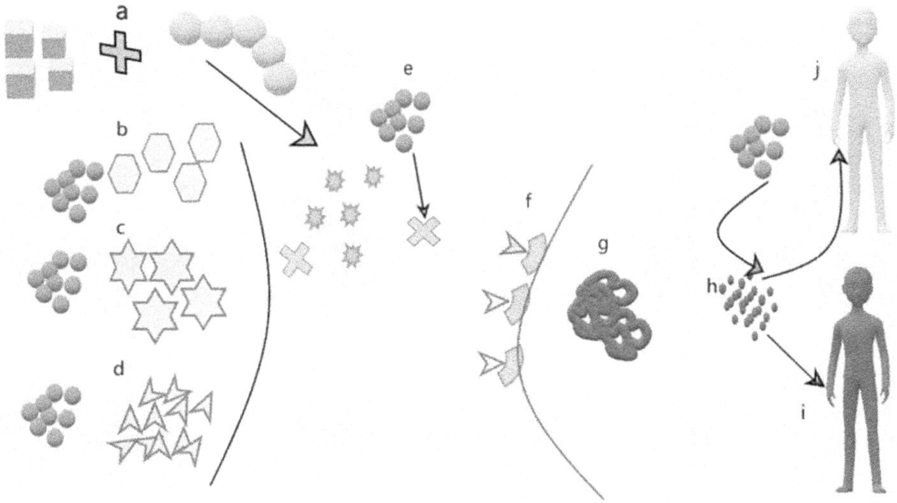

FIGURE 11.5 The putative mechanism of antiglycative action. (a) Sugars in association with amino acids produce the (b) Schiff base, (c) Amadori products, (d) AGEs, and (e) ROS. In addition, RGEs are produced, which are received by RAGE (receptor of RGE) and, as a consequence, (f) RAGE (Receptor for Advanced Glycation End Products) located on the cell membrane, (g) the inflammasome is aggregated with the accumulation of chemokines, (h) these chemokines are then released into surrounding tissues, amplifying the inflammatory response meant for (i) physical injury, (j) When antiglycative agents intervene, they interrupt these harmful biochemical processes, helping to restore cellular balance and strengthening the body's defenses – promoting a protective "Green Belt" of biochemical resilience. Thereby safeguarding human health and ushering in a 'Green Belt' of biochemical protection and cellular resilience.

resveratrol can prevent oxidative stress in cardiac tissues by interacting with proteins which are associated with the induction of stress. The endothelial vessels can also be under the control of polyphenols, thus preventing atherosclerosis. In addition, inflammatory markers can be suppressed in cardiomyocytes by exerting anti-inflammatory responses. Strokes, seizures, and heart failure can also be avoided (Iqbal et al., 2023).

11.2.3 Polyphenols as Antimicrobials

There is a pressing need for alternative drugs or therapeutic agents against the emergence and reemergence of drug-resistant microbial strains in clinical scenarios. Plant-derived polyphenols may be used as antimicrobial agents because of antibacterial, antifungal, and antiviral activities. Polyphenols like norwogonin, ellagic acid, chebulagic acid, chebulinic acid, corilagin, terchebulin, tannins, and flavonoides have shown antibacterial activity against *Acinetobacter baumanii*, *Pseudomonas aeruginosa*, *Staphylococcus aureus*, *Klebsiella pneumoniae*, *Salmonella* spp., and *Escherichia coli*, either tested as in its native form or in combination with conventional antibiotics. Likewise, both topical and systemic fungal strains, namely *Trichophyton mentagrophytes*, *Epidermophyton rubrum*, *Epidermophyton floccosum*, and *Cryptococcus neoformans,* are also restricted in terms of growth and multiplication by polyphenol activities (Manso et al., 2021). The foodborne bacterial strains of

Salmonella, *Pseudomonas*, and *Staphylococcus* are notorious for producing biofilm, which is a collective matrix constituted of exopolysaccharides, extracellular DNA, proteins, and lipids. Biofilm restricts entry of any food preservatives (antibacterial agents) because of its structural barrier. Many of the chemically synthesized antibacterials are unable to dissolve the matrix. The plant-derived polyphenols are reported to have antibiofilm activity (Bae et al., 2022). They can degrade the biofilm matrix components because of their structure, molecular configuration, and chemistry.

Hence, it is understood that polyphenols can exert synergistic activity when tested in combination with antibiotic dosage forms. Polyphenols with dietary supplements can be prescribed for children or patients needing critical care. Polyphenols are a better choice as food preservatives in comparison to synthetic additives (Gutiérrez-del-Río et al., 2018) because the polyphenols have bactericidal activity against food and waterborne bacteria as well as fungal pathogens.

11.2.4 Skin Protection from Polyphenols

Phenolic acids, flavonoides, anthocyanins, and catechins/flavonols (Nichols and Katiyar, 2010) are categorized as skin-protecting agents. The mechanism of action of polyphenols (flavonoides) is multimeric. Polyphenols can transport electrons to free radicals, thus inhibiting the generation of ROS. In addition, polyphenols can chelate metal ions, activate antioxidant-associated enzymes, reduce alpha tocopherol radicals, and inhibit oxidases (Ghazi, 2022).

11.3 LIMITATIONS IN POLYPHENOLIC NUTRACEUTICALS

In the food industry, polyphenols are used as preservatives and processed with food products. As part of the process, thermal processing, UV irradiation, infrared (IR) treatment, microwave, and fermentation are carried out. The polyphenols might not be able to maintain their integrity and undergo degradation, and the degraded products interact with other components in the placebo because polyphenols are polyhydroxyl in nature (Carito et al., 2015).

11.3.1 Poor Bioavailability of Polyphenols

Bioavailability is the amount of optimum concentration of a drug/food supplement in the blood after introduction into the body. It indicates the rate and extent of absorption of drugs and how soon it reaches the circulatory system and is available for systemic distribution (Figure 11.6). The prime cause could be due to limited water solubility (Aatif, 2023). Typical examples of polyphenols exerting poor bioavailability are (a) anthocyanins (1–2%) and black currant juice as reported earlier (Di Lorenzo et al., 2021).

11.3.2 Polyphenols and Dietary Protein/Complex Inhibit or Denature Protein Function

Polyphenols specifically may form complexes with dietary proteins or enzymes using non-specific bonds; thus, proteins are denatured. The structure, solubility,

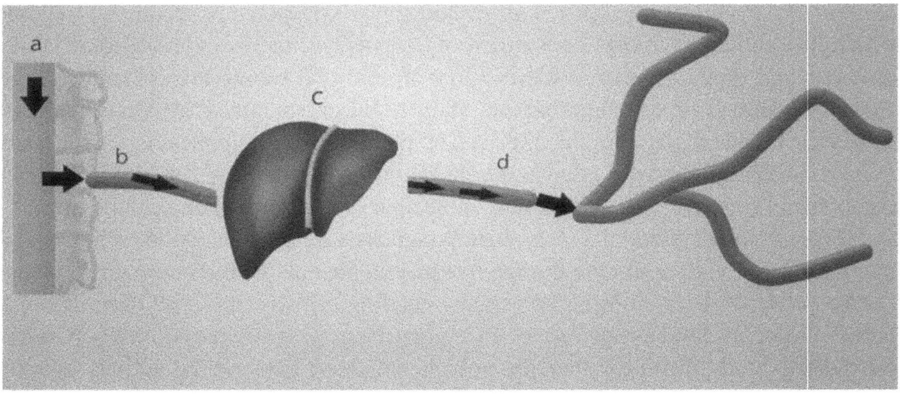

FIGURE 11.6 Drug pharmacokinetics. (a) Entry of drug into the lumen, through tissue, and into the circulatory system. From here it reaches (b) the liver, where first-pass effect is carried out and from which biotransformed drug moves to the circulatory network and is distributed near targets. (c) The bioavailability test is made immediately after the administration of the oral drug dosage form. (d) Polyphenols are reported to have poor bioavailability in this context.

hydrophobicity, thermal stability, and isoelectric points are affected. The catalytic activity of enzymes may be distorted. The amylases, proteases, and lipases seem to be dysfunctional, resulting in impairment of the digestive process (Duda-Chodak and Tarko, 2023).

11.3.3 Phytochemical Integrity of Polyphenols Is Disturbed by Thermal Processing

Thermal processing (Narra et al., 2024) is a key process in food preservation. Heat, either in dry form or steam form, is used to sterilize food. Furthermore, tyndallization, a boiling process, is also practised to keep microorganisms from invading food. UV rays (<300 nm) are also used to kill the microorganisms. However, phenolic compounds are largely affected by thermal processing methods, including boiling, microwave oven heating, and dry heating. The phenolics may be degraded and leached out. The latent heat applied with saturated steam may affect the rotamers and bond angles of phenolics, and the functionality is lost. UV rays can cause mutations in the parent structure, leading to malfunction of candidate phenolics.

11.4 CHALLENGES OF POLYPHENOLIC NUTRACEUTICALS

There are a good number of challenges in the open forum for polyphenolic nutraceuticals. Basically, the challenge lies in every step of the extraction, purification, and isolation of pure compounds using analytical tools like gas chromatography–mass spectrometry (GC-MS), liquid chromatography (LC)-MS, Fourier transform infrared (FTIR), nuclear magnetic resonance (NMR), and x-ray crystallography. The

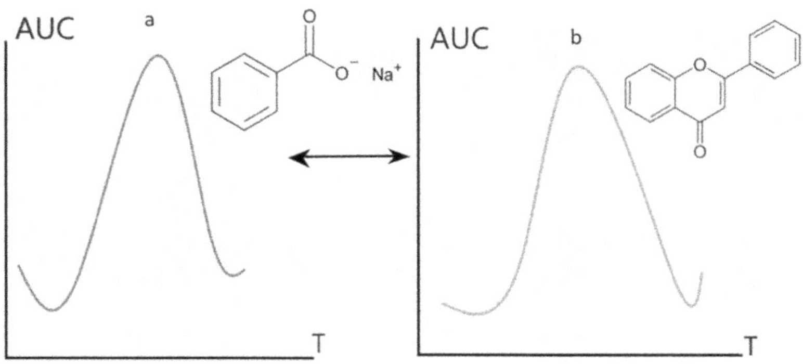

FIGURE 11.7 Biosimilarity test between a conventional preservative, sodium benzoate, and a flavonoid with respect to area under the curve (AUC). This is a graphical representation of the drug concentration at various time periods, T, while being processed in the body.

plants are considered solar-powered factories (Pattnaik, 2023) of enormous amounts of secondary metabolites, including polyphenols. Hence, it is important to extract polyphenols from the plants or their explants using pertinent tools and techniques. There is a great variation in the percentage of polyphenol yield, as several factors like pH, temperature, nutrition, etc., are prime contributory factors. Hence, conditions need to be optimized to receive a better yield. The extracted compounds should be subjected to purification using the previously mentioned analytical tools.

Now the prime task is to placate absorption, distribution, metabolism, and excretion (ADME) processes. The pharmacological tests on animal models should be carried out to have an interaction study between the polyphenols and biological systems operating inside the body.

In addition to these efforts, the toxicity test is a prime objective in polyphenolic-based nutraceuticals. The MD_{50}-based experiments are a prerequisite task to define the polyphenolic nutraceuticals as generally recognized as safe (GRAS). Furthermore, the screening procedure may be upgraded to higher animal models, and finally, case studies may be conducted on volunteers to deliver a concrete statement. Several blind tests are to be carried out to fulfil the maximum criteria assigned for polyphenolic nutraceuticals.

But the great challenge is the 'biosimilarity' test for both conventional food preservatives and polyphenols.

As demonstrated in Figure 11.7, the area under the curve (AUC) generated for each of the drug candidates can provide information about the percentage of biosimilarity between the two drugs. The test also determines the bioefficacy of drugs being tested.

11.4.1 Future of Polyphenolic Nutraceuticals

Currently, metabolomics (Mohapatra and Pattnaik, 2024) approaches are quite conducive in this regard. Metabolomics tools and techniques are contemporary, error free, and require a minimal period. The GC-MS, LC-MS, FTIR, NMR, and

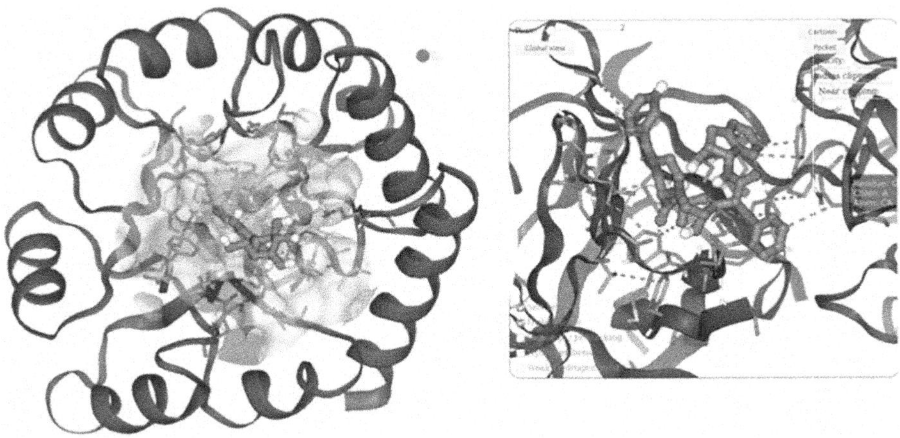

FIGURE 11.8 A flavonoid ($C_{27}H_{30}O_{15}$) is trapped inside the cavity of a protein nanocage, safely forming positive interactions with high energy input. The left image is the nanocage showing its binding pockets where the flavonoid is in a hanging position. The right image depicts the bonds between the nanocage and the flavonoid atoms. Both conventional hydrogen bonds and hydrophobic interactions are noted.

crystallo-graphic dataset available in the public domain can be used as a reference as well as a model of querying polyphenols. Polyphenols may be annotated in terms of structure and physical as well as chemical properties, using both online and stand-alone versions of bioinformatics servers.

In addition, the ADME tests (adding in toxicity tests) can be predicted using computer software. But the in vitro and in silico experiments should be in agreement with each other. Furthermore, it is important to make sure that polyphenol nutraceuticals do not interact with food grains or particles. The polyphenols should be inert towards biological products also. The advanced drug delivery systems are to be exploited so that polyphenols will be delivered to target food particles. The polyphenol-impregnated nano-structured liposomes may be used for this purpose. Polyphenol impregnation may be done using tools of a microfluidics system (Ferreira et al., 2024). Figure 11.8 shows a protein nanocage (PDB ID 5KP9) and a flavonoid (IUPAC name, 5,7-dihydroxy-6–[3,4,5-trihydroxy-6-(hydroxymethyl)oxan-2-yl]-2–[4-[3,4,5-trihydroxy-6-(hydroxymethyl)oxan-2-yl]oxyphenyl]chromen-4-one), syn. It also shows isosaponaretin interaction with the formation of hydrogen bonds using an online docking portal. The flavonoid molecule is trapped inside the cavity of nanocage, although the free energy used for bonding is very high: 44 kcl.

This is an example of the application of -omics technology being used to bring forward polyphenol nutraceutical research.

Hence, liposomes as nanocarriers in appropriate dosage forms can deliver the polyphenols of interest using principles of a target-oriented delivery system. Furthermore, nano-carriers can be used to deliver polyphenols using advanced techniques. Figure 11.9 shows polyphenol-impregnated liposomal coatings for typical food grains.

FIGURE 11.9 Coating of liposome-impregnated polyphenol for the protection of food. The polyphenol impregnation is highlighted in a typical liposome (right panel).

11.5 CONCLUSION

In brief, it is suggested that polyphenols, being plant-derived products, are better than any synthetics as a nutraceuticals, although certain limitations are apparent. With the rise of -omics tools, science now has the means to decode the complex roles of polyphenols and strategically address earlier gaps in nutritional understanding. This progress helps keep our food within the protective orbit of the 'Green Sphere'–a space where health, sustainability, and molecular harmony converge.

REFERENCES

Aatif, M. (2023). Current understanding of polyphenols to enhance bioavailability for better therapies. *Biomedicines*, 11(7), 2078. https://doi.org/10.3390/biomedicines11072078

Bae, J. Y., Seo, Y. H., & Oh, S. W. (2022). Antibacterial activities of polyphenols against food-borne pathogens and their application as antibacterial agents. *Food Science and Biotechnology*, 31(8), 985–997. https://doi.org/10.1007/s10068-022-01058-3

Carito, V., Ceccanti, M., Chaldakov, G., Tarani, L., De Nicolò, S., Ciafrè, S., . . . Fiore, M. (2015). Polyphenols, nerve growth factor, brain-derived neurotrophic factor, and the brain. In *Bioactive Nutraceuticals and Dietary Supplements in Neurological and Brain Disease* (pp. 65–71). Academic Press. https://doi.org/10.1016/B978-0-12-411462-3.00007-2

Cory, H., Passarelli, S., Szeto, J., Tamez, M., & Mattei, J. (2018). The role of polyphenols in human health and food systems: A mini-review. *Frontiers in Nutrition*, 5, 370438. https://doi.org/10.3389/fnut.2018.00087

Di Lorenzo, C., Colombo, F., Biella, S., Stockley, C., & Restani, P. (2021). Polyphenols and human health: The role of bioavailability. *Nutrients*, 13(1), 273. https://doi.org/10.3390/nu13010273

Duda-Chodak, A., & Tarko, T. (2023). Possible side effects of polyphenols and their interactions with medicines. *Molecules*, 28(6), 2536. https://doi.org/10.3390/molecules28062536

Ferreira, M., Carvalho, V., Ribeiro, J., Lima, R. A., Teixeira, S., & Pinho, D. (2024). Advances in microfluidic systems and numerical modeling in biomedical applications: A review. *Micromachines*, 15(7), 873. https://doi.org/10.3390/mi15070873

Ghazi, S. (2022). Do the polyphenolic compounds from natural products can protect the skin from ultraviolet rays?. *Results in Chemistry*, 4, 100428. https://doi.org/10.1016/j.rechem.2022.100428

Grosso, G., Godos, J., Currenti, W., Micek, A., Falzone, L., Libra, M., . . . Galvano, F. (2022). The effect of dietary polyphenols on vascular health and hypertension: current evidence and mechanisms of action. *Nutrients*, 14(3), 545. https://doi.org/10.3390/nu14030545

Gutiérrez-del-Río, I., Fernández, J., & Lombó, F. (2018). Plant nutraceuticals as antimicrobial agents in food preservation: Terpenoids, polyphenols and thiols. *International Journal of Antimicrobial Agents*, 52(3), 309–315. https://doi.org/10.1016/j.ijantimicag.2018.04.024

Iqbal, I., Wilairatana, P., Saqib, F., Nasir, B., Wahid, M., Latif, M. F., . . . Mubarak, M. S. (2023). Plant polyphenols and their potential benefits on cardiovascular health: A review. *Molecules*, 28(17), 6403. https://doi.org/10.3390/molecules28176403

Manso, T., Lores, M., & de Miguel, T. (2021). Antimicrobial activity of polyphenols and natural polyphenolic extracts on clinical isolates. *Antibiotics*, 11(1), 46. https://doi.org/10.3390/antibiotics11010046

Mohapatra, D. D., & Pattnaik, S. (2024). Metabolic profiling technologies for biomarker discovery in ethnomedicinal plants for drug development. In *Ethnomedicinal Plants for Drug Discovery: Current Developments* (pp. 323–343). Singapore: Springer Nature Singapore. https://doi.org/10.1007/9

Narra, F., Piragine, E., Benedetti, G., Ceccanti, C., Florio, M., Spezzini, J., . . . Guidi, L. (2024). Impact of thermal processing on polyphenols, carotenoids, glucosinolates, and ascorbic acid in fruit and vegetables and their cardiovascular benefits. *Comprehensive Reviews in Food Science and Food Safety*, e13426. https://doi.org/10.1111/1541-4337.13426

Nichols, J. A., & Katiyar, S. K. (2010). Skin photoprotection by natural polyphenols: anti-inflammatory, antioxidant and DNA repair mechanisms. *Archives of Dermatological Research*, 302, 71–83. https://doi.org/10.1007/s00403-009-1001-3

Pandey, K. B., & Rizvi, S. I. (2009). Plant polyphenols as dietary antioxidants in human health and disease. *Oxidative Medicine and Cellular Longevity*, 2(5), 270–278. https://doi.org/10.4161/oxim.2.5.9498

Pattnaik, S. (2023). Next-generation sequencing (NGS) for metabolomics study in medicinal plants under stress condition. In *Stress-responsive Factors and Molecular Farming in Medicinal Plants* (pp. 323–343). Singapore: Springer Nature Singapore. https://doi.org/10.1007/9

Yeh, W. J., Hsia, S. M., Lee, W. H., & Wu, C. H. (2017). Polyphenols with antiglycation activity and mechanisms of action: A review of recent findings. *Journal of Food and Drug Analysis*, 25(1), 84–92. https://doi.org/10.1016/j.jfda.2016.10.017

Zhang, Z., Li, X., Sang, S., McClements, D. J., Chen, L., Long, J., Jiao, A., Jin, Z., & Qiu, C. (2022). Polyphenols as plant-based nutraceuticals: health effects, encapsulation, nano-delivery, and application. *Foods*, 11(15), 2189. https://doi.org/10.3390/foods11152189

12 Polyphenol Nutraceuticals Available on the Market and Their Health Benefits

Chandrashekhar Singh and Richa Upadhyay

12.1 INTRODUCTION

Polyphenol nutraceuticals are gaining popularity for their potential health benefits, including antioxidant, anti-inflammatory, and neuroprotective effects (Zhang et al., 2022). These compounds are found in various plant-based foods and are now available in supplement form. Today, a wide range of polyphenol-rich supplements are available on the market, each formulated to deliver specific health benefits from improving cardiovascular function and cognitive performance to promoting gut health and healthy ageing. These products often feature concentrated extracts such as curcumin, resveratrol, grape seed, green tea catechins, and pomegranate polyphenols, either alone or in synergistic blends. This article explores some of the most widely available polyphenol nutraceuticals, their key ingredients, and the scientific evidence supporting their health benefits.

12.1.1 Polyphenol Nutraceuticals: Market Overview

Polyphenol nutraceuticals represent a rapidly expanding segment of the global dietary supplement market, driven by increasing awareness of natural health solutions, ageing populations, and the surge in chronic lifestyle diseases. Polyphenols, a plant-derived compound with antioxidant and anti-inflammatory properties, are being integrated into capsules, tablets, powders, functional foods, and beverages to support holistic health and disease prevention.

12.1.2 Global Trends in Nutraceuticals

The global nutraceuticals industry is undergoing a major transformation, shifting from general wellness supplements to evidence-based, function-specific products. Key trends influencing the market are discussed in the following sections.

DOI: 10.1201/9781003588504-12

12.1.2.1 Personalized Nutrition

Consumers increasingly seek products tailored to individual health needs (e.g., gut health, cognitive support, metabolic balance). Polyphenols, with their diverse biological actions, fit well into this trend.

12.1.2.2 Natural and Clean-Label Products

Demand for plant-based, sustainably sourced, and additive-free supplements has led to growing interest in polyphenol-rich botanicals like green tea, berries, turmeric, and olive leaf.

12.1.2.3 Functional Foods and Beverages

Polyphenols are being added to yoghurts, protein shakes, health drinks, and bars. Brands are capitalising on the dual appeal of nutrition and convenience.

12.1.2.4 Preventive Healthcare

With rising healthcare costs and chronic disease prevalence, consumers are turning to nutraceuticals as proactive health solutions. Polyphenols are widely marketed for their roles in heart health, brain function, and immunity.

12.1.2.5 Regulatory Emphasis on Science-Based Claims

In the United States, Europe, and parts of Asia, there's a push toward substantiating functional claims with clinical evidence. This trend favours polyphenols, which have robust support from epidemiological and interventional studies.

12.2 GROWTH OF THE POLYPHENOL SUPPLEMENT MARKET

The global polyphenol supplement market is projected to grow at a compound annual growth rate (CAGR) of 7–9% from 2023 to 2030, with market valuation expected to surpass USD 3.5 billion by 2030, according to industry reports (https://www.grandviewresearch.com/industry-analysis/polyphenols-market-analysis).

12.2.1 Key Growth Drivers

12.2.1.1 Ageing Population

Older adults are a prime demographic for polyphenol-based products due to concerns over cognitive decline, heart disease, and inflammation.

12.2.1.2 Rising Consumer Health Awareness

Educated consumers are actively seeking antioxidant-rich products, particularly during post-pandemic recovery.

12.2.1.3 Scientific Advancements

New extraction techniques (e.g., liposomal polyphenols, nano-formulations) are enhancing bioavailability, expanding market options.

12.2.1.4 Emerging Economies

Markets like India, China, and Brazil are experiencing rising demand due to a growing middle class, improved access to healthcare, and an increase in lifestyle-related illnesses.

12.2.1.5 Consumer Preferences and Functional Claims

Consumers are now more sophisticated and demanding in terms of what they expect from nutraceuticals. When it comes to polyphenols, their preferences and expectations align with the functional claims discussed next.

12.2.1.6 Indian Polyphenol Market

The Indian polyphenol market is experiencing substantial growth, driven by increasing consumer awareness of health benefits and the demand for natural ingredients. The market is projected to reach USD 221.5 million by 2030, with a CAGR of 7.8%

TABLE 12.1
Key Ingredients and Health Benefits of Polyphenol-Based Supplements

Brand/Product	Key Ingredients	Health Benefits
Pendulum Polyphenol Booster	Pomegranate, grape seed, green tea extracts	Supports gut health, enhances cardiovascular function, promotes healthy ageing
Terry Naturally Polyphenol Defense	Green tea extract, apple extract, BCM-95 curcumin, French grape seed extract, propolis	Provides antioxidant defence; supports heart, brain, liver, and skin health
Polyphenol-C™	Non-GMO vitamin C, fruit extracts (grape, blueberry, cranberry, other berries)	Enhances immune function, supports cardiovascular and brain health, aids in connective tissue regeneration
ProHealth Trans-Resveratrol	Trans-resveratrol, quercetin, polyphenols from red wine and green tea extracts	Supports cardiovascular health, anti-ageing, and antioxidant protection
partiQlar Pure Quercetin	Quercetin from *Sophora japonica* bud extract	Provides antioxidant support, supports immune function, and reduces inflammation
ProHealth Pterostilbene	Trans-pterostilbene	Supports cardiovascular health, anti-ageing, and antioxidant protection
vitalityPRO Curcumin	Curcumin from natural *Curcuma longa* root	Provides anti-inflammatory and antioxidant support, supports joint and brain health
PartiQlar Pure Apigenin	Apigenin	Supports relaxation and sleep, provides antioxidant support
DailyZz by Kemin Industries	Spearmint and green tea extracts	Improves sleep quality and next-day functioning
Hytolive Infinity by Genosa	Hydroxytyrosol from olive fruit extract	Provides antioxidant support, supports cardiovascular and skin health

from 2023 to 2030 (India Polyphenols Market Size & Outlook, 2030). In 2022, India's polyphenol market revenue was valued at 121.7 million USD. The largest revenue-generating product was grape seed in 2022, while peach was the highest profitable product with the fastest growth during the assessment period.

12.3 TOP POLYPHENOL-BASED NUTRACEUTICALS (2025)

12.3.1 Product Overview Table

12.3.2 Detailed Product Reviews

12.3.2.1 Pendulum Polyphenol Booster

Pendulum Polyphenol Booster is a scientifically formulated supplement designed to support the gut microbiome, metabolism, and overall health. Featuring a powerful blend of plant-based polyphenols and precision probiotics, this booster enhances the body's ability to convert polyphenols into beneficial postbiotics, including urolithin A, a compound linked to cellular health, mitochondrial function, and healthy ageing. Thus, overall routine wellness can be elevated with Pendulum Polyphenol Booster.

Key Benefits

Supports Gut Health: Encourages the growth of key gut microbes that help break down polyphenols into bioavailable compounds.

Boosts Metabolic Function: Promotes healthy glucose and lipid metabolism.

Cellular and Mitochondrial Support: Stimulates the production of urolithin A, known to support mitochondrial health and longevity.

Synergistic Formula: Combines targeted polyphenols (like pomegranate and grape extract) with specific strains of probiotics for maximum efficacy.

12.3.2.2 Terry Naturally Polyphenol Defence

Terry Naturally Polyphenol Defence is a premium dietary supplement formulated to provide whole-body antioxidant support. Each capsule delivers a potent blend of polyphenols derived from five natural extracts, totalling 250 mg of polyphenols per serving. This supplement is designed to support various aspects of health, including heart, brain, liver, skin, and immune function.

Key Ingredients and Their Benefits

1. **BCM-95 Curcumin** ***(Curcuma longa)***
   BCM-95 Curcumin has enhanced absorption supporting healthy cellular replication, joint function, liver health, and cognitive well-being.
2. **French Grape Seed Extract VX1** ***(Vitis vinifera)***
   Delivers tannin-free, 100% absorbable oligomeric proanthocyanidins (OPCs) to support cardiovascular health, DNA protection, and healthy blood sugar levels.
3. **EP300 Propolis**
   Supports healthy respiratory function, ear health, and overall immune system function in adults and children alike.

4. **Apple Extract** ***(Malus spp.)***
   Clinically studied support for daily energy, cardiovascular strength, respiratory function, and healthy skin.
5. **Green Tea Extract** ***(Camellia sinensis)***
   Research validates that key polyphenols from green tea support cognitive health, cellular defence, mental well-being, and healthy ageing.

12.3.2.3 Polyphenol-C

Polyphenol-C is a proprietary nutraceutical formulation that combines high-potency polyphenols with vitamin C to enhance antioxidant synergy and immune support. Typically featuring polyphenolic compounds such as green tea catechins, grape seed extract, and citrus bioflavonoids, the blend is designed to provide comprehensive cellular protection against oxidative stress. The inclusion of vitamin C not only boosts immune defence but may also improve the stability and bioavailability of polyphenols. This supplement is often marketed for general wellness, cardiovascular support, and inflammation reduction, appealing to consumers seeking multifunctional antioxidant support in a single formulation.

12.3.2.4 ProHealth Trans-Resveratrol

ProHealth Trans-Resveratrol is a high-potency supplement containing 1000 mg of 99% pure trans-resveratrol per serving. Trans-resveratrol, a polyphenol found in red wine and Japanese knotweed, is known for its antioxidant and anti-inflammatory effects. It's widely studied for potential benefits in cardiovascular health, cellular ageing, and metabolic function. ProHealth emphasises purity and bioavailability, often combining this product with piperine to enhance absorption.

12.3.2.5 VitalityPRO Curcumin

vitalityPRO Curcumin delivers a concentrated dose of curcumin extracted from turmeric (*Curcuma longa*), standardised to contain 95% curcuminoids. This formulation typically includes black pepper extract (piperine) to improve bioavailability. Curcumin has been extensively researched for its anti-inflammatory, antioxidant, and neuroprotective properties, and vitalityPRO positions this supplement for joint support, cognitive health, and systemic inflammation reduction.

12.3.2.6 PartiQlar Pure Quercetin

partiQlar Pure Quercetin provides a pharmaceutical-grade quercetin dihydrate, a flavonol naturally found in onions, apples, and capers. This supplement is marketed for its immune-modulating, antihistamine, and anti-inflammatory benefits. Quercetin may also support cardiovascular health and mitochondrial function. The "Pure" branding suggests minimal excipients and a focus on clean-label formulation.

12.3.2.7 PartiQlar Pure Apigenin

PartiQlar Pure Apigenin offers a standardised form of apigenin, a flavone found in parsley, chamomile, and celery. It is studied for its anxiolytic effects, potential anticancer activity, and role in modulating cellular senescence. As with other partiQlar supplements, the emphasis is on purity and targeted therapeutic use, often in synergy with other longevity-focused compounds.

12.3.2.8 DailyZz by Kemin Industries

DailyZz is a proprietary sleep-support supplement developed by Kemin Industries containing natural polyphenols from spearmint and other botanical sources. It's formulated to improve sleep quality without sedation. Clinical studies have shown improvements in sleep efficiency and reduced nighttime awakenings. DailyZz highlights the emerging role of polyphenols in cognitive and circadian health.

12.3.2.9 Hytolive Infinity (Hydroxytyrosol)

Hytolive Infinity is a high-purity extract of hydroxytyrosol, a potent polyphenol derived from olive fruit. It's recognised for its powerful antioxidant capacity, surpassing many other polyphenols. This supplement is marketed for cardiovascular protection, skin health, and anti-ageing. The "Infinity" branding suggests a premium formulation and long-term health applications, aligning with Mediterranean diet–based wellness approaches.

12.4 MECHANISMS OF ACTION

12.4.1 Free Radical Scavenging

Free radicals (e.g., reactive oxygen species [ROS], reactive nitrogen species [RNS]) are unstable molecules that damage cells by oxidising DNA, proteins, and lipids. Antioxidants (e.g., flavonoids, polyphenols, vitamins C and E) donate an electron to free radicals, stabilising them and terminating chain reactions. This process reduces oxidative stress and protects cells from damage associated with ageing, inflammation, and chronic diseases.

12.4.2 Gene Expression Modulation (e.g., BDNF, CREB)

Certain compounds (e.g., curcumin, resveratrol) activate intracellular signalling pathways such as cAMP/PKA and MAPK/ERK. These pathways lead to phosphorylation of transcription factors like cAMP response element-binding protein (CREB). Activated CREB promotes the transcription of neuroprotective genes such as brain-derived neurotrophic factor (BDNF), enhancing neurogenesis, synaptic plasticity, and cognitive function.

Epigenetic modifications (e.g., histone acetylation) may also up-regulate gene expression.

12.4.3 Enzyme Inhibition (e.g., COX-2, LOX)

Inflammation involves enzymes like cyclooxygenase-2 (COX-2) and lipoxygenase (LOX) that convert arachidonic acid into pro-inflammatory mediators (prostaglandins and leukotrienes). Anti-inflammatory compounds (e.g., non-steroidal anti-inflammatory drugs [NSAIDs], quercetin, curcumin) inhibit COX-2 or LOX activity. This inhibition reduces the production of inflammatory mediators, thereby decreasing pain, swelling, and tissue damage.

12.4.4 Gut Microbiota Interaction

Bioactive compounds (e.g., polyphenols, dietary fibres, prebiotics) are metabolised by gut microbiota into bioactive metabolites (e.g., short-chain fatty acids [SCFAs]). These metabolites influence host metabolism, immune response, and gut barrier function. Simultaneously, compounds may modulate microbiota composition by promoting beneficial bacteria (e.g., *Lactobacillus*, *Bifidobacterium*) and suppressing pathogens. This bidirectional interaction contributes to systemic effects including anti-inflammatory, metabolic, and neuroprotective outcomes.

12.5 BIOAVAILABILITY CHALLENGES

Many bioactive compounds (e.g., curcumin, resveratrol, EGCG) have low water solubility, poor intestinal absorption, rapid metabolism, and quick systemic elimination. Enzymatic degradation in the gastrointestinal (GI) tract and liver (first-pass metabolism) further reduces their systemic concentration. As a result, despite high in vitro efficacy, their bioavailability (fraction of active compound reaching circulation) is limited in vivo, reducing therapeutic impact.

12.6 OVERCOMING BIOAVAILABILITY BARRIERS

12.6.1 Encapsulation and Nanotechnology

Encapsulation involves enclosing the bioactive in protective carriers (e.g., nanoparticles, micelles, polymer matrices) to:

1. Enhance stability in gastric and enzymatic environments.
2. Improve solubility and controlled release at target sites.
3. Use nanotechnology (e.g., nanoemulsions, solid lipid nanoparticles, dendrimers) to reduce particle size to the nanometer scale.
4. Increase surface area and absorption.
5. Enhance cellular uptake via endocytosis.

These systems overcome bioavailability barriers, prolong circulation time, and enable targeted delivery.

12.6.2 Liposomal Polyphenols

Liposomes are spherical vesicles made of phospholipid bilayers that encapsulate both hydrophilic and lipophilic compounds. Polyphenols (e.g., quercetin, catechins) loaded into liposomes gain:

1. Protection from degradation in the GI tract.
2. Improved solubility and intestinal permeability.
3. Enhanced bioavailability and tissue distribution.

Liposomes can also be modified (e.g., PEGylation, targeting ligands) for site-specific delivery, such as to the brain or tumours.

12.6.3 Synergistic Formulations

Combining multiple bioactives (e.g., curcumin + piperine; EGCG + vitamin C) can create synergistic effects, i.e., one compound enhances the absorption, stability, or bioactivity of another. For example, piperine inhibits hepatic glucuronidation, boosting curcumin bioavailability by up to 2000%. These formulations may also target multiple pathways (e.g., antioxidant + anti-inflammatory), enhancing the overall therapeutic effect. Formulating synergists together in single delivery systems (e.g., co-encapsulated nanoparticles) ensures simultaneous release and action.

12.7 CLINICAL STUDIES OF POLYPHENOL PRODUCTS AND SCIENTIFIC EVIDENCE

Over the past five years (2020–2025), numerous clinical studies have investigated the effects of polyphenol products on human health. Next is a detailed overview of recent clinical trials, meta-analyses, and the comparative efficacy of polyphenols in human subjects versus animal models.

12.7.1 Recent Clinical Trials (2020–2025)

12.7.1.1 Bravo Apple Study (Australia, 2025)

Researchers at Murdoch University are conducting an eight-week trial to assess the cardiovascular and metabolic benefits of Bravo apples, a variety rich in polyphenols (Murdoch University, 2024). Participants consume either Bravo apples or a control fruit daily, with outcomes including blood pressure, cholesterol levels, and blood glucose measurements (https://www.murdoch.edu.au/news/articles/help-science-by-eating-apples)

12.7.1.2 Polyphenol Capsules and Cognitive Function (Spain, 2025)

A randomised, double-blind trial involving 92 participants found that daily supplementation with polyphenol-rich capsules led to significant improvements in cognitive functions and elevated levels of BDNF and CREB, both associated with brain health (Carrillo et al., 2025)

12.7.1.3 Sugarcane Extract and Gut Health (Australia, 2024)

Deakin University researchers are investigating the effects of polyphenol-rich sugarcane extract on gut microbiome diversity and chronic disease biomarkers. Participants take daily supplements over six months, with assessments including blood tests and stool samples (Audrey Yow, 2024).

12.7.2 Meta-Analyses and Systematic Reviews

12.7.2.1 Cardiometabolic Risk Factors (2024)

A comprehensive meta-analysis of 281 randomised controlled trials involving 17,126 participants indicated that polyphenol supplementation can improve various cardiometabolic risk markers. Notably, catechins reduced blood pressure, anthocyanins improved lipid profiles, and curcumin enhanced glucose metabolism (Wan et al., 2024)

12.7.2.2 Sarcopenia and Muscle Health (2024)

A systematic review and meta-analysis of seven studies involving 227 individuals with sarcopenia revealed that polyphenol supplementation significantly increased muscle mass but did not affect muscle strength. A near-significant trend was observed for improved physical performance (Medoro et al., 2024).

12.7.2.3 Oxidative Stress and Grape Products (2021)

Analysis of 17 randomised controlled trials with 633 participants found that grape products containing polyphenols significantly increased total antioxidant capacity and enhanced superoxide dismutase and oxygen radical absorbance capacity, suggesting a reduction in oxidative stress (Sarkhosh-Khorasani et al., 2021).

12.7.2.4 Iron Status and Erythropoiesis (2022)

A meta-analysis of 34 randomised controlled trials indicated that polyphenol supplementation reduced serum iron and transferrin saturation but increased haemoglobin concentration, suggesting a potential role in improving erythropoiesis in certain populations (Xu et al., 2021).

12.7.3 Efficacy in Human Subjects versus Animal Models

While animal studies have demonstrated various health benefits of polyphenols, human clinical trials have provided more nuanced insights. For instance, animal models showed that polyphenols can reduce inflammation and oxidative stress, but translating these effects to humans has been complex due to differences in metabolism and bioavailability. Human studies, such as those on grape polyphenol intake, have confirmed antioxidant benefits and improved vascular health but also highlighted variability in individual responses (Rasines-Perea and Teissedre, 2017). Moreover, while animal studies often use higher doses of polyphenols than those achievable through diet alone, human trials typically employ doses that are more reflective of dietary intake. This discrepancy underscores the importance of conducting human clinical trials to assess the practical efficacy and safety of polyphenol-rich interventions.

12.8 FUTURE DIRECTIONS AND RESEARCH OPPORTUNITIES

Future research is poised to uncover a broader spectrum of polyphenols from underexplored natural sources such as wild plants, seaweeds, fungi, and agricultural by-products. Advanced extraction technologies, such as supercritical fluid extraction and green chemistry approaches, are opening avenues for isolating polyphenols with unique bioactivities. Additionally, synthetic biology and metabolic engineering may facilitate the biosynthesis of rare or more potent polyphenol analogues. These novel compounds could offer enhanced therapeutic benefits or more targeted effects against chronic diseases like cancer, cardiovascular conditions, and neurodegenerative disorders.

With the rise of precision medicine and nutrigenomics, there is growing interest in understanding how individual genetic variability affects the metabolism and efficacy of polyphenols. Personalised nutrition models, incorporating genomic, microbiome, and metabolic profiling, could optimise polyphenol-based interventions for disease prevention and health promotion. For instance, interindividual differences in gut microbiota significantly influence the bioavailability and bioactivity of polyphenols. Future studies should focus on gene–diet interactions, microbiome-mediated metabolism, and real-time biomarker monitoring to tailor polyphenol intake to individual needs. The integration of polyphenols into functional foods and beverages represents a promising strategy for enhancing public health through daily dietary habits. Future research should aim to improve the stability, bioavailability, and sensory properties of polyphenol-enriched products. Techniques such as nanoencapsulation, emulsification, and co-formulation with synergistic compounds can enhance delivery and consumer acceptability. Regulatory frameworks, consumer education, and industry collaboration will also play crucial roles in mainstreaming polyphenol-infused products. Moreover, long-term clinical trials are needed to substantiate health claims and ensure efficacy in real-world settings.

12.9 CONCLUSION

This chapter highlights the global and Indian trends of polyphenol nutraceuticals, their current applications, and future possibilities. The integration of polyphenols into functional foods and personalised nutrition models is transforming the landscape of preventive healthcare and nutraceutical development. To translate current knowledge into actionable strategies, several practical recommendations are proposed:

For researchers: Focus on well-designed, long-term human studies to validate health benefits and understand mechanisms of action.

For industry stakeholders: Invest in novel formulation technologies (e.g., nanoencapsulation, emulsification) to improve polyphenol stability and bioavailability in functional products.

For healthcare professionals: Promote dietary patterns rich in polyphenol-containing foods and consider individualised dietary advice based on genetic and microbiome profiles.

For policymakers and regulators: Establish clearer guidelines for health claims, quality control, and labelling of polyphenol-based nutraceuticals and functional foods.

The field of polyphenol research is at a dynamic intersection of food science, medicine, and biotechnology. As consumer demand for natural and effective health solutions grows, polyphenols stand out as key players in next-generation nutraceutical innovation. By embracing a multidisciplinary approach that spans from molecular science to personalised dietary interventions, we can unlock their full potential. Continued collaboration between academia, industry, and regulatory bodies will be essential to drive responsible innovation and bring safe, effective, and evidence-based polyphenol products to global markets.

REFERENCES

Audrey Yow. *Sugarcane Fix: Revolutionary Remote Trial to Explore Health Benefits of Polyphenols in Plant Extract*, 2024 Apr 23. https://www.nutraingredients-asia.com/Article/2024/04/23/Sugarcane-polyphenols-Remote-trial-to-explore-health-benefits.

Carrillo JÁ, Arcusa R, Xandri-Martínez R, Cerdá B, Zafrilla P, Marhuenda J. Impact of Polyphenol-Rich Nutraceuticals on Cognitive Function and Neuroprotective Biomarkers: A Randomized, Double-Blind, Placebo-Controlled Clinical Trial. *Nutrients*. 2025 Feb 7;17(4):601. https://doi.org/10.3390/nu17040601.

Medoro A, Scapagnini G, Davinelli S. Polyphenol Supplementation and Sarcopenia: A Systematic Review and Meta-Analysis of Clinical Trials. *J Frailty Aging*. 2024;13(4):432–440. https://doi.org/10.14283/jfa.2024.73.

Murdoch University. *Help Science by Eating Apples*. Murdoch University, 2024 May 6, https://www.murdoch.edu.au/news/articles/help-science-by-eating-apples.

Polyphenols Market Size, Share & Trends Analysis Report by Product (Grape Seed, Green Tea, Apple, Cocoa), by Application (Beverages, Food, Feed), by Region, And Segment Forecasts, 2023–2030. https://www.grandviewresearch.com/industry-analysis/polyphenols-market-analysis.

Rasines-Perea Z, Teissedre PL. Grape Polyphenols' Effects in Human Cardiovascular Diseases and Diabetes. *Molecules*. 2017 Jan 1;22(1):68. https://doi.org/10.3390/molecules22010068.

Sarkhosh-Khorasani S, Sangsefidi ZS, Hosseinzadeh M. The Effect of Grape Products Containing Polyphenols on Oxidative Stress: A Systematic Review and Meta-analysis of Randomized Clinical Trials. *Nutr J*. 2021 Mar 12;20(1):25.. https://doi.org/10.1186/s12937-021-00686-5. PMID: 33712024; PMCID: PMC7971097.

Wan S, Luo J, Zhu Y, An P, Luo Y, Xing Q. The Effect of Antioxidant Polyphenol Supplementation on Cardiometabolic Risk Factors: A Systematic Review and Meta-Analysis. *Nutrients*. 2024 Dec 5;16(23):4206. https://doi.org/10.3390/nu16234206.

Xu T, Zhang X, Liu Y, Wang H, Luo J, Luo Y, An P. Effects of Dietary Polyphenol Supplementation on Iron Status and Erythropoiesis: A Systematic Review and Meta-analysis of Randomized Controlled Trials. *Am J Clin Nutr*. 2021 Aug 2;114(2):780–793. https://doi.org/10.1093/ajcn/nqab068. PMID: 33871598.

Zhang Z, Li X, Sang S, McClements DJ, Chen L, Long J, Jiao A, Jin Z, Qiu. Polyphenols as Plant-Based Nutraceuticals: Health Effects, Encapsulation, Nano-Delivery, and Application. *Foods*. 2022 Jul 23;11(15):2189. https://doi.org/10.3390/foods11152189. PMID: 35892774; PMCID: PMC9330871.

13 Regulatory Considerations, Safety, and Toxicology of Polyphenol Nutraceuticals

Chandra Prakash, Maneesh Kumar Gupta, and Deepak Sharma

13.1 INTRODUCTION

The word "nutraceuticals" is a combination of "nutrition" and "pharmaceuticals", which are foods or food products providing health and medical benefits, including the prevention and treatment of several chronic diseases (DeFelice 1995). Nutraceuticals are widely known to have physiological benefits or protect against chronic diseases. These products include nutrients, food supplements, genetically modified foods, herbal products, and processed foods like cereals and beverages. Numerous nutraceuticals are naturally occurring bioactive compounds that can be extracted from different food sources, particularly fruits and vegetables.

Polyphenols are plant-derived secondary metabolites consisting of one or more phenol rings and linked hydroxyl groups. Phenolic rings are categorized into many classes based on their strength: phenolic acids, flavonoids, stilbenes, and lignans (Abbas et al. 2017; Cutrim and Cortez 2018). More than half of phenolic compounds exhibit pharmacological properties, e.g., antibacterial, antifungal, anti-inflammatory, and antitumour activities (Rasouli et al. 2017). Although these food components are not essential for the body's metabolism, they can improve consumers' overall health by improving their physiological activity (Cutrim and Cortez 2018). As polyphenolic nutraceuticals are thought to be powerful agents for enhancing human health, there is a surge of research on these molecules (Valdés et al. 2015). Furthermore, polyphenols are natural antioxidants and improve the nutritional value of food by preventing lipid peroxidation; thus adding them to food products can increase oxidative stability and extend their shelf life (Cvejić et al. 2022).

With recent advances in polyphenol nutraceuticals, researchers and medical practitioners are creating frameworks for integrating and evaluating information from clinical studies on complementary and alternative medicines into medical practice

DOI: 10.1201/9781003588504-13

(Keservani et al. 2014). Moreover, given their potential therapeutic roles, understanding polyphenolic nutraceuticals' regulatory considerations, safety, and toxicological aspects is imperative as their market grows.

13.2 REGULATORY CONSIDERATIONS OF POLYPHENOL NUTRACEUTICALS

Nutraceuticals are not typical foods; they contain important components with therapeutic potential or are designed to meet specific nutritional requirements (Puri et al. 2022). Any foodstuff containing nutraceuticals should disclose that these products have different components besides regular food nutrients. Moreover, such food products must specify that the portion is derived from a plant liquid extract, a dry powder, or an extract in the form of ethyl alcohol and hydroalcoholic extract and added in combination or alone (FDA Guidelines 2021). Enzymes, vitamins, minerals, proteins, amino acids, and their metal components should not exceed the limitations (Alagawany et al. 2021). Currently, polyphenols have gained popularity worldwide, and using them as nutraceuticals requires consideration of regulatory aspects. The regulation of these products is affected by geographic areas, each of which has laws regulating dietary supplements and functional foods. Herein, we will look at the regulatory considerations for polyphenols among the key players in terms of the market and consumers.

13.2.1 Regulatory Considerations Around the World

The gross value of the global nutraceuticals market was estimated at USD 419.93 billion in 2023, and it is expected to increase at a compound annual growth rate of 9.95% from USD 457.35 billion in 2024 to USD 976.74 billion in 2032 (Nutraceuticals Market Size, Trends & Growth Report 2024). Moreover, the nutraceutical industry has experienced significant growth over the last two decades, with annual growth rates of 15% and 20%, respectively. Japan, the United States, and the European Union (EU) are the key players in the nutraceuticals industry, accounting for more than 90% of the global market share, including 39%, 31%, and 28%, respectively (Puri et al. 2022). The fast-growing nutraceutical industry, including polyphenols, requires proper regulation for its marketing and manufacturing, ensuring their safety and proper labelling. However, regulatory differences among the nations involved in the pharmaceutical industry pose a major obstacle to the globalization of the polyphenolic nutraceutical industry. The safety and quality of the products are the main concerns of international regulatory bodies, as these products are exported worldwide. Overall, maintaining the safety and quality of polyphenol nutraceuticals is crucial in following the guidelines imposed by the relevant agencies.

In the United States, the Dietary Supplement Health and Education Act was passed in 1994, which regulates the production and marketing of nutraceuticals. According to this act, the producers are responsible for verifying product safety before its marketing (Young and Bass 1995), and the Food and Drug Administration (FDA) reports harmful dietary products once they are on the market (Kennett 2019).

Since 2007, the FDA and the U.S. Department of Health and Human Services have enforced federal Good Manufacturing Practices regulations (21 CFR, Part 111). The regulations state that Indian as well as foreign companies involved in manufacturing, packaging, and labelling of nutraceuticals, including their testing, quality control, and distribution, must ensure the safety of the components used in those products. They should also report major adverse incidences of nutraceuticals, including polyphenols, to the FDA.

The European Food and Safety Authority oversees food regulations in the EU, specifically regulations related to "food supplements" (Patanè et al. 2019). According to the EU, food supplements are concentrated sources of nutrients, including proteins, vitamins, minerals, and other compounds with beneficial effects. The EU's 2002/46/EC directive governs dietary supplements. The EU has compiled a list of authorized vitamin and mineral additives that can be included in dietary supplements to meet specific nutritional requirements. Bourne Partners asserts that the European market is based on these health claims and Europe has greater restrictions on product claims. However, differences have been noticed in the case of product claims and recommended daily allowances for nutraceuticals. To deal with this, European nutraceutical industries have formed a coalition called "Food Supplements Europe" that coordinates with various regulatory bodies to provide the regulatory environment for the future.

Nutraceuticals are very popular in Japan and help a large portion of the global population to maintain health and hygiene. The nutraceutical market was established in Japan back in the 1980s, and at that time it was the first country to do so. In Japan, food and agricultural products must comply with several labelling rules. The Ministry of Agriculture, Forestry, and Fisheries established mandatory quality labelling criteria that should be followed by producers, distributors, and other operators (Japanese Agricultural Standards [JAS]: MAFF 2024). The Ministry of Health, Labour, and Welfare (MHLW) imposes voluntary and statutory labelling regulations for processed foods and beverages, including nutritional and allergen information (Akiyama et al. 2011). The food-sector laws in Japan are classified into two categories: foods for specific health use (FOSHU) and non-FOSHU. The MHLW is authorized to make FOSHU product claims regarding their physiological effects on the body, as they include chemicals with specialized health effects. Furthermore, the selling of FOSHU foods is evaluated for their safety and efficacy and must be certified by the Ministry of Health. FOSHU products are consumed to maintain and promote overall health as well as control specific health issues such as blood pressure and cholesterol. The FOSHU system was first introduced in 1991 with fewer than 20 items and has now increased to more than 1000 products. Non-FOSHU products typically include beneficial substances such as green tea extracts, glucosamine, and fish oil, but they rarely make a claim other than "good for your health". Specific claims are only allowed for FOSHU products, which must carefully show their efficacy (Iwatani and Yamamoto 2019). However, they still lack proper marketing and production regulations for nutraceuticals as is done in the case of pharmaceutical drugs (Siddiqui and Moghadasian 2020).

The Chinese government appointed the "China Health Care Association" to manage the nutraceutical industry. Three major entities shape industry policies. China's

State Food and Drug Administration (SFDA) regulates dietary supplements and awards "blue hat" registrations. The Ministry of Health oversees the SFDA and is responsible for approving new food ingredients in the dietary supplement industry. The Administration of Quality Supervision Inspection and Quarantine controls all imports and exports across China's borders (Liu et al. 2019).

13.2.2 Regulatory Considerations in India

India is currently the world's second-most populous country and is projected to become the most populated by 2027. According to Indian standards, nutraceuticals must include stated ingredients and general qualities. In India, nutraceuticals are classified as either functional foods (fortified foods and beverages) or dietary supplements (vitamins, minerals, protein supplements, and herbal products including polyphenols). Thus, polyphenol nutraceuticals fall under the dietary supplements category (Malve and Bhalerao 2023). Over the years numerous acts have been passed to regulate pharmaceuticals, including nutraceuticals. The Prevention of Food Adulteration Act and the Fruit Product Order were passed in 1954 and 1955, respectively. In 1998, the Prime Minister's Council on Trade and Industry established a food regulatory authority to unify licensing procedures, common application forms and guidelines for food business registration; central licensing; and safety, sanitary, and hygienic conditions (Keservani et al. 2019).

In 2002, a national nonprofit organization was founded to ensure that all food manufacturers submit scientific verification of their products' health claims and safety. In 2003, an expert panel on food legislation from the Ministry of Health recommended that safety testing for food products is necessary. In 2005, the agriculture and other standing committees of Parliament issued their 12th report, recommending the establishment of a unified regulatory agency with integrated legislation. The Indian Food Safety and Standards Bill was approved in 2005, and the president signed it on August 23, 2006. Finally, the Government of India established the Food Safety and Standards Authority of India (FSSAI) in 2006. The Food Safety and Standards Act (FSSA) aimed to harmonize the country's food safety regulations and methodically and scientifically grow the food processing industry, shifting it from a regulated regime to one of self-compliance. The FSSAI Act was passed in 2006, and several rules and regulations concerning nutraceuticals have been in effect since 2008 (Newton 2023). The act has two main goals: (1) to create a single food-related act and (2) to promote scientific progress in the food processing industry. The FSSAI Act 2006 contains 12 chapters, 101 sections, and two schedules. Chapter 4, Article 22 of the act discusses nutraceuticals, functional foods, and dietary supplements, as well as the importance of regulating these products so that they can be manufactured, marketed, distributed, or imported by anybody.

13.3 SAFETY AND TOXICOLOGY OF POLYPHENOL NUTRACEUTICALS

Polyphenols are a large group of compounds comprising phenolic acids, flavonoids, lignans, stilbenes, and, in some classifications, tannins and coumarins (Pandey and

Rizvi 2009; Makarewicz et al. 2021). All of these compounds are available in fruits, vegetables, and legumes and have a broad range of biological properties. Furthermore, many people consume polyphenols in the form of dietary supplements containing either purified polyphenols, e.g., quercetin (Askari et al. 2012) and resveratrol (Ramírez-Garza et al. 2018), or extracts from raw materials high in these compounds (e.g., tea, red wine, grapes, blueberries, pomegranate, etc.) (Martin 2009; Myburgh 2014; Petrisor et al. 2022). Polyphenols gained popularity among scientists and the public as natural medications due to their diverse biological activities and low toxicity. Therefore, they are some time consumed at higher than the prescribed intake limits in the form of natural health products and dietary supplements. Herein, we discuss polyphenols' safety and potential toxicological effects when consumed beyond the recommended intake limits.

13.3.1 Carcinogenic Effects of Polyphenols

Polyphenols have adverse effects on various types of cells by promoting mutagenesis or carcinogenicity. Numerous studies showed that flavonoids interact with topoisomerase IIα and IIβ and can damage DNA through pro-oxidant action. A study demonstrated that genistein accelerates DNA cleavage by human topoisomerase IIα and IIβ (Bandele and Osheroff 2007). Studies also revealed the regulatory aspect of polyphenols by inhibiting topoisomerases through different mechanisms. Polyphenols like epigallocatechin gallate (EGCG) and (-)-epigallocatechin are redox-dependent topoisomerase II toxins, while kaempferol and quercetin are classical toxins. Myricetin uses both methods, and (-)-epicatechin gallate and (-)-epicatechin have no substantial activity (Bandele et al. 2008). Adding -OH groups at C30 and C50 improved the redox activity of the B ring, allowing the polyphenols to behave as a redox-dependent toxin. However, C40-OH is crucial for classical toxicity. This implies that other polyphenols with similar structures could be potential topoisomerase toxins.

The polyphenolic substances have also shown mutagenic properties. A study by Spada et al. (2008) determined the mutagenic effect of a total of 23 frozen fruit samples and found that acai, cashew apple, kiwi fruit, and strawberry pulps exhibit mutagenesis effects at high concentrations (5–15% [wt/vol]) in tested loci (Lys-revertant, His-revertant, and Hom-revertant) of a haploid XV 185–14C strain of *Saccharomyces cerevisiae*. Research studies demonstrated that polyphenols can cause cancer by inducing damage to chromosomes or intercalating with DNA. For instance, quercetin causes carcinogenic effects by H_2O_2-mediated DNA damage, leading to apoptosis or mutations, and luteolin causes apoptosis by topoisomerase II–dependent cleavage of DNA. Additional studies demonstrated that many bioflavonoids can cause chromosomal translocations by either topoisomerase II–independent (luteolin and kaempferol) or topoisomerase II–dependent (myricetin, genistein, and quercetin) pathways. Among hundreds of naturally occurring flavonoids, the flavones C-glycosides vitexin and orientin displayed significant sister chromatid exchange–inducing action, while other compounds had minimal or no activity.

13.3.2 Hepato-Renal Toxic Effects of Polyphenols

The available evidence from in vitro and in vivo studies demonstrated the protective effect of polyphenols on the liver and kidney (Papackova et al. 2018; Levin et al. 2019; Kandemir et al. 2020). The adverse effects of high doses of these compounds have also been reported (Lambert et al. 2010; Levin et al. 2019).

A single dose of EGCG (1500 mg/kg) resulted in enhanced plasma alanine aminotransferase activity and a remarkably decreased survival rate in male CF-1 mice. Moreover, administering 750 mg/kg of EGCG caused moderate to severe hepatic necrosis (Lambert et al. 2010). The consumption of higher amounts of dietary supplements containing green tea or other EGCG-containing products is known to activate inflammatory cascades that ultimately result in hepatotoxicity (James et al. 2018; Hu et al. 2018). Two participants of a double-blind, randomized clinical trial of EGCG were forced to be excluded from the study due to its hepatotoxic effects, while other participants tolerated this compound (Levin et al. 2019), suggesting that the consumption of EGCG of more than 1200 mg/day should be done with caution. In addition, lifestyle and genetic variables could be significant determinants of vulnerability to EGCG-mediated hepatotoxicity.

The nephrotoxic effects of EGCG have also been reported in vivo, as is evident with enhanced serum creatinine, a well-known biomarker of nephropathy. A study by Rasheed et al. (2017) evaluating the protective effect of EGCG (100 mg/kg/day) in diabetic mice reported that the compound causes impairment in the renal tissues. While it is unclear if EGCG-induced nephrotoxicity is dose-dependent, a few studies suggested that duration of administration is a key element in the progression of toxicity. Hence, it can be speculated that though EGCG reduces or prevents diabetes, patients should be cautious while taking it as a nutraceutical.

Another polyphenol that has been linked to nephrotoxicity is quercetin. In 1992, the U.S. National Toxicology Program conducted research on male and female rats fed various amounts of quercetin over 104 weeks. Male rats with higher doses developed chronic nephropathy and kidney adenomas, whereas female rats showed no evident renal damage. Similar results were also observed on re-evaluation of renal histology in quercetin-fed Fischer 344 rats with mild and high quercetin doses (Hard et al. 2007). Moreover, the consumption of a high dose of quercetin in a 6-month study exhibited nephropathy in rats (Andres et al. 2018). Altogether, these studies suggest that consuming higher doses of quercetin may cause kidney damage in animals and humans, despite the absence of adverse effects on the kidney in clinical trials. Therefore, caution should be taken when people with renal inadequacy consume quercetin.

13.3.3 Hormonal Unbalancing Effects of Polyphenols

Isoflavones are structurally similar to oestrogens; therefore, they can act as oestrogens and anti-oestrogens in different tissues. Although these polyphenols are known to be beneficial for postmenopausal women, their effect on reproductive-aged women may be less advantageous. A case report of three women with endometrial disease documented that abnormal uterine bleeding is linked to high soy consumption and removing the diet with soybeans and their products decreased or vanished these issues (Chandrareddy et al. 2008). The adverse effects of isoflavones have also been documented in a placebo-controlled crossover study (Hutchins et al. 2005). The single participant (a 51-year-old, healthy, postmenopausal female) developed high blood pressure (226/117) during the first of four randomly assigned treatments (500 mg vitamin C and 5 mg/kg body weight soy isoflavones), necessitating a medical intervention and terminating her participation in the study.

Accumulating evidence indicates that isoflavones are safe or even helpful for women while affecting the sexual life of males. Adult Sprague-Dawley rats fed on a phytoestrogen-rich diet (containing about 600 µg/g isoflavones) had significantly lower plasma levels of androstenedione and testosterone along with lower body and prostate weights (Weber et al. 2001). Another study documented that a 60-year-old man who consumed soy milk every day experienced bilateral gynecomastia and decreased libido and erectile problems along with four times increase in oestradiol and estrone concentrations (Martinez and Lewi 2008). Contrary to these findings, a meta-analysis including the results of several human studies indicates that soy products and isoflavone supplements do not affect male reproductive hormones (Reed et al. 2021). Thus, it can be predicted that soy products and isoflavones cannot have feminizing effects when taken at amounts equivalent to or even significantly higher than those normal for males.

Soy isoflavones are known to have no significant effect on healthy individuals; however, they inhibit thyroid hormones in persons with iodine insufficiency. An in vivo study on rats showed that genistein reduces thyroid peroxidase (TPO) dose-dependently (Chang and Doerge 2000). Another study found that genistein and daidzein cause microfollicular alterations in thyroid tissue and lower thyroid hormone levels in orchidectomized middle-aged male rats (Šošić-Jurjević et al. 2010). Polyphenols like quercetin, fisetin, kaempferol, naringenin, genistein, and daidzein have been reported to inhibit the activity of TPO (Divi and Doerge 1996; Divi et al. 1997). Individuals with subclinical hypothyroidism supplemented with 16 mg soy phytoestrogens increased the risk of developing overt hypothyroidism threefold while reducing insulin resistance, high-sensitivity C-reactive protein (hsCRP), and both systolic and diastolic blood pressure (Sathyapalan et al. 2011).

A randomized controlled trial involving Chinese postmenopausal women who were equol-producers showed that both soy and daidzein are safe to consume and do not affect thyroid function (Liu et al. 2021). Similarly, soy products or isoflavones did not show any remarkable change in thyroid function in 14 other trials involving individuals who do not have thyroid function as a major health outcome (Messina and Redmond 2006). Therefore, there is little indication that soy products or isoflavones negatively affect thyroid function in euthyroid or iodine-replete people. In contrast, soy meals may raise the dosage of thyroid hormones in hypothyroid patients by preventing absorption.

Polyphenols like genistein, glycitein, and daidzein are known to compete with thyroxine attachment to transthyretin, a primary transport protein for thyroid hormones (Radović et al. 2006). Isoflavones are known to change the concentration of thyroid hormones during development and, in adults, disturb the feedback control of hormonal networks, including the pituitary-thyroid-periphery axis, therefore decreasing tissue availability and metabolism. Isoflavones from soybeans uniquely affect the activity of thyroid hormone receptors (TRs). Genistein and daidzein increase TR-mediated transcription by directly binding to the TR-ligand–binding domain, which attracts steroid receptor coactivators (Ariyani et al. 2018). A study demonstrated that children fed soy-based milk formulas are more likely to develop autoimmune thyroid disorders compared to the healthy controls (Fort et al. 1990), suggesting soy formula feeding during infancy is linked to an autoimmune thyroid disorder. Overall, the findings of these studies are unclear and inconsistent; thus more research is required to determine whether isoflavone use is safe or has negative effects on the human population.

13.3.4 Effects of Polyphenols on Gut Microbiota

Polyphenols have detrimental effects on the digestive system as they can influence the intestinal microbiota and suppress the production of digestive enzymes. Generally, there is a complex association between the digestive tract's bacteria and the dietary polyphenols. Firstly, the bacterial enzymes release polyphenols from the dietary matrix, increasing their bioaccessibility and bioavailability. Next, bacteria convert and degrade these polyphenols, typically by double-bond reduction and hydrolysis processes (Kumar Singh et al. 2019; Sejbuk et al. 2024). Everything transported to the intestine along with food is transformed by intestinal microbiota, as they can catalyse several reactions that are not accessible to human enzymes. Intestinal bacteria use enzymes like β-glucosidase, β-rhamnosidase, β-glucuronidase, β-galactosidase, esterases, and sulfatases to break down glycosides, glucuronides, ester, sulphates, amides, and lactones. In addition, they catalyse aromatic ring cleavage, decarboxylation (decarboxylase), demethylation (demethylase), isomerization (isomerase), reduction (reductases and hydrogenases), and dehydroxylation (dehydroxylase) (Duda-Chodak et al. 2015; Murota et al. 2018). These reactions degrade glycosides (glucosides, galactosides, rhamnosides, rutinosides, etc.) or polymeric polyphenols (such as ellagitannins, tannins, and procyanidins) to simpler molecules, which are then converted to phenolic acids and metabolic products. Polyphenol metabolites have radically different bioactive properties than their parent molecules. Typically, metabolites have lesser antioxidant capabilities; nevertheless, polyphenols acquire biological action only after bacterial biotransformation (Heleno et al. 2015). The well-known examples are the conversions of soya isoflavone daidzein to S-equol and O-desmethylangolensin, ellagitannins to urolithin A, and secoisolariciresinol to entererodiol and enterolactone (Wang et al. 2000; Saha et al. 2016; Mayo et al. 2019).

Polyphenols, on the other hand, have been known for many years to suppress the growth of a variety of bacteria, both gram-positive and gram-negative (Manso et al. 2021). Unfortunately, this inhibitory effect is not exclusive to infections. For example, extracts obtained from cloudberry, raspberry, strawberry, etc., exhibited strong antibacterial action against *Salmonella enterica ser, Typhimurium, Escherichia coli,* and *Lactobacillus rhamnosus* (Puupponen-Pimiä et al. 2001). Strawberry extract is an efficient inhibitor of *Enterococcus faecalis* and *Bifidobacterium lactis*. Myricetin suppressed the growth of all lactic acid bacteria isolated from the human gastrointestinal tract flora, without affecting *Salmonella typhimurium* and *Lactobacillus plantarum*. Luteolin has been shown to have a bacteriostatic effect on several Lactobacillus species, *B. lactis*, and *E. faecalis*, but not on gram-negative bacteria. Apigenin, (+)-catechin, kaempferol, isoquercitrin, and rutin had no effect, whereas phenolic acids suppressed the growth of *E. coli* and *S. enterica*.

13.4 FUTURE PERSPECTIVES

The future of polyphenol nutraceuticals seems promising, with advances in research, technology, and personalized healthcare providing new opportunities for developing more effective and safer products. As the usage of polyphenol nutraceuticals grows, there will be a greater demand for extensive regulatory frameworks to ensure the safety and quality of these products. While current regulatory mechanisms are generally functional, they may need to be changed for accommodating the intricacies of polyphenol nutraceuticals.

One important area of consideration is the creation of standardized testing techniques for polyphenol-containing products. Given the differences in the chemical makeup of polyphenols, based on their sources and extraction process, regulatory authorities are required to develop stringent standards for identifying and quantifying individual polyphenols in supplements. This would help to ensure that consumers get the correct therapeutic dose and that the polyphenolic products remain free of contaminants and adulterants. Furthermore, the safety of polyphenol supplements, particularly those containing high quantities of active components, must be closely checked. Tighter upper intake limits for certain polyphenols may be required to avoid toxicity. Collaborations among researchers, regulators, and manufacturers are required to fully understand the potential of polyphenol nutraceuticals and resolve the safety issues. Hence, in the future, regulatory agencies may need to implement more proactive post-market surveillance systems that can track adverse effects and other safety issues, as they do in the case of pharmaceuticals.

13.5 CONCLUSION

The increasing use of polyphenol nutraceuticals raises both opportunities and risks. While these bioactive compounds have several health benefits, their safe and effective use requires careful consideration of regulatory aspects, doses, bioavailability, and personal health characteristics. To reduce potential health risks, manufacturers must ensure that polyphenol-based products are safe for consumers and thoroughly follow local legislation. To ensure their safety and efficacy, future research should include long-term clinical trials to evaluate the effects of polyphenol supplementation, modify dosage recommendations, and identify potential toxicological risks. As a result, polyphenol nutraceuticals can continue to provide considerable health advantages while protecting people from devastating diseases.

REFERENCES

Abbas M, Saeed F, Anjum FM, et al (2017) Natural Polyphenols: An Overview. *Int J Food Prop* 20:1689–1699. https://doi.org/10.1080/10942912.2016.1220393

Akiyama H, Imai T, Ebisawa M (2011) Japan Food Allergen Labeling Regulation – History and Evaluation. *Adv Food Nutr Res* 62:139–171. https://doi.org/10.1016/B978-0-12-385989-1.00004-1

Alagawany M, Elnesr SS, Farag MR, et al (2021) Nutritional Significance of Amino Acids, Vitamins and Minerals as Nutraceuticals in Poultry Production and Health – A Comprehensive Review. *Vet Q* 41:1–29. https://doi.org/10.1080/01652176.2020.1857887

Andres S, Pevny S, Ziegenhagen R, et al (2018) Safety Aspects of the Use of Quercetin as a Dietary Supplement. *Mol Nutr Food Res* 62:1700447. https://doi.org/10.1002/MNFR.201700447

Ariyani W, Iwasaki T, Miyazaki W, et al (2018) A Possible Novel Mechanism of Action of Genistein and Daidzein for Activating Thyroid Hormone Receptor-Mediated Transcription. *Toxicol Sci* 164:417–427. https://doi.org/10.1093/TOXSCI/KFY097

Askari G, Ghiasvand R, Feizi A, et al (2012) The Effect of Quercetin Supplementation on Selected Markers of Inflammation and Oxidative Stress. *J Res Med Sci* 17:637–641.

Bandele OJ, Clawson SJ, Osheroff N (2008) Dietary Polyphenols as Topoisomerase II Poisons: B Ring and C Ring Substituents Determine the Mechanism of Enzyme-Mediated DNA Cleavage Enhancement. *Chem Res Toxicol* 21:1253–1260. https://doi.org/10.1021/tx8000785

Bandele OJ, Osheroff N (2007) Bioflavonoids as Poisons of Human Topoisomerase IIα and IIβ. *Biochemistry* 46:6097–6108. https://doi.org/10.1021/bi7000664

Chandrareddy A, Muneyyirci-Delale O, McFarlane SI, Murad OM (2008) Adverse Effects of Phytoestrogens on Reproductive Health: A Report of Three Cases. *Complement Ther Clin Pract* 14:132–135. https://doi.org/10.1016/J.CTCP.2008.01.002

Chang HC, Doerge DR (2000) Dietary Genistein Inactivates Rat Thyroid Peroxidase in Vivo without an Apparent Hypothyroid Effect. *Toxicol Appl Pharmacol* 168:244–252. https://doi.org/10.1006/TAAP.2000.9019

Cutrim CS, Cortez MAS (2018) A Review on Polyphenols: Classification, Beneficial Effects and their Application in Dairy Products. *Int J Dairy Technol* 71:564–578. https://doi.org/10.1111/1471-0307.12515

Cvejić J, Atanacković Krstonošić M, Mikulić M, Miljić U (2022) Polyphenols. In: *Nutraceutical and Functional Food Components*. Elsevier, pp 243–312.

DeFelice SL (1995) The Nutraceutical Revolution: Its Impact on Food Industry R&D. *Trends Food Sci Technol* 6:59–61. https://doi.org/10.1016/S0924-2244(00)88944-X

Divi RL, Chang HC, Doerge DR (1997) Anti-Thyroid Isoflavones from Soybean: Isolation, Characterization, and Mechanisms of Action. *Biochem Pharmacol* 54:1087–1096. https://doi.org/10.1016/S0006-2952(97)00301–8

Divi RL, Doerge DR (1996) Inhibition of Thyroid Peroxidase by Dietary Flavonoids. *Chem Res Toxicol* 9:16–23. https://doi.org/10.1021/TX950076M

Duda-Chodak A, Tarko T, Satora P, Sroka P (2015) Interaction of Dietary Compounds, Especially Polyphenols, with the Intestinal Microbiota: A Review. *Eur J Nutr* 54:325–341. https://doi.org/10.1007/s00394-015-0852-y

Fort P, Moses N, Fasano M, et al (1990) Breast and Soy-formula Feedings in Early Infancy and the Prevalence of Autoimmune Thyroid Disease in children. *J Am Coll Nutr* 9:164–167. https://doi.org/10.1080/07315724.1990.10720366

Guidance Documents & Regulatory Information by Topic (Food and Dietary Supplements) | FDA. https://www.fda.gov/food/guidance-regulation-food-and-dietary-supplements/guidance-documents-regulatory-information-topic-food-and-dietary-supplements. Accessed 8 Apr 2025a.

Hard GC, Seely JC, Betz LJ, Hayashi S mo (2007) Re-evaluation of the Kidney Tumors and Renal Histopathology Occurring in a 2-Year Rat Carcinogenicity Bioassay of Quercetin. *Food Chem Toxicol* 45:600–608. https://doi.org/10.1016/J.FCT.2006.10.018

Heleno SA, Martins A, Queiroz MJRP, Ferreira ICFR (2015) Bioactivity of Phenolic Acids: Metabolites Versus Parent Compounds: A Review. *Food Chem* 173:501–513. https://doi.org/10.1016/J.FOODCHEM.2014.10.057

Hu J, Webster D, Cao J, Shao A (2018) The Safety of Green Tea and Green Tea Extract Consumption in Adults – Results of a Systematic Review. *Reg Toxicol Pharmacol* 95:412–433. https://doi.org/10.1016/J.YRTPH.2018.03.019

Hutchins AM, McIver IE, Johnston CS (2005) Hypertensive Crisis Associated with High Dose Soy Isoflavone Supplementation in a Post-menopausal Woman: A Case Report [ISRCTN98074661]. *BMC Women's Health* 5:9. https://doi.org/10.1186/1472-6874-5-9

Iwatani S, Yamamoto N (2019) Functional Food Products in Japan: A Review. *Food Sci Hum Wellness* 8:96–101. https://doi.org/10.1016/J.FSHW.2019.03.011

James KD, Kennett MJ, Lambert JD (2018) Potential Role of the Mitochondria as a Target for the Hepatotoxic Effects of (-)-Epigallocatechin-3-gallate in Mice. *Food and Chemical Toxicology* 111:302–309. https://doi.org/10.1016/J.FCT.2017.11.029

Japanese Agricultural Standards (JAS). MAFF. https://www.maff.go.jp/e/policies/standard/jas/index.html. Accessed 28 Dec 2024c.

Kandemir FM, Yıldırım S, Kucukler S, et al (2020) Protective Effects of Morin against Acrylamide-Induced Hepatotoxicity and Nephrotoxicity: A Multi-biomarker Approach. *Food Chem Toxicol* 138:111190. https://doi.org/10.1016/J.FCT.2020.111190

Kennett G (2019) Time for Change: Stepping Up the FDA's Regulation of Dietary Supplements to Promote Consumer Safety and Awareness. *J Law Health* 33:47–78.

Keservani RK, Jarouliya U, Sharma AK (2019) Nutraceutical and Functional Food Regulations in India. In: *Nutraceutical and Functional Food Regulations in the United States and around the World* (Third Edition). Elsevier, pp 431–444.

Keservani RK, Sharma AK, Ahmad F, Baig ME (2014) Nutraceutical and Functional Food Regulations in India. In: *Nutraceutical and Functional Food Regulations in the United States and around the World* (Second Edition). Elsevier, pp 327–342.

Kumar Singh A, Cabral C, Kumar R, et al (2019) Beneficial Effects of Dietary Polyphenols on Gut Microbiota and Strategies to Improve Delivery Efficiency. *Nutrients* 11:2216. https://doi.org/10.3390/nu11092216

Lambert JD, Kennett MJ, Sang S, et al (2010) Hepatotoxicity of High Oral Dose (−)-Epigallocatechin-3-gallate in Mice. *Food Chem Toxicol* 48:409–416. https://doi.org/10.1016/J.FCT.2009.10.030

Levin J, Maaß S, Schuberth M, et al (2019) Safety and Efficacy of Epigallocatechin Gallate in Multiple System Atrophy (PROMESA): A Randomised, Double-Blind, Placebo-controlled Trial. *Lancet Neurol* 18:724–735. https://doi.org/10.1016/S1474-4422(19)30141–3

Liu Z, Mutukumira AN, Chen H (2019) Food Safety Governance in China: From Supervision to Coregulation. *Food Sci Nutr* 7:4127–4139. https://doi.org/10.1002/FSN3.1281

Liu Z min, Zhang D, Li G, et al (2021) The 6-month Effect of Whole Soy and Purified Isoflavones Daidzein on thyroid function – A Double-blind, Randomized, Placebo Controlled Trial Among Chinese Equol-producing Postmenopausal Women. *Phytother Res* 35:5838–5846. https://doi.org/10.1002/PTR.7244

Makarewicz M, Drożdż I, Tarko T, Duda-Chodak A (2021) The Interactions between Polyphenols and Microorganisms, Especially Gut Microbiota. *Antioxidants* 10:188. https://doi.org/10.3390/antiox10020188

Malve H, Bhalerao P (2023) Past, Present, and Likely Future of Nutraceuticals in India: Evolving Role of Pharmaceutical Physicians. *J Pharm Bioallied Sci* 15:68–74. https://doi.org/10.4103/JPBS.JPBS_96_23

Manso T, Lores M, de Miguel T (2021) Antimicrobial Activity of Polyphenols and Natural Polyphenolic Extracts on Clinical Isolates. *Antibiotics* 11:46. https://doi.org/10.3390/antibiotics11010046

Martin KR (2009) Polyphenols as Dietary Supplements: A Double-edged Sword. *Nutr Diet Suppl* 2:1. https://doi.org/10.2147/NDS.S6422

Martinez J, Lewi JE (2008) An Unusual Case of Gynecomastia Associated with Soy Product Consumption. *Endocr Prac* 14:415–418. https://doi.org/10.4158/EP.14.4.415

Mayo B, Vázquez L, Flórez AB (2019) Equol: A Bacterial Metabolite from The Daidzein Isoflavone and Its Presumed Beneficial Health Effects. *Nutrients* 11:2231. https://doi.org/10.3390/nu11092231

Messina M, Redmond G (2006) Effects of Soy Protein and Soybean Isoflavones on Thyroid Function in Healthy Adults and Hypothyroid Patients: A Review of the Relevant Literature. *Thyroid* 16:249–258. https://doi.org/10.1089/thy.2006.16.249

Murota K, Nakamura Y, Uehara M (2018) Flavonoid Metabolism: The Interaction of Metabolites and Gut Microbiota. *Biosci Biotechnol Biochem* 82:600–610. https://doi.org/10.1080/09168451.2018.1444467

Myburgh KH (2014) Polyphenol Supplementation: Benefits for Exercise Performance or Oxidative Stress? *Sports Med* 44:57–70. https://doi.org/10.1007/s40279-014-0151-4

Newton PI (2023) Food Safety and Standards Act, 2006: Punishment for Unsafe Food in Milk and Milk Products. *Int J Law Manag Hum*anities 6:1950–1965. https://doi.org/10.10000/IJLMH.114256

Nutraceuticals Market Size, Trends & Growth Report [2032]. https://www.fortunebusinessinsights.com/nutraceuticals-market–102530. Accessed 10 Sep 2024b.

Pandey KB, Rizvi SI (2009) Plant Polyphenols as Dietary Antioxidants in Human Health and Disease. *Oxid Med Cell Longev* 2:270–278. https://doi.org/10.4161/OXIM.2.5.9498

Papackova Z, Heczkova M, Dankova H, et al (2018) Silymarin Prevents Acetaminophen-induced Hepatotoxicity in Mice. *PLoS One* 13:e0191353. https://doi.org/10.1371/journal.pone.0191353

Patanè P, Laganà P, Devi P, et al (2019) Polyphenols and Functional Foods from the Regulatory Viewpoint. *J AOAC Int* 102:1373–1377. https://doi.org/10.1093/jaoac/102.5.1373

Petrisor G, Motelica L, Ficai D, et al (2022) New Mesoporous Silica Materials Loaded with Polyphenols: Caffeic Acid, Ferulic Acid and p-Coumaric Acid as Dietary Supplements for Oral Administration. *Materials* 15:7982. https://doi.org/10.3390/ma15227982

Puri V, Nagpal M, Singh I, et al (2022) A Comprehensive Review on Nutraceuticals: Therapy Support and Formulation Challenges. *Nutrients* 14:4637. https://doi.org/10.3390/nu14214637

Puupponen-Pimiä R, Nohynek L, Meier C, et al (2001) Antimicrobial Properties of Phenolic Compounds from Berries. *J Appl Microbiol* 90:494–507. https://doi.org/10.1046/j.1365-2672.2001.01271.x

Radović B, Mentrup B, Köhrle J (2006) Genistein and Other Soya Isoflavones are Potent Ligands for Transthyretin in Serum and Cerebrospinal Fluid. *Br J Nutr* 95:1171–1176. https://doi.org/10.1079/BJN20061779

Ramírez-Garza S, Laveriano-Santos E, Marhuenda-Muñoz M, et al (2018) Health Effects of Resveratrol: Results from Human Intervention Trials. *Nutrients* 10:1892. https://doi.org/10.3390/nu10121892

Rasheed NOA, Ahmed LA, Abdallah DM, El-Sayeh BM (2017) Nephro-toxic Effects of Intraperitoneally Injected EGCG in Diabetic Mice: Involvement of Oxidative Stress, Inflammation and Apoptosis. *Sci Rep* 7:40617. https://doi.org/10.1038/srep40617

Rasouli H, Farzaei MH, Khodarahmi R (2017) Polyphenols and their Benefits: A Review. *Int J Food Prop* 20:1–42. https://doi.org/10.1080/10942912.2017.1354017

Reed KE, Camargo J, Hamilton-Reeves J, et al (2021) Neither Soy Nor Isoflavone Intake Affects Male Reproductive Hormones: An Expanded and Updated Meta-analysis of Clinical Studies. *Reprod Toxicol* 100:60–67. https://doi.org/10.1016/j.reprotox.2020.12.019

Saha P, Yeoh BS, Singh R, et al (2016) Gut Microbiota Conversion of Dietary Ellagic Acid into Bioactive Phytoceutical Urolithin A Inhibits Heme Peroxidases. *PLoS One* 11:e0156811. https://doi.org/10.1371/journal.pone.0156811

Sathyapalan T, Manuchehri AM, Thatcher NJ, et al (2011) The Effect of Soy Phytoestrogen Supplementation on Thyroid Status and Cardiovascular Risk Markers in Patients with Subclinical Hypothyroidism: A Randomized, Double-Blind, Crossover Study. *J Clin Endocrinol Metable* 96:1442–1449. https://doi.org/10.1210/JC.2010-2255

Sejbuk M, Mirończuk-Chodakowska I, Karav S, Witkowska AM (2024) Dietary Polyphenols, Food Processing and Gut Microbiome: Recent Findings on Bioavailability, Bioactivity, and Gut Microbiome Interplay. *Antioxidants* 13:1220. https://doi.org/10.3390/antiox13101220

Siddiqui RA, Moghadasian MH (2020) Nutraceuticals and Nutrition Supplements: Challenges and Opportunities. *Nutrients* 12:1593. https://doi.org/10.3390/nu12061593

Šošić-Jurjević B, Filipović B, Ajdžanović V, et al (2010) Suppressive Effects of Genistein and Daidzein on Pituitary–thyroid Axis in Orchidectomized Middle-aged Rats. *Exp Biol Med* 235:590–598. https://doi.org/10.1258/ebm.2009.009279

Spada PDS, de Souza GGN, Bortolini GV, et al (2008) Antioxidant, Mutagenic, and Antimutagenic Activity of Frozen Fruits. *J Med Food* 11:144–151. https://doi.org/10.1089/jmf.2007.598

Valdés L, Cuervo A, Salazar N, et al (2015) The Relationship between Phenolic Compounds from Diet and Microbiota: Impact on Human Health. *Food Funct* 6:2424–2439. https://doi.org/10.1039/C5FO00322A

Wang L-Q, Meselhy MR, Li Y, et al (2000) Human Intestinal Bacteria Capable of Transforming Secoisolariciresinol Diglucoside to Mammalian Lignans, Enterodiol and Enterolactone. *Chem Pharm Bull* 48:1606–1610. https://doi.org/10.1248/cpb.48.1606

Weber KS, Setchell KDR, Stocco DM, Lephart ED (2001) Dietary Soy-phytoestrogens Decrease Testosterone Levels and Prostate Weight without Altering LH, Prostate 5alpha-reductase or Testicular Steroidogenic Acute Regulatory Peptide Levels in Adult Male Sprague-Dawley Rats. *J Endocrinol* 170:591–599. https://doi.org/10.1677/JOE.0.1700591

Young AL, Bass IS (1995) The Dietary Supplement Health and Education Act. *Food Drug Law J* 50:285–292.

14 Pharmacokinetics of Polyphenols

Moumita Datta, Dwaipayan Sinha, Moumita Chatterjee, and Srijonee Choudhury

14.1 INTRODUCTION

Polyphenols constitute a category of plant-derived functional dietary components prevalent in fruits and vegetables and are increasingly acknowledged for their preventive function in human health, especially within the framework of a diet rich in these foods (Arfaoui 2021; Cory et al. 2018). These chemicals possess antioxidant characteristics that safeguard the body against oxidative reactions, which cause numerous diseases. They are also characterized by diverse chemical structures and various biological actions (Shabbir et al. 2021; Muscolo et al. 2024). Pharmacokinetics of polyphenols include the examination of the mechanisms underlying the absorption and distribution of the consumed compound, its biochemical transformations within the body (through enzymatic processes), and the consequences and pathways of metabolite excretion (Li et al. 2019). The pharmacokinetic ADME framework (absorption, distribution, metabolism, and excretion) often includes both metabolism and excretion as integral components (Garza et al. 2025). The pharmacokinetics of polyphenols depend on their chemical structure (Stromsnes et al. 2021), dietary matrix (Eran Nagar et al. 2020), and gut bacteria (Sejbuk et al. 2024), which affect their bioavailability and subsequent biological effects. Despite their beneficial protective properties, dietary polyphenols exhibit inadequate transport into the body owing to their low bioavailability and metabolic instability (Koudoufio et al. 2020; Bešlo et al. 2023). The nanoformulation of polyphenols can resolve these challenges and enhance the delivery of drugs for the treatment of cancer, neurological disorders, and cardiovascular diseases (Squillaro et al. 2018; Anand et al. 2022). Biobased nanocarriers can be engineered to encapsulate, safeguard, and transport polyphenols, hence enhancing their oral bioavailability, which has extensive applications in health and nutrition formulations, including nutraceuticals (Zhang et al. 2022). The elements affecting bioavailability, methods for improving absorption, and possible interactions with other nutrients highlight the intricacy of translating these chemicals from botanical origins into efficacious nutraceuticals. This chapter will examine several formulations and biobased nanocarriers that can include polyphenols into functional meals and dietary supplements, focusing on stability and shelf-life issues.

DOI: 10.1201/9781003588504-14

14.2 ABSORPTION AND BIOAVAILABILITY

14.2.1 Low Bioavailability

Polyphenols exhibit limited oral bioavailability, indicating that only a minor proportion of consumed polyphenols enter the bloodstream (Manach et al. 2004). Following their administration, some polyphenolic compounds pass through the stomach and are absorbed, while others are absorbed in the small intestine (Spencer 2003; Scalbert et al. 2002). The absorption, distribution, bioavailability, and bioactivity of polyphenols and their metabolites following food consumption have emerged as significant research areas (Scalbert et al. 2002; Manach et al. 2004; Teng and Chen 2019).

Generally, phenolic compounds and their glycosides undergo absorption through the cleavage and release of aglycones, a process facilitated by digestive enzymes and microbial metabolism (Danielsen and Danielsen 2017). The principal enzyme involved is lactase phlorizin hydrolase (LPH), situated in the small intestinal epithelial cells. LPH selectively acts on flavonoid-O-β-glycosides (FBG), liberating aglycones that subsequently diffuse gradually into epithelial cells owing to their increased lipophilicity and closeness to the cell membrane (Kay et al. 2017). Additionally, cytosolic β-glucosidase (CBG) hydrolyzes specific phenolic glycosides once they have crossed the epithelium (Teng and Chen 2019).

Bioavailability is a crucial determinant of the biological efficacy of polyphenols (Manach et al. 2004). Indirect evidence of intestinal absorption includes increased plasma antioxidant capacity after consuming foods containing high polyphenols, which was observed with tea (Serafini et al. 1996; Van Het Hof et al. 1999), red wine (Duthie et al. 1998; Fuhrman et al. 1995; Maxwell et al. 1994; Serafini et al. 1998), and black currant and apple juice (Young et al. 1999). Direct proof has been obtained by quantifying certain phenolic compounds, such as flavan-3-ols, in human plasma and urine following the consumption in either pure form or diets with established quantities of the target ingredient (Di Pede et al. 2023).

14.3 ABSORPTION OF POLYPHENOLS

The molecular structure of polyphenols facilitates high absorption in their free state due to their lipophilic qualities, enabling them to traverse plasmatic membranes via easy diffusion (Gambini et al. 2015). Nonetheless, when they form complexes with other molecules, their absorption properties alter (Stromsnes et al. 2021). Flavonoid absorption from the gut happens through multiple mechanisms. Flavonoid aglycones are readily absorbed by intestinal cells due to their lipophilicity, which aids in traversing the mucosal phospholipid bilayer. LPH, located on the brush border membrane of the small intestinal epithelium, plays a vital role in the absorption of flavonoids with β-glycoside bonds (Suzuki and Hara 2011). LPH was shown to hydrolyze several flavonoid glycosides, including quercetin-4’-glucoside, quercetin-3-glucoside, genistein-7-glucoside, and daidzein-7-glucoside, resulting in the formation of aglycones (Day et al. 2000). The enzymatic conversion of flavonoids to their aglycone forms by LPH enhances their lipid solubility, facilitating their entry into intestinal epithelial cells through passive diffusion (Hu et al. 2025). Conversely, flavonoid monoglucosides may be translocated via sodium-glucose transporter-1 (SGLT-1) located on the brush border membrane of intestinal cells (Walgren et al. 2000; Kwon et al. 2007). It has been shown that SGLT-1 participated in the absorption of quercetin-3-glucoside in rat small intestine cells (Arts et al. 2002). The majority of flavonoid glycosides that enter enterocytes are deglycosylated by β-glucosidases, specifically the broad-specificity CBG (Németh et al. 2003). Flavonoids undergo glucuronidation, sulfation, and methylation in intestinal epithelial cells before entering circulation (Crespy et al. 1999; Piskula and Terao 1998; Radominska-Pandya et al. 1998); however, the degree of conversion is reportedly less in isoflavones and catechins compared to other flavonoids (Clarke et al. 2002; Hollman et al. 1997). The flavonoid conjugates subsequently enter hepatocytes, where they undergo methylation, glucuronidation, or sulfation (O'Leary et al. 2003). These flavonoid conjugates are eliminated in urine and bile fluid, subsequently re-entering the intestinal lumen (Matsukawa et al. 2009). Thereafter, they may be reabsorbed primarily in the large intestine. Subsequent metabolism transpires in the colon, where gut microflora enzymes facilitate the degradation of flavonoids into phenolic acids, which may be absorbed and subsequently metabolized in the liver (Spencer et al. 2008; Scheline 1991). As a result, particular types of flavonoids are found in the small and large intestines as well as in plasma, each capable of eliciting unique biological actions on intestinal epithelial cells (Suzuki and Hara 2011).

14.3.1 Factors Affecting Absorption

14.3.1.1 Chemical Structure

Polyphenols, notwithstanding their exceptional therapeutic qualities, encounter numerous restrictions like low bioavailability; instability in heat, light, and alkaline environments; rapid metabolism; and inadequate membrane permeability, which constrain their clinical use (Oliver et al. 2016; Monfalouti et al. 2024). Various strategies have been explored to enhance their bioavailability and stability. One such approach involves chemical modifications, including glycosylation and esterification, which significantly alter the properties of polyphenols (Ren et al. 2023; Peng and Shahidi 2024).

Glycosylation, for instance, affects the hydrophilicity and membrane permeability of polyphenols. Partition coefficients, which quantify the relative attraction of a substance for aqueous versus organic phases, facilitate the prediction of passive diffusion across biological membranes (Lind et al. 2021). The flavonol quercetin possesses a log octanol/water partition coefficient of 1.2 ± 0.1, but its glycoside variant, quercetin-3-O-rhamnoglucoside, displays a diminished score of 0.37 ± 0.06, signifying enhanced hydrophilicity and decreased membrane permeability (Brown et al. 1998).

Similarly, hydroxycinnamates often undergo esterification with sugars, organic acids, or lipids. Notable examples are the chlorogenic acids, which are esters of quinic acid and a trans-cinnamic acid residue, such as caffeic acid, p-coumaric acid, or ferulic acid, which are abundantly present in coffee (Rojas-González et al. 2022). These ester-linked modifications notably influence the physicochemical and biological behavior of the parent compounds. Nevertheless, the human body is devoid of esterases that can liberate caffeic acid from chlorogenic acid, resulting in negligible absorption and metabolism of chlorogenic acid in the small intestine, as evidenced by animal research (Plumb et al. 1999; Spencer et al. 1999).

To address such limitations, researchers have developed macromolecular systems to stabilize polyphenols and improve their bioavailability. This encompasses the direct polymerization of polyphenol monomers by step-growth, free radical, and enzyme-catalyzed processes, in addition to conjugation with macromolecules through enzyme grafting, free radical grafting, and amidation (Oliver et al. 2016; Zhang et al. 2023; Maraveas et al. 2021). Polymers and other macromolecules are especially effective in enhancing polyphenols' stability and therapeutic efficacy (Parveen et al. 2023).

Furthermore, the degree of polymerization in proanthocyanidins influences their functional properties. Studies have shown that proanthocyanidins with higher degrees of polymerization enhance protein foaming and foam stability due to their improved binding affinity. Their polyhydroxyl structure enables multiple interactions with protein sites, thereby stabilizing food systems more effectively than their lower polymerized counterparts (Prigent et al. 2009; Zhong et al. 2018).

14.3.1.2 Food Matrix

The associations between polyphenols and food matrix components substantially affect their bioavailability by affecting absorption, metabolism, and, ultimately, their biological effects (Di Lorenzo et al. 2021). The food matrix includes proteins, lipids, dietary fibers, and polysaccharides, which can interact with polyphenols and alter their release and uptake (Ben Alaya et al. 2024). When polyphenols are tightly bound to these components, especially in the upper gastrointestinal (GI) tract, their absorption is hindered (Shahidi and Pan 2022). This binding results in less water solubility and restricts the release of polyphenols for optimal absorption in humans (Polia et al. 2022).

Several factors within the food matrix further modulate polyphenol bioaccessibility. For instance, particle size reduction can increase the surface area for polyphenol release, enhancing absorption (Shu et al. 2019). Thermal processing, such as heating, may break down complex structures and improve bioavailability, although it may also degrade sensitive compounds (Pedrosa et al. 2021). Additionally, the presence of certain macronutrients like carbohydrates (Bohn 2014) and proteins (Xue et al. 2024) can either enhance or inhibit polyphenol absorption, depending on the nature of their interactions.

14.3.1.3 Gut Microbiota

The composition of the gut microbiota is essential in influencing the metabolism and bioavailability of polyphenols, bioactive chemicals recognized for their various health advantages (Bié et al. 2023; Lippolis et al. 2023; Marín et al. 2015). Gut microbiota metabolize dietary polyphenols through GI pathways, thereby influencing their absorption, transformation, and overall efficacy in the host (Kumar Singh et al. 2019). This interaction is fundamental to understanding how polyphenols can be effectively utilized for therapeutic purposes.

Enzymatic transformation of polyphenols by gut microbes can significantly affect their bioavailability (Catalkaya et al. 2020; Gupta et al. 2013; Lewandowska et al. 2013; Bohn 2014). These microbial processes include reactions such as C-ring cleavage and dehydroxylation, which modify the molecular structure of polyphenols, often increasing their bioavailability and enhancing biological activity (Stevens and Maier 2016). Certain microbial taxa, including *Adlercreutzia* and *Lactiplantibacillus*, have been identified as key contributors to these beneficial transformations, promoting improved antioxidant properties (Choi et al. 2023).

14.4 TRANSPORTATION OF POLYPHENOLS

Polyphenols are transported in plasma by many processes, including their association with plasma proteins such as albumin and potentially through dedicated transporters. Polyphenols exhibit a notable binding affinity for serum albumins, which are recognized as the primary extracellular proteins with a high concentration in blood plasma (Jaldappagari et al. 2013). Under non-oxidative circumstances, phenolic compounds form reversible complexes with plasma proteins (Diniz et al. 2008; Rohn 2014). These complexes are induced by hydrogen bonding, electrostatic interactions, hydrophobic effects, and van der Waals forces (Tang 2014; Diniz et al. 2008; Rohn 2014). The interactions between polyphenols and proteins are governed by the thermodynamic parameters of enthalpy (ΔH°) and entropy (ΔS°). Positive values for both parameters signify hydrophobic interactions, negative values denote van der Waals forces and hydrogen bonding, and $\Delta S^\circ > 0$ alongside $\Delta H^\circ \approx 0$ suggests that electrostatic forces are crucial in aqueous solutions (Ross and Subramanian 1981). Hydrophobic interactions are predicated on the non-polar aromatic ring of phenolic compounds engaging in hydrophobic interactions with the hydrophobic amino acid residues of proteins, namely alanine, cysteine, glycine, isoleucine, leucine, methionine, phenylalanine, tyrosine, tryptophan, and valine (Zembyla et al. 2019; Baba et al. 2021). Table 14.1 describes the binding pattern of selected polyphenols with proteins and amino acids.

TABLE 14.1
Binding Patterns of Polyphenols with Amino Acids

S. No.	Name of Polyphenol	Name of Protein	Amino Acids to Which Affinity Is Expressed	Reference
1.	Curcumin	Myosin	Hydrophobic zone (residues 971–1410)	(Zhang et al. 2020)
2.	Rosmarinic acid	Bovine serum albumin	Hydrophobic zone (amino acid residues Ser-191, Arg-198, Leu-237, His-241, Leu-259, Ile-263, His-287, Ala-290, and Glu-291)	(Lu et al. 2022)
3.	Chlorogenic acid (catechol part)	β-lactoglobulin	Phe-105	(Liu et al. 2022)

Electrostatic interactions between proteins and phenolics often arise from the deprotonation of low-pKa phenolic acids under neutral pH situations. This deprotonation renders the phenolics negatively charged, enabling interactions with positively charged protein residues, particularly the ε-amino group of lysine (Feng et al. 2023). At physiological pH (7.4), ferulic acid (pKa = 4.58) exists predominantly in its anionic form and interacts electrostatically with basic residues in bovine serum albumin (BSA) (Ojha et al. 2012). Similarly, caffeic, ferulic, and chlorogenic acids exhibit enhanced binding to β-casein due to increased ionization at physiological pH, which amplifies electrostatic attraction (Li et al. 2020).

Hydrogen bonding represents a fundamental mechanism by which polyphenols associate with proteins. Defined as the interaction between a hydrogen atom and electronegative atoms with high electron affinity, such as oxygen or nitrogen, hydrogen bonding is particularly relevant in polyphenol–protein systems (Feng et al. 2023). Polyphenols, through their hydroxyl groups, serve as hydrogen donors and form bonds with carbonyl groups of peptide amide linkages as well as with polar side chains of amino acids, notably hydroxyl and amino functionalities (Feng et al. 2022; Buitimea-Cantúa et al. 2018). Ferulic acid and its methyl ester have been shown to interact with human serum albumin (HSA) and lysozyme via both hydrogen bonding and hydrophobic interactions (Yang et al. 2024). Vanillic acid exhibits affinity for β-lactoglobulin via a combination of four π-alkyl connections and a hydrogen bond within the β-barrel conformation of the monomer (Abdollahi et al. 2021). Quercetin, a flavonoid, binds to transthyretin, a homo-tetrameric protein responsible for transporting thyroxine and retinol-binding protein, by engaging residues from opposing dimers. Notably, hydrogen bonds between Lys-15 and Ser-117 serve as molecular "staples" stabilizing the tetramer, while additional bonding with Thr-119 enhances this structural integrity. The 4′-hydroxyl groups of quercetin's B-ring interact directly with Ser-117, and the 5′-hydroxyl group forms hydrogen bonds with Thr-119 (Boo et al. 2025). Furthermore, hydrogen bonding has been implicated in the interaction between procyanidin B2 and the hydrophobic pocket of α-lactalbumin (Fan et al. 2022).

Van der Waals forces are comparatively weak, non-covalent interactions arising from transient electron oscillations and the resulting dipole–dipole attractions between molecules. Although subtle, these forces contribute to molecular binding and are often observed in concert with stronger associations such as hydrogen bonding and hydrophobic effects (Feng et al. 2023). There are reports that rosamarinic acid is involved in Van der Waals interactions during complex formation with microtubule affinity regulating kinase (Anwar et al. 2020) and human transferrin (Shamsi et al. 2020). Similarly, 4-hydroxybenzoic acid was found to bind with HSA through Van der Waals interactions (Myint et al. 2021). Catechin was also reported to bind with yolk immunoglobulin through Van der Waals interactions (Liu et al. 2023).

14.5 METABOLISM OF POLYPHENOLS

The primary metabolic pathway for polyphenols involves phase II conjugation. Following absorption, polyphenols undergo glucuronidation, sulfation, and methylation (Manach et al. 2004). Glucuronidation and sulfation transform polyphenols into extremely hydrophilic conjugates, whereas methylation produces metabolites that are either similarly or marginally more lipophilic (Liu and Hu 2007). Methylated products are typically further conjugated into glucuronides and sulfates, provided a suitable functional group is available (Chen et al. 2005; Kuhnle et al. 2000). Thus, conjugation enzymes significantly influence the bioavailability of polyphenols (Liu and Hu 2007). Two principal phase II enzyme families, namely UDP–glucuronosyltransferases (UGTs) and sulfotransferases (SULTs), facilitate the synthesis of these hydrophilic conjugates (Runge-Morris and Kocarek 2009).

UGTs are membrane-associated enzymes situated in the endoplasmic reticulum of several organs (Lethe et al. 2024). They facilitate the transfer of glucuronic acid from UDP–glucuronic acid to several substrates, including steroids, bile acids, polyphenols, dietary substances, and xenobiotics (Liu and Hu 2007). In contrast, SULTs are soluble, highly active cytosolic enzymes that transfer sulfate groups from 3′-phosphoadenosine-5′-phosphosulfate to hydroxyl groups on diverse substrates (Klaassen and Boles 1997). Unlike UGTs and SULTs, methyltransferases, such as catechol-O-methyltransferase, catalyze O-methylation of catechol-containing polyphenols, producing metabolites with similar or slightly reduced polarity compared to the parent compounds (Kiss and Soares-da-Silva 2014; Wang et al. 2021).

Alongside phase II enzymes, phase I enzymes – especially cytochrome P450s (CYPs) – play a role in polyphenol metabolism. CYP1A2 and CYP2E1 have been shown to metabolize genistein, with CYP1A2 acting as a prodrug-converting enzyme for less active methylated isoflavones (Hu et al. 2003). CYP1B1 also mediates O-demethylation of biochanin A and formononetin, producing genistein and daidzein, which in turn inhibit CYP1B1 (Roberts et al. 2004). However, CYP-mediated metabolism represents a minor route compared to phase II pathways, as its products are typically rapidly conjugated into hydrophilic glucuronides and sulfates (Jeong et al. 2005).

LPH, a brush border glucosidase located in the small intestine, is crucial to polyphenol metabolism (Wilkinson et al. 2003). In addition to hydrolyzing lactose into glucose and galactose (Han et al. 2024), LPH catalyzes the in vitro hydrolysis of quercetin 3-O-glucoside (Q3G), quercetin 4′-O-glucoside (Q4′G), and monoglucosides of genistein and daidzein, releasing their respective aglycones (Day et al. 2000). Cytosolic β-glucosidase, present in enterocytes, shows broad substrate specificity, hydrolyzing Q4′G, genistin, and daidzin but not Q3G, rutin, or quercetin 3,4′-O-diglucoside, as shown using human intestinal and liver extracts (Day et al. 1998).

A significant population of colonic bacteria participates in polyphenol metabolism (Da C. Pinaffi-Langley et al. 2024). Approximately 90–95% of ingested polyphenols bypass absorption and reach the colon, where microbial enzymes modify their backbone into metabolites of diverse physiological relevance. The colonic microbiota is exposed to two polyphenol sources: unabsorbed dietary glycosides (Cardona et al. 2013) and hydrophilic conjugates – sulfates and glucuronides – secreted by the liver and colon (Liu and Hu 2007). Over 50% of dietary polyphenols accumulate in the lower GI tract, especially the colon, exceeding 90% for polymeric or conjugated forms (Da C. Pinaffi-Langley et al. 2024). Resident microbes use these compounds as substrates, generating more bioavailable metabolites via three primary transformations: (a) cleavage (Stevens and Maier 2016), (b) hydrolysis (Gowd et al. 2020), and (c) reduction (Da C. Pinaffi-Langley et al. 2024). Table 14.2 illustrates the various biochemical processes performed by gut microbiota while acting upon polyphenols.

14.6 FACTORS INFLUENCING THE PHARMACOKINETICS OF POLYPHENOLS

Polyphenols are bioactive, natural compounds widely recognized for their health-promoting effects, including antioxidant, anti-inflammatory, and cardioprotective activities. However, their clinical efficacy mainly depends on their pharmacokinetic action: how they are absorbed, distributed, metabolized, and excreted in the body? Various factors affect the pharmacokinetics of polyphenols, particularly their chemical makeup, solubility, interaction with transporters and metabolic enzymes, the food matrix, gut microbiota activity, and individual physiological differences. Understanding these factors is crucial for developing strategies to enhance polyphenol bioavailability and therapeutic potential.

The bioavailability of dietary polyphenols, defined as the degree and speed of their absorption and utilization in the body, is affected by a complex interaction of physicochemical, biochemical, and physiological variables. A primary determinant is the chemical structure of the polyphenols themselves. Flavonoids such as quercetin and catechins are a particularly diverse group, commonly found in glycosylated forms, which notably influences their solubility and absorption (Rubio et al. 2014; Manach and Donovan 2004). Phenolic acids, such as ferulic acid, are generally more efficiently absorbed when present in their free forms (Rubio et al. 2014; Manach et al. 2004). Stilbenes, including resveratrol, exhibit notable bioactivity but often encounter absorption challenges due to rapid metabolic transformation (Roupe et al. 2006). Lignans, phytoestrogens found in seeds and grains, undergo extensive microbial conversion in the gut, contributing further complexity to their absorption and efficacy (De Silva and Alcorn 2019). The degree of polymerization significantly influences bioavailability, with monomeric forms typically exhibiting greater absorption than their polymeric counterparts (Sánchez-Mundo et al. 2016). Moreover, sugar moieties – glycosides – can either facilitate or hinder absorption depending on their position and structure; these modifications affect solubility and enzymatic hydrolysis in the GI tract (Andreu et al. 2023).

Molecular size is another critical determinant of polyphenol bioavailability. Larger molecules, such as tannins and proanthocyanidins, exhibit markedly lower absorption rates due to limited permeability across the intestinal epithelium (Rubio et al. 2014; Liu et al. 2024; Muntaha et al. 2025). In contrast, smaller molecules such as monomeric catechins can more readily pass through enterocyte membranes, leading to higher plasma concentrations post-ingestion. Lipophilicity also significantly affects absorption. More lipophilic polyphenols tend to integrate into the lipid bilayer of enterocyte membranes, enhancing passive diffusion (Karonen 2022). This is particularly evident for polyphenols like curcumin, which exhibit improved absorption despite its poor water solubility when consumed with dietary fats or in lipophilic formulations (Jäger et al. 2014).

TABLE 14.2
Biochemical Processes Performed by Gut Microbiota on Polyphenols

S. No.	Process	Details	Microbes Involved	Specific Action	Reference
1.	Cleavage	Cleavage of covalent bonds, particularly carbon–carbon and carbon–oxygen bonds, leading to ring opening	*Clostridium butyricum*	Deglycosylation flavonoid O-glucosides in the absence of glucose	(Braune 2025)
			Streptococcus pasteurianus	C-ring cleavage of flavan-3-ol monomers	(Wang et al. 2024)
			Eggerthella lenta	Cleavage of C-ring of (+)-catechin and (-)-epicatechin, followed by p-dehydroxylation of the B-ring	(Jin and Hattori 2012)
			Clostridium sp.	C-ring cleavage of diadzein	H (Hur et al. 2002)
			Enterococcus sp. MRG-2 and *Lactococcus* sp. MRG-IF-4	Deglycosylation: Conversion of isoflavone 7-O-glucosides, glycitin, ononin, and sissotrin to their aglycones: glycitein, formononetin, and biochanin A, respectively	(Kim et al. 2016)
			Enterococcus faecalis	Deglycosylation of flavone C-glycosides	(Zheng et al. 2019)
2.	Hydrolysis	Deconjugation of polyphenols from bigger entities, such as sugars and organic acids	*Bifidobacterium animalis*	Hydrolysis of chlorogenic acid into caffeic acid	(Raimondi et al. 2015)
			Bifidobacterium catenulatum and *Bifidobacterium pseudocatenultum*	Hydrolysis of hesperidin	(Amaretti et al. 2015)
3.	Reduction	Hydrogenation and dehydroxylation	*Enterocloster* sp	Multistep metabolism of dietary ellagitannins to produce urolithin A	(Pidgeon et al. 2025)
			Eggerthella lenta	Dehydroxylation of catecholamine neurotransmitters	(Maini Rekdal et al. 2020)
			Eggerthella lenta	Dehydroxylation of catechol	(Maini Rekdal et al. 2020)

The stability of polyphenols in the GI tract directly affects their availability for absorption. The stomach's acidic pH, the presence of digesting enzymes, and interactions with food components can result in the chemical destruction or complexation of polyphenols, hence diminishing their active concentration in the intestinal lumen (Dzah 2025; Zhang et al. 2021). For instance, anthocyanins are particularly pH-sensitive and degrade quickly under acidic conditions. Furthermore, the solubility of polyphenols in aqueous environments is often limited, which restricts their ability to traverse the unstirred water layer and reach enterocytes. However, various formulation strategies – such as nanoemulsions (Kumar et al. 2019), solid lipid nanoparticles (SLNs) (Arabestani et al. 2024), and self-emulsifying delivery systems (Chavda and Shah 2017) – have been developed to address these limitations and improve intestinal solubility (Sahraeian et al. 2024). Once absorbed, polyphenols undergo extensive metabolism, particularly via the first-pass effect, where enzymes in the intestinal wall and liver rapidly modify the compounds before they reach systemic circulation. Enzymes such as UGTs (Rowland et al. 2013), SULTs (Kurogi et al. 2012), and catechol-O-methyltransferase (COMT) (Sahraeian et al. 2024) play key roles in conjugating polyphenols into more water-soluble forms, facilitating either their excretion or further metabolic processing (Rubio et al. 2014; Li et al. 2025). Phase I reactions – oxidation, reduction, and hydrolysis – precede phase II and are typically less impactful but can alter the chemical structure and influence subsequent conjugation efficiency (McCarthy and Sinal 2005). Although essential for detoxification and elimination, phase II reactions – glucuronidation, sulfation, and methylation – significantly reduce the concentration of bioactive aglycones in circulation, diminishing their pharmacological potential (Phang-Lyn and Llerena 2025).

The intestinal microbiota is crucial for the metabolism of polyphenols, especially for chemicals that are not absorbed in the small intestine. Gut bacteria possess enzymes capable of hydrolyzing glycosidic bonds, deconjugating metabolites, and generating smaller phenolic acids that are more readily absorbed and often retain or even enhance biological activity (Dzah 2025; Kawabata et al. 2019; Pasinetti et al. 2018). The composition and activity of the gut microbiota vary significantly among individuals due to factors such as diet (Leeming et al. 2019), age (Conlon and Bird 2014), antibiotic use (Jandhyala 2015), and overall health (Hasan and Yang 2019), leading to significant interindividual differences in polyphenol bioavailability.

Transport proteins also influence polyphenol absorption and distribution. Efflux transporters like P-glycoprotein (P-gp), breast cancer resistance protein (BCRP), and various multidrug resistance proteins (MRPs) actively pump polyphenols and their conjugates back into the intestinal lumen, thereby limiting their net absorption (Scheepens et al. 2010; Bart et al. 2004; Liu and Hu 2007). These proteins, part of the ATP-binding cassette (ABC) transporter family, are highly expressed in the intestinal lining and regulated by genetic and environmental factors (Rubio et al. 2014). Interactions with dietary components can modulate the activity of these transporters, either inhibiting or enhancing their function.

The food matrix significantly affects polyphenol bioavailability. Co-ingestion with fats can enhance the solubility and incorporation of lipophilic polyphenols into micelles, facilitating their absorption. Conversely, dietary fibers may bind polyphenols and form insoluble complexes, reducing bioaccessibility (Palafox-Carlos et al. 2011).

Proteins may also interact with polyphenols to form precipitates, further decreasing the absorption potential (Oliveira et al. 2018). Digestive enzymes such as LPH can cleave glycosidic bonds in the intestinal brush border, releasing more lipophilic aglycones and leading to better absorption (Wilkinson et al. 2003). These enzymatic processes are essential for transforming many dietary polyphenols into their absorbable forms.

The pH along the GI tract also modulates polyphenol solubility and stability. While some polyphenols degrade in acidic conditions (e.g., in the stomach) (Bešlo et al. 2023), others may be more stable and absorbable in the slightly alkaline environment of the small intestine (Xiao 2022). Age is another factor influencing polyphenol pharmacokinetics. Elderly individuals often experience slower gastric emptying, reduced digestive enzyme activity, altered gut microbiota, and impaired hepatic metabolism, all of which can lead to reduced bioavailability and altered therapeutic outcomes (Mangoni and Jackson 2004).

Gender differences in polyphenol metabolism have also been observed. Hormonal variations, particularly involving estrogen, can influence the expression of metabolic enzymes and transport proteins, resulting in sex-dependent pharmacokinetics (Soldin and Mattison 2009). Similarly, health status plays a crucial role. Diseases such as diabetes, obesity, and liver dysfunction can impair the enzymatic machinery involved in polyphenol metabolism and alter transporter activity, thereby modifying absorption and elimination pathways (Xiao and Hogger 2014).

Genetic polymorphisms in the genes encoding for metabolic enzymes (e.g., UGTs, SULTs, COMT) and transporters (e.g., ABC transporters) further contribute to interindividual variability in polyphenol bioavailability (Umamaheswaran et al. 2014). These polymorphisms may lead to altered enzyme activity or expression levels, significantly affecting the extent and duration of polyphenol exposure in different individuals. Advances in nutrigenomics are helping to elucidate these genotype–phenotype correlations, potentially paving the way for personalized dietary recommendations (Singar et al. 2024).

Formulation and delivery systems have emerged as effective strategies to improve polyphenol stability, absorption, and bioavailability. Technologies such as nanoparticle encapsulation, liposomes, micelles, or co-crystals protect polyphenols from degradation in the GI tract and facilitate controlled release and targeted delivery (Lu et al. 2025). For instance, encapsulating curcumin in lipid nanoparticles significantly enhances oral bioavailability and systemic exposure. Moreover, the dose and frequency of administration also influence bioavailability (Scalbert and Williamson 2000). Higher doses may saturate enzymatic or transporter systems, leading to non-linear pharmacokinetics, while repeated dosing may induce adaptive responses such as up-regulation of efflux transporters (Chow et al. 2003; Feliciano et al. 2017).

The route of administration is a fundamental factor in determining polyphenol bioavailability. Oral administration, the most common method for dietary polyphenols, subjects compounds to digestive processes and first-pass metabolism, resulting in reduced systemic availability (Scheepens et al. 2010). In contrast, intravenous or sublingual administration bypasses the GI tract and hepatic first-pass metabolism, leading to higher bioavailability, although such routes are less practical for routine dietary intake (Rubio et al. 2014). Finally, polyphenols and their metabolites are primarily excreted via the bile into feces or via the kidneys into urine (Guerreiro et al. 2022). The extent and route of excretion depend on the nature of conjugates formed during metabolism. Enterohepatic circulation, where conjugates secreted into bile are hydrolyzed by intestinal microbiota and reabsorbed, can prolong the half-life and biological activity of polyphenols (Roberts et al. 2002).

In summary, the bioavailability of polyphenols is determined by a multifaceted interplay of intrinsic and extrinsic factors, including chemical structure, solubility, metabolism, microbiota composition, genetic polymorphisms, and formulation strategies. A comprehensive understanding of these variables is essential for optimizing the health benefits associated with dietary polyphenols and developing effective polyphenol-based nutraceuticals.

14.7 TECHNOLOGICAL STRATEGIES TO ENHANCE BIOAVAILABILITY

Polyphenols possess numerous health benefits, but their poor bioavailability limits their therapeutic potential. Factors such as low water solubility, instability during digestion, and rapid metabolism reduce their effectiveness in the body. Several technological strategies have been developed to overcome these challenges to improve the absorption, stability, and bioefficacy of polyphenols, enabling their more effective use in health promotion and disease prevention. Table 14.3 illustrates selected technological approaches to enhance the efficacy of polyphenols.

14.8 CONCLUSION

Polyphenols, a varied group of plant-derived chemicals, have attracted significant scientific attention due to their extensive health advantages, which encompass antioxidant, anti-inflammatory, cardioprotective, and anticancer effects. The translation of these potential bioactivities into therapeutic efficacy is hindered by their complex pharmacokinetic profile, primarily marked by low bioavailability, restricted absorption, extensive metabolism, and rapid systemic clearance. This chapter has methodically examined the multifaceted nature of polyphenol pharmacokinetics and underscored essential physiological, biochemical, and technical factors affecting their disposition in the human body.

The absorption of polyphenols is significantly affected by their chemical structure, encompassing molecular size, polymerization degree, glycosylation, and lipophilicity. Aglycone forms are often more lipophilic and hence more easily absorbed by the intestinal epithelium, whereas glycosides and conjugated forms necessitate enzymatic breakdown before absorption. Enzymes like LPH and CBG are crucial for releasing absorbable aglycones from dietary polyphenols, whereas the structural characteristics of these compounds influence membrane permeability and transporter interactions.

Upon absorption, polyphenols experience significant phase II metabolism, including glucuronidation, sulfation, and methylation, predominantly in enterocytes and hepatocytes, resulting in water-soluble conjugates with diminished bioactivity. These

TABLE 14.3
Different Technological Strategies and Their Efficacy

Procedure	Explanation	Example/Outcome of Study	Reference
Nanoencapsulation	Encapsulating polyphenols in nanoparticles to protect from degradation and enhance solubility, stability, and absorption	• Lipid nanoparticles containing curcumin and resveratrol enhance the efficacy of pharmaceuticals utilized in the prevention and treatment of disorders linked to oxidative stress.	(Coradini et al. 2014)
		• Curcumin encapsulated in casein nanoparticles exhibited stronger antioxidant and cytotoxic effects than unencapsulated curcumin. • The researchers devised a cost-effective, energy-efficient encapsulation technique that is devoid of organic solvents, utilizing curcumin's pH-dependent solubility and the self-assembling characteristics of sodium caseinate.	(Rashidinejad et al. 2023)
		• Resveratrol was encapsulated in two types of SLNs, which are commonly used for hydrophobic compounds due to their compatible properties.	(Subroto et al. 2023)
Lipid-Based Delivery Systems	Formulating polyphenols with lipids (liposomes, micelles, SEDDS) to improve solubility, protect from metabolism, and promote lymphatic absorption	• Liposomal quercetin enhances absorption and offers an innovative strategy for diabetes management.	(Rocha et al. 2024)
		• Silica coating markedly improved the stability of flexible liposomes, with silica-coated liposomes containing curcumin exhibiting a 3.31-fold rise in bioavailability over curcumin-loaded flexible liposomes lacking the silica coating.	(Li et al. 2012)
		• The integration of the phytosomal variant of silymarin into a liposomal carrier system improved its oral bioavailability and conferred greater hepatoprotective and anti-inflammatory properties relative to a silymarin suspension.	(Kumar et al. 2014)
Co-Administration with Bioenhancers	Using natural compounds to inhibit enzymes and efflux pumps, reducing metabolism and increasing polyphenol bioavailability	• Piperine with curcumin increases bioavailability by 2000%.	(Prasad et al. 2014)
		• Another study indicated that the co-administration of piperine (20 mg/kg orally) with curcumin (2 g/kg orally) enhanced the bioavailability of curcumin by as much as 20-fold in epileptic rats.	(Sharma et al. 2010)
		• A study highlighted quercetin's potential as a bioenhancer for curcumin in the treatment of ischemia-reperfusion–induced myocardial toxicity in rats.	(Chakraborty et al. 2018)

(Continued)

TABLE 14.3 (*Continued*)
Different Technological Strategies and Their Efficacy

Procedure	Explanation	Example/Outcome of Study	Reference
Structural Modifications	Chemical modification like glycosylation or esterification to improve solubility, stability, and permeability	• Glycosylated flavonoids show better intestinal absorption.	(Crespy et al. 2001)
		• Esterifying EGCG by substituting its hydroxyl groups with fatty acid chains increased its lipophilicity and enhanced its antioxidant activity by improving its ability to donate hydrogen atoms.	(Zhong and Shahidi 2011)
		• Protein–polyphenol conjugates are frequently utilized to improve the stability of food emulsions owing to their robust emulsifying capabilities and antioxidant characteristics.	(Quan et al. 2019)
Solid Dispersion Techniques	Dispersing polyphenols in solid matrices with polymers to enhance solubility and dissolution rate	• Solid-state amorphous dispersions of curcumin with hydroxypropyl methyl cellulose and polyvinyl pyrrolidine enhanced its solubility, rate of dissolution, and physicochemical evaluations, especially at a 1:4 ratio.	Ishtiaq et al. 2022)
		• At pH 6.8, PVP demonstrated the highest ability to solubilize ellagic acid (EA) from amorphous solid dispersions (ASDs) (≈92%), followed by hydroxypropyl-methylcellulose acetate and carboxymethylcellulose acetate butyrate, with polyvinyl pyrrolidinone allowing rapid but unstable release across pH levels, while HPMCAS enabled pH-triggered release, making it a practical option for enhancing EA solubility.	(Li et al. 2013)
		• A study was undertaken to develop curcumin solid dispersions to enhance its solubility.	(Onoue et al. 2010)
Microencapsulation	Encapsulating polyphenols within micro-sized particles to protect from environmental factors and mask taste	• Microencapsulated green tea polyphenols maintain antioxidant activity, determined by the scavenging power of cationic free radicals.	(Rodrigues Silva et al. 2021)
		• Microencapsulation enables the controlled release of polyphenols, thereby improving their bioavailability.	(Mehta et al. 2022)
		• Garcinia fruit extract, rich in hydroxycitric acid, was microencapsulated and incorporated into bread, resulting in increased volume, softer crumb texture, improved color and sensory qualities, and a higher concentration of free hydroxycitric acid (HCA).	(Ezhilarasi et al. 2013)

Prebiotics and Gut Microbiota Modulation	Enhancing polyphenol metabolism by promoting beneficial gut bacteria capable of transforming polyphenols into more absorbable forms	• Synbiotic combinations of polyphenols, probiotics, and prebiotics serve as therapeutic agents for treating inflammation-related depression.	(Westfall and Pasinetti 2019)
		• Orange peels are rich in dietary fiber and polyphenols, which are transformed by gut microbiota into more absorbable forms, enhancing their health benefits in the human body.	(Ademosun et al. 2025)
		• Mice subjected to a high-fat diet exhibited an augmentation in gut microbiota after the ingestion of green tea polyphenols.	(Ma et al. 2019)
Emulsion-Based Systems	Emulsions (O/W, W/O/W) were used to solubilize lipophilic polyphenols and facilitate their intestinal absorption.	• In vivo pharmacokinetic studies in mice showed that curcumin delivered via nanoemulsion had a 9-fold higher oral bioavailability compared to unformulated curcumin.	(Yu and Huang 2012)
		• Caseinate-stabilized emulsions provided the highest total curcumin availability for absorption, making them the most suitable as excipient emulsions.	Zou et al. 2015)
		• Encapsulating catechin and curcumin in an emulsion markedly improved their stability in simulated gastrointestinal fluid, resulting in a 4-fold improvement in bioaccessibility compared to their freely floating solutions.	(Aditya et al. 2015)

conjugates are either expelled or re-enter the intestinal lumen through bile, facilitating enterohepatic recirculation and possible reabsorption. A substantial fraction of polyphenols arrives in the colon unabsorbed, where gut microbes facilitate their biotransformation into smaller phenolic metabolites with improved bioavailability and occasionally superior biological activity. The interaction between host enzymatic pathways and microbial metabolism is a crucial factor influencing polyphenol effectiveness.

The transport and distribution of polyphenols are influenced by their binding affinity to plasma proteins and interactions with efflux transporters, including P-gp and BCRP, which actively reduce systemic exposure by facilitating intestinal excretion. Furthermore, individual variability – encompassing age, sex, genetic polymorphisms, health condition, and gut flora composition – complicates pharmacokinetic findings, rendering interindividual responses to dietary polyphenols highly variable.

Innovative technological techniques have emerged to address these pharmacokinetic issues. Nanoencapsulation, lipid-based delivery methods, structural alterations, solid dispersions, and microencapsulation techniques are designed to improve the solubility, stability, and absorption of polyphenols. Furthermore, the concurrent use of bioenhancers and dietary prebiotics, as well as the modification of gut microbiota, presents intriguing strategies to enhance polyphenol bioefficacy. These interventions highlight the significance of formulation science and host–microbe interactions in enhancing polyphenol-based therapeutics.

The clinical translation of polyphenols relies on both their inherent biological potential and a thorough comprehension of their pharmacokinetics. Future research ought to concentrate on individualized nutraceutical approaches, utilizing advancements in -omics technologies and delivery mechanisms to optimize therapeutic results. Integrating molecular insights with practical technologies enables the full health-promoting potential of polyphenols to be utilized in combating chronic diseases and enhancing human wellness.

REFERENCES

Abdollahi, Kourosh, Lloyd Condict, Andrew Hung, and Stefan Kasapis. 2021. "Binding Parameters and Molecular Dynamics of β-Lactoglobulin-Vanillic Acid Complexation as a Function of pH – Part A: Acidic pH." *Food Chemistry* 360 (October): 130059. https://doi.org/10.1016/j.foodchem.2021.130059.

Ademosun, Ayokunle Olubode, Olufunke Florence Ajeigbe, Mary Tosin Ademosun, Omodesola Oluwafisayo Ogunruku, and Ganiyu Oboh. 2025. "Improving Gut Microbiome through Diet Rich in Dietary Fibre and Polyphenols: The Case for Orange Peels." *Human Nutrition & Metabolism* 39 (March): 200295. https://doi.org/10.1016/j.hnm.2024.200295.

Aditya, N.P., Sheetal Aditya, Hanjoo Yang, Hye Won Kim, Sung Ook Park, and Sanghoon Ko. 2015. "Co-Delivery of Hydrophobic Curcumin and Hydrophilic Catechin by a Water-in-Oil-in-Water Double Emulsion." *Food Chemistry* 173 (April): 7–13. https://doi.org/10.1016/j.foodchem.2014.09.131.

Amaretti, Alberto, Stefano Raimondi, Alan Leonardi, Andrea Quartieri, and Maddalena Rossi. 2015. "Hydrolysis of the Rutinose-Conjugates Flavonoids Rutin and Hesperidin by the Gut Microbiota and Bifidobacteria." *Nutrients* 7 (4): 2788–800. https://doi.org/10.3390/nu7042788.

Anand, Santosh, Ramachandregowda Sowbhagya, Mohammad Azam Ansari, Mohammad A. Alzohairy, Mohammad N. Alomary, Asiyah I. Almalik, Wasim Ahmad, Takshashila Tripathi, and Abozer Y. Elderdery. 2022. "Polyphenols and their Nanoformulations: Protective Effects against Human Diseases." *Life* 12 (10): 1639. https://doi.org/10.3390/life12101639.

Andreu, Alicia, Marija Ćorović, Carla Garcia-Sanz, A. Sofia Santos, Ana MilivojeviĆ, Clara Ortega-Nieto, Cesar Mateo, Dejan Bezbradica, and Jose M. Palomo. 2023. "Enzymatic Glycosylation Strategies in the Production of Bioactive Compounds." *Catalysts* 13 (10): 1359. https://doi.org/10.3390/catal13101359.

Anwar, Saleha, Anas Shamsi, Mohd Shahbaaz, Aarfa Queen, Parvez Khan, Gulam Mustafa Hasan, Asimul Islam, et al. 2020. "Rosmarinic Acid Exhibits Anticancer Effects via MARK4 Inhibition." *Scientific Reports* 10 (1): 10300. https://doi.org/10.1038/s41598-020-65648-z.

Arabestani, Mohammad Reza, Ashkan Bigham, Farideh Kamarehei, Mahya Dini, Fatemeh Gorjikhah, Aref Shariati, and Seyed Mostafa Hosseini. 2024. "Solid Lipid Nanoparticles and their Application in the Treatment of Bacterial Infectious Diseases." *Biomedicine & Pharmacotherapy* 174 (May): 116433. https://doi.org/10.1016/j.biopha.2024.116433.

Arfaoui, Leila. 2021. "Dietary Plant Polyphenols: Effects of Food Processing on their Content and Bioavailability." *Molecules* 26 (10): 2959. https://doi.org/10.3390/molecules26102959.

Arts, Ilja C.W., Aloys L.A. Sesink, and Peter C.H. Hollman. 2002. "Quercetin-3-Glucoside is Transported by the Glucose Carrier SGLT1 across the Brush Border Membrane of Rat Small Intestine." *The Journal of Nutrition* 132 (9): 2823. https://doi.org/10.1093/jn/132.9.2823.

Baba, Waqas N., David Julian McClements, and Sajid Maqsood. 2021. "Whey Protein–Polyphenol Conjugates and Complexes: Production, Characterization, and Applications." *Food Chemistry* 365 (December): 130455. https://doi.org/10.1016/j.foodchem.2021.130455.

Bart, J., H. Hollema, H.J.M. Groen, E.G.E. De Vries, N.H. Hendrikse, D.T. Sleijfer, T.D. Wegman, W. Vaalburg, and W.T.A. Van Der Graaf. 2004. "The Distribution of Drug-Efflux Pumps, P-Gp, BCRP, MRP1 and MRP2, in the Normal Blood–Testis Barrier and in Primary Testicular Tumours." *European Journal of Cancer* 40 (14): 2064–70. https://doi.org/10.1016/j.ejca.2004.05.010.

Ben Alaya, Ikbel, Gilberto Alves, João Lopes, and Luís R. Silva. 2024. "Use of Encapsulated Polyphenolic Compounds in Health Promotion and Disease Prevention: Challenges and Opportunities." *Macromol* 4 (4): 805–42. https://doi.org/10.3390/macromol4040048.

Bešlo, Drago, Nataša Golubić, Vesna Rastija, Dejan Agić, Maja Karnaš, Domagoj Šubarić, and Bono Lučić. 2023. "Antioxidant Activity, Metabolism, and Bioavailability of Polyphenols in the Diet of Animals." *Antioxidants* 12 (6): 1141. https://doi.org/10.3390/antiox12061141.

Bié, Joaquim, Bruno Sepodes, Pedro C.B. Fernandes, and Maria H.L. Ribeiro. 2023. "Polyphenols in Health and Disease: Gut Microbiota, Bioaccessibility, and Bioavailability." *Compounds* 3 (1): 40–72. Multidisciplinary Digital Publishing Institute. https://doi.org/10.3390/compounds3010005.

Bohn, Torsten. 2014. "Dietary Factors Affecting Polyphenol Bioavailability." *Nutrition Reviews* 72 (7): 429–52. https://doi.org/10.1111/nure.12114.

Boo, Hye Joon, Danbi Yoon, Yujeong Choi, Younghyun Kim, Jeong Seok Cha, and Jiho Yoo. 2025. "Quercetin: Molecular Insights into Its Biological Roles." *Biomolecules* 15 (3): 313. https://doi.org/10.3390/biom15030313.

Braune, Annett. 2025. "Flavonoid-Converting Capabilities of Clostridium Butyricum." *Applied Microbiology and Biotechnology* 109 (1): 53. https://doi.org/10.1007/s00253-025-13434-0.

Brown, E. Jonathan, Hicham Khodr, C. Robert Hider, and Catherine A. Rice-Evans. 1998. "Structural Dependence of Flavonoid Interactions with Cu2+ Ions: Implications for their Antioxidant Properties." *Biochemical Journal* 330 (3): 1173–78. https://doi.org/10.1042/bj3301173.

Buitimea-Cantúa, Nydia E., Janet A. Gutiérrez-Uribe, and Sergio O. Serna-Saldívar. 2018. "Phenolic–Protein Interactions: Effects on Food Properties and Health Benefits." *Journal of Medicinal Food* 21 (2): 188–98. https://doi.org/10.1089/jmf.2017.0057.

Cardona, Fernando, Cristina Andrés-Lacueva, Sara Tulipani, Francisco J. Tinahones, and María Isabel Queipo-Ortuño. 2013. "Benefits of Polyphenols on Gut Microbiota and Implications in Human Health." *The Journal of Nutritional Biochemistry* 24 (8): 1415–22. https://doi.org/10.1016/j.jnutbio.2013.05.001.

Catalkaya, Gizem, Koen Venema, Luigi Lucini, Gabriele Rocchetti, Dominique Delmas, Maria Daglia, Anna De Filippis, et al. 2020. "Interaction of Dietary Polyphenols and Gut Microbiota: Microbial Metabolism of Polyphenols, Influence on the Gut Microbiota, and Implications on Host Health." *Food Frontiers* 1 (2): 109–33. https://doi.org/10.1002/fft2.25.

Chakraborty, Manodeep, Mohammed Gulzar Ahmed, and Ananya Bhattacharjee. 2018. "Effect of Quercetin on Myocardial Potency of Curcumin against Ischemia Reperfusion Induced Myocardial Toxicity." *Synergy* 7 (December): 25–29. https://doi.org/10.1016/j.synres.2018.09.001.

Chavda, Vivek P., and Dhaval Shah. 2017. "Self-Emulsifying Delivery Systems: One Step Ahead in Improving Solubility of Poorly Soluble Drugs." In *Nanostructures for Cancer Therapy*, 653–718. Elsevier. https://doi.org/10.1016/B978-0-323-46144-3.00025-8.

Chen, Jun, Huimin Lin, and Ming Hu. 2005. "Absorption and Metabolism of Genistein and Its Five Isoflavone Analogs in the Human Intestinal Caco-2 Model." *Cancer Chemotherapy and Pharmacology* 55 (2): 159–69. https://doi.org/10.1007/s00280-004-0842-x.

Choi, Se Rin, Hyunji Lee, Digar Singh, Donghyun Cho, Jin-Oh Chung, Jong-Hwa Roh, Wan-Gi Kim, and Choong Hwan Lee. 2023. "Bidirectional Interactions between Green Tea (GT) Polyphenols and Human Gut Bacteria." *Journal of Microbiology and Biotechnology* 33 (10): 1317–28. https://doi.org/10.4014/jmb.2306.06014.

Chow, H.-H. Sherry, Yan Cai, Iman A. Hakim, James A. Crowell, Farah Shahi, Chris A. Brooks, Robert T. Dorr, Yukihiko Hara, and David S. Alberts. 2003. "Pharmacokinetics and Safety of Green Tea Polyphenols after Multiple-Dose Administration of Epigallocatechin Gallate and Polyphenon E in Healthy Individuals." *Clinical Cancer Research: An Official Journal of the American Association for Cancer Research* 9 (9): 3312–19.

Clarke, Don B., Antony S. Lloyd, Nigel P. Botting, Mark F. Oldfield, Paul W. Needs, and Helen Wiseman. 2002. "Measurement of Intact Sulfate and Glucuronide Phytoestrogen Conjugates in Human Urine Using Isotope Dilution Liquid Chromatography-Tandem Mass Spectrometry with [13C3]Isoflavone Internal Standards." *Analytical Biochemistry* 309 (1): 158–72. https://doi.org/10.1016/S0003-2697(02)00275–0.

Conlon, Michael, and Anthony Bird. 2014. "The Impact of Diet and Lifestyle on Gut Microbiota and Human Health." *Nutrients* 7 (1): 17–44. https://doi.org/10.3390/nu7010017.

Coradini, K., F.O. Lima, C.M. Oliveira, P.S. Chaves, M.L. Athayde, L.M. Carvalho, and R.C.R. Beck. 2014. "Co-Encapsulation of Resveratrol and Curcumin in Lipid-Core Nanocapsules Improves their in Vitro Antioxidant Effects." *European Journal of Pharmaceutics and Biopharmaceutics* 88 (1): 178–85. https://doi.org/10.1016/j.ejpb.2014.04.009.

Cory, Hannah, Simone Passarelli, John Szeto, Martha Tamez, and Josiemer Mattei. 2018. "The Role of Polyphenols in Human Health and Food Systems: A Mini-Review." *Frontiers in Nutrition* 5 (September): 87. https://doi.org/10.3389/fnut.2018.00087.

Crespy, Vanessa, Christine Morand, Catherine Besson, Claudine Manach, Christian Démigné, and Christian Rémésy. 2001. "Comparison of the Intestinal Absorption of Quercetin,

Phloretin and their Glucosides in Rats." *The Journal of Nutrition* 131 (8): 2109–14. https://doi.org/10.1093/jn/131.8.2109.

Crespy, Vanessa, Christine Morand, Claudine Manach, Catherine Besson, Christian Demigne, and Christian Remesy. 1999. "Part of Quercetin Absorbed in the Small Intestine is Conjugated and Further Secreted in the Intestinal Lumen." *American Journal of Physiology-Gastrointestinal and Liver Physiology* 277 (1): G120–26. https://doi.org/10.1152/ajpgi.1999.277.1.G120.

Da C. Pinaffi-Langley, Ana Clara, Stefano Tarantini, Norman G. Hord, and Andriy Yabluchanskiy. 2024. "Polyphenol-Derived Microbiota Metabolites and Cardiovascular Health: A Concise Review of Human Studies." *Antioxidants* 13 (12): 1552. https://doi.org/10.3390/antiox13121552.

Danielsen, E. Thomas, and E. Michael Danielsen. 2017. "Glycol Chitosan: A Stabilizer of Lipid Rafts in the Intestinal Brush Border." *Biochimica et Biophysica Acta (BBA) – Biomembranes* 1859 (3): 360–67. https://doi.org/10.1016/j.bbamem.2016.12.017.

Day, Andrea J., F. Javier Cañada, Juan C. Díaz, Paul A. Kroon, Russell Mclauchlan, Craig B. Faulds, Geoff W. Plumb, Michael R.A. Morgan, and Gary Williamson. 2000. "Dietary Flavonoid and Isoflavone Glycosides Are Hydrolysed by the Lactase Site of Lactase Phlorizin Hydrolase." *FEBS Letters* 468 (2–3): 166–70. https://doi.org/10.1016/S0014-5793(00)01211-4.

Day, Andrea J., M. Susan DuPont, Saxon Ridley, Mike Rhodes, Michael J.C. Rhodes, Michael R.A. Morgan, and Gary Williamson. 1998. "Deglycosylation of Flavonoid and Isoflavonoid Glycosides by Human Small Intestine and Liver B-glucosidase Activity." *FEBS Letters* 436 (1): 71–75. https://doi.org/10.1016/S0014-5793(98)01101-6.

De Silva, S. Franklyn, and Jane Alcorn. 2019. "Flaxseed Lignans as Important Dietary Polyphenols for Cancer Prevention and Treatment: Chemistry, Pharmacokinetics, and Molecular Targets." *Pharmaceuticals* 12 (2): 68. https://doi.org/10.3390/ph12020068.

Di Lorenzo, Chiara, Francesca Colombo, Simone Biella, Creina Stockley, and Patrizia Restani. 2021. "Polyphenols and Human Health: The Role of Bioavailability." *Nutrients* 13 (1): 273. https://doi.org/10.3390/nu13010273.

Di Pede, Giuseppe, Pedro Mena, Letizia Bresciani, Mariem Achour, Rosa M. Lamuela-Raventós, Ramon Estruch, Rikard Landberg, et al. 2023. "Revisiting the Bioavailability of Flavan-3-Ols in Humans: A Systematic Review and Comprehensive Data Analysis." *Molecular Aspects of Medicine* 89 (February): 101146. https://doi.org/10.1016/j.mam.2022.101146.

Diniz, Andréa, Laura Escuder-Gilabert, Norberto P. Lopes, Rosa María Villanueva-Camañas, Salvador Sagrado, and María José Medina-Hernández. 2008. "Characterization of Interactions between Polyphenolic Compounds and Human Serum Proteins by Capillary Electrophoresis." *Analytical and Bioanalytical Chemistry* 391 (2): 625–32. https://doi.org/10.1007/s00216-008-2046-4.

Duthie, G.G., M.W. Pedersen, P.T. Gardner, P.C. Morrice, A.M.E. Jenkinson, D.B. McPhail, and G.M. Steele. 1998. "The Effect of Whisky and Wine Consumption on Total Phenol Content and Antioxidant Capacity of Plasma from Healthy Volunteers." *European Journal of Clinical Nutrition* 52 (10): 733–36. https://doi.org/10.1038/sj.ejcn.1600635.

Dzah, Courage Sedem. 2025. "How to Increase Dietary Polyphenol Bioavailability: Understanding the Roles of Modulating Factors, Lifestyle and Gut Microbiota." *Food and Humanity* 4 (May): 100602. https://doi.org/10.1016/j.foohum.2025.100602.

Eran Nagar, Eden, Zoya Okun, and Avi Shpigelman. 2020. "Digestive Fate of Polyphenols: Updated View of the Influence of Chemical Structure and the Presence of Cell Wall Material." *Current Opinion in Food Science* 31 (February): 38–46. https://doi.org/10.1016/j.cofs.2019.10.009.

Ezhilarasi, P.N., D. Indrani, B.S. Jena, and C. Anandharamakrishnan. 2013. "Freeze Drying Technique for Microencapsulation of Garcinia Fruit Extract and Its Effect on Bread

Quality." *Journal of Food Engineering* 117 (4): 513–20. https://doi.org/10.1016/j.jfoodeng.2013.01.009.

Fan, Yuting, Qingyu He, Chao Gan, Zhen Wen, and Jiang Yi. 2022. "Investigation of Binding Interaction between Bovine α-Lactalbumin and Procyanidin B2 by Spectroscopic Methods and Molecular Docking." *Food Chemistry* 384 (August): 132509. https://doi.org/10.1016/j.foodchem.2022.132509.

Feliciano, Rodrigo, Charlotte Mills, Geoffrey Istas, Christian Heiss, and Ana Rodriguez-Mateos. 2017. "Absorption, Metabolism and Excretion of Cranberry (Poly)Phenols in Humans: A Dose Response Study and Assessment of Inter-Individual Variability." *Nutrients* 9 (3): 268. https://doi.org/10.3390/nu9030268.

Feng, Yulin, Xuejia Feng, Shuchang Liu, Huijuan Zhang, and Jing Wang. 2022. "Interaction Mechanism between Cereal Phenolic Acids and Gluten Protein: Protein Structural Changes and Binding Mode." *Journal of the Science of Food and Agriculture* 102 (15): 7387–96. https://doi.org/10.1002/jsfa.12107.

Feng, Yulin, Chengming Jin, Shihao Lv, Huijuan Zhang, Feiyue Ren, and Jing Wang. 2023. "Molecular Mechanisms and Applications of Polyphenol-Protein Complexes with Antioxidant Properties: A Review." *Antioxidants* 12 (8): 1577. https://doi.org/10.3390/antiox12081577.

Fuhrman, B., A. Lavy, and M. Aviram. 1995. "Consumption of Red Wine with Meals Reduces the Susceptibility of Human Plasma and Low-Density Lipoprotein to Lipid Peroxidation." *The American Journal of Clinical Nutrition* 61 (3): 549–54. https://doi.org/10.1093/ajcn/61.3.549.

Gambini, J., M. Inglés, G. Olaso, R. Lopez-Grueso, V. Bonet-Costa, L. Gimeno-Mallench, C. Mas-Bargues, et al. 2015. "Properties of Resveratrol: *In Vitro* and *In Vivo* Studies about Metabolism, Bioavailability, and Biological Effects in Animal Models and Humans." *Oxidative Medicine and Cellular Longevity* 2015: 1–13. https://doi.org/10.1155/2015/837042.

Garza, Aaron Z., Sharon B. Park, and Remek Kocz. 2025. "Drug Elimination." In *StatPearls*. Treasure Island (FL): StatPearls Publishing. http://www.ncbi.nlm.nih.gov/books/NBK547662/.

Gowd, Vemana, Lianghua Xie, Chongde Sun, and Wei Chen. 2020. "Phenolic Profile of Bayberry Followed by Simulated Gastrointestinal Digestion and Gut Microbiota Fermentation and Its Antioxidant Potential in HepG2 Cells." *Journal of Functional Foods* 70 (July): 103987. https://doi.org/10.1016/j.jff.2020.103987.

Guerreiro, Íris, Cíntia Ferreira-Pêgo, Diogo Carregosa, Cláudia N. Santos, Regina Menezes, Ana S. Fernandes, and João G. Costa. 2022. "Polyphenols and their Metabolites in Renal Diseases: An Overview." *Foods* 11 (7): 1060. https://doi.org/10.3390/foods11071060.

Gupta, Apoorva, Lalit D. Kagliwal, and Rekha S. Singhal. 2013. "Biotransformation of Polyphenols for Improved Bioavailability and Processing Stability." In *Advances in Food and Nutrition Research*, 69: 183–217. Elsevier. https://doi.org/10.1016/B978-0-12-410540-9.00004-1.

Han, Sihao, Jiemin Yao, Hajime Yamazaki, Samantha A. Streicher, Jianyu Rao, Roch A. Nianogo, Zuofeng Zhang, and Brian Z. Huang. 2024. "Genetically Determined Circulating Lactase/Phlorizin Hydrolase Concentrations and Risk of Colorectal Cancer: A Two-Sample Mendelian Randomization Study." *Nutrients* 16 (6): 808. https://doi.org/10.3390/nu16060808.

Hasan, Nihal, and Hongyi Yang. 2019. "Factors Affecting the Composition of the Gut Microbiota, and Its Modulation." *PeerJ* 7 (August): e7502. https://doi.org/10.7717/peerj.7502.

Hollman, Peter C.H., Lilian B.M. Tijburg, and Chung S. Yang. 1997. "Bioavailability of Flavonoids from Tea." *Critical Reviews in Food Science and Nutrition* 37 (8): 719–38. https://doi.org/10.1080/10408399709527799.

Hu, Lei, Yiqing Luo, Jiaxin Yang, and Chunsong Cheng. 2025. "Botanical Flavonoids: Efficacy, Absorption, Metabolism and Advanced Pharmaceutical Technology for Improving Bioavailability." *Molecules* 30 (5): 1184. https://doi.org/10.3390/molecules30051184.

Hu, Ming, Kristopher Krausz, Jun Chen, Xia Ge, Jianqi Li, Harry L. Gelboin, and Frank J. Gonzalez. 2003. "Identification of Cyp1a2 as the Main Isoform for the Phase I Hydroxylated Metabolism of Genistein and a Prodrug Converting Enzyme of Methylated Isoflavones." *Drug Metabolism and Disposition* 31 (7): 924–31. https://doi.org/10.1124/dmd.31.7.924.

Hur, Hor-Gil, Richard Beger, Thomas Heinze, Jackson Lay, James Freeman, Joel Dore, and Fatemeh Rafii. 2002. "Isolation of an Anaerobic Intestinal Bacterium Capable of Cleaving the C-Ring of the Isoflavonoid Daidzein." *Archives of Microbiology* 178 (1): 8–12. https://doi.org/10.1007/s00203-002-0414-6.

Ishtiaq, Memoona, Sajid Asghar, Ikram Ullah Khan, Muhammad Shahid Iqbal, and Syed Haroon Khalid. 2022. "Development of the Amorphous Solid Dispersion of Curcumin: A Rational Selection of Polymers for Enhanced Solubility and Dissolution." *Crystals* 12 (11): 1606. https://doi.org/10.3390/cryst12111606.

Jäger, Ralf, Ryan P. Lowery, Allison V. Calvanese, Jordan M. Joy, Martin Purpura, and Jacob M. Wilson. 2014. "Comparative Absorption of Curcumin Formulations." *Nutrition Journal* 13 (1): 11. https://doi.org/10.1186/1475-2891-13-11.

Jaldappagari, Seetharamappa, Sandhya Balakrishnan, Ashwini H. Hegde, Nagappa L. Teradal, and Prashanth S. Narayan. 2013. "Interactions of Polyphenols with Plasma Proteins: Insights from Analytical Techniques." *Current Drug Metabolism* 14 (4): 456–73. https://doi.org/10.2174/1389200211314040009.

Jandhyala, Sai Manasa. 2015. "Role of the Normal Gut Microbiota." *World Journal of Gastroenterology* 21 (29): 8787. https://doi.org/10.3748/wjg.v21.i29.8787.

Jeong, E., X. Liu, X. Jia, J. Chen, and M. Hu. 2005. "Coupling of Conjugating Enzymes and Efflux Transporters: Impact on Bioavailability and Drug Interactions." *Current Drug Metabolism* 6 (5): 455–68. https://doi.org/10.2174/138920005774330657.

Jin, Jong-Sik, and Masao Hattori. 2012. "Isolation and Characterization of a Human Intestinal Bacterium *Eggerthella* Sp. CAT-1 Capable of Cleaving the C-Ring of (+)-Catechin and (−)-Epicatechin, Followed by *p*-Dehydroxylation of the B-Ring." *Biological and Pharmaceutical Bulletin* 35 (12): 2252–56. https://doi.org/10.1248/bpb.b12-00726.

Karonen, Maarit. 2022. "Insights into Polyphenol–Lipid Interactions: Chemical Methods, Molecular Aspects and their Effects on Membrane Structures." *Plants* 11 (14): 1809. https://doi.org/10.3390/plants11141809.

Kawabata, Kyuichi, Yasukiyo Yoshioka, and Junji Terao. 2019. "Role of Intestinal Microbiota in the Bioavailability and Physiological Functions of Dietary Polyphenols." *Molecules* 24 (2): 370. https://doi.org/10.3390/molecules24020370.

Kay, Colin D., Gema Pereira-Caro, Iziar A. Ludwig, Michael N. Clifford, and Alan Crozier. 2017. "Anthocyanins and Flavanones Are More Bioavailable than Previously Perceived: A Review of Recent Evidence." *Annual Review of Food Science and Technology* 8 (1): 155–80. https://doi.org/10.1146/annurev-food-030216-025636.

Kim, Mihyang, Nayoung Kim, and Jaehong Han. 2016. "Deglycosylation of Flavonoid O-Glucosides by Human Intestinal Bacteria Enterococcus Sp. MRG-2 and Lactococcus Sp. MRG-IF–4." *Applied Biological Chemistry* 59 (3): 443–49. https://doi.org/10.1007/s13765-016-0183-6.

Kiss, László E., and Patrício Soares-da-Silva. 2014. "Medicinal Chemistry of Catechol *O*-Methyltransferase (COMT) Inhibitors and their Therapeutic Utility." *Journal of Medicinal Chemistry* 57 (21): 8692–717. https://doi.org/10.1021/jm500572b.

Klaassen, Curtis D., and James W. Boles. 1997. "The Importance of 3'-phosphoadenosine 5'-phosphosulfate (PAPS) in the Regulation of Sulfation." *The FASEB Journal* 11 (6): 404–18. https://doi.org/10.1096/fasebj.11.6.9194521.

Koudoufio, Mireille, Yves Desjardins, Francis Feldman, Schohraya Spahis, Edgard Delvin, and Emile Levy. 2020. "Insight into Polyphenol and Gut Microbiota Crosstalk: Are their Metabolites the Key to Understand Protective Effects against Metabolic Disorders?" *Antioxidants* 9 (10): 982. https://doi.org/10.3390/antiox9100982.

Kuhnle, Gunter, Jeremy P.E. Spencer, Hagen Schroeter, Baskar Shenoy, Edward S. Debnam, S. Kaila S. Srai, Catherine Rice-Evans, and Ulrich Hahn. 2000. "Epicatechin and Catechin Are O-Methylated and Glucuronidated in the Small Intestine." *Biochemical and Biophysical Research Communications* 277 (2): 507–12. https://doi.org/10.1006/bbrc.2000.3701.

Kumar, Manish, Ram Singh Bishnoi, Ajay Kumar Shukla, and Chandra Prakash Jain. 2019. "Techniques for Formulation of Nanoemulsion Drug Delivery System: A Review." *Preventive Nutrition and Food Science* 24 (3): 225–34. https://doi.org/10.3746/pnf.2019.24.3.225.

Kumar, Nitesh, Amita Rai, Neetinkumar D. Reddy, P. Vasanth Raj, Prateek Jain, Praful Deshpande, Geetha Mathew, N. Gopalan Kutty, Nayanabhirama Udupa, and C. Mallikarjuna Rao. 2014. "Silymarin Liposomes Improves Oral Bioavailability of Silybin besides Targeting Hepatocytes, and Immune Cells." *Pharmacological Reports* 66 (5): 788–98. https://doi.org/10.1016/j.pharep.2014.04.007.

Kumar Singh, Amit, Célia Cabral, Ramesh Kumar, Risha Ganguly, Harvesh Kumar Rana, Ashutosh Gupta, Maria Rosaria Lauro, Claudia Carbone, Flávio Reis, and Abhay K. Pandey. 2019. "Beneficial Effects of Dietary Polyphenols on Gut Microbiota and Strategies to Improve Delivery Efficiency." *Nutrients* 11 (9): 2216. https://doi.org/10.3390/nu11092216.

Kurogi, Katsuhisa, Adnan Alazizi, Ming-Yih Liu, Yoichi Sakakibara, Masahito Suiko, Takuya Sugahara, and Ming-Cheh Liu. 2012. "Concerted Actions of the Catechol O-Methyltransferase and the Cytosolic Sulfotransferase SULT1A3 in the Metabolism of Catecholic Drugs." *Biochemical Pharmacology* 84 (9): 1186–95. https://doi.org/10.1016/j.bcp.2012.08.009.

Kwon, Oran, Peter Eck, Shenglin Chen, Christopher P. Corpe, Je-Hyuk Lee, Michael Kruhlak, and Mark Levine. 2007. "Inhibition of the Intestinal Glucose Transporter GLUT2 by Flavonoids." *The FASEB Journal* 21 (2): 366–77. https://doi.org/10.1096/fj.06-6620com.

Leeming, Emily R., Abigail J. Johnson, Tim D. Spector, and Caroline I. Le Roy. 2019. "Effect of Diet on the Gut Microbiota: Rethinking Intervention Duration." *Nutrients* 11 (12): 2862. https://doi.org/10.3390/nu11122862.

Lethe, Mary Caroline L., Vincent Paris, Xiaoqiang Wang, and Clement T.Y. Chan. 2024. "Similarities in Structure and Function of UDP-Glycosyltransferase Homologs from Human and Plants." *International Journal of Molecular Sciences* 25 (5): 2782. https://doi.org/10.3390/ijms25052782.

Lewandowska, Urszula, Karolina Szewczyk, Elżbieta Hrabec, Anna Janecka, and Sylwia Gorlach. 2013. "Overview of Metabolism and Bioavailability Enhancement of Polyphenols." *Journal of Agricultural and Food Chemistry* 61 (50): 12183–99. https://doi.org/10.1021/jf404439b.

Li, Bin, Kim Harich, Lindsay Wegiel, Lynne S. Taylor, and Kevin J. Edgar. 2013. "Stability and Solubility Enhancement of Ellagic Acid in Cellulose Ester Solid Dispersions." *Carbohydrate Polymers* 92 (2): 1443–50. https://doi.org/10.1016/j.carbpol.2012.10.051.

Li, Chong, Yan Zhang, Tingting Su, Lianlian Feng, Yingying Long, and Zhangbao Chen. 2012. "Silica-Coated Flexible Liposomes as a Nanohybrid Delivery System for Enhanced Oral Bioavailability of Curcumin." *International Journal of Nanomedicine* (December): 5995. https://doi.org/10.2147/IJN.S38043.

Li, Ti, Xin Li, Taotao Dai, Peng Hu, Xiaoqin Niu, Chengmei Liu, and Jun Chen. 2020. "Binding Mechanism and Antioxidant Capacity of Selected Phenolic Acid – β-Casein Complexes." *Food Research International* 129 (March): 108802. https://doi.org/10.1016/j.foodres.2019.108802.

Li, Xiaopeng, Enjun Xie, Shumin Sun, Jie Shen, Yujin Ding, Jiaqi Wang, Xiaoyu Peng, Ruting Zheng, Mohamed A. Farag, and Jianbo Xiao. 2025. "Flavonoids for Gastrointestinal Tract Local and Associated Systemic Effects: A Review of Clinical Trials and Future Perspectives." *Journal of Advanced Research* (January): S2090123225000335. https://doi.org/10.1016/j.jare.2025.01.014.

Li, Yuhua, Qiang Meng, Mengbi Yang, Dongyang Liu, Xiangyu Hou, Lan Tang, Xin Wang, et al. 2019. "Current Trends in Drug Metabolism and Pharmacokinetics." *Acta Pharmaceutica Sinica B* 9 (6): 1113–44. https://doi.org/10.1016/j.apsb.2019.10.001.

Lind, Christoffer, Poonam Pandey, Richard W. Pastor, and Alexander D. MacKerell. 2021. "Functional Group Distributions, Partition Coefficients, and Resistance Factors in Lipid Bilayers Using Site Identification by Ligand Competitive Saturation." *Journal of Chemical Theory and Computation* 17 (5): 3188–3202. https://doi.org/10.1021/acs.jctc.1c00089.

Lippolis, Tamara, Miriam Cofano, Giusy Rita Caponio, Valentina De Nunzio, and Maria Notarnicola. 2023. "Bioaccessibility and Bioavailability of Diet Polyphenols and their Modulation of Gut Microbiota." *International Journal of Molecular Sciences* 24 (4): 3813. https://doi.org/10.3390/ijms24043813.

Liu, Jiayuan, Gongshuai Song, Yawen Yuan, Like Zhou, Danli Wang, Tinglan Yuan, Ling Li, et al. 2022. "Ultrasound-Assisted Assembly of β-Lactoglobulin and Chlorogenic Acid for Non Covalent Nanocomplex: Fabrication, Characterization and Potential Biological Function." *Ultrasonics Sonochemistry* 86 (May): 106025. https://doi.org/10.1016/j.ultsonch.2022.106025.

Liu, Lili, Xiaodan Zhang, Mapinyi Zhang, Mengjun Zhang, Weiwei Cheng, and Baocheng Xu. 2023. "Effect of Catechin on Yolk Immunoglobulin Structure and Properties: A Polyphenol–Protein Interaction Approach." *Foods* 12 (3): 462. https://doi.org/10.3390/foods12030462.

Liu, Yichen, Jianjun Deng, Tong Zhao, Xiaojie Yang, Juntao Zhang, and Haixia Yang. 2024. "Bioavailability and Mechanisms of Dietary Polyphenols Affected by Non-Thermal Processing Technology in Fruits and Vegetables." *Current Research in Food Science* 8: 100715. https://doi.org/10.1016/j.crfs.2024.100715.

Liu, Zhongqiu, and Ming Hu. 2007. "Natural Polyphenol Disposition via Coupled Metabolic Pathways." *Expert Opinion on Drug Metabolism & Toxicology* 3 (3): 389–406. https://doi.org/10.1517/17425255.3.3.389.

Lu, Yingcong, Ru Zhao, Ce Wang, Xiaoge Zhang, and Cuina Wang. 2022. "Deciphering the Non-Covalent Binding Patterns of Three Whey Proteins with Rosmarinic Acid by Multi-Spectroscopic, Molecular Docking and Molecular Dynamics Simulation Approaches." *Food Hydrocolloids* 132 (November): 107895. https://doi.org/10.1016/j.foodhyd.2022.107895.

Lu, Yuxuan, Kai Wang, and Lin Hu. 2025. "Advancements in Delivery Systems for Dietary Polyphenols in Enhancing Radioprotection Effects: Challenges and Opportunities." *Npj Science of Food* 9 (1): 51. https://doi.org/10.1038/s41538-025-00419-6.

Ma, Hui, Bowei Zhang, Yaozhong Hu, Jin Wang, Jingmin Liu, Renbing Qin, Shiwen Lv, and Shuo Wang. 2019. "Correlation Analysis of Intestinal Redox State with the Gut Microbiota Reveals the Positive Intervention of Tea Polyphenols on Hyperlipidemia in High Fat Diet Fed Mice." *Journal of Agricultural and Food Chemistry* 67 (26): 7325–35. https://doi.org/10.1021/acs.jafc.9b02211.

Maini Rekdal, Vayu, Paola Nol Bernadino, Michael U. Luescher, Sina Kiamehr, Chip Le, Jordan E. Bisanz, Peter J. Turnbaugh, Elizabeth N. Bess, and Emily P. Balskus. 2020. "A Widely Distributed Metalloenzyme Class Enables Gut Microbial Metabolism of Host- and Diet-Derived Catechols." *eLife* 9 (February): e50845. https://doi.org/10.7554/eLife.50845.

Manach, Claudine, and Jennifer L. Donovan. 2004. "Pharmacokinetics and Metabolism of Dietary Flavonoids in Humans." *Free Radical Research* 38 (8): 771–85. https://doi.org/10.1080/10715760410001727858.

Manach, Claudine, Augustin Scalbert, Christine Morand, Christian Rémésy, and Liliana Jiménez. 2004. "Polyphenols: Food Sources and Bioavailability." *The American Journal of Clinical Nutrition* 79 (5): 727–47. https://doi.org/10.1093/ajcn/79.5.727.

Mangoni, A.A., and S.H.D. Jackson. 2004. "Age-related Changes in Pharmacokinetics and Pharmacodynamics: Basic Principles and Practical Applications." *British Journal of Clinical Pharmacology* 57 (1): 6–14. https://doi.org/10.1046/j.1365-2125.2003.02007.x.

Maraveas, Chrysanthos, Ilker S. Bayer, and Thomas Bartzanas. 2021. "Recent Advances in Antioxidant Polymers: From Sustainable and Natural Monomers to Synthesis and Applications." *Polymers* 13 (15): 2465. https://doi.org/10.3390/polym13152465.

Marín, Laura, Elisa M. Miguélez, Claudio J. Villar, and Felipe Lombó. 2015. "Bioavailability of Dietary Polyphenols and Gut Microbiota Metabolism: Antimicrobial Properties." *BioMed Research International* 2015: 1–18. https://doi.org/10.1155/2015/905215.

Matsukawa, Noriko, Megumi Matsumoto, and Hiroshi Hara. 2009. "High Biliary Excretion Levels of Quercetin Metabolites after Administration of a Quercetin Glycoside in Conscious Bile Duct Cannulated Rats." *Bioscience, Biotechnology, and Biochemistry* 73 (8): 1863–65. https://doi.org/10.1271/bbb.90031.

Maxwell, Simon, Alison Cruickshank, and Gary Thorpe. 1994. "Red Wine and Antioxidant Activity in Serum." *The Lancet* 344 (8916): 193–94. https://doi.org/10.1016/S0140-6736(94)92795-2.

McCarthy, Tanya C., and Christopher J. Sinal. 2005. "Biotransformation." In *Encyclopedia of Toxicology*, 299–312. Elsevier. https://doi.org/10.1016/B0-12-369400-0/00137-X.

Mehta, Nitin, Pavan Kumar, Akhilesh K. Verma, Pramila Umaraw, Yogesh Kumar, Om Prakash Malav, Awis Qurni Sazili, Rubén Domínguez, and José M. Lorenzo. 2022. "Microencapsulation as a Noble Technique for the Application of Bioactive Compounds in the Food Industry: A Comprehensive Review." *Applied Sciences* 12 (3): 1424. https://doi.org/10.3390/app12031424.

Monfalouti, Hanae El, Badr Eddine Kartah, Hanae El Monfalouti, and Badr Eddine Kartah. 2024. *Enhancing Polyphenol Bioavailability through Nanotechnology: Current Trends and Challenges*. IntechOpen. https://doi.org/10.5772/intechopen.1005764.

Muntaha, Sidra Tul, Allah Rakha, Hina Rasheed, Iraj Fatima, Masood Sadiq Butt, Gholam-Reza Abdi, and Rana Muhammad Aadil. 2025. "Polyphenol-Protein Particles: A Nutraceutical Breakthrough in Nutrition and Food Science." *Journal of Agriculture and Food Research* 19 (March): 101641. https://doi.org/10.1016/j.jafr.2025.101641.

Muscolo, Adele, Oliva Mariateresa, Torello Giulio, and Russo Mariateresa. 2024. "Oxidative Stress: The Role of Antioxidant Phytochemicals in the Prevention and Treatment of Diseases." *International Journal of Molecular Sciences* 25 (6): 3264. https://doi.org/10.3390/ijms25063264.

Myint, Ohnmar, Sakornniya Wattanapongpitak, Benjamaporn Supawat, Suchart Kothan, Chatchanok Udomtanakunchai, Singkome Tima, and Montree Tungjai. 2021. "Protein Binding of 4-Hydroxybenzoic Acid and 4-Hydroxy-3-Methoxybenzoic Acid to Human Serum Albumin and their Anti-proliferation on Doxorubicin-sensitive and Doxorubicin-resistant Leukemia Cells." *Toxicology Reports* 8: 1381–88. https://doi.org/10.1016/j.toxrep.2021.07.001.

Németh, Kitti, Geoff W. Plumb, Jean-Guy Berrin, Nathalie Juge, Ralf Jacob, Hassan Y. Naim, Gary Williamson, Dallas M. Swallow, and Paul A. Kroon. 2003. "Deglycosylation by Small Intestinal Epithelial Cell?-Glucosidases is a Critical Step in the Absorption and Metabolism of Dietary Flavonoid Glycosides in Humans." *European Journal of Nutrition* 42 (1): 29–42. https://doi.org/10.1007/s00394-003-0397-3.

Ojha, Himanshu, Krishnanand Mishra, Mohammad I. Hassan, and Nabo K. Chaudhury. 2012. "Spectroscopic and Isothermal Titration Calorimetry Studies of Binding Interaction of Ferulic Acid with Bovine Serum Albumin." *Thermochimica Acta* 548 (November): 56–64. https://doi.org/10.1016/j.tca.2012.08.016.

O'Leary, Karen A., Andrea J. Day, Paul W. Needs, Fred A. Mellon, Nora M. O'Brien, and Gary Williamson. 2003. "Metabolism of Quercetin-7- and Quercetin-3-Glucuronides by an in Vitro Hepatic Model: The Role of Human β-Glucuronidase, Sulfotransferase, Catechol-O-Methyltransferase and Multi-Resistant Protein 2 (MRP2) in Flavonoid Metabolism." *Biochemical Pharmacology* 65 (3): 479–91. https://doi.org/10.1016/S0006-2952(02)01510–1.

Oliveira, Ana, Ana L. Amaro, and Manuela Pintado. 2018. "Impact of Food Matrix Components on Nutritional and Functional Properties of Fruit-Based Products." *Current Opinion in Food Science* 22 (August): 153–59. https://doi.org/10.1016/j.cofs.2018.04.002.

Oliver, Susan, Orazio Vittorio, Giuseppe Cirillo, and Cyrille Boyer. 2016. "Enhancing the Therapeutic Effects of Polyphenols with Macromolecules." *Polymer Chemistry* 7 (8): 1529–44. https://doi.org/10.1039/C5PY01912E.

Onoue, Satomi, Haruki Takahashi, Yohei Kawabata, Yoshiki Seto, Junya Hatanaka, Barbara Timmermann, and Shizuo Yamada. 2010. "Formulation Design and Photochemical Studies on Nanocrystal Solid Dispersion of Curcumin with Improved Oral Bioavailability." *Journal of Pharmaceutical Sciences* 99 (4): 1871–81. https://doi.org/10.1002/jps.21964.

Palafox-Carlos, Hugo, Jesús Fernando Ayala-Zavala, and Gustavo A. González-Aguilar. 2011. "The Role of Dietary Fiber in the Bioaccessibility and Bioavailability of Fruit and Vegetable Antioxidants." *Journal of Food Science* 76 (1). https://doi.org/10.1111/j.1750-3841.2010.01957.x.

Parveen, Shama, Pratishtha Gupta, Saurabh Kumar, and Monisha Banerjee. 2023. "Lipid Polymer Hybrid Nanoparticles as Potent Vehicles for Drug Delivery in Cancer Therapeutics." *Medicine in Drug Discovery* 20 (December): 100165. https://doi.org/10.1016/j.medidd.2023.100165.

Pasinetti, Giulio Maria, Risham Singh, Susan Westfall, Francis Herman, Jeremiah Faith, and Lap Ho. 2018. "The Role of the Gut Microbiota in the Metabolism of Polyphenols as Characterized by Gnotobiotic Mice." *Journal of Alzheimer's Disease* 63 (2): 409–21. https://doi.org/10.3233/JAD–171151.

Pedrosa, Mercedes M., Eva Guillamón, and Claudia Arribas. 2021. "Autoclaved and Extruded Legumes as a Source of Bioactive Phytochemicals: A Review." *Foods* 10 (2): 379. https://doi.org/10.3390/foods10020379.

Peng, Han, and Fereidoon Shahidi. 2024. "Metabolic, Toxicological, Chemical, and Commercial Perspectives on Esterification of Dietary Polyphenols: A Review." *Critical Reviews in Food Science and Nutrition* 64 (21): 7465–7504. https://doi.org/10.1080/10408398.2023.2185589.

Phang-Lyn, Simone, and Valerie A. Llerena. 2025. "Biochemistry, Biotransformation." In *StatPearls*. Treasure Island (FL): StatPearls Publishing. http://www.ncbi.nlm.nih.gov/books/NBK544353/.

Pidgeon, Reilly, Sacha Mitchell, Michael Shamash, Layan Suleiman, Lharbi Dridi, Corinne F. Maurice, and Bastien Castagner. 2025. "Diet-Derived Urolithin A is Produced by a Dehydroxylase Encoded by Human Gut Enterocloster Species." *Nature Communications* 16 (1): 999. https://doi.org/10.1038/s41467-025-56266-2.

Piskula, Mariusz K., and Junji Terao. 1998. "Accumulation of (−)-Epicatechin Metabolites in Rat Plasma after Oral Administration and Distribution of Conjugation Enzymes in Rat Tissues." *The Journal of Nutrition* 128 (7): 1172–78. https://doi.org/10.1093/jn/128.7.1172.

Plumb, Geoff W., Maria T. Garcia-Conesa, Paul A. Kroon, Mike Rhodes, Saxon Ridley, and Gary Williamson. 1999. "Metabolism of Chlorogenic Acid by Human Plasma, Liver, Intestine and Gut Microflora." *Journal of the Science of Food and Agriculture* 79 (3). Wiley-Blackwell: 390–92. https://doi.org/10.1002/(SICI)1097-0010(19990301)79:3<390::AID-JSFA258>3.0.CO;2-0.

Polia, Franck, Marta Pastor-Belda, Alberto Martínez-Blázquez, Marie-Noelle Horcajada, Francisco A. Tomás-Barberán, and Rocío García-Villalba. 2022. "Technological and Biotechnological Processes to Enhance the Bioavailability of Dietary (Poly)Phenols in Humans." *Journal of Agricultural and Food Chemistry* 70 (7): 2092–107. https://doi.org/10.1021/acs.jafc.1c07198.

Prasad, Sahdeo, Amit K. Tyagi, and Bharat B. Aggarwal. 2014. "Recent Developments in Delivery, Bioavailability, Absorption and Metabolism of Curcumin: The Golden Pigment from Golden Spice." *Cancer Research and Treatment* 46 (1): 2–18. https://doi.org/10.4143/crt.2014.46.1.2.

Prigent, S.V.E., A.G.J. Voragen, G.A. van Koningsveld, A. Baron, C.M.G.C. Renard, and H. Gruppen. 2009. "Interactions between Globular Proteins and Procyanidins of Different Degrees of Polymerization." *Journal of Dairy Science* 92 (12): 5843–53. https://doi.org/10.3168/jds.2009-2261.

Quan, Nguyen, Tran Xuan, Hoang-Dung Tran, Nguyen Thuy, Le Trang, Can Huong, Yusuf Andriana, and Phung Tuyen. 2019. "Antioxidant, α-Amylase and α-Glucosidase Inhibitory Activities and Potential Constituents of Canarium Tramdenum Bark." *Molecules* 24 (3): 605–605. https://doi.org/10.3390/molecules24030605.

Radominska-Pandya, Anna, Joanna M. Little, Jay T. Pandya, Thomas R. Tephly, Christopher D. King, Gary W. Barone, and Jean-Pierre Raufman. 1998. "UDP-Glucuronosyltransferases in Human Intestinal Mucosa." *Biochimica et Biophysica Acta (BBA) – Lipids and Lipid Metabolism* 1394 (2–3): 199–208. https://doi.org/10.1016/S0005-2760(98)00115-5.

Raimondi, Stefano, Andrew Anighoro, Andrea Quartieri, Alberto Amaretti, Francisco A. Tomás-Barberán, Giulio Rastelli, and Maddalena Rossi. 2015. "Role of Bifidobacteria in the Hydrolysis of Chlorogenic Acid." *MicrobiologyOpen* 4 (1): 41–52. https://doi.org/10.1002/mbo3.219.

Rashidinejad, Ali, Matthijs Nieuwkoop, Harjinder Singh, and Geoffrey Jameson. 2023. "Assessment of Various Food Proteins as Structural Materials for Delivery of Hydrophobic Polyphenols Using a Novel Co-precipitation Method." *Molecules* 28 (8): 3573. https://doi.org/10.3390/molecules28083573.

Ren, Jie, Caleb Don Barton, and Jixun Zhan. 2023. "Engineered Production of Bioactive Polyphenolic O-Glycosides." *Biotechnology Advances* 65 (July): 108146. https://doi.org/10.1016/j.biotechadv.2023.108146.

Roberts, Dean W., Daniel R. Doerge, Mona I. Churchwell, Gonçalo Gamboa Da Costa, M. Matilde Marques, and William H. Tolleson. 2004. "Inhibition of Extrahepatic Human Cytochromes P450 1A1 and 1B1 by Metabolism of Isoflavones Found in *Trifolium Pratense* (Red Clover)." *Journal of Agricultural and Food Chemistry* 52 (21): 6623–32. https://doi.org/10.1021/jf049418x.

Roberts, Michael S., Beatrice M. Magnusson, Frank J. Burczynski, and Michael Weiss. 2002. "Enterohepatic Circulation: Physiological, Pharmacokinetic and Clinical Implications." *Clinical Pharmacokinetics* 41 (10): 751–90. https://doi.org/10.2165/00003088-200241100-00005.

Rocha, Sónia, M. Luísa Corvo, Marisa Freitas, and Eduarda Fernandes. 2024. "Liposomal Quercetin: A Promising Strategy to Combat Hepatic Insulin Resistance and Inflammation in Type 2 Diabetes Mellitus." *International Journal of Pharmaceutics* 661 (August): 124441. https://doi.org/10.1016/j.ijpharm.2024.124441.

Rodrigues Silva, Francisca Mayla, Larissa Morais Ribeiro Da Silva, Antoniella Souza Gomes Duarte, Carlos Eduardo Da Silva Monteiro, Adriana Rolim Campos, Dayse Karine

Rodrigues Holanda, Ana Cristina Silva De Lima, Luiz Bruno De Sousa Sabino, and Raimundo Wilane De Figueiredo. 2021. "Microencapsulation of Green Tea Extract (Camellia Sinensis Var Assamica) to Increase the Bioaccessibility of Bioactive Compounds and Gastroprotective Effects." *Food Bioscience* 42 (August): 101190. https://doi.org/10.1016/j.fbio.2021.101190.

Rohn, Sascha. 2014. "Possibilities and Limitations in the Analysis of Covalent Interactions between Phenolic Compounds and Proteins." *Food Research International*, 7th World Congress on Polyphenols Applications, 65 (November): 13–19. https://doi.org/10.1016/j.foodres.2014.05.042.

Rojas-González, Alexis, Claudia Yuritzi Figueroa-Hernández, Oscar González-Rios, Mirna Leonor Suárez-Quiroz, Rosa María González-Amaro, Zorba Josué Hernández-Estrada, and Patricia Rayas-Duarte. 2022. "Coffee Chlorogenic Acids Incorporation for Bioactivity Enhancement of Foods: A Review." *Molecules* 27 (11): 3400. https://doi.org/10.3390/molecules27113400.

Ross, Philip D., and S. Subramanian. 1981. "Thermodynamics of Protein Association Reactions: Forces Contributing to Stability." *Biochemistry* 20 (11): 3096–102. https://doi.org/10.1021/bi00514a017.

Roupe, Kathryn A., Jaime A. Yáñez, Xiao Wei Teng, and Neal M. Davies. 2006. "Pharmacokinetics of Selected Stilbenes: Rhapontigenin, Piceatannol and Pinosylvin in Rats." *Journal of Pharmacy and Pharmacology* 58 (11): 1443–50. https://doi.org/10.1211/jpp.58.11.0004.

Rowland, Andrew, John O. Miners, and Peter I. Mackenzie. 2013. "The UDP-Glucuronosyltransferases: Their Role in Drug Metabolism and Detoxification." *The International Journal of Biochemistry & Cell Biology* 45 (6): 1121–32. https://doi.org/10.1016/j.biocel.2013.02.019.

Rubio, Laura, Alba Macia, and Maria-Jose Motilva. 2014. "Impact of Various Factors on Pharmacokinetics of Bioactive Polyphenols: An Overview." *Current Drug Metabolism* 15 (1): 62–76. https://doi.org/10.2174/1389200214666131210144115.

Runge-Morris, Melissa, and Thomas A. Kocarek. 2009. "Regulation of Sulfotransferase and UDP-Glucuronosyltransferase Gene Expression by the PPARs." Edited by John Y.L. Chiang. *PPAR Research* 2009 (1): 728941. https://doi.org/10.1155/2009/728941.

Sahraeian, Shahriyar, Ali Rashidinejad, and Mohammad-Taghi Golmakani. 2024. "Recent Advances in the Conjugation Approaches for Enhancing the Bioavailability of Polyphenols." *Food Hydrocolloids* 146 (January): 109221. https://doi.org/10.1016/j.foodhyd.2023.109221.

Sánchez-Mundo, M.L., V.M. Escobedo-Crisantes, S. Mendoza-Arvizu, and M.E. Jaramillo-Flores. 2016. "Polymerization of Phenolic Compounds by Polyphenol Oxidase from Bell Pepper with Increase in their Antioxidant Capacity." *CyTA – Journal of Food* 14 (4): 594–603. https://doi.org/10.1080/19476337.2016.1181672.

Scalbert, Augustin, Christine Morand, Claudine Manach, and Christian Rémésy. 2002. "Absorption and Metabolism of Polyphenols in the Gut and Impact on Health." *Biomedicine & Pharmacotherapy* 56 (6): 276–82. https://doi.org/10.1016/S0753-3322(02)00205-6.

Scalbert, Augustin, and Gary Williamson. 2000. "Dietary Intake and Bioavailability of Polyphenols." *The Journal of Nutrition* 130 (8): 2073S–85S. https://doi.org/10.1093/jn/130.8.2073S.

Scheepens, Arjan, Kee Tan, and James W. Paxton. 2010. "Improving the Oral Bioavailability of Beneficial Polyphenols through Designed Synergies." *Genes & Nutrition* 5 (1): 75–87. https://doi.org/10.1007/s12263-009-0148-z.

Scheline, Ronald R. 1991. "Metabolism of Oxygen Heterocyclic Compounds." In *Handbook of Mammalian Metabolism of Plant Compounds*. CRC Press.

Sejbuk, M., I. Mirończuk-Chodakowska, S. Karav, and A. M. Witkowska. 2024. "Dietary Polyphenols, Food Processing and Gut Microbiome: Recent Findings on Bioavailability,

Bioactivity, and Gut Microbiome Interplay." *Antioxidants (Basel)* 13 (10): 1220. Published 2024 Oct 10. https://doi.org/10.3390/antiox13101220

Serafini, M., A. Ghiselli, and A. Ferro-Luzzi. 1996. "In Vivo Antioxidant Effect of Green and Black Tea in Man." *European Journal of Clinical Nutrition* 50 (1): 28–32.

Serafini, Mauro, Giuseppe Maiani, and Anna Ferro-Luzzi. 1998. "Alcohol-Free Red Wine Enhances Plasma Antioxidant Capacity in Humans." *The Journal of Nutrition* 128 (6): 1003–7. https://doi.org/10.1093/jn/128.6.1003.

Shabbir, Umair, Akanksha Tyagi, Fazle Elahi, Simon Okomo Aloo, and Deog-Hwan Oh. 2021. "The Potential Role of Polyphenols in Oxidative Stress and Inflammation Induced by Gut Microbiota in Alzheimer's Disease." *Antioxidants* 10 (9): 1370. https://doi.org/10.3390/antiox10091370.

Shahidi, Fereidoon, and Yao Pan. 2022. "Influence of Food Matrix and Food Processing on the Chemical Interaction and Bioaccessibility of Dietary Phytochemicals: A Review." *Critical Reviews in Food Science and Nutrition* 62 (23): 6421–45. https://doi.org/10.1080/10408398.2021.1901650.

Shamsi, Anas, Saleha Anwar, Mohd Shahbaaz, Taj Mohammad, Mohamed F. Alajmi, Afzal Hussain, Imtaiyaz Hassan, Faizan Ahmad, and Asimul Islam. 2020. "Evaluation of Binding of Rosmarinic Acid with Human Transferrin and Its Impact on the Protein Structure: Targeting Polyphenolic Acid-Induced Protection of Neurodegenerative Disorders." Edited by German Gil. *Oxidative Medicine and Cellular Longevity* 2020 (November): 1–14. https://doi.org/10.1155/2020/1245875.

Sharma, V., Nehru, B., Munshi, A., and Jyothy, A. 2010. "Antioxidant Potential of Curcumin against Oxidative Insult Inducedby Pentylenetetrazol in Epileptic Rats." *Methods and Findings in Experimental and Clinical Pharmacology* 32 (4): 227. https://doi.org/10.1358/mf.2010.32.4.1452090.

Shu, Yang, Jie Li, Xiaoping Yang, Xinyang Dong, and Xujie Wang. 2019. "Effect of Particle Size on the Bioaccessibility of Polyphenols and Polysaccharides in Green Tea Powder and Its Antioxidant Activity after Simulated Human Digestion." *Journal of Food Science and Technology* 56 (3): 1127–33. https://doi.org/10.1007/s13197-019-03573-4.

Singar, Saiful, Ravinder Nagpal, Bahram H. Arjmandi, and Neda S. Akhavan. 2024. "Personalized Nutrition: Tailoring Dietary Recommendations through Genetic Insights." *Nutrients* 16 (16): 2673. https://doi.org/10.3390/nu16162673.

Soldin, Offie P., and Donald R. Mattison. 2009. "Sex Differences in Pharmacokinetics and Pharmacodynamics:" *Clinical Pharmacokinetics* 48 (3): 143–57. https://doi.org/10.2165/00003088-200948030-00001.

Spencer, Jeremy P.E. 2003. "Metabolism of Tea Flavonoids in the Gastrointestinal Tract." *The Journal of Nutrition* 133 (10): 3255S–61S. https://doi.org/10.1093/jn/133.10.3255S.

Spencer, Jeremy P.E., Manal M. Abd El Mohsen, Ann-Marie Minihane, and John C. Mathers. 2008. "Biomarkers of the Intake of Dietary Polyphenols: Strengths, Limitations and Application in Nutrition Research." *British Journal of Nutrition* 99 (1): 12–22. https://doi.org/10.1017/S0007114507798938.

Spencer, Jeremy P.E, George Chowrimootoo, Ruksana Choudhury, Edward S. Debnam, S. Kaila Srai, and Catherine Rice-Evans. 1999. "The Small Intestine Can Both Absorb and Glucuronidate Luminal Flavonoids." *FEBS Letters* 458 (2): 224–30. https://doi.org/10.1016/S0014-5793(99)01160–6.

Squillaro, T., A. Cimini, G. Peluso, A. Giordano, and M.A.B. Melone. 2018. "Nano-Delivery Systems for Encapsulation of Dietary Polyphenols: An Experimental Approach for Neurodegenerative Diseases and Brain Tumors." *Biochemical Pharmacology* 154 (August): 303–17. https://doi.org/10.1016/j.bcp.2018.05.016.

Stevens, Jan F., and Claudia S. Maier. 2016. "The Chemistry of Gut Microbial Metabolism of Polyphenols." *Phytochemistry Reviews* 15 (3): 425–44. https://doi.org/10.1007/s11101-016-9459-z.

Stromsnes, Kristine, Rudite Lagzdina, Gloria Olaso-Gonzalez, Lucia Gimeno-Mallench, and Juan Gambini. 2021. "Pharmacological Properties of Polyphenols: Bioavailability, Mechanisms of Action, and Biological Effects in in Vitro Studies, Animal Models, and Humans." *Biomedicines* 9 (8): 1074. https://doi.org/10.3390/biomedicines9081074.

Subroto, Edy, Robi Andoyo, and Rossi Indiarto. 2023. "Solid Lipid Nanoparticles: Review of the Current Research on Encapsulation and Delivery Systems for Active and Antioxidant Compounds." *Antioxidants* 12 (3): 633. https://doi.org/10.3390/antiox12030633.

Suzuki, Takuya, and Hiroshi Hara. 2011. "Role of Flavonoids in Intestinal Tight Junction Regulation." *The Journal of Nutritional Biochemistry* 22 (5): 401–8. https://doi.org/10.1016/j.jnutbio.2010.08.001.

Tang, Lin. 2014. "Comparison of the Interaction between Three Anthocyanins and Human Serum Albumins by Spectroscopy." *Journal of Luminescence* 153: 54–63. Amsterdam: Elsevier [etc.].

Teng, Hui, and Lei Chen. 2019. "Polyphenols and Bioavailability: An Update." *Critical Reviews in Food Science and Nutrition* 59 (13): 2040–51. https://doi.org/10.1080/10408398.2018.1437023.

Umamaheswaran, Gurusamy, Dhakchinamoorthi Krishna Kumar, and Chandrasekaran Adithan. 2014. "Distribution of Genetic Polymorphisms of Genes Encoding Drug Metabolizing Enzymes & Drug Transporters – a Review with Indian Perspective." *The Indian Journal of Medical Research* 139 (1): 27–65.

Van Het Hof, Karin H., Sheila A. Wiseman, Chung S. Yang, and Lilian B.M. Tijburg. 1999. "Plasma and Lipoprotein Levels of Tea Catechins Following Repeated Tea Consumption." *Proceedings of the Society for Experimental Biology and Medicine* 220 (4): 203–9. https://doi.org/10.1046/j.1525-1373.1999.d01–34.x.

Walgren, R.A., J.T. Lin, R.K. Kinne, and T. Walle. 2000. "Cellular Uptake of Dietary Flavonoid Quercetin 4'-Beta-Glucoside by Sodium-Dependent Glucose Transporter SGLT1." *The Journal of Pharmacology and Experimental Therapeutics* 294 (3): 837–43.

Wang, Fang-Yuan, Ping Wang, Dong-Fang Zhao, Frank J. Gonzalez, Yu-Fan Fan, Yang-Liu Xia, Guang-Bo Ge, and Ling Yang. 2021. "Analytical Methodologies for Sensing Catechol-O-Methyltransferase Activity and their Applications." *Journal of Pharmaceutical Analysis* 11 (1): 15–27. https://doi.org/10.1016/j.jpha.2020.03.012.

Wang, Li, Ruonan Liu, Fangfang Yan, Wanbing Chen, Mo Zhang, Qun Lu, Bijun Huang, and Rui Liu. 2024. "A Newly Isolated Intestinal Bacterium Involved in the C-Ring Cleavage of Flavan-3-Ol Monomers and the Antioxidant Activity of the Metabolites." *Food & Function* 15 (2): 580–90. https://doi.org/10.1039/D3FO03601D.

Westfall, Susan, and Giulio Pasinetti. 2019. "Design of a Novel Synbiotic Formulation to Optimize Gut-Derived Phenolic Acid Mediated Gut-Brain Axis Signals for the Treatment of Stress-Induced Depression and Anxiety (OR23-03-19)." *Current Developments in Nutrition* 3 (June): nzz040.OR23-03-19. https://doi.org/10.1093/cdn/nzz040.OR23-03-19.

Wilkinson, A.P., J.M. Gee, M.S. Dupont, P.W. Needs, F.A. Mellon, G. Williamson, and I.T. Johnson. 2003. "Hydrolysis by Lactase Phlorizin Hydrolase is the First Step in the Uptake of Daidzein Glucosides by Rat Small Intestine in Vitro." *Xenobiotica* 33 (3): 255–64. https://doi.org/10.1080/0049825021000058088.

Xiao, Jianbo. 2022. "Recent Advances on the Stability of Dietary Polyphenols." *eFood* 3 (3): e21. https://doi.org/10.1002/efd2.21.

Xiao, Jianbo, and Petra Hogger. 2014. "Influence of Diabetes on the Pharmacokinetic Behavior of Natural Polyphenols." *Current Drug Metabolism* 15 (1): 23–29. https://doi.org/10.2174/1389200214666131210142614.

Xue, Hongkun, Junyao Feng, Yingqi Tang, Xuewu Wang, Jintian Tang, Xu Cai, and Huaqiang Zhong. 2024. "Research Progress on the Interaction of the Polyphenol–Protein–Polysaccharide Ternary Systems." *Chemical and Biological Technologies in Agriculture* 11 (1): 95. https://doi.org/10.1186/s40538-024-00632-7.

Yang, Ying, Shuqin Wang, Xingyan Liu, Wenbin Zhang, Wenhua Tong, Huibo Luo, and Liming Zhao. 2024. "Interactions of Ferulic Acid and Ferulic Acid Methyl Ester with Endogenous Proteins: Determination Using the Multi-Methods." *Heliyon* 10 (2): e24605. https://doi.org/10.1016/j.heliyon.2024.e24605.

Young, Jette F., Salka E. Nielsen, Jóhanna Haraldsdóttir, Bahram Daneshvar, Søren T. Lauridsen, Pia Knuthsen, Alan Crozier, Brittmarie Sandström, and Lars O. Dragsted. 1999. "Effect of Fruit Juice Intake on Urinary Quercetin Excretion and Biomarkers of Antioxidative Status." *The American Journal of Clinical Nutrition* 69 (1): 87–94. https://doi.org/10.1093/ajcn/69.1.87.

Yu, Hailong, and Qingrong Huang. 2012. "Improving the Oral Bioavailability of Curcumin Using Novel Organogel-Based Nanoemulsions." *Journal of Agricultural and Food Chemistry* 60 (21): 5373–79. https://doi.org/10.1021/jf300609p.

Zembyla, Morfo, Brent S. Murray, Stewart J. Radford, and Anwesha Sarkar. 2019. "Water-in-Oil Pickering Emulsions Stabilized by an Interfacial Complex of Water-Insoluble Polyphenol Crystals and Protein." *Journal of Colloid and Interface Science* 548 (July): 88–99. https://doi.org/10.1016/j.jcis.2019.04.010.

Zhang, Leilei, Pascual García-Pérez, Erika Martinelli, Gianluca Giuberti, Marco Trevisan, and Luigi Lucini. 2023. "Different Fractions from Wheat Flour Provide Distinctive Phenolic Profiles and Different Bioaccessibility of Polyphenols Following in Vitro Digestion." *Food Chemistry* 404 (March): 134540. https://doi.org/10.1016/j.foodchem.2022.134540.

Zhang, Li, Peng Wang, Zongyun Yang, Feifei Du, Zhen Li, Changling Wu, Aihu Fang, Xinglian Xu, and Guanghong Zhou. 2020. "Molecular Dynamics Simulation Exploration of the Interaction between Curcumin and Myosin Combined with the Results of Spectroscopy Techniques." *Food Hydrocolloids* 101 (April): 105455. https://doi.org/10.1016/j.foodhyd.2019.105455.

Zhang, Wenwen, Suzhen Qi, Xiaofeng Xue, Yahya Al Naggar, Liming Wu, and Kai Wang. 2021. "Understanding the Gastrointestinal Protective Effects of Polyphenols Using Foodomics-Based Approaches." *Frontiers in Immunology* 12 (July): 671150. https://doi.org/10.3389/fimmu.2021.671150.

Zhang, Zhiheng, Xiaojing Li, Shangyuan Sang, David Julian McClements, Long Chen, Jie Long, Aiquan Jiao, Zhengyu Jin, and Chao Qiu. 2022. "Polyphenols as Plant-Based Nutraceuticals: Health Effects, Encapsulation, Nano-delivery, and Application." *Foods* 11 (15): 2189. https://doi.org/10.3390/foods11152189.

Zheng, Shiqi, Di Geng, Shuangyue Liu, Qingqing Wang, Siqi Liu, and Rufeng Wang. 2019. "A Newly Isolated Human Intestinal Bacterium Strain Capable of Deglycosylating Flavone C-Glycosides and Its Functional Properties." *Microbial Cell Factories* 18 (1): 94. https://doi.org/10.1186/s12934-019-1144-7.

Zhong, Huairong, Yong Xue, Xiaoyuan Lu, Qiang Shao, Yuelei Cao, Zhaoxia Wu, and Gao Chen. 2018. "The Effects of Different Degrees of Procyanidin Polymerization on the Nutrient Absorption and Digestive Enzyme Activity in Mice." *Molecules (Basel, Switzerland)* 23 (11): 2916. https://doi.org/10.3390/molecules23112916.

Zhong, Ying, and Fereidoon Shahidi. 2011. "Lipophilized Epigallocatechin Gallate (EGCG) Derivatives as Novel Antioxidants." *Journal of Agricultural and Food Chemistry* 59 (12): 6526–33. https://doi.org/10.1021/jf201050j.

Zou, Liqiang, Wei Liu, Chengmei Liu, Hang Xiao, and David Julian McClements. 2015. "Designing Excipient Emulsions to Increase Nutraceutical Bioavailability: Emulsifier Type Influences Curcumin Stability and Bioaccessibility by Altering Gastrointestinal Fate." *Food & Function* 6 (8): 2475–86. https://doi.org/10.1039/C5FO00606F.

15 Emerging Role of Polyphenols in Nutraceutical Development

Current Insights and Future Directions

Manoj Kumar Sharma

15.1 INTRODUCTION

Polyphenols, abundant in medicinal plants, constitute a diverse group of natural compounds. Known for their antioxidant properties, polyphenols assume a crucial role in safeguarding cells against the deleterious effects of oxidative stress. Plants are utilized in herbal medicine for a variety of spiritual purposes, such as making libations, offering sacrifices, pleasing the gods, invoking, divining, preparing food, and treating illnesses. Herbal medicine is made from whole plants or plant parts, and its effectiveness depends on the bioactive ingredients found in the plant and its parts. Memory loss, cardiovascular problems, arthritis, osteoarthritis, digestive and respiratory disorders, reproductive issues, skin diseases, neurological disorders, diabetes, various cancers, hypertension, and many more conditions have been effectively treated with herbal medicine. Herbal medicines can be used in liquid or powder form, and depending on the condition, they can be either administered topically to the affected areas of the body or consumed. However, dosages for herbal medicines are sometimes vague or non-specific (Lokhande, 2018; Izah *et al.*, 2023).

Their anti-inflammatory and cardiovascular preventive properties, among other potential health advantages, have drawn a lot of attention in natural medicine (Pandey *et al.*, 2009). The present exposition underscores the significance of polyphenols within medicinal plants and elucidates their prospective therapeutic applications. Medicinal plants harbour a rich reservoir of polyphenols, encompassing flavonoids, phenolic acids, and polyphenolic amides. These bioactive compounds are recognized for their potent antioxidant capabilities, crucial in mitigating oxidative stress and maintaining cellular health (Tungmunnithum *et al.*, 2018).

Polyphenols exhibit diverse properties, including anti-inflammatory and cardiovascular benefits, making them valuable components in natural medicine. This

DOI: 10.1201/9781003588504-15

introduction sets the stage for exploring the multifaceted roles of polyphenols in medicinal plants and their potential contributions to human health and well-being (Cory *et al.,* 2018). Within the realm of medicinal plants, polyphenols emerge as versatile compounds, embracing flavonoids, phenolic acids, and polyphenolic amides. Renowned for their robust antioxidant attributes, polyphenols play a pivotal role in safeguarding cells against oxidative stress. Their multifaceted properties extend to anti-inflammatory effects and support for cardiovascular health, positioning them as key players in the landscape of natural medicine. This introduction underscores the richness of polyphenols in medicinal plants and foreshadows their potential impact on human health, paving the way for further exploration and understanding (Iqbal *et al.,* 2023).

15.2 SOURCES OF POLYPHENOLS

Polyphenols constitute a heterogeneous category of naturally existing compounds prevalent in plants, recognized for their anti-oxidative attributes and prospective health-promoting effects (Rudrapal *et al.,* 2022). Here are some common sources of polyphenols:

1. **Fruits:**
   - Apples
   - Berries (blueberries, strawberries, raspberries, blackberries)
   - Cherries
   - Citrus fruits (oranges, lemons, grapefruits)
   - Cranberries
   - Grapes
   - Pomegranate
   - Quercetin-rich foods
2. **Vegetables:**
   - Artichokes
   - Broccoli
   - Green leafy vegetables
   - Onions
   - Purple and red cabbage
   - Purple-fleshed potatoes
   - Red and yellow peppers
   - Spinach
   - Tomatoes
3. **Nuts and seeds:**
   - Almonds
   - Flaxseeds
   - Sunflower seeds
   - Walnuts
4. **Whole grains:**
   - Barley
   - Brown rice

- Buckwheat
- Oats
- Whole wheat

5. **Tea/herbal teas:**
   - Black tea
   - Chamomile tea
   - Green tea
   - Hibiscus tea
   - Peppermint tea
   - White tea
6. **Coffee:**
   - Coffee beans contain chlorogenic acids
7. **Herbs and spices:**
   - Cinnamon
   - Cloves
   - Cumin
   - Ginger
   - Rosemary
   - Thyme
   - Turmeric
8. **Dark chocolate:**
9. **Red Wine:**
   - Red wine contains various polyphenols, such as resveratrol.
10. **Olives and extra virgin olive oil:**
11. **Legumes:**
    - Black beans
    - Chickpeas
    - Lentils
    - Red beans
12. **Seaweed:**
    - Certain types of seaweed, like brown seaweed
13. **Cocoa and cocoa products:**
    - Cocoa powder and dark chocolate
14. **Flaxseeds:**
    - Besides being a source of omega-3 fatty acids, flaxseeds contain lignans, a type of polyphenol.
15. **Cherimoya:**
    - Cherimoya, also known as custard apple, is a tropical fruit containing polyphenols.
16. **Rose hips:**
    - Rose hips, the fruit belonging to the rose plant, are high in vitamin C and other polyphenols.

It is important to note that a range of factors, including plant variety, growing conditions, and processing techniques, might affect the amount of polyphenols in these foods. A varied diet that includes whole grains, nuts, seeds, fruits, vegetables, and other plant-based foods will help you consume more polyphenols. Incorporating a variety of these polyphenol-rich foods into your diet can contribute to a more balanced and nutritious intake of these beneficial compounds. Keep in mind that cooking and processing methods can affect the polyphenol content in foods, so choosing minimally processed options is often recommended (Arfaoui *et al.*, 2021).

15.3 HEALTH BENEFITS OF POLYPHENOLS

Polyphenols, prevalent in diverse medicinal plant sources, confer a wide range of health advantages attributable to their antioxidant, anti-inflammatory, and other bioactive properties. Here are some explanations of the health advantages of polyphenols obtained from therapeutic plants.

15.3.1 Polyphenols as Antioxidants

Polyphenols encompass a diverse array of natural compounds present in plants, drawing notable attention owing to their antioxidant attributes. Antioxidants serve as agents that facilitate the neutralization of free radicals within the body – exceptionally reactive molecules capable of inducing cellular damage and contributing to diverse diseases, including cancer and cardiovascular disorders (Lobo *et al.*, 2010). Polyphenols obtained from medicinal plants provide anti-oxidative properties that are essential for protecting cellular and tissue structures from the damaging effects of oxidative stress (Rudrapal *et al.*, 2022). Flavonoids, a prominent polyphenolic class renowned for their anti-oxidative capabilities, encompass subgroups such as flavanones, flavones, flavanols, isoflavones, and anthocyanins. The anti-oxidative attributes inherent in flavonoids stem from their capacity to neutralize free radicals and modulate cellular signalling pathways (Panche *et al.*, 2016; Dias *et al.*, 2021). Phenolic acids, exemplified by potent antioxidants like caffeic acid and ferulic acid, are prevalent in various plants. Their role in averting oxidative damage involves the neutralization of free radicals and the inhibition of reactive oxygen species (ROS) production (Kumar and Goel, 2019). Lignans, classified as polyphenolic compounds possessing antioxidant attributes, are present in seeds, grains, and vegetables. Their capacity to shield cells from oxidative stress involves the scavenging of free radicals and augmentation of antioxidant enzyme activity (Pandey and Rizvi, 2009). Resveratrol, a recognized stilbene, is notably present in red grapes, red wine, and select berries. Scientific research has extensively investigated its antioxidant qualities and prospective health advantages, including anti-inflammatory and anticancer activities (Koushki *et al.*, 2018). The substantial anti-oxidative properties exhibited by polyphenols sourced from medicinal plants play a crucial role in fostering their potential health-promoting effects. Integration of a diverse range of plant-based foods, abundant in polyphenols, into the dietary regimen, holds promise for mitigating oxidative stress and potentially lowering the risk of various chronic diseases (Pandey and Rizvi, 2009; Lokhande, 2018). Nevertheless, it is imperative to acknowledge that the bioavailability and effectiveness of polyphenols exhibit variability, necessitating ongoing research efforts to comprehensively elucidate their mechanisms and explore potential therapeutic applications (Cory *et al.*, 2018).

15.3.1.1 Antioxidant Mechanisms

Polyphenols manifest their antioxidant effects through diverse mechanisms, encompassing free radical scavenging, metal ion chelation, and modulation of antioxidant enzyme activity. Their ability to donate electrons stabilizes free radicals, thwarting their potential to induce cellular damage (Pandey and Rizvi, 2009; Lobo *et al.,* 2010; Liu *et al.,* 2023). It is well known that polyphenols have strong antioxidant qualities, thereby playing a role in mitigating ageing processes and potentially dropping the risk of various chronic diseases. By scavenging free radicals, polyphenols help protect cells and tissues from oxidative stress (Pandey and Rizvi, 2009; Lobo *et al.,* 2010; Liu *et al.,* 2023).

15.3.2 Anti-Inflammatory Effects

Apart from their anti-oxidative characteristics, polyphenols frequently demonstrate anti-inflammatory effects. Chronic inflammation, intimately associated with oxidative stress, is mitigated by polyphenols, thereby contributing to the overall well-being of cellular processes (Hussain *et al.,* 2016; Lobo *et al.,* 2010; Liu *et al.,* 2023). Polyphenols manifest anti-inflammatory characteristics by modulating inflammatory pathways and diminishing the production of inflammatory molecules. The association between chronic inflammation and various diseases, including cardiovascular diseases, arthritis, and specific cancers, underscores the potential significance of these anti-inflammatory effects. Polyphenols' ability to mitigate inflammation contributes to overall health (Lobo *et al.,* 2010; Lokhande, 2018; Liu *et al.,* 2023; Arias-Sánchez *et al.,* 2023). Emerging evidence indicates that dietary polyphenols mitigate oxidative and inflammatory markers across various diseases. Through the modulation of diverse cell signalling pathways and gene expression, they play a role in ameliorating chronic conditions, including metabolic and neurodegenerative diseases (Fanaro *et al.,* 2023).

15.3.3 Cardio Protective Effects

Polyphenols have been correlated with cardiovascular health advantages. Their potential to ameliorate lipid profiles, augment endothelial function, and diminish oxidative stress within blood vessels suggests a capacity to mitigate the possibility of cardiovascular diseases (Khurana *et al.,* 2013; Alotaibi *et al.,* 2021; Iqbal *et al.,* 2023). Numerous studies suggest that polyphenols may contribute to cardiovascular health. They can improve endothelial function, reduce blood pressure, and lower low-density lipoprotein (LDL) cholesterol levels.

Collectively, these effects may contribute to a decrease in the risk of heart diseases, encompassing conditions like atherosclerosis and coronary artery disease (Khurana *et al.,* 2013; Alotaibi *et al.,* 2021; Iqbal *et al.,* 2023).

15.3.4 Neuroprotective Effects

Certain polyphenols, notably flavonoids, have undergone scrutiny for their neuroprotective attributes. Their potential to deter neurodegenerative diseases is attributed to their capability to hunt free radicals, diminish inflammation, and foster neuronal survival (Lobo *et al.,* 2010; Liu *et al.,* 2023; Arias-Sánchez *et al.,* 2023). Certain polyphenols, such as those found in green tea and berries, have neuroprotective properties. Polyphenols may contribute to the safeguarding of brain cells against oxidative stress, the attenuation of inflammation within the brain, and the prospective reduction of the risk related to neurodegenerative diseases such as Parkinson's and Alzheimer's (Lobo *et al.,* 2010; Liu *et al.,* 2023; Arias-Sánchez *et al.,* 2023). The bidirectional communication between the gut microbiota and the gut-brain axis (GBA) plays a crucial role in brain function. Various intrinsic and extrinsic factors influence this connection, with diet, particularly polyphenols, being prominent among them. Polyphenols, known for their anti-inflammatory properties and antioxidants, have been connected with beneficial properties on brain ageing by modulating both gut microbiota and the GBA (Sarubbo *et al.,* 2023).

15.3.5 Cancer Prevention

The anti-oxidative attributes of polyphenols have been associated with potential cancer prevention. Through the neutralization of free radicals and inhibition of oxidative DNA damage, polyphenols play a role in mitigating the risk of cancer development (Pandey and Rizvi, 2009). Certain kinds of cancer may be prevented or have their progression slowed down by polyphenols. The anti-oxidative and anti-inflammatory attributes exhibited by these substances play a role in hindering the proliferation of cancer cells and lowering the risk of developing diverse cancers, including those affecting the breast, prostate, and colon (Pandey and Rizvi, 2009; Lobo *et al.,* 2010; Liu *et al.,* 2023; Arias-Sánchez *et al.,* 2023). Flavonoids, categorized as polyphenols, exhibit notable antioxidant and anti-inflammatory effects and have recently demonstrated strong anticancer capabilities (Asnaashari *et al.,* 2023). Epidemiological studies have linked the anti-oxidative potential of polyphenols to a decreased risk of cancer. These properties also include the ability to prevent pro-carcinogen activation, cancer cell growth, metastasis, and angiogenesis, as well as to promote the inhibition or down-regulation of active drug efflux transporters. Furthermore, polyphenols can alter immune reactions, affect inflammatory cascades, and cause cancer cells to undergo apoptosis (Farhan, 2023).

15.3.6 Improved Blood Sugar Regulation

Certain polyphenols, notably those existing in foods *viz,* green tea, cinnamon, and berries, exhibit the potential to regulate blood sugar levels. Their capacity to enhance insulin sensitivity, alleviate insulin resistance, and enhance glucose metabolism has potential benefits for people with diabetes or at risk of acquiring the disease (Aryaeian *et al.,* 2017). The pivotal action of oxidative stress in diabetes and its complications underscores the importance of nutritional antioxidants from dietary sources in enhancing antidiabetic therapy. This review explores polyphenols, examining their plant origins, human metabolism, and their significance in the antioxidant mechanism. The emphasis is on how they might improve antidiabetic outcomes by reducing oxidative stress, improving insulin sensitivity, and maintaining β-cell function (Krawczyk *et al.,* 2023). Emerging research indicates that dietary polyphenols play a role in diminishing oxidative and inflammatory indicators associated with various diseases. This is accomplished by regulating a variety of gene expression and cell signalling pathways, which improves some chronic illnesses, such as metabolic and neurological disorders (Fanaro *et al.*, 2023).

15.3.7 Gut Health Promotion

Polyphenols possess the capacity to exert positive effects on the gut microbiota by fostering the growth of advantageous bacteria while inhibiting the proliferation of detrimental microbes. This modulation helps establish a positive gut environment, ultimately enhancing processes such as digestion and nutrient absorption (Wang *et al.*, 2022). In a symbiotic connection with the host, billions of bacteria known as the intestinal microbiota find the ideal home in a healthy gut. This microbiota is highly approachable to the long-term diet and produces cytokines, circulating metabolites, and hormones that are essential for metabolism. The gut-heart axis is a unique, recently developed idea that is based on mounting evidence that dysbiosis, or a disturbed gut microbiota, contributes to the pathophysiology of cardiovascular disease (Bianchi *et al.*, 2022). The gut microbial genome encodes enzymes shaping the bioavailability of polyphenols, revealed by high-throughput sequencing and -omics data. While not strictly prebiotics, polyphenols exhibit direct prebiotic effects. Certain probiotic bacteria produce bioactive metabolites from polyphenols. On phenolic substrates, polyphenol-transforming bacteria flourish in vivo, supporting gut symbionts linked to health, such as *Akkermansia muciniphila*. To characterize polyphenols that modulate the gut microbiota through prebiotic and antimicrobial activities, we suggest the term "duplibiotic." This may help to mitigate metabolic disorders and gut dysbiosis for therapeutic purposes (Rodríguez-Daza *et al.*, 2021).

15.3.8 Weight Management

Some polyphenols, such as those in green tea, may aid in weight management. They can enhance metabolism, increase fat oxidation, and potentially reduce body weight and fat accumulation (Aloo *et al.*, 2023).

15.3.9 Joint Health

Polyphenols with anti-inflammatory properties may benefit individuals with joint disorders, such as osteoarthritis and rheumatoid arthritis. By reducing inflammation, polyphenols may help alleviate joint pain and improve mobility (Hussain *et al.*, 2016; Lobo *et al.*, 2010; Sirše 2022; Aloo *et al.*, 2023; Liu *et al.*, 2023).

15.4 SKIN HEALTH

The antioxidant properties of polyphenols contribute to skin health by protecting against oxidative stress caused by factors like ultraviolet (UV) radiation and pollution. Polyphenols may help maintain skin elasticity, reduce signs of ageing, and support overall skin health (Michalak, 2022).

15.5 ANTIMICROBIAL EFFECTS

Some polyphenols possess antimicrobial properties, helping to combat infections and support the immune system. This can be particularly beneficial in preventing and managing various infectious diseases (Pandey and Rizvi, 2009).

Incorporating a varied array of polyphenol-rich foods, including fruits, vegetables, tea, coffee, and herbs, into a well-balanced diet can provide these health benefits. However, it is crucial to note that each person's reaction may be different, and further research is needed to fully understand the mechanisms and determine the right dosages for particular health outcomes.

15.6 BIOAVAILABILITY AND HEALTH BENEFITS

Bioavailability pertains to the degree and speed at which a substance is assimilated and becomes accessible in the bloodstream or reaches the designated tissue following administration. The analysis goes deeper into the topic of polyphenol bioavailability, emphasizing important elements such as intestinal absorption and the impact of the food matrix, chemical structure (such as polymerization, glycosylation, and esterification), and the process of outflow back into the intestinal lumen. Additionally, information is given about the role that microflora plays in the breakdown of polyphenols and the production of particular active metabolites (Manach *et al.*, 2004; D'Archivio *et al.*, 2010). Comprehending these factors is vital for optimizing the potential health benefits of polyphenols. The subsequent section provides detailed insights into the bioavailability of polyphenols:

15.6.1 Factors Affecting Bioavailability

15.6.1.1 Chemical Structure

The chemical structure of polyphenols significantly influences their bioavailability. Factors such as molecular size, polarity, and the presence of glycosylation affect absorption. For example, smaller and less polar polyphenols are generally better absorbed than larger and more polar ones (D'Archivio *et al.*, 2010; Marín *et al.*, 2015).

15.6.1.2 Gastrointestinal Digestion

Polyphenols undergo extensive transformations during gastral digestion. They are subject to enzymatic degradation, and some may form conjugates with other molecules.

The degree to which polyphenols undergo degradation and captivation in the stomach and small intestine is subject to variation contingent upon their specific chemical characteristics (Pandey and Rizvi, 2009; Wang *et al.*, 2022).

15.6.1.3 Food Matrix

The food matrix in which polyphenols are ingested plays a pivotal role in influencing their bioavailability. It has long been known that antioxidants found in newly harvested plant materials might differ significantly from antioxidants obtained through food. This disparity is frequently a result of employing food processing techniques, including both thermal and non-thermal treatments (Toydemir *et al.*, 2022). Polyphenols are frequently situated within intricate matrices of fruits, vegetables, and other plant-derived foods. The existence of dietary fibres, proteins, and fats can affect the liberation and assimilation of polyphenols. Notably, the concurrent intake of polyphenols with lipids has the potential to augment their solubility and absorption (Aryaeian *et al.*, 2017; Aloo *et al.*, 2023; Fernandes *et al.*, 2023).

15.6.1.4 Gut Microbiota

The biochemical breakdown of polyphenols is significantly influenced by the gut bacteria. Certain polyphenols undergo microbial metabolism, transforming into bioactive compounds that potentially exhibit enhanced bioavailability. The composition and diversity of an individual's gut microbiota can vary, influencing the degree of polyphenol metabolism and absorption (Pandey and Rizvi, 2009; Wang *et al.*, 2022).

15.6.1.5 Phase II Metabolism

Polyphenols undergo phase II metabolism in the liver, where they undergo coupling with molecules such as glucuronic acid and sulphate. This process prepares them for elimination via bile or urine. The efficiency of this metabolism can affect the overall bioavailability of polyphenols (Liu and Hu, 2007).

15.6.1.6 Transport Proteins

Specific transport proteins in the intestinal epithelium are responsible for facilitating the absorption of polyphenols. The expression and activity of these proteins may vary among individuals, affecting the absorption rates of different polyphenols (Scott *et al.*, 2022).

15.6.1.7 Competition with Other Nutrients

The presence of other nutrients in the digestive system can compete for absorption sites and influence the bioavailability of polyphenols. For example, minerals like calcium and iron may interfere with the absorption of certain polyphenols (Piskin *et al.*, 2022).

15.6.1.8 Food Processing

Food processing can affect the bioavailability of polyphenols. For example, cooking may induce structural changes in polyphenols, influencing their accessibility for absorption. Conversely, specific food processing methods may facilitate the release of polyphenols from the food matrix (Arfaoui, 2021).

15.6.1.9 Individual Variability

Individuals can exhibit variability in their ability to absorb and metabolize polyphenols. Genetic factors, age, and overall health status can influence how efficiently polyphenols are absorbed and utilized in the body (Scott *et al.*, 2022).

15.6.1.10 Formulation and Delivery Systems

Scientists are exploring novel approaches to augment the bioavailability of polyphenols by devising innovative formulations and delivery systems, including the utilization of nanoparticles or encapsulation technologies (Yang *et al.*, 2020).

The bio-accessibility of polyphenols derived from medicinal plants is influenced by numerous factors, resulting in a complex process. The health benefits associated with polyphenols depend on their bioavailability, which is generally limited and subject to significant variability among individuals. Factors such as low water solubility, presence in foods as polymers or glycosylated forms, and strong binding to food matrices primarily contribute to the restricted absorption of polyphenols in the upper gastrointestinal tract (Polia *et al.*, 2022). Enhancing the intake of foods abundant in polyphenols within a varied and well-balanced diet, while considering the food matrix and acknowledging individual variances in gut microbiota, genetics, and metabolism that affect how these compounds are processed, making personalized dietary approaches, can augment the potential health advantages, including antioxidant, anti-inflammatory, and cardioprotective effects, linked with these bioactive substances.

15.7 FUTURE DIRECTIONS AND RESEARCH GAPS

Prospective avenues and research voids in the examination of polyphenols from plants underscore domains necessitating additional exploration to augment comprehension and optimize the potential benefits of these bioactive compounds. Empirical evidence has established the diverse biological activities of the bioactive compounds existing in these sources. Consequently, there exists an imperative for scientific exploration of eco-friendly traditional and advanced methodologies distinguished by economical production costs. This investigation should focus on prioritizing the extraction, isolation, and purification of phytochemical compounds derived from agricultural residues, aiming to achieve high yields and optimal quality outcomes (Doria *et al.*, 2021).

A brief overview follows.

15.7.1 Mechanistic Understanding

Scientific investigation ought to delve into the intricate molecular mechanisms through which polyphenols exert their effects. An in-depth comprehension of the distinct pathways and targets influenced by various polyphenols will significantly contribute to the development of targeted therapeutic applications.

15.7.2 Bioavailability Enhancement Strategies

There is a pressing need for research into strategies aimed at augmenting the bioavailability of polyphenols. Enhancing stability, absorption, and delivery systems has the potential to amplify the efficacy of interventions based on polyphenols. Current research endeavours are dedicated to investigating the therapeutic capabilities of marine polyphenols and other micronutrients present in algae, fish, and crustaceans. Characterized by their distinct chemical compositions, these compounds showcase diverse biological functionalities, encompassing anti-inflammatory, antioxidant, antimicrobial, and antitumour properties. Owing to these attributes, marine polyphenols are currently being scrutinized as prospective therapeutic interventions for a wide array of ailments, spanning from cardiovascular diseases, diabetes, and neurodegenerative disorders to cancer (Pereira and Cotas, 2023).

15.7.3 Standardization of Extracts

Standardization of polyphenol-rich extracts is crucial for ensuring consistency in research and product development. Establishing standardized methods for extraction and analysis will facilitate accurate comparisons across studies.

15.7.4 Interaction with the Gut Microbiota

Further investigation into the interface between polyphenols and the gut microbiota is imperative. Gaining insight into the processes by which gut bacteria metabolize polyphenols and contribute to their bioactivity holds the potential to unveil novel avenues for promoting health.

15.7.5 Dose-Response Relationships

Research should focus on establishing optimal dosages of polyphenols for various health outcomes. Understanding dose-response relationships is crucial for designing effective interventions and avoiding potential adverse effects.

15.7.6 Clinical Trials and Human Studies

While many pre-clinical studies have shown promise, more well-designed clinical trials are needed to validate the health benefits of polyphenols in humans. Research should address factors such as dosage, duration, and population-specific effects.

15.7.7 Synergistic Effects

Investigating the potential synergistic effects of polyphenols when combined with other bioactive compounds or therapies could provide insights into enhanced health benefits.

15.7.8 Long-Term Effects and Safety

Long-term studies assessing the safety and potential adverse effects of sustained polyphenol consumption are needed. Understanding the long-term impact on different populations will guide recommendations for dietary and therapeutic interventions.

15.7.9 Genetic Variability

Considering individual genetic variability in responding to polyphenols is essential. Research should explore how genetic factors influence the metabolism and effectiveness of polyphenols in diverse populations.

15.7.10 Exploration of Novel Sources

Exploring less studied plants for their polyphenol content and bioactivity has the potential to unveil novel compounds exhibiting unique health-promoting properties.

15.7.11 Influence on Ageing and Age-Related Pathologies

Scientific inquiry should investigate the potential of polyphenols in alleviating age-related diseases and fostering healthy ageing. Unravelling the mechanisms by which polyphenols influence cellular processes linked to ageing represents a promising avenue for research.

15.7.12 Integration into Public Health Strategies

Translating research findings into public health strategies and recommendations is crucial. Identifying target populations, optimal dietary patterns, and feasible interventions will help integrate polyphenols into preventive healthcare practices.

Through the examination of these research voids, scientists can advance our comprehension of polyphenols and establish the foundation for their effective utilization in health enhancement and disease prevention. A study explores recent progressions in environmentally sustainable extraction techniques for natural compounds, particularly in the context of Indonesia, evaluating their possible practical implementations (Putra *et al.*, 2023). Collaborative initiatives spanning various disciplines and geographical regions will enrich our holistic understanding of the involvement of polyphenols in human health.

15.8 CONCLUSION

Polyphenols and antioxidant-rich nutraceuticals are foundational in comprehending the bioactive nature of these compounds and their potential health benefits. Present in various plant sources, polyphenols have garnered significant interest owing to their antioxidant properties, which are essential for counteracting free radicals and reducing oxidative stress. This introductory overview highlights the diverse subclasses of polyphenols, including flavonoids, phenolic acids, lignans, and stilbenes, each distinguished by unique chemical structures and health-promoting attributes (Dias *et al.*, 2021; Puri *et al.*, 2022).

The association between oxidative stress and a range of chronic diseases underscores the critical role of polyphenols in preventive healthcare. Their antioxidant, anti-inflammatory, and other bioactive properties position them as promising candidates for nutraceutical applications. Acting as antioxidants, polyphenols contribute to cellular health by protecting against oxidative damage, promoting cardiovascular well-being, and potentially reducing the risk of conditions such as cancer and neurodegenerative diseases.

The introduction also accentuates the significance of comprehending the bioavailability of polyphenols, a determinant in their efficacy for health promotion. Factors including chemical structure, food matrix, gut microbiota, and interactions with other compounds contribute to the intricate landscape of polyphenol bioavailability.

As the nutraceutical industry explores the applications of polyphenols, future research directions and potential gaps come to the fore. Areas such as mechanistic understanding, strategies for enhancing bioavailability, and long-term effects necessitate further investigation to unlock the complete therapeutic potential of polyphenols.

In summary, this introduction lays the groundwork for a thorough exploration of polyphenols and antioxidant-rich nutraceuticals, paving the way for subsequent discussions on their diverse applications, health benefits, and the evolving scientific landscape in this field.

REFERENCES

Aloo S-O, Ofosu FK, Kim N-H, Kilonzi SM, Oh D-H. (2023). Insights on Dietary Polyphenols as Agents against Metabolic Disorders: Obesity as a Target Disease. *Antioxidants.* 12:416. https://doi.org/10.3390/antiox12020416

Alotaibi BS, Ijaz M, Buabeid M, Kharaba ZJ, Yaseen HS, Murtaza G. (2021). Therapeutic Effects and Safe Uses of Plant-Derived Polyphenolic Compounds in Cardiovascular Diseases: A Review. *Drug Des Dev Ther.* Nov 20;15:4713–4732. https://doi.org/10.2147/DDDT.S327238. PMID: 34848944; PMCID: PMC8619826.

Arfaoui L. (2021). Dietary Plant Polyphenols: Effects of Food Processing on their Content and Bioavailability. *Molecules.* May 16;26(10):2959. https://doi.org/10.3390/molecules26102959. PMID: 34065743; PMCID: PMC8156030.

Arias-Sánchez RA, Torner L, Fenton Navarro B. (2023). Polyphenols and Neurodegenerative Diseases: Potential Effects and Mechanisms of Neuroprotection. *Molecules.* Jul 14;28(14):5415. https://doi.org/10.3390/molecules28145415. PMID: 37513286; PMCID: PMC10385962.

Aryaeian N, Sedehi SK, Arablou T. (2017). Polyphenols and their Effects on Diabetes Management: A Review. *Med J Islam Repub Iran.* Dec 26;31:134. https://doi.org/10.14196/mjiri.31.134. PMID: 29951434; PMCID: PMC6014790.

Asnaashari S, Amjad E, Sokouti B. (2023). Synergistic Effects of Flavonoids and Paclitaxel in Cancer Treatment: A Systematic Review. *Cancer Cell Int.* 23(1):211.

Bianchi F, Cappella A, Gagliano N, Sfondrini L, Stacchiotti A. (2022). Polyphenols-Gut-Heart: An Impactful Relationship to Improve Cardiovascular Diseases. *Antioxidants.* 11(9). https://doi.org/10.3390/antiox11091700

Cory H, Passarelli S, Szeto J, Tamez M, Mattei J. (2018). The Role of Polyphenols in Human Health and Food Systems: *A Mini-Review. Front Nutr.* 21;5:87. https://doi.org/10.3389/fnut.2018.00087. PMID: 30298133; PMCID: PMC6160559.

D'Archivio M, Filesi C, Varì R, Scazzocchio B, Masella R. (2010). Bioavailability of the Polyphenols: Status and Controversies. *Int J Mol Sci.* Mar 31;11(4):1321–1342. https://doi.org/10.3390/ijms11041321. PMID: 20480022; PMCID: PMC2871118.

Dias MC, Pinto DCGA, Silva AMS. (2021). Plant Flavonoids: Chemical Characteristics and Biological Activity. *Molecules.* 4;26(17):5377. https://doi.org/10.3390/molecules26175377. PMID: 34500810; PMCID: PMC8434187.

Doria E, Boncompagni E, Marra A, Dossena M, Verri M, Buonocore D. (2021). Polyphenols Extraction from Vegetable Wastes Using a Green and Sustainable Method. *Front Sustain Food Syst.* 5:690399.

Fanaro GB, Marques MR, Calaza KDC, *et al.* (2023). New Insights on Dietary Polyphenols for the Management of Oxidative Stress and Neuroinflammation in Diabetic Retinopathy. *Antioxidants.* 12(6). https://doi.org/10.3390/antiox12061237

Farhan MJB. (2023). Insights on the Role of Polyphenols in Combating Cancer Drug Resistance. *Biomedicines.* 11(6):1709.

Fernandes A, Mateus N, de Freitas V. (2023). Polyphenol-Dietary Fiber Conjugates from Fruits and Vegetables: Nature and Biological Fate in a Food and Nutrition Perspective.

Foods. Mar 1;12(5):1052. https://doi.org/10.3390/foods12051052. PMID: 36900569; PMCID: PMC10000549.

Hussain T, Tan B, Yin Y, Blachier F, Tossou MC, Rahu N. (2016). Oxidative Stress and Inflammation: What Polyphenols Can Do for Us? *Oxid Med Cell Longev*. 2016;2016:7432797. https://doi.org/10.1155/2016/7432797. Epub Sep 22. PMID: 27738491; PMCID: PMC5055983.

Iqbal I, Wilairatana P, Saqib F, Nasir B, Wahid M, Latif MF, Iqbal A, Naz R, Mubarak MS. (2023). Plant Polyphenols and their Potential Benefits on Cardiovascular Health: A Review. *Molecules*. Sep 1;28(17):6403. https://doi.org/10.3390/molecules28176403. PMID: 37687232; PMCID: PMC10490098.

Izah SC, Ogidi OI, Ogwu MC, *et al.* (2023). *Historical Perspectives and Overview of the Value of Herbal Medicine Herbal Medicine Phytochemistry: Applications and Trends*. Springer, pp. 1–33.

Khurana S, Venkataraman K, Hollingsworth A, Piche M, Tai TC. (2013). Polyphenols: Benefits to the Cardiovascular System in Health and in Aging. *Nutrients*. Sep 26;5(10):3779–3827. https://doi.org/10.3390/nu5103779. PMID: 24077237; PMCID: PMC3820045.

Koushki M, Amiri-Dashatan N, Ahmadi N, Abbaszadeh HA, Rezaei-Tavirani M. (2018). Resveratrol: A Miraculous Natural Compound for Diseases Treatment. *Food Sci Nutr*. Oct 26;6(8):2473–2490. https://doi.org/10.1002/fsn3.855. PMID: 30510749; PMCID: PMC6261232.

Krawczyk M, Burzynska-Pedziwiatr I, Wozniak LA, Bukowiecka-Matusiak M. (2023). Impact of Polyphenols on Inflammatory and Oxidative Stress Factors in Diabetes Mellitus: Nutritional Antioxidants and their Application in Improving Antidiabetic Therapy. *Biomolecules*. 13(9):1402.

Kumar N., Goel N. (2019). Phenolic acids: Natural versatile molecules with promising therapeutic applications. *Biotechnol. Rep.* 24, e00370. 10.1016/j.btre.2019.e00370.

Liu H-M, Cheng M-Y, Xun M-H, Zhao Z-W, Zhang Y, Tang W, Cheng J, Ni J, Wang W. (2023). Possible Mechanisms of Oxidative Stress-Induced Skin Cellular Senescence, Inflammation, and Cancer and the Therapeutic Potential of Plant Polyphenols. *Int J Mol Sci*. 24:3755. https://doi.org/10.3390/ijms24043755

Liu Z, Hu M. (2007). Natural Polyphenol Disposition via Coupled Metabolic Pathways. *Expert Opin Drug Metab Toxicol*. Jun;3(3):389–406. https://doi.org/10.1517/17425255.3.3.389. PMID: 17539746; PMCID: PMC2777985.

Lobo V, Patil A, Phatak A, Chandra N. (2010). Free Radicals, Antioxidants and Functional Foods: Impact on Human Health. *Pharmacogn Rev*. Jul;4(8):118–126. https://doi.org/10.4103/0973-7847.70902. PMID: 22228951; PMCID: PMC3249911.

Lokhande SS. (2018). Role of Nutraceuticals in Various Diseases: A Comprehensive Review. *Asian J Pharm Res*. 8(4):236–240. https://doi.org/10.5958/2231-5691.2018.00040.0

Manach C, Scalbert A, Morand C, Rémésy C, Jiménez L. (2004). Polyphenols: Food Sources and Bioavailability. *Am J Clin Nutr*. 79(5):727–747. https://doi.org/10.1093/ajcn/79.5.727

Marín L, Miguélez EM, Villar CJ, Lombó F. (2015). Bioavailability of Dietary Polyphenols and Gut Microbiota Metabolism: Antimicrobial Properties. *Biomed Res Int*. 2015:905215. https://doi.org/10.1155/2015/905215. Epub 2015 Feb 23. PMID: 25802870; PMCID: PMC4352739.

Michalak M. (2022). Plant-Derived Antioxidants: Significance in Skin Health and the Ageing Process. *Int J Mol Sci*. Jan 6;23(2):585. https://doi.org/10.3390/ijms23020585. PMID: 35054770; PMCID: PMC8776015.

Panche AN, Diwan AD, Chandra SR. (2016). Flavonoids: An Overview. *J Nutr Sci*. Dec 29;5:e47. https://doi.org/10.1017/jns.2016.41. PMID: 28620474; PMCID: PMC5465813.

Pandey KB, Rizvi SI. (2009). Plant Polyphenols as Dietary Antioxidants in Human Health and Disease. *Oxid Med Cell Longev*. Nov–Dec;2(5):270–278. https://doi.org/10.4161/oxim.2.5.9498. PMID: 20716914; PMCID: PMC2835915.

Pereira L, Cotas J (2023) Therapeutic Potential of Polyphenols and Other Micronutrients of Marine Origin. *Marine Drugs*. 21(6). https://doi.org/10.3390/md21060323

Piskin E, Cianciosi D, Gulec S, Tomas M, Capanoglu E. (2022). Iron Absorption: Factors, Limitations, and Improvement Methods. *ACS Omega*. Jun 10;7(24):20441–20456. https://doi.org/10.1021/acsomega.2c01833. PMID: 35755397; PMCID: PMC9219084.

Polia F, Pastor-Belda M, Martínez-Blázquez A, *et al.* (2022). Technological and Biotechnological Processes to Enhance the Bioavailability of Dietary (Poly) Phenols in Humans. *J Agric*. 70(7):2092–2107.

Puri V, Nagpal M, Singh I, Singh M, Dhingra GA, Huanbutta K, Dheer D, Sharma A, Sangnim T. (2022). A Comprehensive Review on Nutraceuticals: Therapy Support and Formulation Challenges. *Nutrients*. Nov 3;14(21):4637. https://doi.org/10.3390/nu14214637. PMID: 36364899; PMCID: PMC9654660.

Putra NR, Yustisia Y, Heryanto B, *et al.* (2023). Advancements and Challenges in Green Extraction Techniques for Indonesian Natural Products: A Review. *South Afr J Chem Eng*. 46:88–98.

Rodríguez-Daza MC, Pulido-Mateos EC, Lupien-Meilleur J, Guyonnet D, Desjardins Y, Roy D. (2021). Polyphenol-Mediated Gut Microbiota Modulation: Toward Prebiotics and Further. *Front Nutr*. 8:689456.

Rudrapal M, Khairnar SJ, Khan J, Dukhyil AB, Ansari MA, Alomary MN, Alshabrmi FM, Palai S, Deb PK, Devi R. (2022). Dietary Polyphenols and their Role in Oxidative Stress-Induced Human Diseases: Insights into Protective Effects, Antioxidant Potentials and Mechanism(s) of Action. *Front Pharmacol*. 13:806470. https://doi.org/10.3389/fphar.2022.806470

Sarubbo F, Moranta D, Tejada S, Jiménez M, Esteban S. (2023). Impact of Gut Microbiota in Brain Ageing: Polyphenols as Beneficial Modulators. *Antioxidants*. 12(4):812.

Scott MB, Styring AK, McCullagh JSO. (2022). Polyphenols: Bioavailability, Microbiome Interactions and Cellular Effects on Health in Humans and Animals. *Pathogens*. Jul 5;11(7):770. https://doi.org/10.3390/pathogens11070770. PMID: 35890016; PMCID: PMC9324685.

Siršе M. (2022). Effect of Dietary Polyphenols on Osteoarthritis-Molecular Mechanisms. *Life*. Mar 16;12(3):436. https://doi.org/10.3390/life12030436. PMID: 35330187; PMCID: PMC8955436.

Toydemir G, Subasi BG, Hall RD, Beekwilder J, Boyacioglu D, Capanoglu E. (2022). Effect of Food Processing on Antioxidants, their Bioavailability and Potential Relevance to Human Health. *Food Chem X*. 18;14:100334. https://doi.org/10.1016/j.fochx.2022.100334. PMID: 35712535; PMCID: PMC9194584.

Tungmunnithum D, Thongboonyou A, Pholboon A, Yangsabai A. (2018). Flavonoids and Other Phenolic Compounds from Medicinal Plants for Pharmaceutical and Medical Aspects: An Overview. *Medicines*. Aug 25;5(3):93. https://doi.org/10.3390/medicines5030093. PMID: 30149600; PMCID: PMC6165118.

Wang X, Qi Y, Zheng H. (2022). Dietary Polyphenol, Gut Microbiota, and Health Benefits. *Antioxidants*. Jun 20;11(6):1212. https://doi.org/10.3390/antiox11061212. PMID: 35740109; PMCID: PMC9220293.

Yang B, Dong Y, Wang F, Zhang Y. (2020). Nano-formulations to Enhance the Bioavailability and Physiological Functions of Polyphenols. *Molecules*. Oct 10;25(20):4613. https://doi.org/10.3390/molecules25204613. PMID: 33050462; PMCID: PMC7587200.

Index

D

E

F

G

H

I

J

K

L

M

N

O

P

Q

R

For Product Safety Concerns and Information please contact our EU
representative GPSR@taylorandfrancis.com
Taylor & Francis Verlag GmbH, Kaufingerstraße 24, 80331 München, Germany

www.ingramcontent.com/pod-product-compliance
Ingram Content Group UK Ltd.
Pitfield, Milton Keynes, MK11 3LW, UK
UKHW022215190926
475702UK00003B/376
* 9 7 8 1 0 3 2 9 6 1 9 8 9 *